SEMICONDUCTOR MATERIALS

SEMICONDUCTOR MATERIALS

Lev I. Berger

CRC Press
Boca Raton   New York   London   Tokyo

Associate Editor, Engineering: Felicia Shapiro
Assistant Managing Editor, EDP: Paul Gottehrer
Marketing Manager: Arline Massey
Cover Design: Dawn Boyd
Prepress: Gary Bennett

Library of Congress Cataloging-in-Publication Data

Berger, Lev I. (Lev Isaakovich)
    Semiconductor materials / Lev I. Berger.
      p.   cm.
    Includes bibliographical references and index.
    ISBN 0-8493-8912-7 (alk. paper)
    1. Semiconductors. 2. Organic semiconductors. I. Title.
QC611.B46   1997
537.6′22--dc20   96-41739
                                                                                  CIP

This book contains information obtained from authentic and highly regarded sources. Reprinted material is quoted with permission, and sources are indicated. A wide variety of references are listed. Reasonable efforts have been made to publish reliable data and information, but the author and the publisher cannot assume responsibility for the validity of all materials or for the consequences of their use.

Neither this book nor any part may be reproduced or transmitted in any form or by any means, electronic or mechanical, including photocopying, microfilming, and recording, or by any information storage or retrieval system, without prior permission in writing from the publisher.

CRC Press, Inc.'s consent does not extend to copying for general distribution, for promotion, for creating new works, or for resale. Specific permission must be obtained in writing from CRC Press for such copying.

Direct all inquiries to CRC Press, Inc., 2000 Corporate Blvd., N.W., Boca Raton, Florida 33431.

© 1997 by CRC Press, Inc.

No claim to original U.S. Government works
International Standard Book Number 0-8493-8912-7
Library of Congress Card Number 96-41739
Printed in the United States of America   1 2 3 4 5 6 7 8 9 0
Printed on acid-free paper

TO MY TEACHERS AND TO MY STUDENTS

Acknowledgments

This book has benefited greatly from the many suggestions, comments, and corrections made by Dr. H. P. R. Frederikse (NIST). I would like to thank Prof. L. J. Burnett and Prof. D. Rehfuss (SDSU), Dr. A. Borshchevsky (JPL), and Dr. S. E. Gould (US EPA) for valuable advice and information I received from them that was relevant to the content of this book. The permanent attention of the Editor of this series, Dr. D. R. Lide (NIST-CRC), to its content and his stimulating comments are greatly appreciated. I would like especially to express my gratitude to Dr. C. L. Musgrove (UCR) for detailed analysis of the entire text of this book and the comments the value of which is impossible to overestimate. I am grateful to Mr. P. Gottehrer, CRC Press, Assistant Managing Editor, EDP, for scrupulous reading of the manuscript and refining of the text. Last, but certainly not least, it is a pleasure to thank the collective of the Science Library of the University of California at Riverside for the help I received during my work upon the manuscript of this book.

Preface

The goal of this book is to provide information on semiconductor materials. The text includes the systemization of semiconductors and the history of their discovery, investigation, and application. An introduction to Semiconductor Physics in which the stress is on the understanding of physical phenomena in semiconductors consists of an overview of the electrical conductivity in semiconductors and thermoelectric phenomena in them, recombination of charge carriers, photoconductivity and photo-e.m.f., and phenomena at the interface of n-type and p-type semiconductors. Because this book is written, first of all, for engineers specializing in materials science and technology as well as students in these fields, the mathematical apparatus that would be a necessity for a physicist is used only for better understanding of the physical meaning of the effects in semiconductors, and always in the most elemental form.

The section of the book pertinent to the experimental methods of evaluation of the basic properties of semiconductors is meant to introduce the reader to the fundamental principles of measurements. Detailed information regarding equipment for measurements may be, as a rule, obtained from the original references with descriptions of the methods or from the references containing the results of measurements of the properties of particular materials.

The bulk of the text is devoted to the comprehensive description of the chemical, thermochemical, thermodynamic, thermal, and crystallographic parameters of elemental and compound semiconductors; their band structure, electrical and thermal transport properties, including the thermoelectric parameters; dielectric and magnetic properties; their optical and photoelectric characteristics; and mechanical properties, such as elastic moduli and hardness. Special attention is paid to the description of the materials which have been known and investigated for a long time (among them diamond; boron; and pnictogenides, oxides, and chalcogenides of transition metals, and in particular, the VIIIA group metals) that are currently moving to positions in the forefront of research and development projects. Because the first semiconductors utilized in radio electronics were minerals, very brief information regarding the natural semiconductors which are known in mineralogy is also included in connection with particular groups of materials. Attention is also paid to the description of binary, ternary, and quaternary semiconductor solid solutions, in particular, correlations between the energy gap and lattice parameters of the solutions and their composition.

The section devoted to the principles of the semiconductor devices includes analysis of the processes at the p-n junction and the Schottky barrier. It includes description of the p-n junction and Schottky diodes, varistor and varactor, Zener and avalanche diodes, tunnel diode, solar cell, light-emitting diode and semiconductor laser, unipolar and bipolar devices, metal-insulator diode, and charge-coupled devices.

A portion of this book is written on the basis of the lecture course on semiconductor materials and devices taught by the author by invitation of the Department of Electrical and Computer Engineering of San Diego State University. The material of the course is substantially extended to include information important for solid state and semiconductor chemists and materials scientists.

The author would like to hope that this book attracts attention of not only engineers, but also physicists and physical chemists, and particularly students specializing in the fields of semiconductor synthesis, crystal and film growth, and semiconductor device design.

The volume of the published information relevant to semiconductor materials and devices currently is so great that it is necessary to use a very strict selection of the data to keep it in the limits of a relatively small book. It is well known that any selection is subjective. The author has to apologize to the specialists in the field of semiconductors whose subjects of specialization are not considered in this book in sufficient detail. Under any circumstances, this text cannot be considered as a comprehensive one. Nevertheless, the author hopes that he has been able to present information on the most important features of Semiconductor

Physics, Semiconductor Chemistry, and Semiconductor Materials Science, and (to a somewhat lesser degree) of the topic of semiconductor devices. The limitation of the volume of the material included in this book the author tried to compensate for by inclusion of a great number of references on the original publications and reviews so this book can be considered as a sort of handbook in the field and a comprehensive source of the reference material.

The SI units are used predominantly in this book, in accordance with Special Publication No. 811 (1995) of the U.S. National Institute of Standards and Technology.

As with any publication which includes a large bulk of information, this book is not free of defects and inaccuracies. The author will be grateful to the readers for any ideas and comments relevant to the book content which may be sent to the address of the Publisher: CRC Press, Inc., 2000 Corporate Blvd., N.W., Boca Raton, FL 33431.

The Author

Lev I. Berger is one of the leading specialists in the field of electronic materials synthesis and evaluation. Having been trained in a doctorate program in the Dept. of Physics at Moscow Institute of Non-Ferrous Metals and Gold, he received his Ph.D in semiconductor physics from Belorussian State University in 1959. The topic of investigation was thermal properties of III-V and III_2-VI_3 semiconductors. In 1968, he received a Professor-Doctor of Technical Sciences degree from Moscow Institute of Steel and Alloys for a cycle of investigations of ternary adamantine semiconductors. By decision of the Supreme Certification Commission of the USSR in 1971, he was awarded the academic title of Professor in the Field of Electronic and Electro-Vacuum Materials.

While in Russia, Dr. Berger was a Docent and Professor of Physics and Director of research groups in Moscow Institute of Reagents and Special-Purity Substances and Moscow Research Institute of Introscopy. After immigrating to the USA in 1978 (naturalized in 1985), he worked as a lecturer in Physics and research associate for the Brooklyn College of CUNY, New England Research Center, San Diego State University and the University of San Diego. He is a member of several standard-controlling committees of the American Society for Testing and Materials (ASTM), member of the American Physical Society, Materials Research Society, Society for Advancement of Material and Process Engineering (Executive Board), and American Association for Crystal Growth.

Dr. Berger has written over 150 published articles and four books, mainly in physics and physical chemistry of semiconductor materials; he is a holder of many patents in semiconductor materials and device technology and evaluation.

Currently, Dr. Berger is the CEO of the California Institute of Electronics and Materials Science, a research organization specializing in the synthesis of electronic materials and evaluation of physical and physico-chemical properties of solids.

Table of Contents

Chapter 1
Semiconductor Materials: Definition, History, Systematization 1

Chapter 2
Introduction to Semiconductor Physics 15
   2.1. Chemical Bonds in Semiconductor Crystals 15
   2.2. Electrical Conductivity in Semiconductors 18
   2.3. Thermoelectric Phenomena in Semiconductors 24
   2.4. Recombination of Electrons and Holes 25
   2.5. Photoconductivity and Photo-emf 26
   2.6. Phenomena at the Interface of n-type and p-type Semiconductors 28
   2.7. Brief Description of the Methods of Evaluation of the Basic Properties of Semiconductors 29
        2.7.1. Electrical Conductivity, Hall Effect, Nernst Effect 29
        2.7.2. Generation/Recombination Parameters 31
        2.7.3. Thermal and Thermoelectric Measurements 32

Chapter 3
Elemental Semiconductors 37
   3.1. Boron 37
   3.2. Diamond 43
   3.3. Silicon 56
   3.4. Germanium 65
   3.5. Gray Tin 74
   3.6. Group V — Phosphorus 81
   3.7. Selenium 86
   3.8. Tellurium 88
   3.9. Chemical Interaction between the Elemental Semiconductors: Solid Solutions and Compounds 91
        3.9.1. Boron-Carbon 91
        3.9.2. Boron-Silicon 92
        3.9.3. Boron-Phosphorus 92
        3.9.4. Carbon-Silicon 92
        3.9.5. Silicon-Germanium 92
        3.9.6. Silicon-Tin and Germanium-Tin 94

Chapter 4
Binary IV-VI and III-V Semiconductors 105
   4.1. IVb-IVb Compound — Silicon Carbide 105
   4.2. IIIb-Vb Compounds 110
        4.2.1. Crystal Structure 110
        4.2.2. Boron Nitride 112
        4.2.3. Boron Phosphide 116
        4.2.4. Boron Arsenide 121
        4.2.5. Aluminum Nitride 123
        4.2.6. Aluminum Phosphide 124
        4.2.7. Aluminum Arsenide 125
        4.2.8. Aluminum Antimonide 128
        4.2.9. Gallium Nitride 131

	4.2.10.	Gallium Phosphide	134
	4.2.11.	Gallium Arsenide	138
	4.2.12.	Gallium Antimonide	145
	4.2.13.	Indium Nitride	152
	4.2.14.	Indium Phosphide	155
	4.2.15.	Indium Arsenide	160
	4.2.16.	Indium Antimonide	165

Chapter 5
Binary II-VI and I-VII Tetrahedral Semiconductors 183

5.1.	Binary II-VI Compounds	183
	5.1.1. Zinc Oxide	183
	5.1.2. Zinc Sulfide	186
	5.1.3. Zinc Selenide	187
	5.1.4. Zinc Telluride	193
	5.1.5. Cadmium Sulfide	198
	5.1.6. Cadmium Selenide	201
	5.1.7. Cadmium Telluride	203
	5.1.8. Mercury Sulfide	207
	5.1.9. Mercury Selenide	210
	5.1.10. Mercury Telluride	210

Chapter 6
Ternary Adamantine Semiconductors 229

6.1.	Ternary Analogs of the III-V Adamantine Semiconductors	230
	6.1.1. The II-IV-V_2 Compounds (Ternary Pnictides)	230
	6.1.2. The I-IV_2-V_3 Compounds	242
6.2.	Ternary Analogs of II-VI Adamantine Semiconductors	246
	6.2.1. I-III-VI_2 Compounds	246
	6.2.2. I_2-IV-VI_3 Compounds	256
	6.2.3. I_3-V-VI_4 Compounds	266
6.3.	Defect Adamantine Semiconductors	271
	6.3.1. IV_3-V_4 Compounds	271
	6.3.2. III_2-VI_3 Compounds	273
	6.3.3. Defect Ternary Compounds	281
6.4.	III_2-IV-VI Compounds	284

Chapter 7
Non-adamantine Semiconductors and Variable-Composition Semiconductor Phases 295

7.1.	IA-IB Semiconductors	295
7.2.	Semiconductors in the I-V Binary Systems	297
	7.2.1. IA-VB Compounds	297
	7.2.2. $(IA)_3$-VB Compounds	297
7.3.	I-VI Compounds	302
	7.3.1. Copper and Silver Oxides	303
	7.3.2. Copper and Silver Chalcogenides	307
7.4.	II_2-IV Compounds with Antifluoride Structure	313
7.5.	IV-VI Galenite Type Compounds	317
7.6.	V_2-VI_3 Compounds	338
7.7.	Binary Compounds of the Group VIIIA Elements	351

7.8.	Semiconductor Solid Solutions		355
	7.8.1.	Alloys of Elemental Semiconductors	355
	7.8.2.	III-V/III-V Semiconductor Alloys	362
	7.8.3.	II-VI/II-VI Solid Solutions	367
	7.8.4.	Solid Solutions of II-VI and III-V Compounds	380
	7.8.5.	Organic Semiconductors	380

Chapter 8
Semiconductor Devices 411

8.1.	p-n Junction and Schottky Barrier		411
	8.1.1.	p-n Junction	411
	8.1.2.	p-n Junction Diode	413
	8.1.3.	Metal-Semiconductor Rectifying Contact. Schottky Barrier Diode	415
8.2.	Ohmic Contact		417
8.3.	p-n-Junction and Schottky-Barrier Two-Electrode Devices		420
	8.3.1.	Varistors and Varactors	422
	8.3.2.	Zener and Avalanche Diodes	423
	8.3.3.	Tunnel Diode	424
	8.3.4.	Photodiode (Solar Cell)	427
	8.3.5.	Unijunction Transistor	433
8.4.	Multiterminal (Multibarrier) Devices		433
	8.4.1.	Bipolar Devices	434
		8.4.1.1. Bipolar Junction Transistor	434
		8.4.1.2. Thyristor	436
	8.4.2.	Unipolar Devices	438
		8.4.2.1. Field-Effect Transistors	439
		8.4.2.2. Metal-Insulator-Semiconductor Diode and Charge-Coupled Device	444
8.5.	Conclusion		445

Index 449

CHAPTER 1

Semiconductor Materials: Definition, History, Systematization

The transition of electricity through solids may be performed either by ions or by electrons. Thus, the solid materials may be divided into two classes: (1) solid electrolytes and (2) electronic materials, although there is a large group of substances of an intermediate type. The majority of solid salts and glasses, as well as some polymers, are electrolytes. A common feature for electrolytes is the current decrease in a sample almost to zero with time (at constant applied voltage) and existence of a substantial time-dependent reverse voltage. This feature specifies use of solid electrolytes in electrical and electronic applications.

The class of solids with electronic charge transport is traditionally divided into (1) insulators with resistivity higher than 10^{12} Ohm·cm, (2) metals with conductivity 10^5 to 10^6 S/cm, and (3) a gigantic group of materials with conductivity between 10^5 and 10^{-11} S/cm which are called semiconductors. Virtually all inorganic nature surrounding us, as well as many organic compounds, belong to the latter category. A relatively small group of solids, among them arsenic, antimony, and bismuth, has electrical resistivity 10^2 to 10^3 times higher than that of metals, but does not possess certain properties, typical for intrinsic semiconductors, such as high sensitivity of electrical properties to radiation and temperature, etc. These materials form a class of solids called semimetals.

The most noticeable difference between metals and nonmetal solids consists of the presence in the former of a substantial electrical conductivity which increases with cooling up to the lowest magnitudes of temperature (and many of them transfer at low temperatures into the superconducting state), whereas all nonmetal solids behave like perfect insulators at low temperatures and increase their conductivity with heating. The difference between semiconductors and insulators is essentially only quantitative; namely, semiconductors are the nonmetal solids which possess a noticeable electrical conductivity at room temperature, whereas insulators need to be heated to very high temperatures to acquire appreciable thermally enhanced conductivity. The better an insulator is, the higher the temperature must be at which a noticeable electric current can be developed in it. An "ideal" solid insulator is a material in which it is impossible to have a perceptible electric current at any temperature below its melting point if the applied potential drop is below its breakdown voltage. (For liquid nonmetals, the characteristic temperature which separates semiconductors and insulators is, apparently, their boiling point.) All solids, depending on their electrical properties, may be distributed in the following sequence: insulators — semiconductors — semimetals — metals.

While metals and insulators have a long history of study and application, through the times of William Gilbert (1544–1603), Otto von Guericke (1602–1686), Benjamin Franklin (1706–1790), Luigi Galvani (1737–1798), Alessandro Volta (1745–1827), Georg Ohm

(1787–1854), Michael Faraday (1791–1867), and Thomas Edison (1847–1931), the era of semiconductors started, apparently, in the 1870s, when the selenium photoconductivity was discovered[1.1] and the first attempts to use selenium in the devices for the visual image transport by telegraph had been undertaken, although the first researcher who discovered (1833) the negative temperature coefficient of resistivity (in silver sulfide) was Faraday.[1.2] The first solid-state rectifiers of the alternating current based on selenium were made by Fritts,[1.3] but his work did not attract much attention until Grondahl and Geiger[1.4] designed and manufactured a technically feasible rectifier based on cuprous oxide (1927). This event and successful use of silicon and galenite crystals in the electromagnetic wave detectors (initially patented by Bose[1.5]) for radio receivers enhanced interest in the study and application of semiconductors, but the invention of the electron tubes (Edison, Fleming, De Forest) and their wide use had pushed the semiconductor materials and devices on to the back pages of the science and engineering textbooks.

Defense needs, on the threshold of World War II, were a cause of the return to the crystal devices with their much smaller interelectrode capacitance, in comparison with electric tubes, for application in radiolocators in the centimeter wavelength range. In the search for materials with improved parameters, the methods of material purification and single crystal growth, first of all for silicon and germanium, were developed, and the properties of semiconductor materials and devices became a subject of a large number of basic and applied research efforts. These efforts led a group of the Bell Telephone Laboratories specialists, W. B. Shockley (1910–1989), J. Bardeen (1980–1992), W. Brattain (1902–1987), and their colleagues, to the discovery of the amplification effect in semiconductors and to the invention (1949) of the transistor.[1.6] This work, added to the building of the computer by Vannevar Bush and development of cybernetics by Norbert Wiener, may be considered as the true cornerstone of the modern Information Revolution and the beginning of the triumphal march of semiconductors into virtually all branches of modern life and history.

A substantial development in the search for materials with new combinations of physical and chemical properties was reached when A. Ioffe predicted (1949) and his colleagues proved[1.7] that gray tin, which has a crystal structure similar to that of silicon and germanium, is a semiconductor. The next step was the investigation of the electrical properties of a large number of binary compounds with components that belong to the groups of the periodic table equidistant from group IV. The similarity of their crystal structure with that of silicon and germanium has been known since 1929.[1.8] These compounds belong to one of three types, III-V, II-VI, or I-VII. The average number of valence electrons per atom in compounds of these types is equal to four, the same as in the group IV elements; each atom in the crystal lattice is surrounded by and bonded with four nearest neighbors (tetrahedral coordination). These compounds possess a crystal structure similar to either one of the cubic mineral, sphalerite, or to hexagonal wurtzite. There are fifteen III-V compounds, eighteen II-VI compounds, and four I-VII compounds with this structure and all are semiconductors.

Investigation of properties of the III-V and II-VI compounds and a very successful application of some of them in semiconductor devices, such as fectifiers, photodiodes, Gunn diodes,[1.9] etc., increased interest in the tetrahedral compounds. Using an empirical approach developed (1959) by Goodman[1.10] and ideas of Huggins[1.15] and Grimm and Sommerfeld,[1.16] Goryunova[1.11] developed a method of prediction of composition of chemical compounds with the tetrahedral and octahedral (NaCl-type) coordination of atoms in their crystal lattice. This method is based on consideration of the number of valence electrons in the elements, which is assumed to be equal to the number of the group of the periodic table they belong to (so-called full or normal valency). The goal is to combine a set of (two, three, etc.) representatives of different groups in a composition which has an average number of valence electrons per atom equal to their number in germanium or silicon, i.e., four. Later, Berger and Prochukhan[1.12] suggested a mathematical presentation of Goryunova's method in the form of the equation system

$$\sum_{i=1}^{A} B_i x_i = 4$$

$$\sum_{i=1}^{n} B_i x_i = \sum_{i=n+1}^{A} (8 - B_i) x_i \quad (1.1)$$

$$\sum_{i=1}^{A} x_i = 1$$

for calculating the content of the representatives of different groups of the Periodic table in a compound containing A components, n of which are considered as cations. Here, B_i is the number of the group of the Periodic table to which the i-th component of the compound belongs, and x_i is its concentration in the compound. The first equation of the three represents the condition that the average number of valence electrons per atom is equal to four (tetrahedral rule). The second equation states that in the compound, the number of valence electrons which "cations" give to "anions" is equal to the number of electrons the anion needs to form the octet (the condition of the normal valency). The quotation marks are used to remind us that in the solids with dominating covalent bond, which factually form the class of semiconductors, the ion concept is quite conditional.

Based on the above-discussed mathematical criteria, it is possible to calculate all probable types of tetrahedral substances, taking into consideration only the location of their components in one or another group of the Periodic table. This work has been done[1.11-1.13] and its results are presented later in this book.

In the process of analysis of the normal valency compounds, Goryunova has pointed out that not only tetrahedral phases, but all compounds of this category have to be grouped, first of all, in accordance with the average valence electron content per atom of a compound, Z. Table 1.1 contains all possible types of the binary valence compounds. One can see from the table that if A and B are the cation valencies in a binary compound (or A and B are the numbers of the groups of the Periodic table to which the cation and anion belong), the atomic ratio of components in a valence compound $A_x B_y$ is equal to

$$x/y = (8 - B)/A \quad (1.2)$$

and the valence electron concentration per atom in the binary compound is

$$Z = \frac{8A}{8 - (B - A)} = 8A/(8 + A - B) \quad (1.3)$$

as follows from Equations 1.1.

Table 1.1 shows that there are 21 types of valence binary compounds. From analysis of the available experimental data, Goryunova[1.11] (p. 7) concluded that the phases which possess semiconductor properties are ones with Z between two and six. The phases with Z smaller than two are either metal alloys or the compounds with typical metallic crystal structure and properties. Although there are known semiconductors among the compounds with Z greater than six, e.g., BI_3, SnI_4, or $SbCl_5$ (below 276 K), in these materials, the covalent-ionic forces typical for semiconductor crystals act predominantly within the limits of molecular dimensions, whereas the van der Waals bonds dominate intermolecular interaction. Using as a basis for calculation the magnitudes of Z between two and six in a table similar to Table 1.1, Goryunova showed that there are 148 types of ternary compounds which are the electron

Table 1.1 The Types of the Valence Binary Compounds[a]

Cation valency	Cation to anion content ratio (x/y) for the anion valency					
	II	III	IV	V	VI	VII
I	6:1 (1.14)	5:1 (1.33)	4:1 (1.60)	3:1 (2.00)	2:1 (2.67)	1:1 (4.00)
II	—	5:2 (2.29)	2:1 (2.67)	3:2 (3.20)	1:1 (4.00)	1:2 (5.33)
III	—	—	4:3 (3.43)	1:1 (4.00)	2:3 (4.80)	1:3 (6.00)
IV	—	—	—	3:4 (4.57)	1:2 (5.33)	1:4 (6.40)
V	—	—	—	—	2:5 (5.71)	1:5 (6.67)
VI	—	—	—	—	—	1:6 (6.86)

[a] The numbers in parentheses show the magnitude of Z for the compound type.

analogs of the binary valence compounds (Table 1.2). There is a substantial number of known chemical compounds that belong to the types listed in the table, and among them there is a large group of single-anion and single-cation diamond-like ternary compounds. Some of these compounds possess properties advantageous for their practical applications, but it has to be mentioned that our knowledge regarding these materials is still quite superficial.

Analysis of Equation 1.3 shows, in accordance with data of Table 1.1, that there is not any combination of integers for A and B which gives magnitude of Z equal to three or five. The same conclusion emerges from application of Equation 1.1 to a binary combination with the right-hand side of the first of the equations equal to three or five. Thus, there are binary electron analogs of the elements of second, fourth, and six groups of the Periodic table presented in Table 1.1. This may be an explanation of why Goryunova excluded from consideration possible ternary analogs of the third and fifth groups elements.

Application of Equation system 1.1 for calculation of composition of the ternary electron analogs of the elements of third and fourth groups (the right-hand side of the first of Equations 1.1 is replaced by three or four, respectively) showed[1,13] that there are 20 possible types of ternary electron analogs of the third group elements and 19 types of ternary analogs of the fifth group elements (Table 1.3). Some of the representatives of these types, e.g., compounds $I_9.III VI_6$ type, such as Ag_9GaS_6, are known and the available information regarding them will be presented later.

Equations 1.1 limit the number of components of a compound that belong to different groups of the periodic table — the maximum number is three. This is in very good agreement with the majority of known experimental data: the valence compounds with the number of elemental components of different groups greater than three may be considered as (and in many cases are) the linear combinations of ternary and binary valence compounds. For example, $CuGaGeSe_2$ may be considered as a phase of the (quasi)binary system $CuGaSe_2$-Ge; $Cu_2ZnGeSe_4$ is a combination of two diamond-like compounds Cu_2GeSe_3 and ZnSe, etc. The purely mathematical criteria used in consideration of the limits of existence of the valence compounds cannot predict the character of the chemical interaction of the components that belong to the different periods of the Periodic table. They may be individual phases (berthollides or daltonides), ordered or disordered solid solutions, eutectics, and so on. Formation of semiconductor phases by the chemical elements of different groups and periods of the Periodic table is determined by certain criteria, such as the ability of the atoms in a compound to utilize all or a part of their valence electrons in formation of the interatomic bonds, difference of electronegativities between participating atoms, etc. The list of elemental, binary, and ternary semiconductors that includes only entities with the ratio of components determined

SEMICONDUCTOR MATERIALS: DEFINITION, HISTORY, SYSTEMATIZATION

Table 1.2   Ternary Analogs of the Binary Valence Compounds

Group numbers	Ratio of the components' atomic concentrations in a ternary compound with valence electron concentration is equal to										
	2.00	2.29	2.67	3.20	3.43	4.00	4.57	4.80	5.33	5.71	6.00
I-II-III	1:4:1										
I-II-IV	2:1:1	2:3:2									
I-II-V		4:1:2	1:1:1								
I-II-VI	12:2:3	20:1:7	2:1:2	2:2:3							
I-II-VII	15:2:3	25:3:7	10:1:4	15:1:9	20:1:14		2:1:4	1:1:3			
I-III-IV	5:1:2	7:3:4	1:1:1	1:5:4							
I-III-V		9:1:4	3:1:2	3:3:4	3:5:6						
I-III-VI	9:1:2	15:1:5		5:1:4	3:1:3	1:1:2	2:5:8				
I-III-VII	6:1:1	20:3:5	8:1:3	12:1:7	16:1:11		5:1:8	3:1:6	1:1:4	1:3:10	
I-IV-V		14:1:6	5:1:3	2:1:2	7:5:9	1:2:3					
I-IV-VI	6:1:1	10:1:3		8:1:6	10:2:9	2:1:3	4:5:12	2:4:9			
I-IV-VII	9:2:1	5:1:1	6:1:2	9:1:5	12:1:8		8:1:12	5:1:9	2:1:6	1:1:5	1:2:9
I-V-VI		5:1:1		11:1:8	7:1:6	3:1:4	7:5:16	1:1:3	1:3:8		
I-V-VII		10:3:1	4:1:1	6:1:3	8:1:5		11:1:16	7:1:12	3:1:8	5:3:20	1:1:6
I-VI-VII				3:1:1	4:1:2		14:1:20	9:1:15	4:1:10	7:3:25	3:2:15
II-III-IV				1:2:2							
II-III-V			4:1:1		3:1:3						
II-III-VI			6:2:1	9:2:4	12:2:7		1:2:4				
II-III-VII			3:3:1	3:1:1	16:5:7	4:1:3	6:1:7	8:1:11		1:1:5	
II-IV-V					7:1:6	1:1:2					
II-IV-VI				3:1:1	4:1:2		2:1:4	1:1:3			
II-IV-VII				9:4:2	12:5:4	3:1:2	9:2:10	6:1:8		3:1:10	1:1:6
II-V-VI					4:2:1		7:2:12	4:2:9	1:2:6		
II-V-VII					8:5:1	2:1:1	3:1:3	4:1:5		5:1:15	2:1:9
II-VI-VII							3:2:2	2:1:2		7:1:20	3:1:12
III-IV-V											
III-IV-VI						2:1:1	6:1:7				
III-IV-VII						3:2:1	9:5:7	2:1:2	3:1:5	6:1:14	
III-V-VI						3:1:3		1:1:4			
III-V-VII						6:5:3	4:3:3	2:1:3		4:1:9	
III-VI-VII								1:1:1	2:1:4		
IV-V-VI								2:2:1			
IV-V-VII							4:5:1	1:1:1		4:3:7	2:1:5
IV-VI-VII										4:3:2	1:1:2
V-VI-VII											1:2:1

by their valency ("true chemical", stoichiometric compounds, or daltonides) has about 1100 entries. The list of the semiconductor materials with variable composition (e.g., bertollides, solid solutions of the semiconductor elements, and compounds) would be virtually endless.

The physical and chemical properties of the vast majority of known semiconductors are, by now, known only superficially, with the exception of the Group IV elements and a few of III-V, II-VI, and IV-VI compounds, while the list of new semiconductor phases is continuously growing, in particular, for the account of the compounds containing elements with multiple valency, e.g., transition metals. The huge number of available semiconductor materials makes it a very difficult task to choose the best possible candidates for a particular application, if the materials are not systematized in accordance with their fundamental features. The basic feature of a chemical entity is, apparently, the location of its elemental components in the Periodic table. The available experimental data indicate that semiconductor properties, such as the energy gap, mobility of the current carriers, Seebeck coefficient, etc., of a material strongly correlate with the atomic numbers of the elements used as its components, their electron structures, ionization potential, electron affinity and electronegativity, which certainly obey the Periodic law.

Table 1.3 Ternary Electron Analogs of the Third and Fifth Group Elements

Group numbers	Ratio of the components' atomic concentrations in a ternary compound with valence electron concentration (electron/atom) equal to	
	3.00	5.00
I-II-V	1:4:3	—
I-II-VI	4:1:3	—
I-II-VII	25:2:13	1:2:5
I-III-IV	3:7:6	—
I-III-V	3:2:3	—
I-III-VI	9:1:6	—
I-III-VII	10:1:5	2:1:5
I-IV-V	11:4:9	—
I-IV-VI	14:1:9	2:7:15
I-IV-VII	15:2:7	7:2:15
I-V-VI	19:1:12	5:7:20
I-V-VII	5:4:2	5:1:10
I-VI-VII	5:2:1	13:2:25
II-III-IV	3:2:3	—
II-III-V	10:1:5	—
II-III-VI	15:4:5	—
II-III-VII	20:7:5	12:1:19
II-IV-V	5:1:2	—
II-IV-VI	5:2:1	1:2:5
II-IV-VII	15:7:2	9:1:14
II-V-VI	—	5:4:15
II-V-VII	—	6:1:9
II-VI-VII	—	3:1:4
III-IV-VII	—	9:4:11
III-V-VI	—	5:1:10
III-V-VII	—	3:2:3
III-VI-VII	—	3:4:1
IV-V-VI	—	3:2:3
IV-V-VII	—	6:7:3

In spite of numerous attempts to find principles, which may be used for a reliable prediction of the chemical composition of a semiconductor with a desirable combination of its properties, these attempts until now have not produced a solution. The search for materials with a certain combination of properties for a particular application is continuing to be an art based to some extent on the mathematical criteria described above and the analysis of properties of already known materials and cautious change of their qualitative and quantitative composition, observing the direction in which the properties of interest are changing. In some cases, this process leads to formation of a solid solution of some elemental, binary, or ternary semiconductors; in others, a solid solution of a dopant in the semiconductor matrix. Sometimes, the dopant is a product of a very carefully chosen deviation from the stoichiometric composition of a semiconductor compound.

The list of elemental, binary, and ternary semiconductors, which are considered in this book, is presented in the following tables, in accordance with their or their components' location in the Periodic table.

From Tables 1.4 and 1.5, one can see that while there are only eight elemental semiconductors, the number of known binary inorganic semiconductor compounds exceeds 600 (among them there are more than 250 oxides and chalcogenides and about 200 halides). Even such a fundamental characteristic of a semiconductor as the magnitude of its energy gap is apparent only for less than half of them. The majority of these materials have never been synthesized in the single-crystal or epitaxial layer form; therefore the available (if any) data on their electrical and thermal transport properties and optical characteristics cannot be considered reliable. It will take, apparently, decades of investigation of these materials to estimate their application potential.

Table 1.4 Elemental Semiconductors

Period	Group			
	IIIb	IVb	Vb	VIb
2	B	$C_{diamond}$		
3		Si	P	
4		Ge		Se
5		Sn (gray)		Te

Table 1.5 Binary Semiconductors

"Cation" group	"Anion" group						
	I	II	III	IV	V	VI	VII
I	CsAu		CuB_{24}	Li_2C_2	NaSb	CuO	CuCl
	RbAu		Cu_3B_2	Li_6Si_2	KSb	Cu_2O	CuBr
	NaH			Na_2C_2	CsSb	Cu_2S	CuI
	CuH			Cu_4Si	Li_3Sb	Cu_2Se	AgI
	CuD			Ag_2C_2	Li_3Bi	Cu_2Te	$CuBr_2$
					Na_3Sb	Ag_2O	$CuCl_2$
					K_3Sb	Ag_2S	AuBr
					Rb_3Sb	Ag_2Se	$AuBr_3$
					Cs_3Sb	Ag_2Te	AuCl
					Cu_3As	CuS	$AuCl_2$
					Cu_5As_2	Cu_4Te_3	AuI
					Cu_3P	Au_2S	AuI_3
					Au_2P_3	Au_2S_3	
					H_2As_2	$AuTe_2$	
II		BeH_2	MgB_4	Be_2C	Mg_3As_2	BeO	$MgBr_2$
		MgH_2	CaB_6	Mg_2Si	Mg_3Sb_2	BeS	$SrCl_2$
		CaH_2	SrB_6	Mg_2Ge	Mg_3Bi_2	BeSe	SrI_2
		BaH_2	BaB_6	Mg_2Sn	Mg_3N_2	BeTe	$CdBr_2$
				Mg_2Pb	Mg_3P_2	BePo	$CdCl_2$
				Ca_2C	$Ca(N_3)_2$	MgTe	CdI_2
				Ca_2Si	Ca_3N_2	CaS	$HgCl_2$
				Ca_2Sn	Ca_3P_2	CaSe	$HgBr_2$
				Ca_2Pb	Zn_3Sb_2	CaTe	HgI_2
				Sr_2Ge	Zn_3As_2	ZnO	
				Ba_2C	Zn_3N_2	ZnS	
				Ba_2Si	Zn_3P_2	ZnSe	
				Ba_2Ge	Cd_3As_2	ZnTe	
				Ca_2Si	Cd_3P_2	ZnPo	
				Sr_2C	Sr_3N_2	SrO	
					Ba_3N_2	SrS	
					Ba_3As_2	SrSe	
					Hg_3N_2	SrTe	
						CdO	
						CdS	
						CdSe	
						CdTe	
						CdPo	
						BaO	
						BaO_2	
						BaS	
						BaSe	
						BaTe	

Table 1.5 (continued)   Binary Semiconductors

"Cation" group	"Anion" group						
	I	II	III	IV	V	VI	VII
						HgO HgS Hg$_2$S HgSe HgTe	
III			AlB$_{12}$ AlB$_2$ CeB$_4$ CeB$_6$ SmB$_6$ SmB$_{66}$ EuB$_6$ GdB$_{66}$ LaB$_6$ YbB$_6$ YbB$_{66}$	B$_4$C B$_2$Si B$_6$Si Al$_4$C$_3$ YC$_2$ LaC$_2$ CeC$_2$ CeSi$_2$ PrC$_2$	BN BP B$_6$P BAs AlN AlP AlAs AlSb GaN GaP GaAs GaSb InN InP InAs InSb	B$_2$S$_3$ B$_2$S$_5$ B$_2$Se$_3$ GaS GaSe GaTe InS InSe InTe TlS TlSe TlTe In$_6$S$_7$ In$_4$Se$_3$ In$_6$Se$_7$ In$_4$Te$_3$ Ti$_5$Te$_3$ Al$_2$S$_3$ Al$_2$Se$_3$ Ga$_2$O$_3$ Ga$_2$O Ga$_2$S$_3$ Ga$_2$Se$_3$ Ga$_2$Te$_3$ In$_2$S$_3$ In$_2$Se$_3$ In$_2$Te$_3$ Y$_2$S$_3$ La$_2$S$_3$ La$_2$Te$_3$ Ce$_2$S$_3$ SmS SmSe SmTe EuO EuS EuSe EuTe TmTe YbS YbSe YbTe Nd$_2$S$_3$ Sm$_3$S$_4$ Eu$_3$S$_4$ Gd$_2$S$_3$ Dy$_2$S$_3$ Ho$_2$S$_3$ PrO$_2$ PrO$_3$ Eu$_2$O$_3$ Tb$_2$O$_3$ Tb$_4$O$_7$	AlF$_3$ GaBr$_3$ GaI$_3$ GaBr$_3$ GaI$_3$ ScBr$_3$ YBr$_3$ LaCl$_3$ LaI$_3$ TbF$_3$ TbBr$_3$ TbI$_3$ DyF$_3$ DyCl$_3$ DyBr$_3$ DyI$_3$ HoF$_3$ HoCl$_3$ HoBr$_3$ HoI$_3$ ErF$_3$ ErI$_3$ TmF$_3$ TmBr$_3$ TmI$_3$ YbF$_2$ YbCl$_2$ YbBr$_2$ YbI$_2$ YbF$_3$ YbBr$_3$ YbI$_3$ PrCl$_2$ PrBr$_3$ PrI$_3$ LuF$_3$ LuCl$_3$ LuBr$_3$ LuI$_3$

Table 1.5 (continued) Binary Semiconductors

"Cation" group	"Anion" group					
	I	II	III	IV	V	VI	VII
IV	TiH$_2$		TiB$_2$ ZrB$_2$	SiC TiC ZrC ZrSi$_2$ HfC	SiP SiAs GeP GeAs SiP$_2$ GeAs$_2$ Ge$_3$N$_2$ Ge$_3$N$_4$ SnP SnP$_3$ Sn$_4$P$_3$ TiN TiP ZrN ZrP$_2$ HfN	Ho$_2$O$_3$ Er$_2$O$_3$ Tm$_2$O$_3$ Yb$_2$O$_3$ Lu$_2$O$_3$ GeS GeSe GeTe SnS SnSe SnTe PbO PbS PbSe PbTe GeS$_2$ GeO$_2$ GeO SiS$_2$ SiS SnS$_2$ SnTe$_2$ TiS Ti$_2$S$_3$ TiS$_2$ TiSe$_2$ TiO Ti$_2$O$_3$ TiO$_2$ Ti$_3$O$_5$ Ti$_4$O$_7$ Ti$_5$O$_9$ Ti$_6$O$_{11}$ Ti$_8$O$_{15}$ ZrS$_2$ ZrSe$_2$ ZrS$_3$ ZrSe$_3$ ZrO$_2$ HfS$_2$ HfSe$_2$ HfS$_3$ VS	C$_2$Br$_6$ CBr$_4$ C$_2$Cl$_6$ GeBr$_2$ GeBr$_4$ SnBr$_2$ PbF$_2$ PbCl$_2$ PbBr$_2$ PbI$_2$ TiCl$_2$ TiBr$_2$ TiI$_2$ TiF$_3$ TiCl$_3$ TiF$_4$ TiCl$_4$ TiBr$_4$ TiI$_4$ ZrCl$_2$ ZrBr$_2$ ZrCl$_3$ ZrBr$_3$ ZrF$_4$ ZrCl$_4$ ZrBr$_4$ ZrI$_4$ HfF$_4$ HfCl$_4$ HfBr$_4$ HfI$_4$
V	NbH		VB$_2$ NbB TaB$_2$	VC VSi$_2$ V$_2$Si NbC TaC	VN NbN TaN	As$_2$S$_3$ As$_2$Se$_3$ As$_2$Te$_3$ Sb$_2$S$_3$ Sb$_2$Se$_3$ Sb$_2$Te$_3$ Bi$_2$O$_3$ Bi$_2$S$_3$ Bi$_2$Se$_3$ Bi$_2$Te$_3$ As$_4$S$_4$ VS V$_2$S$_5$ V$_2$S$_3$ VO$_2$	AsI$_3$ SbI$_3$ BiI$_3$ VCl$_2$ VI$_2$ VF$_3$ VCl$_3$ VBr$_3$ VF$_4$ VCl$_4$ VF$_5$ NbF$_5$ NbCl$_5$ NbBr$_5$ TaF$_5$

Table 1.5 (continued)   Binary Semiconductors

"Cation" group	\multicolumn{7}{c}{"Anion" group}
	I	II	III	IV	V	VI	VII
						V_2O_3	$TaCl_5$
						V_2O_5	$TaBr_5$
						V_3O_5	
						Nb_2O_3	
						NbO_2	
						Nb_2O_5	
						TaO_2	
						Ta_2O_5	
						TaS_2	
VI			CrB	$CrSi_2$	$CrSb_2$	Cr_2O_3	TeI_2
			MoB_2	Cr_3Si_2	$CrAs$	Cr_2S_3	TeI_4
			MoB	Cr_3C_2	CrN	Cr_3Se_4	CrF_2
			Mo_2B	MoC	CrP	CrS	$CrCl_2$
			WB_2	Mo_2C	MoP_2	CrO_2	$CrBr_2$
				$MoSi_2$	MoP	MoS_2	CrI_2
				WSi_2	WAs_2	$MoSe_2$	CrF_3
				WC	WN_2	$MoTe_2$	$CrCl_3$
				W_2C	WP	Mo_2S_3	$CrBr_3$
					WP_2	MoS_3	CrI_3
					W_2P	MoS_4	$MoCl_2$
						MoO_2	$MoBr_2$
						Mo_2O_3	MoI_2
						Mo_2O_5	$MoCl_3$
						WO_2	$MoBr_3$
						WO_3	$MoCl_4$
						W_2O_5	$MoBr_4$
						WS_2	MoI_4
						WSe_2	$MoCl_5$
							WCl_2
							WBr_2
							WI_2
							WCl_4
							WBr_4
							WI_4
							WCl_5
							WBr_5
							WF_6
							WCl_6
							WBr_6
VII			MnB	Mn_3C	MnP	MnO	MnF_2
			MnB_2	$MnSi_2$	Mn_3P_2	Mn_3O_4	$MnCl_2$
				$MnSi$	MnP_4	Mn_2O_3	$MnBr_2$
				Mn_3Si	$MnAs$	MnO_2	MnI_2
				ReC_3	Mn_2As	MnS	MnF_3
				$ReSi_2$	Mn_2As_3	$MnSe$	$MnCl_3$
						$MnTe$	$ReBr_3$
						MnS_2	ReF_4
						TcS_2	$ReCl_4$
						$TcSe_2$	$ReCl_5$
						ReO_2	ReF_6
						ReO_3	
						Re_2O_7	
						ReS_2	
						$ReSe_2$	
						Re_2S_7	
VIII			FeB	Fe_3C	Fe_3P	FeO	FeF_2
			CoB	$FeSi$	Fe_2P	Fe_3O_4	$FeBr_2$
			NiB	$FeSi_2$	FeP_2	Fe_2O_3	FeI_2

Table 1.5 (continued) Binary Semiconductors

"Cation" group	"Anion" group						
	I	II	III	IV	V	VI	VII
				Ru_2Si_3	FeP_4	FeS	FeF_3
				RuSi	$FeAs_2$	Fe_2S_3	$FeCl_3$
				Ru_2Ge	FeAs	RuS_2	$FeBr_3$
				Ru_2Ge_3	Fe_2N	$RuSe_2$	$RuCl_3$
				CoSi	$RuAs_2$	$RuTe_2$	RuF_5
				$CoSi_2$	RuP_2	OsS_2	$OsCl_2$
				Co_2Si	RuP_4	$OsTe_2$	$OsCl_3$
				Ni_3C	$RuSb_2$	CoO	OsF_4
				Ni_2Si	OsP_2	Co_3O_4	$OsCl_4$
				PdSi	OsP_4	Co_2O_3	OsI_4
					$OsAs_2$	CoS	CoF_2
					$OsSb_2$	CoS_2	$CoCl_2$
					CoP_2	Co_3S_4	$CoBr_2$
					CoP_3	Co_2S_3	CoI_2
					Co_2P	CoSe	CoF_3
					Co_2As	RhS_3	$CoCl_3$
					$CoAs_2$	RhS	RhF_3
					$CoAs_3$	Rh_2S_3	$RhCl_3$
					RhP_2	$RhSe_2$	RhI_3
					$RhAs_2$	$RhSe_3$	$IrCl_2$
					IrP_2	IrS	$IrCl_3$
					$IrAs_2$	Ir_2S_3	IrI_3
					Ni_5P_2	IrS_2	IrF_6
					Ni_2P	IrS_3	NiF_2
					Ni_3P_2	$IrSe_2$	$NiCl_2$
					NiP_2	$IrSe_3$	$NiBr_2$
					$NiAs_2$	$IrTe_3$	NiI_2
					NiAs	Ni_3S_2	PdF_2
					PdP	NiS	$PdCl_2$
					PdP_2	Ni_3S_4	$PdBr_2$
					PtP_2	NiS_2	PdI_2
					$PtAs_2$	NiSe	PdF_3
					$PtSb_2$	PdS	$PtCl_2$
						PdS_2	$PtBr_2$
						PdSe	PtI_2
						$PdSe_2$	$PtCl_3$
						$PdTe_2$	PtI_3
						PtS	PtF_4
						Pt_2S_3	$PtCl_4$
						PtS_2	$PtBr_4$
						$PtSe_2$	PtI_4
						$PtSe_3$	
						$PtTe_2$	

Table 1.6 contains the list of known ternary inorganic semiconductor materials also presented in accordance with location of their components in the Periodic table.

The distribution of the ternary compounds in Table 1.6 was done strictly in accordance with location of the elemental components in the Periodic table. This may create some doubts regarding the logic of the choice. For example, from the 8A-group elements, only ruthenium and osmium possess valency equal to eight. We found, nevertheless, that this classification is better than any one available in the reference literature, in spite of a certain inconvenience connected to the fact that the elements of A and B subgroups are not separated. Tables 1.5 and 1.6 do not contain solid solutions, with a few exceptions of the ordered variable composition phases in which the component atoms' arrangement is the same through the crystal, so the material may be considered as a "true" chemical compound.

Table 1.6 Ternary Inorganic Semiconductors

Component groups	Compounds					
I-I-IV	KHC_2	$CsHC_2$				
I-I-V	K_2CsSb	K_2RbSb	Na_2CsSb	Na_2KSb	Na_2RbSb	Rb_2CsSB
I-I-VI	Li_3CuO_3	$LiHS$	KHS	$NaHS$		
I-I-VII	KHF_2	K_2RuCl_6	$KAuBr_4$	$KAuCl_4$	$KAuI_4$	$CsAuCl_4$
I-II-VII	K_2BeF_4	$KCaCl_3$	$KHgI_3$	K_2HgI_4	Cu_2HgI_4	Ag_2HgI_4
I-III-I	$LiBH_4$	$LiGaH_4$	$NaBH_4$	KBH_4	$RbBH_4$	
I-III-III	$LiAlB_{14}$					
I-III-V	Li_3GaN_2	$CsBN_4$				
I-III-VI	$LiBO_2$	$Li_2B_4O_7$	KBO_2	$NaBO_2$	$NaAlO_2$	$KAlO_2$	$CuAlS_2$
	$CuAlSe_2$	$CuAlTe_2$	$CuGaS_2$	$CuGaSe_2$	$Cu_2Ga_4Se_7$	$CuGaTe_2$	$CuInS_2$
	$CuInSe_2$	$Cu_2In_4Se_7$	$Cu_3In_5Se_9$	$CuInTe_2$	$CuIn_3Te_5$	$Cu_2In_4Te_7$	$AgAlS_2$
	$AgAlSe_2$	$AgAlTe_2$	$AgGaS_2$	Ag_9GaS_6	$AgGaSe_2$	Ag_9GaSe_6	$AgGaTe_2$
	$AgInS_2$	$AgIn_5S_8$	$AgInSe_2$	$Ag_3In_5Se_9$	$AgInTe_2$	$CuTlS_2$	$CuTlSe_2$
	Cu_3ScS_3	Cu_3YS_3	Cu_3SmS_3	Cu_3GdS_3	Cu_3TbS_3	Cu_3DyS_3	Cu_3HoS_3
	Cu_3YbS_3	Cu_3LuS_3	Cu_5HoS_4	Cu_5LuS_4	Cu_3ScSe_3	Cu_3YSe_3	Cu_3SmSe_3
	Cu_3GdSe_3	Cu_3TbSe_3	Cu_3DySe_3	Cu_3HoSe_3	Cu_3YbSe_3	Cu_5GdSe_4	Cu_5TbSe_4
	Cu_5YbSe_4	Cu_5LuSe_4	Cu_3YTe_3	Cu_3SmTe_3	Cu_3GdTe_3	Cu_3TbTe_3	Cu_3DyTe_3
	Cu_3HoTe_3	Cu_3ErTe_3	Cu_3TmTe_3				
I-III-VII	$NaBF_4$	Na_3AlF_6	$RbBF_6$	$CsBF_4$		
I-IV-V	$NaCN$	$RbCN$	$CsCN$	$CuCN$	$CuSi_2P_3$	$CuGe_2P_3$	$AgGe_2P_3$
	$AuCN$						
I-IV-VI	Li_2CO_3	Li_2SiO_3	Li_4SiO_4	Li_2GeO_3	$Na_2C_2O_4$	Na_2CO_3	$Na_2Si_2O_5$
	Na_4SiO_4	$Na_2Ti_2O_7$	$K_2Si_2O_5$	K_2SiO_3	K_2SiS_3	$Cs_2C_2O_4$	Cu_2CO_3
	Cu_2SiS_3	Cu_2SiTe_3	Cu_2GeS_3	Cu_2GeSe_3	Cu_2GeTe_3	Cu_2SnS_3	Cu_2SnSe_3
	Cu_2SnTe_3	$Cu_4Ge_3S_5$	Cu_8GeS_6	$Cu_4Ge_3Se_5$	Cu_4SnS_4	$Cu_4Sn_3Se_5$	Ag_8SiSe_6
	Ag_8GeS_6	Ag_8GeSe_6	Ag_8GeTe_6	Ag_8SnS_6	Ag_8SnSe_6	$Ag_3Ge_8Se_9$	Na_2GeO_3
I-IV-VII	Na_2SiF_6	Na_2GeF_6	K_2SiF_6	K_2SnCl_6	K_2SnBr_6	K_2ZrF_6	Rb_2SiF_6
	Rb_2GeF_6	Cs_2SiF_6	Cs_2GeF_6	Cs_2SnCl_6	Cu_2SiF_6		
I-V-I	$LiNH_2$	$NaNH_2$	MNH_2				
I-V-VI	$LiNO_3$	$LiPO_3$	$LiVO_3$	$NaNO_2$	$Na_4P_2O_7$	$Na_5P_3O_{10}$	$NaVO_3$
	Na_3VO_4	$Na_4V_2O_7$	KNO_2	K_3AsO_4	$KAsO_2$	Ka_3AsO_3	K_3AsS_3
	Ka_2AsS_4	KVO_3	$KNbO_3$	$NaNbO_3$	$KTaO_3$	$NaAsO_3$	$NaAsO_2$
	Na_4AsO_7	$NaBiO_3$	Li_3PO_4	Li_3AsO_4	$RbNO_3$	Cu_3PS_4	Cu_3AsS_4
	Cu_3AsSe_4	Cu_3SbS_4	Cu_3SbSe_4	Cu_3VS_4	Cu_3NbS_4	Cu_3TaS_4	$NaSbSe_2$
	$CuSbS_2$	$CuSbSe_2$	$CuSbTe_2$	$CuBiSe_2$	$CuBiTe_2$	Cu_3AsS_3	Cu_3SbS_3
	Ag_3AsS_3	Ag_3SbS_3	$AgAsS_2$	$AgAsSe_2$	$AgAsTe_2$	$AgSbS_2$	$AgSbSe_2$
	$AgSbTe_2$	$AgBiS_2$	$AgBiSe_2$	$AgBiTe_2$			
I-V-VII	$NaSbF_6$	KPF_6	K_2TaF_7				
I-VI-VI	$Cu_2Cr_2O_4$	$CuTeO_3$	Ag_2TeO_3				
I-VIII-VI	$AgFeSe_2$	$AgFeTe_2$	$KFeS_2$	K_2RuO_4	$CuFe_2O_4$	$CuFeS_2$	$CuFeSe_2$
	$CuFeTe_2$						
I-VIII-VII	Na_3RhCl_6	Na_2PtCl_6	K_2RuCl_6	K_2OsCl_6	K_2RhCl_5	K_2IrCl_4	K_2IrCl_6
	K_2PdCl_6	K_2PdCl_4	K_2PtCl_6	K_2PtCl_4	K_2PtI_6	$CsPtCl_6$	
II-III-VI	$ZnAl_2S_4$	$ZnAl_2Se_4$	$ZnAl_2Te_4$	$ZnGa_2S_4$	$ZnGa_2Se_4$	$ZnGa_2Te_4$	$ZnIn_2S_4$
	$ZnIn_2Se_4$	$ZnIn_2Te_4$	$CdAl_2S_4$	$CdAl_2Se_4$	$CdAl_2Te_4$	$CdGa_2S_4$	$CdGa_2Se_4$
	$CdGa_2Te_4$	$CdIn_2S_4$	$CdIn_2Se_4$	$CdIn_2Te_4$	$HgAl_2S_4$	$HgAl_2Se_4$	$HgAl_2Te_4$
	$HgGa_2S_4$	$HgGa_2Se_4$	$HgGa_2Te_4$	$HgIn_2Te_4$	$Hg_5Ga_2Te_8$	$Hg_3In_2Te_6$	$HgTlS_2$
	$HgTlSe_2$	$HgTlTe_2$	$CdTlS_2$	$CdTlSe_2$	$CdTlTe_2$	$ZnTlS_2$	$ZnTlSe_2$
	$ZnTlTe_2$	$CdInS_2$	$CdInSe_2$	$CdInTe_2$	$Zn_2In_2S_5$	$Zn_3In_2S_3$	$ZnSc_2S_4$
	$ZnTm_2S_4$	$ZnYb_2S_4$	$ZnLu_2S_4$	$CdLa_2S_4$	$CdCe_2S_4$	$CdCe_2Se_4$	$CdPr_2S_4$
	$CdGd_2S_4$	$CdSc_2S_4$	$CdDy_2S_4$	$CdEr_2S_4$	$CdTm_2S_4$	$CdYb_2S_4$	
II-IV-V	$CaSiN_2$	$ZnGeN_2$	$ZnGeP_2$	$ZnGeAs_2$	$ZnSnP_2$	$ZnSnAs_2$	$ZnSnSb_2$
	$CdSiP_2$	$CdSiAs_2$	$CdGeP_2$	$CdGeAs_2$	$CdGeSb_2$	$CdSiSb_2$	$CdSnSb_2$
	$ZnSiSb_2$	$ZnGeSb_2$	$CdSnP_2$	$CdSnAs_2$	$BeSiN_2$		
II-IV-VI	$BaTiO_3$	$CaTiO_3$	$CaZrO_3$	$CaSiO_3$	Ca_2SiO_4	Ca_2PbO_4	$CaSiS_3$
	$MgSiO_3$	Mg_2SiO_4	$SrTiO_3$	$SrZrO_3$	$SrHfO_3$	$ZnSiO_3$	Zn_2SiO_4
	Be_2SiO_4	$ZnGeS_3$	Zn_2GeS_4	$CdCO_3$	CdC_2O_4	$CdSnO_3$	Cd_2SnO_4
	$Hg_2C_2O_4$						

Table 1.6 (continued) Ternary Inorganic Semiconductors

Component groups	Compounds						
II-V-VI	BeAsO$_4$	CaVO$_3$	Mg$_2$P$_2$O$_7$	Ba$_2$V$_2$O$_7$	Hg$_3$PS$_3$	Hg$_3$PS$_4$	
II-V-VII	Cd$_4$P$_2$Cl$_3$	Cd$_4$P$_2$Br$_3$	Cd$_4$P$_2$I$_3$	Cd$_4$As$_2$Cl$_3$	Cd$_4$As$_2$Br$_3$	Cd$_4$As$_2$I$_3$	
II-VI-VI	MgCr$_2$O$_4$	MgW$_2$O$_4$	CaCr$_2$O$_4$	CaMo$_2$O$_4$	SrSeO$_4$	CdCr$_2$O$_4$	Hg$_2$WO$_4$
	Hg$_3$TeO$_6$						
II-VII-VI	BaMnO$_4$						
II-VII-VII	HgBrI						
II-VIII-VI	MgFe$_2$O$_4$	ZnFe$_2$O$_4$					
III-IV-VI	La$_2$GeSe$_5$	La$_2$SnSe$_5$	Ce$_2$GeSe$_5$	Ce$_2$SnSe$_5$	Pr$_2$GeSe$_5$	Pr$_2$SnSe$_5$	Nd$_2$GeSe$_5$
	Nd$_2$SnSe$_5$	Sm$_2$GeSe$_5$	Sm$_2$SnSe$_5$	Gd$_2$GeSe$_5$	Gd$_2$SnSe$_5$	GdTiO$_3$	TbTiO$_3$
	ErTiO$_3$	YbTiO$_3$	Al$_2$CO	Al$_2$GeSe	Al$_2$GeTe	Al$_2$SnTe	In$_2$GeSe
	In$_2$GeTe	Tl$_2$CO					
III-V-VI	BAsO$_4$	AlPO$_4$	Ga$_6$Sb$_5$Te	In$_6$Sb$_5$Te	In$_7$SbTe$_6$	TlNO$_3$	Tl$_3$PO$_4$
	TlAsS$_2$	TlSbS$_2$	TlBiS$_2$	TlBiSe$_2$	TlBiTe$_2$	TlVO$_3$	Tl$_4$V$_2$O$_7$
III-VI-VI	La$_{10}$S$_{14}$O	NdCrO$_3$	SmCrO$_3$	HoCrO$_3$	YbCrO$_3$		
III-VI-VII	CeOCl						
IV-IV-VI	ZrSiO$_4$	SnGeS$_3$	PbTiO$_3$	PbCO$_3$	Pb$_2$Si$_2$O$_7$	PbGeS$_3$	PbSnS$_3$
IV-V-VI	GeSb$_2$Te$_4$	GeSb$_2$Te$_7$	GeBi$_2$Te$_4$	GeBi$_2$Te$_7$	Sn$_2$P$_2$O$_7$	Sn$_2$As$_2$O$_7$	Pb$_2$As$_2$O$_7$
	Pb$_2$Sb$_2$O$_7$	PbSb$_2$S$_4$	PbBi$_4$Te$_7$				
IV-VI-VI	ZrOS	CSTe	CSeS				
V-VI-VII	VOCl	NbOBr$_3$					
VII-III-VI	MnGa$_2$S$_4$						
VII-V-VI	MnSb$_2$S$_4$	MnVO$_3$	Mn$_2$P$_2$O$_7$				
VII-VI-VI	MnCr$_2$O$_4$						
VIII-V-V	RuPAs	IrAsSb	PdPAs	PtPAs			
VIII-V-VI	FeAsS	FeAsSe	FePS	FeSbTe	OsPS	OsPSe	OsAsS
	OsSbS	CoPS	CoAsS	CoAsSe	CoSbS	Co$_3$V$_2$O$_8$	
VIII-VI-VI	FeCr$_2$S$_4$						

In addition to the inorganic semiconductors, there is a substantial number of organic materials which possess semiconductor properties. These materials will be dealt with in later chapters.

REFERENCES

1.1. W. Smith, *J. Soc. Telegr. Eng.,* 2, 31, 1873.
1.2. M. Faraday, *Exp. Res. Electr.,* Ser. 4, 433, 1883; *Beibl. Ann. Phys.,* 31, 25, 1834.
1.3. C. E. Fritts, *Am. J. Sci.,* 26, 465, 1883.
1.4. L. O. Grondahl and P. H. Geiger, *Trans. AIEE,* 46, 357, 1927.
1.5. J. C. Bose, U.S. Patent 755,840, 1904.
1.6. W. Shockley, *Electrons and Holes in Semiconductors,* Academic Press, New York, 1950.
1.7. A. I. Blum and N. A. Goryunova, *Sov. Phys. Dokl.,* 75, 367, 1950.
1.8. V. M. Goldschmidt, *Sov. Phys. Usp.,* 9, 811, 1929.
1.9. J. B. Gunn, *Solid State Commun.,* 1, 88, 1963.
1.10. C. H. L. Goodman, *Phys. Chem. Solids,* 6, 305, 1958.
1.11. N. A. Goryunova, *The Chemistry of Diamond-Like Semiconductors,* MIT Press, Cambridge, 1965.
1.12. L. I. Berger and V. D. Prochukhan, *Ternary Diamond-Like Semiconductors,* Plenum Press, New York, 1969.
1.13. L. I. Berger, 5th American Conf. on Crystal Growth, Fallen Leaf Lake, CA, Abstr., 1980, 35.
1.14. T. Caillat, A. Borshchevsky, and J.-P. Fleurial, *Proc. 10th Symp. Nucl. Power Propulsion Albuquerque NM,* 2, 771, 1993; Proc. 12th Int. Conf. on Thermoelectricity, 1993.
1.15. M. L. Huggins, *Phys. Rev.,* 27, 286, 1926.
1.16. H. Grimm and A. Sommerfeld, *Z. Phys.,* 36, 36, 1926.
1.17. N. A. Goryunova, Slozhnyie Almazopodobnyie Poluprovodniki (Multinary Diamond-Like Semiconductors), Soviet Radio, Moscow, 1968.

CHAPTER 2

Introduction to Semiconductor Physics

In this chapter, the basic principles of band theory of solids, and semiconductors in particular, are presented in such a way that the mathematical apparatus is used only in the cases where it helps to understand the physical picture. The readers who wish to consider the detailed paths in development of the semiconductor theory are referred to comprehensive reviews, e.g., by Seeger[2.1] and Böer,[2.2] and excellent introductions to physics of semiconductor materials (Hummel[2.3]) and devices (Shur[2.4]).

2.1. CHEMICAL BONDS IN SEMICONDUCTOR CRYSTALS

Atoms in solids are connected by forces of chemical bonding which may be a result of (1) formation of electronic pairs by valence electrons that belong to neighboring atoms (covalent bond, binding energy about 10 eV/atom), (2) formation of cations and anions by transition of valence electrons from one part of atoms forming the crystal to another (ionic bond, binding energy about 5 eV/atom), (3) mutual polarization of neutral atoms and attraction of the formed dipoles (molecular or van der Waals bond, bonding energy about 0.1 eV/atom), and (4) attraction of all free electrons of a solid to all its positive ions (metallic bond, bonding energy about 3 eV/atom). The bond of the first type gives to a crystal high strength, hardness, and brittleness (an example: diamond); ionic crystals (an example: rock salt) with Coulomb forces of interatomic interaction in them are less strong with low hardness (but brittle); metallic bonding makes the material strong and hard, but ductile. The van der Waals bonds dominate in organic solids which have typically low strength and hardness, and low melting temperature. As a rule, the bonding of atoms in semiconductor crystals is performed by a combination of different types of bond, but the covalent bond prevails.

At temperatures closest to 0 K, all solids are either insulators, which possess no free electrons, or metals that contain an amount of free electrons in unit volume of the order of number of atoms in it. At room temperature, a part of insulators, in which the bond energy between a valence electron and the atom it belongs to is under 3 to 5 eV, possesses a noticeable number of free valence electrons. These electrons are called the conduction electrons. This provides an electrical conductivity of the material of the order of 1 mS/cm. This part of insulators forms a vast class of materials called semiconductors. The higher the temperature, the greater is the concentration of free electrons and the greater is electrical conductivity. In metals, the electrical conductivity is decreasing with the temperature growth because the number of free electrons in them does not depend on temperature, whereas the probability of dissipation of electrons by positive ions at higher temperatures is greater. This decreases the magnitude of current in the metal (at a given voltage) and, consequently, the electrical conductivity. While the dependence of electron dissipation on temperature has, in general, the same character in semiconductors, its effect is exceeded by the much greater increase of the number of conduction electrons with the temperature increase.

It makes sense to note that the ionic solids also are insulators at low temperatures, and their electrical conductivity increases with temperature because of the growing probability for the ions in the crystal to break loose from their locations in the crystal lattice and to move in the direction of an applied electric field force. The higher the temperature, the higher is the kinetic energy of the ions in the lattice relative to the potential energy of attraction to their neighbors in the lattice.

The electronic and ionic semiconductors have specific features in the dependence of their properties; first of all, the mobility of the current carriers, on the content of lattice defects and temperature. The mobility of electrons is very high in a crystal without defects and decreases with the lattice imperfection increase. The mobility of ions is greater in a material with imperfections. The mobility of electrons decreases with the temperature growth and especially sharply decreases with transition into the liquid state. The mobility of ions in a salt increases with temperature and is substantially greater in the melt than in the crystal. Based on these features, these two kinds of materials are divided into (electronic) semiconductors and solid electrolytes. This book is devoted to the electronic semiconductors which will simply be called semiconductors.

Specifics of semiconductors, in comparison with metals and insulators, are explained on the basis of the quantum theory of solids. The main feature of the theory is the energy band concept. As we know from atomic physics, the electrons of an atom may occupy only certain orbits, and the parameters of the orbits, such as radius, eccentricity, orbital angular and magnetic momenta, and spin angular and magnetic momenta, are quantized. The maximum number of electrons in a quantum state is determined by Pauli's exclusion principle. In a gas, each atom is an independent quantum system. The distribution of electrons upon the orbits in an atom of the gas does not depend upon the electron distribution in the other atoms surrounding it, because the distance between the atoms is by several orders of magnitude greater than the distances between the electrons and nucleus in an atom.

The electrons occupy one or another of "allowed" states with certain orbit radii, kinetic and potential energy, angular momentum, etc. (energy levels). Electrons may change their energy only by certain portions (quanta) when they jump from one of the allowed states to another. Consequently, the electrons in an atom possess a certain set of the energy magnitudes (levels) available for them, and this set is called the energy spectrum, the same for all atoms of the gas.

If the gas is condensed as a result of the temperature and/or volume decrease to such a degree that the distances between atoms are comparable with the radii of the electron orbits in the atoms (so-called condensed matter), the distribution of electrons upon the energy levels of an atom depends on the distribution of electrons in all other atoms of the (liquid or solid) body. The electrons in a solid material behave in such a way as if they were a part of a gigantic molecule. Of course, the electrons of the innermost atomic orbits are much closer to their own nuclei and so they do not change their energy spectrum, but the outermost, valence electrons experience a substantial change of their energy spectrum. Because the Pauli's exclusion principle does not permit more than one electron in a quantum system to be with a particular set of parameters, each of the individual electron energy levels splits into a band of levels, the number of which is equal to the total number of atoms in the body which is of the order of about 10^{22} to 10^{23}. If the ionization energy of an individual atom (i.e., energy of transition of a valence electron of the atom into the free state) is of the order of 5 to 20 eV, the energy needed for transition of the valence electron of the solid into a free state is zero in metals, between 0.1 and 3 eV in the majority of semiconductors, and about 6 to 8 eV in insulators, at room temperature. This is the result of expansion of the energy levels into the bands, because this expansion decreases the ranges of the forbidden magnitudes of energy between the individual "allowed" electron energy levels of a free atom into the relatively narrow forbidden energy zones.

As was mentioned before, the valence electrons of a solid that left their location around "their" atoms as a result of the action of either the thermal motion of atoms or the bombardment of the solid with photons are called conduction electrons because they can increase their kinetic energy under action of an external electric field, i.e., these electrons provide the electrical conductivity of the solid. In metals, the energy bands of the valence electrons and conduction electrons overlap, or the highest energy of the valence electrons is equal to the lowest energy of conduction electrons. This means that the valence and conduction electrons are indistinguishable and even a small external electric force may accelerate the electrons which are located near the upper "border" of the valence band in the metal in the direction of the force. At any temperature, including very low, metals have a large number of electrons which are able to participate in electric current.

A general expression for electrical conductivity, σ, is

$$\sigma = ne\mu \qquad (2.1)$$

where n is the number of conduction electrons in unit volume, e is the electron charge, and μ is the mobility of the current carriers which is equal to the average velocity the carriers acquire in the unit electric field along the direction of the field. For metals, n is independent from temperature, but mobility is changing antibatically with temperature. At the relatively high temperatures, σ is reverse proportional to the absolute temperature; at very low temperatures it strongly depends on the lattice defects and the presence of impurities, whereas the influence of the thermal motion of atoms and electrons on conductivity is quite small. In semiconductors and insulators the magnitude of n depends exponentially on temperature, so the effect of the temperature dependence of mobility on the temperature dependence of conductivity is small.

In a semiconductor at 0 K, all electrons occupy energy levels of the valence band; the conduction band is empty. When the temperature increases, some of the valence electrons with energy near the top of the valence band may get an additional energy (the source of energy — the thermal oscillations of the crystal lattice) if the energy is equal or greater than the energy gap between the valence and conduction bands. These electrons are responsible for the crystal conductivity. The vacant positions in the valence band may be occupied by the remaining valence electrons. While the electrons of the valence band move in the direction opposite to the direction of the outer electric field, the vacant electron positions are moving in the opposite direction as if they had the positive charge. The number of the electrons in the valence band is very large in comparison with the number of the vacancies; therefore, the motion of these electrons is limited and has such a complicated character that it cannot be described in the traditional terms of mechanics. The result of motion of the valence electrons in an electric or magnetic field is totally equivalent to the result of motion of the relatively small number of the vacant electron positions, if they are considered having a positive charge equal in the absolute magnitude to the electron charge. This motion may be described using common terms of mechanics. The conductivity created by the electrons of the conduction band is called the n-type (negative) conductivity; the conductivity which is the result of the motion of the vacant electron positions of the valence band is called the p-type (positive) conductivity. The vacant positions acquire a certain acceleration and average velocity under action of the electric forces; therefore, they behave as if they had a certain mass (m = F/a) and a certain mobility. These "particles" are called holes (Defektelektronen, in German literature).

The greater is the number of electrons which are transferred from the valence to the conduction band at a given temperature (free electrons and holes generation), the greater is the number of electrons which move back from the conduction to the valence band, giving their energy to the crystal lattice or releasing it in the form of electromagnetic radiation (recombination). As a result, the concentration of the conduction electrons, n, and holes, p, reaches a dynamic equilibrium magnitude which in a semiconductor crystal without defects and impurities (ideal semiconductor) may be expressed as

$$n_i = p_i = AT^{3/2} \exp(-E_g/2k_BT) \qquad (2.2)$$

where A is a constant, E_g is the energy gap, and k_B is the Boltzmann constant. The conductivity formed by these electrons and holes is called intrinsic.

2.2. ELECTRICAL CONDUCTIVITY IN SEMICONDUCTORS

The analysis of motion of the electrons and holes in a crystal under the action of an external electric field is complicated by the fact that these particles, on their way through the crystal, will interact with the internal fields of atoms located along their trajectories. The analysis may be simplified, if we assume that the particles are free (like in vacuum), but their masses differ from the free electron mass. These "artificial" charged particles are called quasiparticles and their masses (m_n^* and m_p^*) are called effective masses. It is possible to show that in an intrinsic semiconductor

$$n_i = p_i = 2(2\pi k_BT/h^2)^{3/2} (m_n^* m_p^*/m_o^2)^{3/4} \exp(-E_g/2k_BT) \qquad (2.3)$$

or

$$n_i = p_i = 4.82 \cdot 10^{15} T^{3/2} (m_n^* m_p^*/m_o^2)^{3/4} \exp(-E_g/2k_BT)$$

where h is Planck's constant and m_o is the free nonrelativistic electron mass. It makes sense to note that the effective mass of electrons and holes in a crystal depends on the direction of their motion in the crystal.

In a real crystal, especially at low temperatures, the concentration of electrons and holes depends on the quantity and chemical nature of the impurities in the crystal. If the binding energy of a valence electron of an impurity atom with this atom is small in comparison with E_g of the semiconductor crystal that hosts the impurity, the atom will release the electron into the conduction band at such a low temperature at which the generation of conduction electrons from the semiconductor atoms is negligibly small. The minimum energy of a conduction electron, independently of origin, is the lower border of the conduction band. It means that the energy level of the impurity valence electron has to be above the upper edge of the semiconductor valence band, i.e., the level has to be in the energy gap close to the lower edge of the conduction band. The impurities of this kind are called donors or the n-type impurities, because they generate conducting electrons without producing an equal number of holes in the valence band. The binding energy of the valence electron of an impurity is reverse proportional to the square of the semiconductor dielectric constant. The typical location of the donor levels for, e.g., germanium is about 10 meV below the bottom of the conduction band (typical donor impurities are As, Bi, Li, O, P, Sb).

Another kind of impurities are atoms which have a tendency to capture the electrons of the valence band, thus turning themselves into the negative ions. The result of this process is the generation of holes in the valence band without simultaneous generation of electrons in the conduction band. The energy levels of these impurities are located in the energy gap,

INTRODUCTION TO SEMICONDUCTOR PHYSICS

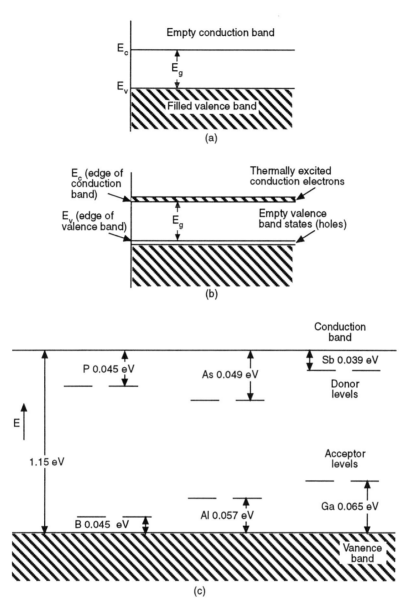

Figure 2.1 Conduction band, energy gap, and valence band of semiconductors. Conductor and valence bands of a pure semiconductor (a) at 0 k and (b) at room temperature; (c) the silicon energy gap with some impurity levels. (After D.H. Navon, *Electronic Materials and Devices,* Houghton Mifflin, Boston, 1975. With permission.)

about 10 to 20 meV above the top of the valence band. These impurities are called acceptors or p-type impurities (typical acceptors for germanium are Al, B, Ga, In, and Ta). In semiconductor compounds, the excessive component atoms may act as donors or acceptors. The band structures of an ideal semiconductor and a semiconductor with the donor and acceptor impurities are presented in Figure 2.1.

In a real semiconductor crystal, when the temperature increases to about 300 K, virtually all donor impurity atoms turn themselves into positive ions, releasing their valence electrons into the conduction band; and all acceptor atoms are negative ions, getting electrons from the valence band and producing holes there. The following increase of temperature results in the growth of the electron concentration in the conduction band for the account of electrons

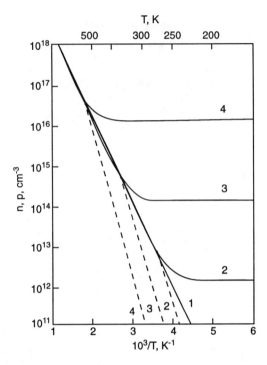

Figure 2.2 Typical temperature dependence of the electron (n) and hole (p) concentration for Ge (solid curves — electrons, broken lines — holes). Curves 1, 2, 3, and 4 are for the samples with $N_D = N_A$, $N_D - N_A = 10^{12}$ cm^{-3}, $N_D - N_A = 10^{14}$ cm^{-3}, and $N_D - N_A = 10^{16}$ cm^{-3}, respectively. (After R.A. Smith, *Semiconductors*, Cambridge University Press, New York, 1959. With permission.)

that transfer there from the valence band with formation in the latter an equal number of hole. Then, the dependence of the concentration of current carriers on temperature becomes the same as if the crystal were an intrinsic semiconductor (Figure 2.2). The temperature of transition from the impurity to intrinsic conductivity is determined by conditions p = N_D or n = $2N_D$ and may be written in the form

$$T_{trans} = E_g / k_B \cdot \ln\left[N_c(T_{trans}) N_v(T_{trans}) / 2N_D^2\right] \tag{2.4}$$

where $N_c(T_{trans})$ and $N_v(T_{trans})$ are the (effective) numbers of states in the conduction and valence bands, respectively, at the temperatures close to T_{trans}.

If the donor impurities dominate in the crystal, the concentration of conduction electrons, n, will be greater than that of holes, p, at any temperature. This crystal is called one with the electron or n-type conductivity. If the concentration of the acceptor impurities, N_A, is greater, the holes will be the majority current carriers and the crystal possesses the hole or p-type conductivity. The holes in the first case and the electrons in the second are called the minority carriers.

The principle of the electrical neutrality of the semiconductor, which states that the total charge of holes and positive donor ions must be equal to the total charge of the conduction electrons and the negative acceptor ions, leads to an important conclusion that independently on the degree of impurity of a semiconductor (unless it is so high that the semiconductor behaves as a metal),

$$np = n_i p_i = n_i^2 \tag{2.5}$$

If all impurity atoms are ionized,

$$n - p = N_D - N_A \tag{2.6}$$

Combining Equations 2.5 and 2.6 gives for $N_D \gg N_A$,

$$n = N_D \left[1 + \left(1 + 4n_i^2/N_D^2\right)^{1/2}\right]/2 \tag{2.7a}$$

and

$$p = 2n_i^2 \left[1 + \left(1 + 4n_i^2/N_D^2\right)^{1/2}\right]/N_D \tag{2.7b}$$

If $N_D \gg N_A$ and $N_D \gg n_i$,

$$n \cong \left(N_D^2 + n_i^2\right)/N_D \tag{2.8a}$$

and

$$p \cong n_i^2/N_D \tag{2.8b}$$

When $N_A \gg N_D$,

$$n = 2n_i^2 \Big/\left[1 + \left(1 + 4n_i^2/N_A^2\right)^{1/2}\right]N_A \tag{2.9a}$$

and

$$p = N_A \left[1 + \left(1 + 4n_i^2/N_A^2\right)^{1/2}\right]/2 \tag{2.9b}$$

If $N_A \gg N_D$ and $N_A \gg n_i$,

$$n \cong n_i^2/N_A \tag{2.10a}$$

and

$$p \cong N_A + n_i^2/N_A \tag{2.10b}$$

Equations 2.7 to 2.10 may be applied for calculation of n and p in a semiconductor crystal using Equation 2.2.

Analysis of motion of electrons and holes under action of an external electric field may be performed using F. Bloch's solution of Schrödinger's equation for an electron moving in the periodic field of an ideal (an infinite and defectless) crystal. The instant velocity of the electron varies depending on distribution of the atomic electric fields in the crystal in the direction of its motion. Its average velocity is equal to

$$\vec{v} = \hbar^{-1} \text{grad } E(\vec{k}) \tag{2.11}$$

where E is the electron energy and $\vec{k}$ is the wave vector of the electron ($\vec{k} = \vec{p}/\hbar$); $\vec{p}$ is the electron quasiimpulse. If an external force $\vec{F} = \vec{F}(\vec{r})$ ($\vec{r}$ is the radius-vector) acts on the electron, it changes its energy with time in accordance with the well-known expression

$$dE/dt = \vec{F} \cdot \vec{v} \qquad (2.12)$$

The electron acceleration ($d\vec{v}/dt = \vec{F}/m$) may be calculated using Equations 2.11 and 2.12:

$$d\vec{v}/dt = \hbar^{-2} \mathrm{grad}\left[\mathrm{grad}\, E(\vec{k}) \cdot \vec{F}\right] \qquad (2.13)$$

In the Descarte's coordinates, Equation 2.13 gives:

$$\partial v_x/\partial t = \hbar^{-2}\left(F_x \frac{\partial^2 E}{\partial k_x^2} + F_y \frac{\partial^2 E}{\partial k_x \cdot \partial k_y} + F_z \frac{\partial^2 E}{\partial k_x \cdot \partial k_z}\right)$$

$$\partial v_y/\partial t = \hbar^{-2}\left(F_x \frac{\partial^2 E}{\partial k_y \cdot \partial k_x} + F_y \frac{\partial^2 E}{\partial k_y^2} + F_z \frac{\partial^2 E}{\partial k_y \cdot \partial k_z}\right) \qquad (2.14)$$

and

$$\partial v_z/\partial t = \hbar^{-2}\left(F_x \frac{\partial^2 E}{\partial k_z \cdot \partial k_x} + F_y \frac{\partial^2 E}{\partial k_z \cdot \partial k_y} + F_z \frac{\partial^2 E}{\partial k_z^2}\right)$$

Equations 2.14 may be presented as

$$\partial v_x/\partial t = F_x/m_{xx} + F_y/m_{xy} + F_z/m_{xz}$$
$$\partial v_y/\partial t = F_x/m_{yx} + F_y/m_{yy} + F_z/m_{yz} \qquad (2.15)$$
$$\partial v_z/\partial t = F_x/m_{zx} + F_y/m_{zy} + F_z/m_{zz}$$

It is possible to show that, if x, y, and z axes coincide with the main axes of the crystal, only m_{xx}, m_{yy}, and m_{zz} in Equations 2.15 are not equal to zero. Then

$$\partial v_x/\partial t = F_x/m_{xx} = F_x/m_{1n}^*$$
$$\partial v_y/\partial t = F_y/m_{yy} = F_y/m_{2n}^* \qquad (2.16)$$
$$\partial v_z/\partial t = F_z/m_{zz} = F_z/m_{3n}^*$$

The tensors m_1^*, m_2^*, and m_3^* are called the electron effective mass tensors. In the case of the primitive cubic lattice, $m_{1n}^* = m_{2n}^* = m_{3n}^*$, and denoting them as m_n^* we may write

$$dv/dt = F/m_n^* \qquad (2.17)$$

From Equations 2.11 and 2.17, it is possible to show that energy of electrons relative to the bottom of the conduction band is

$$E(k) = \hbar^2 k^2 / 2 m_n^* \qquad (2.18)$$

which means that in the coordinate system k_x, k_y, k_z, the loci of the equienergy points are spheres; otherwise,

$$E(k) = \frac{\hbar^2 k_x^2}{2 m_{1n}^*} + \frac{\hbar^2 k_y^2}{2 m_{2n}^*} + \frac{\hbar^2 k_z^2}{2 m_{3n}^*} \qquad (2.19)$$

and the loci of the equienergy points in the k coordinate system (k-space) are ellipsoids. The behavior and properties of holes in the valence band may be described in the same terms. The effective mass of holes, m_p^*, is not equal to m_n^* and both, as a rule, are smaller than m_o (Equation 2.3).

If an electric field is applied to the crystal, the electrons and holes will move under its action. Atoms in their way oscillate; consequently, the electrons and holes will be scattered from time to time. The average time between two sequential acts of scattering, τ_n, τ_p, is called the carrier free path time. If the energy of current carriers in the crystal follows Equation 2.18, the direction of the current density vector, $\vec{i}$, is collinear with the electric field vector, $\vec{E}_e$, and Equation 2.1 may be written in the form

$$\sigma = n e \mu_n + p e \mu_p \qquad (2.20)$$

then Ohm's law may be presented as

$$\vec{i} = \sigma \vec{E}_e = e \vec{E}_e \left(n \mu_n + p \mu_p \right) \qquad (2.21)$$

with

$$\mu_n = e \tau_n / m_n^* \qquad (2.22)$$

and

$$\mu_p = e \tau_p / m_p^* \qquad (2.23)$$

In a semiconductor with nonspherical isoenergy surfaces (Equation 2.19), σ is a tensor with components

$$\begin{aligned} \sigma_{xxn} &= n e \mu_{1n} & \sigma_{xxp} &= p e \mu_{1p} \\ \sigma_{yyn} &= n e \mu_{2n} & \sigma_{yyp} &= p e \mu_{2p} \\ \sigma_{zzn} &= n e \mu_{3n} & \sigma_{zzp} &= p e \mu_{3p} \end{aligned} \qquad (2.24)$$

all nondiagonal tensor components are equal to zero, and, in general, $\vec{i}$ is not parallel to $\vec{E}_e$ with the exception where $\vec{E}_e$ is parallel to the main crystallographic axes.

The τ in Equations 2.22 and 2.23 and, consequently, the mobility depend on the character of interaction of the current carriers with atoms in the crystal. If the carriers are scattered by the longitudinal acoustic oscillations of the crystal lattice, τ and μ are proportional to $T^{-3/2}$. If carriers scatter on charged impurities, the mobility is proportional to $T^{-3/2}$. If carriers scatter on charged impurities, the mobility is proportional to E_c^2 and $T^{3/2}$; scattering of the carriers on the neutral impurity atoms, in cubic crystals, gives $\mu \propto 1/(NE_e)$, where N is the impurity concentration.

2.3. THERMOELECTRIC PHENOMENA IN SEMICONDUCTORS

Three thermoelectric phenomena — the Seebeck effect, the Peltier effect, and the Thomson effect — are related to each other and present a substantial theoretical and practical interest. The Seebeck effect consists of generation of a potential difference between any two points of a conductor, ΔV, if these points have different temperatures. The Peltier effect consists of release or absorption of the heat on the junction of two conductors of different materials when electric current passes through the junction in one or another direction. The Thomson effect consists of the heat absorption or release in a conductor which occurs when an electric current flows in one or another direction through the conductor, if the conductor has a temperature gradient along the current direction.

The Seebeck effect may be expressed in the form

$$\Delta V = S \, \Delta T \qquad (2.25)$$

where S is called the Seebeck coefficient or the absolute thermal electromotive force. S depends on the conductor electronic properties and temperature. The Peltier effect is described by

$$\Delta Q = \Pi \, I \qquad (2.26)$$

where Π is the Peltier coefficient.

If an electric circuit is formed by two conductors with Seebeck coefficients S_1 and S_2 and the junctions have temperatures T_1 and T_2, the current in this circuit is

$$I = \frac{S_{1-2}(T_1 - T_2)}{R} = \frac{S_{1-2} \Delta T}{R} \qquad (2.27)$$

where S_{1-2} is the relative Seebeck coefficient which is a function of S_1 and S_2, and R is the total electrical resistance of the circuit. If the junctions of this circuit have the same temperature, T, and a current is developed in the circuit by an external source (e.g., a battery), the current will absorb heat from one of the junctions and release it on the other, thus forming a temperature difference between the junctions. It is possible to show by the thermodynamic treatment that

$$\Pi_{1-2} = S_{1-2} T \qquad (2.28)$$

In metals, the Seebeck effect is caused by migration of electrons from the part of the conductor with higher temperature to the part where temperature is lower, forming an electric field inside of it, but because the concentration of free electrons is very high and does not depend on temperature, the electric field is relatively small and S has a magnitude of about 0.5 to 10 µV/K. The Seebeck coefficient for a metal may be described by the formula

$$S = -\frac{\pi^2}{3} \frac{k_B}{e} \frac{k_B T}{E_F} \qquad (2.29)$$

INTRODUCTION TO SEMICONDUCTOR PHYSICS

where E_F is called Fermi energy, the maximum energy the electrons may possess in the metal at 0 K.

In semiconductors, the Seebeck coefficient is usually much higher because (1) the concentration of free charges is much smaller and their displacement forms relatively strong electric fields and (2) the concentration of free charges exponentially depends on temperature. Presence of both electrons and holes in semiconductors decreases S because the charge carriers compensate each other. But, as a rule, electrons and holes have different mobility, and S even in an intrinsic semiconductor is relatively high. To increase S, it makes sense to introduce certain impurities in the semiconductor, which produce additional current carriers of a particular sign.

For semiconductors with two kinds of carriers, S may be expressed as

$$S = -\frac{k_B}{\sigma}\left\{n\mu_n\left[A + \ln\frac{2(2\pi m_n^* k_B T)^{3/2}}{h^3 n}\right] - p\mu_p\left[A + \ln\frac{2(2\pi m_p^* k_B T)^{3/2}}{h^3 p}\right]\right\} \quad (2.30)$$

where A is a parameter which determines the charge free path (A is equal to 1, 2, 2.5, or 4 for amorphous semiconductors, for acoustic scattering, for neutral impurity scattering, or for ionized impurity scattering, respectively). For semiconductors with one dominating type of free charges, either one or another item in the right-hand side of Equation 2.30 may be omitted. In semiconductors, S is up to three or even four orders of magnitude higher than that of metals.

2.4. RECOMBINATION OF ELECTRONS AND HOLES

Equations 2.3, 2.7, and 2.9 are based on the assumption that the semiconductor is in a state of thermodynamic equilibrium. If the equilibrium concentrations, n and p, at initial moment (t = 0) are changed by magnitudes Δn and Δp, the n and p will return to the same initial magnitudes at a certain moment after the cause of the change (an external influence) was eliminated. The rates of generation, g, and recombination, r, of the nonequilibrium electrons and holes are important characteristics of a semiconductor, especially for its use in certain semiconductor devices. In a neutral ($\Delta n = \Delta p$) semiconductor,

$$dn/dt \equiv d(\Delta n)/dt = g - n/\tau_n \quad (2.31)$$

where τ_n is the conducting electron's lifetime. At equilibrium, $dn/dt = 0$ and

$$g = n_o/\tau_{no} \quad (2.32)$$

Equations 2.31 and 2.32 give

$$dn/dt = n_o/\tau_{no} - n/\tau_n \quad (2.33)$$

The magnitude of τ_n is reverse proportional to p. A similar expression may be written for holes:

$$d(\Delta p)/dt \equiv dp/dt = p_o/\tau_{po} - p/\tau_p \quad (2.34)$$

with τ_p being reverse proportional to n.

In an n-type semiconductor, $n \gg p$; the assumption that $\Delta p \leq p$ gives $\Delta n \ll n$ and $n \simeq n_o$. Then $\tau_{po} \simeq \tau_p$ and Equation 2.34 gives

$$d(\Delta p)/dt = (p_o - p)/\tau_p = -\Delta p/\tau_p,$$

$$d(\Delta p)/p = -dt/\tau_p$$

and

$$\Delta p = \Delta p_o \exp(-t/\tau_p) \qquad (2.35)$$

Equation 2.35 shows that the number of minority charges decreases e times in the time period τ_p, which is called the lifetime of the excess minority charge carriers. Because the condition of neutrality of the semiconductor requires $\Delta p = \Delta n$, the magnitude τ_p characterizes the average lifetime of the excess electron-hole pairs. In the case of a p-type semiconductor, the lifetime of the excess electron-hole pairs is equal to τ_n.

2.5. PHOTOCONDUCTIVITY AND PHOTO-EMF

Photoconductivity of a substance is the change of its electrical conductivity as a result of absorption of electromagnetic radiation. The absorption of a photon may transfer an electron from the valence to conduction band with simultaneous generation of a hole in the valence band (intrinsic photoconductivity). The extrinsic photoconductivity results in transition of the donor electrons to the conduction band or transition of holes from acceptor impurities to the valence band. These electrons and/or holes change the conductivity of the crystal which the crystal has at the same temperature without irradiation (dark conductivity).

The more free charges per unit of time are generated in a unit of the semiconductor volume by irradiation, the greater is the rate of their recombination. Because photoconductivity in an extrinsic semiconductor is the result of the change of the majority carrier concentration, its magnitude strongly depends on the dark (or thermal) concentration of these carriers. In the intrinsic semiconductors, photoconductivity is ambipolar, i.e., the number of (photo)electrons, Δn, is equal to the number of (photo)holes, Δp. The total electrical conductivity is

$$\sigma = e(n_{dark} + \Delta n)\mu_n + e(p_{dark} + \Delta p)\mu_p \qquad (2.36)$$

$$= e(n_{dark}\mu_n + p_{dark}\mu_p) + e(\Delta n\mu_n + \Delta p\mu_p)$$

Hence, the photoconductivity is

$$\sigma_{ph} = e(\Delta n\mu_n + \Delta p\mu_p) \qquad (2.37)$$

For an intrinsic semiconductor,

$$\sigma_{ph} = e\,\Delta n(\mu_n + \mu_p) \qquad (2.38)$$

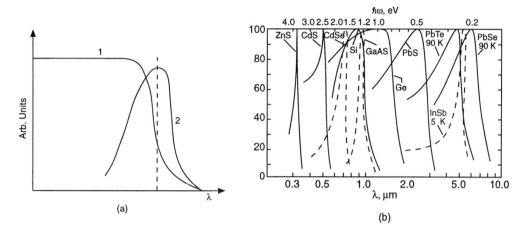

Figure 2.3 Optical and photoelectric properties of semiconductors. (a) Typical spectral distribution of the optical absorption (curve 1) and photoconductivity (curve 2); (b) normalized photoconductivity of some semiconductor materials. (After P.S. Kireev, *Semiconductor Physics*, Mir Publishers, Moscow, 1978. With permission.)

The photoconductivity effect takes place under the condition that the energy of bombarding photons is equal to or greater than the energy gap

$$\hbar\omega \geq E_g \tag{2.39}$$

Experiments show that in many cases the energy of photons responsible for photoconductivity (photoionization energy) differs from the thermal ionization energy, E_g. This can be explained by the great difference between the impulse of photons and impulse of phonons — the quanta of thermal energy of oscillations of the crystal lattice. In accordance with the selection rules, the azimutal quantum number, L, of an electron when it changes its state must be changed only by 1 or –1, so not only energy of the emerging photon, $\hbar\omega$, but its impulse, $\hbar\omega/c$, must meet the transition criteria. A phonon of the same energy, $\hbar\omega$, has an impulse equal to $\hbar\omega/v$, where v is the speed of sound in the crystal, which is about 10^5 times smaller than the speed of light, c. Then, the phonon impulse is so great in comparison with that of photons that the selection rules are not applicable: the electrons have equal probability of transition in any state.

The condition (2.39) implies that the greater the incident photon energy, the greater will be the photoconductivity, but the experimental data show that the photoconductivity exists only in the long-wave part of the absorption spectral area near the wavelength

$$\lambda = 2\pi\hbar c/E_g$$

(see Figure 2.3). This effect may be explained by dependence of the number of recombining electrons and holes on the number of free charges. When total number of the free charges is increasing, the rate of its change is decreasing (see Equation 2.31). At a certain concentration of the electrons and holes released by photons, their recombination will take place so fast that they will not produce virtually any change in the electric conductivity. Decrease of the total number of the free charge, e.g., by decreasing the temperature of a semiconductor crystal, increases its photoconductivity (or so-called photoresponse).

If one of the surfaces of an intrinsic homogeneous semiconductor crystal is illuminated with light of a proper wavelength, the concentration of the free electrons and holes near this surface increases and they diffuse into the crystal. But, if their mobility is different, they will have different diffusion coefficient, D, in accordance with Einstein's equation

$$D = \mu k_B T/e \tag{2.40}$$

and will migrate with the different rate. If the electron mobility is greater, the lighted crystal surface will acquire a positive charge relative to the dark surface. The migrated particles form an electric field inside of the crystal which will oppose the charge motion. At equilibrium, a potential drop of about 0.1 to 0.2 V appears between the lighted and dark surfaces. This effect is called the Dember effect. The direction of the Dember field is such that it enhances the diffusion of the less mobile charge carriers and hinders the diffusion of the more mobile ones. If the electron and hole mobilities are equal, the Dember voltage drop is zero.

2.6. PHENOMENA AT THE INTERFACE OF n-TYPE AND p-TYPE SEMICONDUCTORS

If two pieces of semiconductors having conductivity of different types are pressed to each other, the interface is called a p-n junction. Of course, the ohmic electrical resistance of the junction can be much lower if instead of contacting two crystals, we diffuse donor impurities and acceptor impurities through the opposite surfaces of an initially pure crystal with intrinsic conductivity, forming in the crystal two contacting parts, with n- and p-type conductivity, without formation of the high-resistance area in the region where conductivity changes its type. The effects at p-n junction are utilized in many semiconductor devices. The most important feature of the p-n junction is that its resistance may be varied; this is the place at which the flux of electrons that diffuse from the n-type part, where they are the majority charges, meets the flux of holes which diffuse from the p-type part. They recombine, and the number of the free charge carriers per unit of volume at the p-n junction decreases. This recombination charges n-type part positively and p-part negatively. The resulting electric field inside of the crystal increases its magnitude until the charge migration ceases. Thus, the crystal with a p-n junction acquires the feature of a capacitor consisting of the positively charged n-part of high conductivity, negatively charged p-part, also of high conductivity, and a layer between them (depletion layer) of the p-n junction with low content of free charges and, consequently, low conductivity, that behaves as the insulator in the capacitor. The total resistance of this region is directly proportional to the thickness of the depletion layer. If we connect the n-part of the crystal to the negative (low potential) pole of the battery and the p-part to the positive pole, the electrons from the n-part and the holes from the p-part will migrate in greater number toward the p-n junction; the thickness of the depletion layer will decrease and so its resistance. Then the electric current of a greater magnitude can pass through this electric circuit. This configuration is called "forward bias". If the positive pole of the current source is connected to the n-part and the negative pole to the p-part, the electrons of the n-part and the holes of the p-part will go away from the p-n junction, increasing the thickness of the depletion layer and its resistance. The current in this circuit will decrease to some very small magnitude. This configuration is called "reverse bias". The dependence of current on voltage in this circuit may be described by the Shockley equation

$$I = I_s \left[\exp(\beta eV/k_B T) - 1 \right] \tag{2.41}$$

where β is a constant ($0 < \beta \le 1$) and V is the voltage of the current source, positive for the forward bias and negative for the reverse bias. I_s is called the saturation current, which depends on diffusion characteristics of electrons and holes in the crystal. At $V > 0$ and relatively low temperatures, the first term in the right-hand side of Equation 2.41 grows very fast with V increase, so the second term is negligibly small in comparison with it and the current has an exponential dependence on voltage. With V < 0, the exponential term in the right-hand side becomes negligibly small with the voltage increase, and $I = -I_s$ = constant for a certain range of voltage. The described property of p-n junction is utilized in the semiconductor rectifiers. The same property can be used to make a variable capacitor which capacitance is controlled by applied voltage. Formation in a crystal of two reverse-oriented p-n junctions permits to make a solid-state amplifier of electric signals called a transistor. Intensive recombination of electrons and holes in the depletion layer at the forward bias, resulting in release of energy in the form of electromagnetic radiation, is used in the light-emitting diodes and semiconductor lasers. Specially designed diodes similar to the light-emitting diodes are used for transformation of solar energy in the energy of electric current.

2.7. BRIEF DESCRIPTION OF THE METHODS OF EVALUATION OF THE BASIC PROPERTIES OF SEMICONDUCTOR MATERIALS

The main parameters of a semiconductor which determine its practical applications are sign, concentration, and mobility of the current carriers and their effective mass, the energy gap and location of the donor and acceptor impurity levels in the energy gap, quantity and quality of the crystal lattice defects, and character and content of the surface defects. The majority of these characteristics may be evaluated from the data of measurements of electrical conductivity, thermal emf, galvano- and thermomagnetic effects such as the Hall and Nernst effects, dielectric constant, optical absorption and refraction coefficients, magnetic susceptibility, lattice parameters, elastic constants, thermal conductivity, thermal expansion coefficient, and specific heat.

2.7.1. Electrical Conductivity, Hall Effect, Nernst Effect

The traditional and, apparently, most accurate methods of measurement of electrical conductivity of semiconductor crystals are based on formation and measurement of electric current in a sample of the shape of a cylinder or a prism, resulting from application of a uniform electric field in the sample along the current direction. Measurement of the potential drop between two points in this direction permits us to calculate resistance between the points or (averaged) volume resistivity, ρ, of the sample material in the sample part adjacent to the points, using the well-known formula

$$\rho = VA/IL$$

where A is the sample cross section in the current direction and L is the distance between the potential probes. This method does not provide sufficient information relevant to the impurity distribution in the sample and to the lattice defects. The accuracy of this method strongly depends on the dimensions and electrical properties of the contacts between the sample and the components of the electric circuit used in the test.

There are two methods that may be considered the most convenient for evaluation of the resistivity distribution along semiconductor surface and, to some extent, the distribution of dopants in the sample: the four-point method[2.5] and the van der Pauw method.[2.6] The four-probe method is based on the use of an array of four needle-like electrodes isolated from each other, which are located in a probe in a plane parallel to each other and with equal (0.3 to 1.2 mm) distance between the adjacent needles. Being pressed to a surface, the electrode ends leave four imprints along a straight line. The outermost needles are used usually (but not always[2.7]) to create an electric current in the sample; the inner needles are used to measure the potential drop between the points of their contact with the sample. The resistivity of the sample at the probe location is equal to

$$\rho = F(2\pi s V/I) \qquad (2.42)$$

where F is a dimensionless correction factor, the magnitude of which depends on location of the probe relative to the sample edge,[2.5] dimensions of the sample surface under the test,[2.8] and the sample thickness.[2.9] A description of the measurement technique relevant to semiconductors is presented in Reference 2.8.

The van der Pauw method[2.6] is based on formation of four (or more) point contacts along the circumference of an arbitrarily shaped (but uniform in thickness) sample, provided that the sample does not have isolated holes through it. If the contacts form a sequence A, B, C, and D, the test procedure consists of formation of a current I_{AB} between the contacts marked A and B and measurement of voltage drop V_{DC} between the contacts C and D; then the current is formed between contacts B and C (I_{BC}), and voltage drop is measured between contacts D and A (V_{DA}). Denoting $V_{CD}/I_{AB} = R_{AB,CD}$ and $V_{DA}/I_{BC} = R_{BC,DA}$, one may calculate the volume resistivity of the sample as

$$\rho = \frac{\pi d}{\ln 2} \frac{R_{AB,CD} + R_{BC,DA}}{2} \cdot f\left(\frac{R_{AB,CD}}{R_{BC,DA}}\right) \qquad (2.43)$$

where f is a dimensionless function of the ratio $R_{AB,CD}/R_{BC,DA}$ (f is equal to 1.0, 0.7, 0.4, and 0.26 when the ratio is equal to 1, 10, 10^2, and 10^3, respectively).

Strong dependence of semiconductor resistivity on temperature requires the application of measuring devices that dissipate thermal energy in the sample on the level of nanowatts and even picowatts to diminish the error related to temperature uncertainty. The application of the direct electric field for measurement may be complicated by the electrical polarization of the sample which may lead to erroneous results. Several four-terminal alternating current bridges were designed to diminish the role of these two sources of error. Some of these bridges are also successfully used in the semiconductor resistor thermometry (see, e.g., References 2.11 to 2.13), where these sources of error are especially important.

Combination of data on electrical resistivity with those on the Hall effect, Seebeck effect, and Nernst effect provides quite comprehensive information on basic properties of the charge carriers in a semiconductor material.

Electrical conductivity in the case of simultaneous existence of electrons in the conduction band and holes in the valence band is expressed by Equation 2.21. Hall constant, R_H, in the case of mixed conductivity, may be expressed as

$$R_H = \frac{r_H}{e\sigma^2}\left(p\mu_p^2 - n\mu_n^2\right) \qquad (2.44)$$

where r_H is a dimensionless parameter equal to 1.18 for acoustic scattering of charge carriers and equal to 1.93 for scattering on ionized impurities. Measurements of the thermoelectric effect give the magnitude of the Seebeck coefficient, S, (Equation 2.30) which correlates with the charge concentration and effective mass.

If a semiconductor sample with dimensions l, w, and d is placed in a magnetic field B (B∥d, B⊥l, B⊥w), and the temperature gradient $\Delta T/l$ exists in the l-direction, an electric field V_N/w appears in the sample, parallel to w. This is called the Nernst thermomagnetic effect; it may be described by

INTRODUCTION TO SEMICONDUCTOR PHYSICS

$$\frac{V_N}{w} = Q_N \frac{\Delta T}{l} \qquad (2.45)$$

where Q_N is the Nernst coefficient which depends on the material properties and the sample temperature. For a material with the ambipolar conductivity

$$Q_N = \frac{k_B}{\sigma}\left[r\left(n\mu_n^2 + p\mu_p^2\right) + \frac{n_i^2 \mu_n \mu_p}{\sigma}\left(\mu_n + \mu_p\right)\left(2r_H + 5 + E_g/k_B T\right)\right] \qquad (2.46)$$

From the above discussion, it is possible to conclude that the measurements of the temperature dependence of electrical conductivity, Hall constant, Seebeck coefficient, and Nernst coefficient (Equations 2.3, 2.21, 2.23, 2.29, 2.30, and 2.46) may be used for evaluation of the parameters, such as the energy gap, the sign, concentration, mobility, effective mass, and lifetime of the current carriers.

2.7.2. Generation/Recombination Parameters

The information regarding the electron and hole lifetime and the diffusion length is important for many semiconductor applications, in particular, for light-emitting diodes and semiconductor lasers. There are several recombination mechanisms which control the recombination lifetime. For semiconductors with the direct band gap, such as AlN or GaAs, the recombination results in photon release (radiative recombination). In the indirect band gap semiconductors, a dominating recombination process is one in which the energy of the recombining electron-hole pairs is transferred to phonons, i.e., to the lattice vibrations. This process is called the multiphonon recombination (see e.g., References 2.14 to 2.16). In the semiconductors with narrow energy gap, such as HgCdTe, and in ones with high dopant content, the Auger recombination mechanism may dominate. In the process of Auger recombination,[2.17-2.19] a recombining electron-hole pair gives its energy either to the electrons of the conduction band or to the holes in the valence band.

The total rate, N_{tot}, of recombination (the number of electron-hole pairs recombining in the unit of time) is the sum of contributions from the three mechanisms: radiative, N_r, multiphonon, N_m, and Auger, N_{Au}:

$$N_{tot} = N_r + N_m + N_{Au} \qquad (2.47)$$

Because N is reverse proportional to the lifetime, from Equation 2.47 it follows that

$$\frac{1}{\tau_{tot}} = \frac{1}{\tau_r} + \frac{1}{\tau_m} + \frac{1}{\tau_{Au}} \qquad (2.48)$$

or

$$\tau_{tot} = \left(\tau_r^{-1} + \tau_m^{-1} + \tau_{Au}^{-1}\right)^{-1} \qquad (2.49)$$

i.e., the total lifetime is determined by the shortest of the three lifetimes. Generation of the electron-hole pairs also has the three mechanisms: radiative, by absorption of photons; thermal (multiphonon), by the lattice vibrations (or absorption of phonons); and the impact ionization generation which is a counterpart of the Auger recombination.

One of the first experimental methods used for evaluation of the recombination lifetime is the photoconductive decay method based on the use of Equation 2.35. The application of this method for silicon and germanium was reported by Stevenson and Keyes.[2.19] If a semiconductor sample is connected in series with a constant voltage source and a load resistor, the voltage drop, V_L, on the resistor will be proportional to the sample conductivity if the sample resistance is much greater than that of the load:

$$V_L \propto \sigma$$

If a specially chosen and prepared surface of the sample is lighted by a short light impulse, V_L will sharply increase followed by a relatively slow decrease. The magnitude of V_L change, ΔV, is proportional to $\Delta\sigma$ which, in turn, depends on Δn. If we assume $\Delta n = \Delta p$, then $\Delta\sigma = \Delta n e(\mu_n + \mu_p)$ and

$$\Delta V_{Lt} = \Delta V_{Lo} \exp(-t/\tau_o) \tag{2.50}$$

and τ_o can be calculated from the record of $\Delta V_L(t)$.

There are several other methods of the lifetime evaluation, but their description is outside the limits of this book (see a review of these methods compiled by Maracas and Schröder.[2.20]

2.7.3. Thermal and Thermoelectric Measurements

Methods of measurements of the thermal conductivity, specific heat, and thermal expansion of semiconductors are similar to the methods used for other solids and are described in numerous articles, reviews (see, e.g., Reference 2.13), and industrial standards.[2.21]

Thermal conductivity, κ, of a material is a parameter which is measured by the amount of thermal energy that passes through a unit area of a sample of the material in a unit of time, $(dQ/dt)/A$, in the direction, z, normal to the area, A, if this energy flux is created by the unit temperature gradient along this direction

$$\kappa = -\left(\frac{dQ}{dt \cdot A}\right) \Big/ \left(\frac{\Delta T}{\Delta z}\right) \tag{2.51}$$

Another parameter that characterizes thermal transport through a material is its thermal diffusivity, K. These two parameters are connected by the formula

$$K = \frac{\kappa}{c_p \cdot D} \tag{2.52}$$

where c_p is the specific heat of the material at constant pressure and D is its density (the product $c_p D$ is the volume specific heat of the material). Thermal diffusivity is an important parameter for the heat propagation processes in which the temperature gradients are not uniform in the volume of the sample. A method of thermal diffusivity measurement suitable for semiconductors is described in Reference 2.22.

Among the thermal conductivity measurement methods, the stationary methods, in which a constant temperature difference is developed between the opposite surfaces of a sample by the steady thermal flux from a heat source, connected to one surface, and a heat sink, connected to the opposite surface, are the most accurate, but the measurement consumes more time in comparison with the nonstationary methods (the latter give magnitude of K, from which the thermal conductivity may be calculated using Equation 2.52). The main sources of error are measurements of the thermal energy passing through the sample and the temperature drop, ΔT, between (two) points of the sample along the thermal flux direction. Usually, ΔT for semiconductors is small, so the relative accuracy of its measurement is poor. Increase of ΔT by the increase of the amount of heat passing through the sample leads to additional thermal losses in the surrounding sample media and, hence, increases the error in the estimation of the heat flux in the sample under test.

An improvement of the test accuracy was reported by Berger,[2.23] who used for the thermal conductivity measurement a thermal analog of the electrical Wheatstone bridge in which a thermal flux from a heat source passes through the test sample and a reference sample, connected in series, to a heat sink. Another flux from the same source to the same sink passes through an auxiliary uniform sample. A differential thermocouple with one junction at the contact between the test sample and the reference sample and another junction which can be moved along the auxiliary sample in the thermal contact with it was used. The location of the moving junction at the auxiliary sample, at which the thermocouple reading is zero, shows the ratio of the thermal conductivities of the test and reference samples. Because this is a zero method, the sensitivity of the instrument used with the thermocouple is unlimited, whereas the range of its measurement is not important, and the temperature difference between the heat source and sink may be as small as a fraction of 1 K.

Measurement of the specific heat is performed by the calorimetry methods. Because the specific heat of semiconductors is usually measured in very wide temperature ranges (from 1 to 10 K up to the melting point), the most appropriate method appears to be one or another variant of the Nernst vacuum calorimeter method. In this method, a test sample of a material in an evacuated and sealed refractory vessel (e.g., fused silica, SiO_2) surrounded by a small electric heater, which quite often consists of a few turns of a high-resistance wire, is placed into a regulated thermostat or a cryostat. After the thermostat (and the sample) has reached a chosen temperature, a current is passed through the heater during a certain time period. This increases the sample temperature. The heat the sample received from the heater, the change of the sample temperature, and the sample mass are used to calculate the specific heat of the sample material at the chosen temperature. The correction factor related to the heat capacity of the vessel with the heater and a temperature sensor is measured and/or calculated prior to the sample test. At high temperature, a substantial part of the energy from the heater may be lost through radiation. To decrease the losses, either the sample or the calorimeter inner surface has to be covered by a thin layer of a reflective material, such as silver or palladium.

The coefficient of thermal expansion, α, is a very important parameter which determines compatibility of the parts of a semiconductor device. Dilatometers used for the measurements are divided into two groups: absolute and relative. The more precise but time-consuming absolute measurements are performed with (1) the optical interference dilatometers with very high elongation sensitivity (up to 20 to 30 nm) and (2) the comparator dilatometers in which two microscopes, focused at the ends of a sample located inside of a heater (a furnace), measure the change of the distance between the sample ends. The most common relative method is utilized in the quartz (fused silica, vitreous silica) dilatometer, in which a test sample in the shape of a cylinder is installed upon a silica pedestal inside of a (vertical) silica tube sealed at the lower end; a light silica rod is installed on the upper end of the sample, and the upper end of the rod is connected to the elongation measuring instrument fixed relative to the upper end of the silica tube. The part of the tube with the sample is installed either in a thermostat or a cryostat. When the temperature of this part is changed by ΔT, both the silica parts and the sample change their length. The displacement of the end of the silica rod resulting from the thermal expansion is equal to the difference between the elongation of the sample, ΔL_{sample}, and the elongation, ΔL_{silica}, of the part of the silica tube surrounding the sample and equal to the sample in the length, L. (The other parts of the tube expand equally with the silica pedestal and rod.)

The reading of the measuring instrument is

$$\Delta L = \Delta L_{sample} - \Delta L_{silica} \quad (2.53)$$

Dividing both sides of Equation 2.53 by the product ΔTL, we get

$$\frac{\Delta L}{\Delta TL} = \frac{\Delta L_{sample}}{\Delta TL} - \frac{\Delta L_{silica}}{\Delta TL}$$

or

$$\frac{\Delta L}{\Delta TL} = \alpha_{sample} - \alpha_{silica}$$

and

$$\alpha_{sample} = \frac{\Delta L}{\Delta TL} + \alpha_{silica} \quad (2.54)$$

Semiconductors usually have α of the order of 10^{-6} K^{-1}, whereas $\alpha_{silica} = 5 \cdot 10^{-7}$ K^{-1} at 300 K.[2.24,2.25] At temperature about 200 K, α_{silica} passes through zero and is negative at lower temperatures, reaching the magnitude of $-7 \cdot 10^{-7}$ K^{-1} at about 80 K. Some semiconductors (see following chapters) at these temperatures have $|\alpha|$ of the same order of magnitude, so the error of measurement connected to the properties of fused silica may be quite large. In addition, fused silica, being a glass, slowly changes its coefficient of thermal expansion with repetitious heating and cooling. This complicated behavior of the silica thermal expansion requires a very thorough calibration of the fused silica dilatometers[2.25] for precision α measurements.

The thermal emf test of a semiconductor crystal consists of measuring a potential difference, ΔV, between two metal contacts on the crystal surface, if the temperature of the contacts is different. It is important to note that the measurement has to be done in such a way that the electric current through the crystal is equal to zero (potentiometric method of the voltage drop measurement), because a current through the sample may create an additional temperature difference (the Peltier effect) and an error in the thermal emf evaluation.

Table 2.1 Absolute Thermal emf of Some Metals and Alloys

Material	α (μV/K)	Material	α (μV/K)	Material	α (μV/K)
Sb	+43[a]	Pb	0.0	Ni	−20.8
Fe	+15.0	Sn	0.0	Bi	−68.0
Mo	+7.6	Mg	−0.2	Chromel	+24
Cd	+4.6	Al	−0.4	Nichrom	+18
W	+3.6	Hg	−4.4	Pt-10% Rd	+2
Cu	+3.2	Pt	−4.4	Alumel	−17.3
Zn	+3.1	Na	−6.5	Constantan	−38
Au	+2.9	Pd	−8.9		
Ag	+2.7	K	−13.8		

[a] Sign + means that the hot part of a sample of the material has a positive and cold part a negative potential.

The magnitude of ΔV depends on properties of both semiconductor and contact metal, but, as was mentioned before, the thermal emf of semiconductors is up to 10^3 times higher than that of metals, and, in many cases, the metal part of the total thermal emf may be considered equal to zero. For precision measurements, the contacting metal's thermal emf has to be taken into account. The average magnitudes of the thermal emf of some metals and metal alloys for the temperature range of 273 to 373 K are presented in Table 2.1.

One can see from Table 2.1 that the best contact materials for the thermal emf measurements on semiconductors are lead and tin. Unfortunately, low melting temperature, phase transformations (Sn), and some problems with contact forming limit their application in the thermal emf measuring devices.

REFERENCES

2.1. K. Seeger, *Semiconductor Physics,* 3rd ed., Springer-Verlag, New York, 1985.
2.2. K. W. Böer, *Survey of Semiconductor Physics,* Van Nostrand Reinhold, New York, 1990 (Vol. 1), 1992 (Vol. 2).
2.3. R. E. Hummel, *Electronic Properties of Materials. An Introduction for Engineers,* Springer-Verlag, New York, 1985.
2.4. M. Shur, *Physics of Semiconductor Devices,* Prentice-Hall, Englewood Cliffs, NJ, 1990.
2.5. L. B. Valdes, *Proc. IRE,* 42, 420, 1954.
2.6. L. J. van der Pauw, *Phillips Res. Rep.,* 13, 1, 1958.
2.7. R. Rymaszewski, *J. Phys. E. Sci. Instrum.,* 2, 170, 1969.
2.8. M. P. Albert and J. F. Combs, *IEEE,* ED-11, 148, 1964.
2.9. J. Albers and H. L. Bercowitz, *J. Electrochem. Soc.,* 132, 2453, 1985.
2.10. ASTM Standards F43-88 and F84-88, ASTM 1992 Annual Book of Standards, Vol. 10.05.
2.11. J. W. Ekin and D. K. Wagner, *Rev. Sci. Instrum.,* 41, 1109, 1970.
2.12. P. K. Mukhopadhyay and A. K. Raychaudhuri, *J. Phys. E Sci. Instrum.,* 19, 207, 1986.
2.13. G. K. White, *Experimental Techniques in Low Temperature Physics,* Clarendon Press, Oxford, 1979.
2.14. R. H. Hall, *Phys. Rev.,* 87, 387, 1952.
2.15. W. Shockley and W. T. Read, *Phys. Rev.,* 87, 835, 1952.
2.16. V. S. Vavilov, *Sov. Phys. Usp.,* 68, 247, 1959.
2.17. R. H. Hall, *Proc. IEE,* 106B, 923, 1960.
2.18. J. Dzewior and W. Schmid, *Appl. Phys. Lett.,* 31, 346, 1977.
2.19. D. T. Stevenson and R. J. Keyes, *J. Appl. Phys.,* 26, 190, 1955.
2.20. G. N. Maracas and D. K. Schröder, *Characterization of Semiconductor Materials. Principles and Methods,* Vol. 1, G.E. McGuire, Ed., Noyes Publication, Park Ridge, NJ, 1989, 1.
2.21. ASTM Standards C518-91 and E1225, ASTM 1992 Annual Book of Standards, Vol. 04.06 and 14.02.
2.22. ASTM Standard C714-85, ASTM 1992 Annual Book of Standards, Vol. 15.01.
2.23. L. I. Berger, in Electronic Materials: Technology Here and Now, Proc. 5th Int. SAMPE Electronic Conf. 1991, Vol. 5, R. Keyson, D. Basiulis, and B. Wong, Eds., 1991, 615.
2.24. T. A. Hahn and R. K. Kirby, *Thermal Expansion-1971, AIP Conf. Proc.,* M.G. Graham and H.E. Hagy, Eds., Amer. Inst. of Physics, New York, 1972, 13.
2.25. J. Valentich, *J. Thermal Anal.,* 11, 387, 1977.
2.26. U.S. NIST Standard Reference Material 739, NIST SRM Catalog 1990–91, 104.

CHAPTER 3

Elemental Semiconductors

This chapter contains a description of the elemental materials which possess semiconductor properties. These materials are boron, carbon (diamond), silicon, germanium, gray tin, phosphorus, selenium, and tellurium. They belong to B-subgroups of third, fourth, fifth, and sixth groups of the Periodic table. Investigation of electrical and photoelectrical properties of some elemental substances by Moss[3.1] showed that in thin films some elements, such as arsenic, antimony, and iodine, behave as semiconductors, but detailed studies of their properties have indicated that the elements are either semimetals (As, Sb) or insulators (I) and therefore are not included in this chapter.

3.1. BORON

Boron is the only element of IIIB group which possesses semiconductor properties. As an element, boron was first isolated by Joseph Louis Gay-Lussac and Louis Jacques Thenard (reported June 21, 1808) by treating born sesquioxide with potassium. Sir Humphry Davy announced isolation of boron 10 days later.

The atomic number of boron is 5; the atomic weight is 10.811 in the units of 1/12 of the atomic weight of isotope ^{12}C. Natural boron consists of two stable isotopes, ^{10}B (19.9[2]%) and ^{11}B (80.1[2]%). Known radioactive isotopes have a half-life between 10^{-23} sec and 0.8 sec. Capture of a neutron by a ^{10}B nucleus is followed by reaction $^{10}B(n,\alpha)^7Li$ which is used for neutron registration in the boron counters. The thermal neutron cross section for natural boron is 760 barn, whereas for ^{10}B isotopes it is close to 3800 barn. This strong ability of boron for neutron capture makes it a very undesirable impurity for materials used in nuclear fuel and in construction of nuclear reactors.

The electronic shell configuration of boron is $1s^22s^22p$. Ionization potential of boron is 8.30 eV (first) and 25.15 eV (second). Atomic radius of boron is 0.086 nm. Boron crystallizes in rhombohedral and tetragonal lattices. The basic units of the majority of known allotropic modifications of boron and its compounds are either a B_{12} icosahedron or a B_{12} cube-octahedron[3.2] (Figure 3.1). The B_{12} units combining with individual atoms (or forming a bond through interstitials) form a crystal lattice (mainly rhombohedral[3.3,3.4] or tetragonal[3.5]). The B_{12} units may also combine with each other, forming larger units, B_{84} or $(B_{12})_7$.[3.6] This unit consists of a B_{12} icosahedron and 12 pentagonal pyramids connected with their vertices to the vertices of the central icosahedron. Each of these pyramids is one half of a B_{12} icosahedron

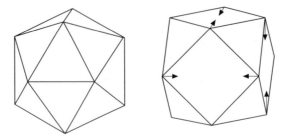

Figure 3.1 Icosahedron (left) and cube-octahedron (right). (After V.I. Matkovich and J. Economy, *Boron and Refractory Borides,* V.I. Matkovich, Ed., Springer-Verlag, New York, 1977, 78. With permission.)

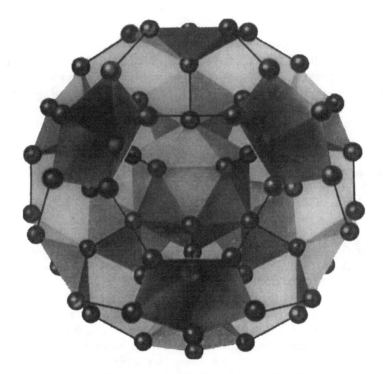

Figure 3.2 Ideal arrangement of boron atoms in the B_{84} polyatomic unit. The arrangement consists of a central B_{12} icosahedron and 12 pentagonal pyramids centered radially at the vertices of the icosahedron. The 60 peripheral atoms occupy vertices of a truncated icosahedron (12 pentagonal and 20 hexagonal faces). (After V.I. Matkovich and J. Economy, *Boron and Refractory Borides,* V.I. Matkovich, Ed., Springer-Verlag, New York, 1977, 78. With permission.)

(Figure 3.2)*. The atoms on the surface of a B_{84} polyatomic unit form a surface similar to the famous bukmasterfullerene unit, C_{60}.

In spite of the high strength of the intra-icosahedral bonds, the boron crystals cannot be considered as molecular ones because the intericosahedral bonds are comparable in strength or even stronger to those inside of the icosahedra.[3.7] A common characteristic of boron crystals is the presence of holes in the crystal lattice framework between the B_{12} or B_{84} structure elements. These holes are so large that the boron crystal can accommodate a large number of single atoms (or their groups[3.2]) in them, forming solid solutions in the boron matrix or chemical compounds without noticeable changes in the crystal lattice dimensions.

---

* Another approach to the B_{84} unit is to consider it as a combination of 12 pentagonal pyramids with their apical atoms forming the central icosahedron and with one atom in the center of each pyramid.[3.7]

Pure boron crystallizes in four modifications: alpha- and beta-rhombohedral and alpha- and beta-tetragonal (see, e.g., Reference 3.9). The alpha-rhombohedral boron (a-rB) lattice is a cubic close-packing of B_{12} icosahedra which generates a simple rhombohedral structure. It is worth mentioning[3.10] that there is no known boron allotrope with hexagonal structure, although a hexagonal close-packed arrangement of B_{12} icosahedra would have the same near-neighbor distribution as in a-rB.

The crystal structure of a-rB was first reported by Decker and Kasper.[3.4] Morosin et al.[3.11] reinvestigated the structure to R = 0.062 on 349 hkl's, including anisotropic temperature factor. A following study of the a-rB by Switendick and Morosin[3.12] showed that the unit cell parameters are a = 0.49075(9) nm and c = 1.2559(3) nm. There is a noticeable variation in the lattice parameter data for a-rB (from a = 0.49 nm to 0.51 nm) presented by different research groups. This variation is, apparently, related to the purity of the primary material and differences in crystal growth techniques. The authors of a compilation (Reference 3.13, p. 4) based on the data taken from Reference 3.9 assume that the "most reliable" parameters for a-rB are a = 0.5057 nm and α = 58.06°. The number of atoms per unit cell is 12, the point group 3m, and the smallest distance between the centers of B_{12} icosahedra in an a-rB rhombohedral cell is close to 0.506 nm. Hence, the lattice constant of the (distorted) cubic face-centered unit that generates the a-rB lattice is about 0.716 nm.

The beta-rhombohedral boron, b-rB, crystal lattice is formed by the B_{84} entities. A description of its structure was presented first by Hoard et al.[3.7] and later by Geist et al.[3.14] According to Callmer's analysis of the b-rB crystal lattice,[3.15] each unit cell has two boron sites which are randomly and partially occupied.

A cubic close-packing of the B_{84} boron entities generates a rhombohedral unit cell with 105 atoms in it. The difference between 84 atoms in a B_{84} "ball" and the number of atoms in a unit cell may be explained by the presence of condensed fragments of icosahedra (B_{10}-B-B_{10}) in the very large octahedral interstices. A study of b-rB crystal structure by Hughes et al.[3.16] led to lattice constants of a = 1.0944 and c = 2.3811 nm with c/a = 2.1757 < $6^{1/2}$. This correlates fairly well with the "most reliable" parameters a = 1.0145 nm and α = 65.17° from Reference 3.13 (p. 4).

Two forms of boron with tetragonal crystal lattice can be grown under certain conditions. The crystal structure of alpha-tetragonal boron, a-tB, was first described by Hoard et al.[3.5] (see also References 3.30 and 3.31). The unit cell consists of four nearly regular icosahedra B_{12} centered at (1/4, 1/4, 1/4), (1/4, 3/4, 1/4), (3/4, 1/4, 3/4), and (3/4, 3/4, 3/4); the interstitial boron atoms are at (0, 0, 0) and (1/2, 1/2, 1/2). The structure as viewed along the c axis is shown in Figure 3.3.[3.17,3.18] The lattice parameters[3.5] are a = 0.875 nm and c = 0.508 nm. The structural formula of this lattice is $(B_{12})_4B_2$ with 50 atoms in the unit cell, but the results of Ploog and Amberger[3.19] and Will and Kossobutzki[3.20] have shown that a-tB is probably a phase of composition $(B_{12})_4B_2C_2$ in which carbon atoms act as lattice stabilizers. The question of the stabilizing role of some trace impurities in the a-tB is still unanswered (see, e.g., Reference 3.21); the situation is similar to the one relevant to the fullerene cubic close-packed C_{60} lattice.[3.22]

The second tetragonal form of boron was first synthesized by Tally et al.[3.6] and was initially thought of as a phase based on B_{84} polyhedra with lattice parameters a = 1.012 nm and c = 1.414 nm (see, e.g., Reference 3.2, p. 82). Later, Vlasse et al.[3.23] described the beta-tetragonal boron, b-tB, unit cell as a tetrahedron based on nonintersecting (but bonded) chains of B_{12} icosahedra. The specifics of this cell consist of existence of (B_{21}) groups in the gaps between the B_{12} chains (Figure 3.4). The (B_{21}) entity consists of two icosahedra twinned across one shared face. There are four (B_{21}) groups in the unit cell. In addition to (B_{12}) and (B_{21}) groups, each unit cell contains ten individual boron atoms. Eight of them are said to satisfy bonding requirements of (B_{21}) groups, the other two occupy random interstitial positions. The distance between adjacent (B_{12}) chains along c axis is c/4. Per Vlasse et al.,[3.23] lattice parameters of b-tB are a = 1.014 nm and c = 1.417 nm. A description of the crystal structure of a tetragonal

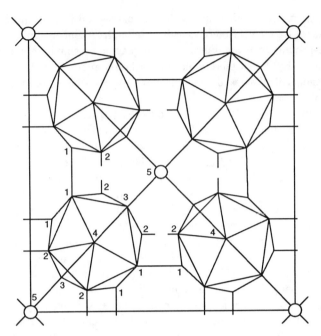

Figure 3.3 Structure of tetragonal boron as viewed in the direction of the c axis. One unit cell is shown. Two of the icosahedral groups (light lines) are centered at z = 1/4, and the other two (heavy lines) at z = 3/4. The interstitial boron atoms (open circles) are at (0,0,0) and (1/2,1/2,1/2). The various structurally nonequivalent boron atoms are identified by numbers. All the extraicosahedral boron nearest-neighbor interactions are shown with the exception of $B_4 - B_4$, which is parallel to the c axis from each icosahedron to the icosahedra in cells directly above and below it. (After L. Pauling and Z.S. Herman, *Advances in Boron and the Boranes*, Vol. 5, J.F. Liebman, A. Greenberg, and R.E. Williams, Eds., VCH Press, 1988, 97. With permission.)

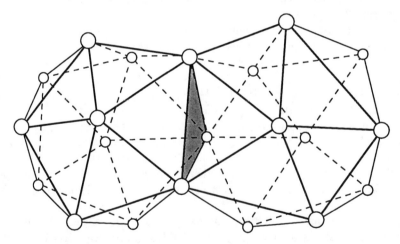

Figure 3.4 (B_{21}) face-sharing icosahedra. (After H.L. Yakel, *Boron-Rich Solids*, AIP Conf. Proc. 140, American Institute of Physics, 1986, 97. With permission.)

boron modification with structural formula $(B_{12})_8B_4$ and lattice parameters a = 0.875 nm and c = 1.015 nm was given by Gorski.[3.28]

Methods of the elemental boron synthesis were reviewed by Bower.[3.24] The most thermodynamically stable allotrope is, apparently, beta-rhombohedral boron with melting point 2450 ± 20 K.[3.25] Holcombe et al.[3.26] found a melting point of 2365 K; the authors of compilation in Reference 3.13 recommend a range between 2347 and 2392 K. The alpha-rhombohedral boron is a low-temperature allotrope which transforms with heating into b-rB near 1370 K[3.27] through a series of intermediate (tetragonal) phases.[3.24]

Many procedures of boron synthesis (see, e.g., Reference 3.29) show that the lowest temperatures (up to 1300 to 1400 K) are favorable for a-rh.B deposition, the range between 1250 and 1550 K is the tetragonal boron area, and b-rh.B is mainly deposited above 1550 to 1600 K. Different technologies are related to somewhat different temperature ranges of the boron allotropes, but sequence remains virtually unchanged.

A substantial number of original reports and several comprehensive reviews contain results of investigation of electrical transport and other electrical properties of boron: the temperature dependence of electrical conductivity and Hall constant, thermoelectric power, and dielectric constant, as well as optical absorption and reflectance. The data on electrical resistivity of boron were first published by Weintraub.[3.32,3.33] About a decade later, Warth[3.34] published results of electrical resistivity measurements, and in 1934 Freymann and Stieber[3.35] reported their results. All these measurements were performed on polycrystalline samples of boron, as well as measurements of thermoelectric power and other electrical properties by Lagrenaudie.[3.36,3.37] First results of measurements on very small (about 5×10^{-6} g) single crystals of tetragonal boron were published by Shaw et al.,[3.38,3.39] who measured resistivity between 200 and 1000 K, Hall effect between 300 and 500 K, thermoelectric effect between 330 and 900 K, and photoconductivity. The band gap reported was 1.55 ± 0.05 eV at temperatures above 800 K, which was substantially higher than that published by Moss[3.1] (1.0 to 1.25 eV) and Lagrenaudie[3.37] (1.2 eV). The large majority of crystals had resistivity about 1.7 MOhm $\times$ cm at 300 K. At lower temperatures, samples had n-type conductivity which changed to p-type at temperatures above 500 K. The authors of Reference 3.39 found that mobility of holes in boron is greater than that of electrons (the latter was of 0.7 ± 0.3 cm^2/V·sec). Later, Jauman and Schnell[3.40] and Anderson et al.[3.41] found the hole to electron mobility ratio close to 2.9.

The boron intrinsic band gap, in accordance with published results, has magnitudes from 1.00 to 1.60 eV from the resistivity measurements,[3.39,3.42-3.48] and from 1.30 to 1.62 eV from optical measurements.[3.24,3.48,3.49] The Hall effect measurements performed by Brungs[3.50] gave the magnitude of 1.38 eV. As could be expected, the bulk of data relates to the measurements on beta-tetrahedral boron.

Investigation of the boron thermoelectric power, S, by different research groups has led to a wide spectrum of data correlated to the samples' technology. Shaw et al.[3.39] measured S on (tetragonal) microcrystals and found its magnitude close to +0.60 mV/K and practically independent on temperature in the range from 500 to 700 K for the samples with typical resistivity between 1 kOhm·cm and 1 Ohm·cm, respectively. King et al.[3.46] found that maximum magnitude of S measured on a beta-rh.B single crystal equal to 0.76 mV/K around 370 K. Dietz and Herman[3.45] found decreases from 0.63 to 0.33 mV/K with a temperature increase from 450 to 900 K. Very close S data were reported by Jaumann and Schnell[3.40] for the temperature range 570 to 970 K. Anderson et al.,[3.41] who measured S on polycrystalline samples of a-tetragonal and b-rhombohedral boron, found its magnitude increasing with temperature from 0.15 to 0.66 mV/K in the range from 300 to about 600 K. The measurements were performed under 44-MPa pressure.

Thermal conductivity of elemental boron reported by different authors[3.51-3.54] has certain similarities, namely, at low temperatures it shows T^3-dependence, it passes through a maximum near 50 K with a magnitude of 3 W/cm·K,[3.54] and decreases with raise of temperature above 50 K. Thompson and McDonald[3.54] found thermal conductivity magnitude at 300 K close to 0.6 W/cm·K. Slack[3.51] reported a maximum of about 4 W/cm·K also near 50 K. Golikova et al.[3.53] found a magnitude of about 0.1 W/cm·K at 1000 K. All cited results were received on samples of b-rh.B. Measurements (see, e.g., a review by Türkes et al.[3.55]) showed that thermal conductivity along c-axis of b-rh.B is only slightly higher than its magnitude across c-axis in the range of temperatures 3 to 300 K. This may be considered as an indication that the main sources of phonon scattering are impurities as well as one- and two-dimensional lattice defects (see discussion in References 3.52 and 3.56). Polycrystalline samples of b-rh.B have a maximum of thermal conductivity close to 0.4 W/cm·K at about 200 K, but at room temperature and up to 1000 K the thermal conductivity of single crystals is lower than that of polycrystalline samples. This feature of boron may be of interest for thermoelectric applications.

Specific heat of (beta) boron, C, was first measured by Johnston et al.[3.56] Later, Bilir et al.[3.57] reported the results of C measurements at low temperatures. Türkes et al.[3.55] measured C between 1.5 and 100 K and found that at low temperatures it approaches a T^3-dependence. From these data, they found Debye temperature θ_D = 1520 K, whereas Silvestrova et al.[3.58] found θ_D = 1370 K from the speed of sound measurements at room temperature. Wysnyi and Pijanowski[3.59] reported C_p (298.15 K) = 11.090 J/mol·K (1.026 J/g·K) in fair agreement with data presented in Reference 3.55.

Photoconductivity of boron was first described by Freymann and Stieber[3.35] and later by Moss.[3.1] A detailed investigation of photoconductivity of beta-rh. boron was performed by Neft and Seiler,[3.60] Gaulé et al.,[3.61] and Nadolny et al.[3.62] Their data are in satisfactory agreement with the results of investigations of dependence of electrical resistivity on temperature, as well as with the results of the optical properties study.

The transmission of light by b-rh.B is limited by a transparency window between 900 and 8000 nm (in the temperature range from 290 to 870 K) with maximum transmission near 3500 nm and an absorption edge at about 800 nm.[3.40,3.46,3.61] Alpha-rhomb. boron is transparent for visible red light, in accordance with Amberger and Dietze[3.63] and Newkirk.[3.64] Refractive index of b-rh.B depends on the technology of sample preparation and has a magnitude from 3.29 to 3.50 in the range 410 to 650 nm.[3.65,3.66] Its magnitude in the range 2 to 12 μm is between 2.6 and 2.7.[3.40]

Boron possesses outstanding mechanical properties. Herring[3.67] reported the elasticity modulus of boron up to 0.48 TPa at room temperature, which is at least twice as high as its magnitude for steel. Wawner[3.68] found tensile strength up to 4.8 GPa. This is at least 30 times higher than tensile strength of the water-quenched and tempered SAE 1340 steel at 640 K. Speerschneider and Sartel,[3.69] studying the response of boron to tensile stress, found that the dimensional response of boron samples to stress was entirely elastic. The samples tested returned to their initial dimensions after being strained; no dislocation movement was noticed.

The hardness of boron is 9.5 in the original Mohs' scale. Knoop hardness number for beta-rhombohedral boron is between 2110 and 2580 and Vickers hardness number is 5000.[3.64,3.70]

Elemental boron may be prepared by chemical or electrolytic reduction of its halides, hydrides, or oxides, as well as by direct thermal decomposition of boron hydrides and boron triiodide. Purification of elemental boron may be performed by chemical methods, by zone refining, and by evaporation of boron under high vacuum using heating or electron bombardment, with a following condensation of boron on a substrate. The problem of preparation and purification of boron was reviewed by Newkirk,[3.64,3.71] Bower,[3.24] Starks et al.,[3.72] and Becher and Mattes.[3.73] Current price of elemental boron is $280 per gram of polycrystalline, 4- to 6-mm-diameter rods with purity of 99.999%, and $150 per gram of boron powder with purity of 99.9995%.[3.74] Boron with purity of 99.5% is much less expensive ($2.69/gram). The price of boron filament of technical purity is about $1500 per kilogram.[3.24]

Practical applications of elemental boron are based, first of all, on its outstanding mechanical properties. Boron filaments are successfully used as reinforcement in composites with epoxy resin or aluminum as the matrix material (see, e.g., Reference 3.68). One of the applications of boron in electronics is based on a large difference in thermal neutron capture cross section of ^{10}B and ^{11}B.[3.75] A thermal neutron detector consisting of a pair of thermistors, one made from ^{10}B and another from ^{11}B, was designed and a prototype was successfully tested.[3.77] Resistors based on a mixture of boron with $MoSi_2$ and designed for use in potentiometers and rheostats were patented by Ruben.[3.78]

The factor which slows down the use of semiconductor properties of boron is the specific of its crystal structure with a large number of "interstitial" vacancies that complicate the purification of boron and single crystal growth. This specific also causes high chemical activity of boron at elevated temperatures, resulting in absorption of impurities from substrates, containers, etc. Another particularity of boron is the existence of many crystallographic modifications which are stable in several ranges of temperature. This creates difficulties in the use of traditional methods of crystal growth and purification, such as Czochralski's and Pfann's processes.

Similar problems existed in the diamond technology, but they were essentially solved when the needs of semiconductor electronics forced technologists to find a solution. It is possible to assume that boron will find some effective applications in electronics in the future, among which solar energy devices, thermistors, tensometers, and high-pressure manometers may be listed in addition to the applications mentioned above.

3.2. DIAMOND

Diamond is an allotrope of carbon in the crystal lattice of which every atom is surrounded by four atoms located in the vertices of a tetrahedron. A vast number of solids with similar arrangements of atoms — and the majority of them are semiconductors — are called diamond-like materials.

Current applications of diamond utilize the extreme properties it possesses, namely, uniquely high hardness and thermal conductivity in combination with exceptional optical transparency, thermal stability, chemical inertness, and low friction coefficient.[3.79-3.81,3.183] Relatively high carrier mobility and intrinsic (indirect) energy gap close to 5.5 eV[3.82] make diamond a promising material for high-temperature semiconducting and optical devices. Interest in diamond as a material for electronic devices was enhanced by the indications that epitaxial diamond films may be grown on diamond and some other substrates[3.83] (see, also, Reference 3.84).

Carbon basically exists in two allotrope modifications: hexagonal graphite and tetragonal diamond. The p-T diagram of carbon as was presented by Bundy[3.85] is shown in Figure 3.5. From the diagram, it is clear that synthetic diamond may be produced by metallurgical methods only at very high pressures and temperatures which technically allow growth of relatively small crystals. By now and apparently in the foreseeable future, the cost of synthetic gem-quality diamonds is substantially higher than that of native diamonds, and many physical parameters of natural diamonds important for their applications in electronic and optical devices are superior to those of synthetic diamonds.

Natural diamonds were divided by Robertson et al.[3.86] into two groups in accordance with certain similarities of their properties. Those groups were called Type I and II. Later, Custer[3.87] suggested a division of Type II diamonds into two subgroups, Type IIa and IIb. Type IIb diamonds have an appreciable electrical conductivity, whereas the diamonds of other types are insulators. Type II diamonds are, in general, more pure than the ones of Type I and are transparent for UV light. Currently, it is common to divide natural diamonds into four types: Ia, Ib, IIa, and IIb depending on their optical and luminescence properties which depend, in their turn, on quality and quantity of impurities. Type I diamonds are UV absorbent. Some natural diamonds are of "intermediate" types.

The most common impurity, presented in the diamonds of all types, is nitrogen.[3.89,3.90] The high conductivity and semiconducting behavior of the Type IIb diamonds are, apparently, a result of the presence of the substitutional boron in the diamond crystal lattice.[3.91]

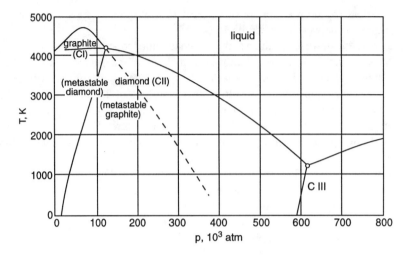

Figure 3.5 Pressure-temperature diagram for carbon. (After F.P. Bundy, *Science*, 137, 1055, 1962. With permission.)

Electrical properties of insulating diamond were reported first by Wartenberg,[3.92] and later by Mendelson and Dember.[3.93] At room temperature, its resistivity is of the order of 10^8 Ohm·cm. The electrical properties of semiconducting natural diamond crystals were measured at different conditions by Austin and Wolfe,[3.94] Wedepohl,[3.95,3.96] Kemmey,[3.97] and later by Williams[3.98] and Collins and Williams[3.99] (see, also, Collins and Lightowler[3.100]). The majority of the reported data on temperature dependence of resistivity, ρ, by different authors shows a noticeable similarity (Figures 3.6 and 3.7), namely, the temperature increase from about 150 to 200 K is followed by the decrease of ρ close to linear in coordinates log ρ vs. 1/T. The resistivity reaches a minimum near 600 to 700 K and is increasing with further temperature growth. This minimum is, apparently, the result of saturation of the free current carriers (holes) concentration with the temperature increase. The maximum free hole concentration, p, in the extrinsic part of conductivity is close to the difference between the acceptor, N_A, and donor, N_D, concentrations. The slope of the ln ρ vs. 1/T curve gives the acceptor ionization energy, E_A, magnitude ranging from 0.32 to 0.38 eV. A thorough investigation of five carefully selected diamonds led Collins and Williams[3.98,3.99] to an E_A average magnitude of 368 ± 1.5 meV.

The Hall effect measurements (Figures 3.8 and 3.9) show a decrease of the Hall constant and saturation of p with the temperature increase. Collins and Williams found the magnitudes of N_A and N_D in the diamonds investigated close to 6×10^{16} and 2.5×10^{15} cm^{-3}, respectively. As was mentioned before, the most probable candidate for the acceptor impurity in natural diamonds is substitutional boron and the donor impurity is nitrogen. The confirmation of this can be found in the results of the electrical properties study of synthetic semiconducting diamonds. Williams et al.[3.101] carried out electrical transport measurements on a series of synthetic diamonds in the range of temperatures between 80 and 450 K. An assumption that aluminum may be an electrically active impurity in natural diamonds was proven erroneous (see, e.g., Chrenko[3.91,3.102]). An additional proof of electrical activity of boron can be found in the results of experiments with ion implantation into diamond (Vavilov[3.104]).

Thermal properties of diamonds were studied, first of all, for calculation of the graphite-diamond equilibrium conditions which, in their turn, could be used for the artificial diamond synthesis (see, e.g., References 3.105 to 3.107). The equilibrium curve may be calculated using the equations

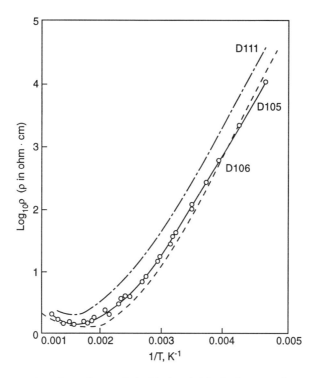

Figure 3.6 The temperature dependence of electrical resistivity in three specimens of diamond. (After P.J. Kemmey and P.T. Wedepohl, *Physical Properties of Diamond*, R. Berman, Ed., Clarendon Press, Oxford, 1965, 325. With permission.)

$$G = F + pV$$
$$F = U - TS$$
$$H = U + pV$$

or

$$G = U + pV - TS = H - TS \tag{3.1}$$

where G is thermodynamic potential (Gibbs energy), U is internal energy, F is free energy, p, V, T, S, and H are pressure, volume, temperature, entropy, and enthalpy, respectively. A two-phase system (e.g., diamond-graphite) is in equilibrium when the thermodynamic potentials of both phases are equal. Therefore, from Equation (3.1)

$$H_g - TS_g = H_d - TS_d$$

or

$$\Delta H - T\Delta S = 0 \tag{3.2}$$

where subscripts g and d mean graphite and diamond, respectively, ΔH is the heat of the transition graphite-diamond, and ΔS can be derived from the specific heat at constant pressure. The pressure of equilibrium can be calculated from the data on volume thermal expansion coefficient and compressibility.

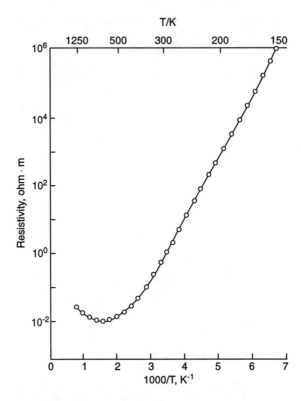

Figure 3.7 The temperature dependence of electrical resistivity of a diamond specimen with $N_A = 8.3 \cdot 10^{16}$ cm^{-3}, $N_D = 0.42 \cdot 10^{16}$ cm^{-3}, and $E_A = 368.1$ meV. (After A.T. Collins and E.C. Lightowler, *The Properties of Diamond*, J.E. Field, Ed., Academic Press, New York, 1979, 79. With permission.)

The heat of transition, ΔH, from graphite to diamond at standard conditions can be determined from the difference in the heat of combustion of graphite and diamond and then can be corrected for any temperature using the differences in their specific heat at any chosen pressure (e.g., standard pressure).

$\Delta H_{25°C}$ was derived by Prosen et al.[3.108] Its magnitude (453 ± 75 cal/mol) is in satisfactory agreement with the data of Hawtin et al.[3.109] (1872 ± 75 J/mol). The data on specific heat of graphite were reported by DeSorbo and Tyler,[3.110] Bergenlid et al.,[3.111] Keeson and Pearlman,[3.112] and Bowman and Krumhansl[3.113] below room temperature, and by Lucks et al.[3.114] and Rasor and McClelland[3.115] above room temperature up to 3900 K. The specific heat of diamond below room temperature was measured by DeSorbo,[3.116] Burk and Friedburg,[3.117] and Desnoyers and Morrison.[3.118] Measurements above room temperature up to 2600 K were performed by Victor,[3.119] McDonald,[3.120] and West and Ishinara.[3.121]

Thermal expansion and molar volume of graphite were evaluated by Nelson and Riley,[3.122] Steward and Cook,[3.123] Steward et al.,[3.124] and Kellett and Richards[3.125] (between 77 and 3300 K). Thermal expansion and molar volume of diamond were investigated by Straumanis and Aka,[3.126] Thewlis and Davey,[3.127] Skinner,[3.128] Kaiser and Bond,[3.129] Novikova,[3.130] Mykolajewycz et al.,[3.131] and Wright.[3.132] A detailed analysis of diamond thermal expansion based on a large series of measurements was performed by Slack and Bartram.[3.133] These data, together with data on the bulk modulus of graphite and diamond (Drickamer et al.[3.134]), were used by Berman[3.107] to calculate the equilibrium pressure at temperatures from 298 to 3000 K.

Thermal conductivity, κ, of diamond is of particular interest because its magnitude in the wide range of temperature is higher than or comparable with that of metals such as copper

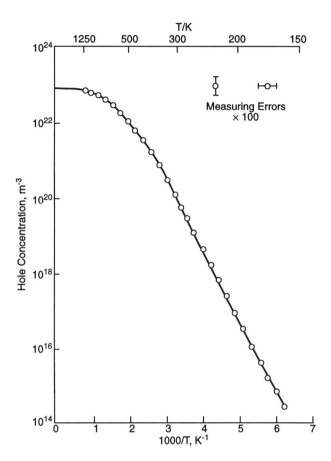

Figure 3.8 The temperature dependence of the hole concentration for the diamond specimen described in the caption to Figure 3.7. (After A.T. Collins and E.C. Lightowler, *The Properties of Diamond*, J.E. Field, Ed., Academic Press, New York, 1979, 79. With permission.)

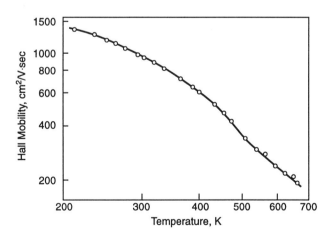

Figure 3.9 The temperature dependence of the Hall mobility for a p-type diamond specimen. (After A.T. Collins and E.C. Lightowler, *The Properties of Diamond*, J.E. Field, Ed., Academic Press, New York, 1979, 79. With permission.)

(maximum of 196 W/cm·K at about 10 K; 3.98 W/cm·K at 300 K) and silver (maximum of 193 W/cm·K at about 7 K; 4.27 W/cm·K at 300 K).[3.135] At room temperature, κ of diamonds is about an order of magnitude higher than κ of Ag and Cu. Semiconductor devices based on diamond and those with diamond heat sinks operate at substantially higher current densities than the devices based, say, on silicon with conventional metal (copper) heat sinks. The thermal conductivity of diamonds of all types and that of synthetic diamonds passes through its maximum in the range of temperatures between 70 and 90 K (see Figures 3.10 to 3.13). Type IIa diamonds possess κ at room temperature greater than any other known substance, including diamonds of other types (at least seven times higher than that of silver and copper). Data on κ of diamonds were reported first by Eucken[3.136] and later by DeHaas and Biermasz,[3.137] but their results were later shown to have inaccuracies related to the test methods used or to the dimensions of the samples used.[3.140] The κ of semiconducting Type IIb diamond between 20 and 300 K was measured by Berman et al.[3.140] Their results at room temperature were substantially lower than data reported by Burgemeister and Seal,[3.143] who measured κ between 310 and 420 K.

The κ of Type IIa diamonds was measured by Berman et al.[3.140] and Berman and Martinez.[3.144] It is interesting to note that the magnitudes of κ measured by Slack[3.142] on two very pure synthetic diamonds are, in common features, very close to the κ for most natural Type IIa diamonds.

Type I diamonds, in general, have lower thermal conductivity in comparison with Type II ones. The reason for it is apparently in the presence in the former a substantial amount of impurities in the form either of the point or planar defects (platelets). At its peak, κ of the Type I diamonds is about three times lower than that of the Type II diamonds (Figure 3.10a). At low temperatures, κ shows a noticeable dependence on the cross-sectional dimensions of the specimens (see Figure 3.10b). The difference between κ of the diamonds of two types may be even greater depending on the impurity content. Kaiser and Bond[3.145] found a linear correlation between the absorption coefficient, K, in diamond at 7.8 μm and the nitrogen content, n, in the form

$$K(cm^{-1}) = 1.7 \times 10^{-19} n \ (cm^{-3})$$

Martinez[3.146] showed (see Figure 3.11) the existence of very clearly visible correlation between κ and K of the Type I diamonds.

The elastic properties of diamonds were measured by different methods, and the known results have some deviations from each other. Bulk modulus data were reported first by Adams[3.147] and Williamson.[3.148] Bhagavantam and Bhimasenachar[3.149] were, apparently, the first who measured the diamond elastic constants by studying the velocity of the acoustic wave propagation through crystals. Prince and Wooster[3.150] used diffuse reflection of X-rays for determination of elastic constants of diamond crystals. The measurements of McSkimin and Bond[3.151] gave for second-order elastic moduli magnitudes $c_{11} = 1.07$ TPa, $c_{12} = 0.125$ TPa, and $c_{44} = 0.576$ TPa. Later, McSkimin and Andreatch[3.152] measured the dependence of the elastic moduli on pressure and temperature. Their results are presented in Table 3.1. The later data published by Grimsditch and Ramdas[3.153] ($c_{11} = 1.0764$ TPa, $c_{12} = 0.1252$ TPa, $c_{44} = 0.5774$ TPa) are in good agreement with the elastic moduli magnitudes from Reference 3.152.

One can see that diamond does not obey the Cauchy relations for cubic crystals ($c_{12} = c_{44}$). On the other hand, the anisotropy factor

$$A = 2c_{44}/(c_{11} - c_{12})$$

ELEMENTAL SEMICONDUCTORS

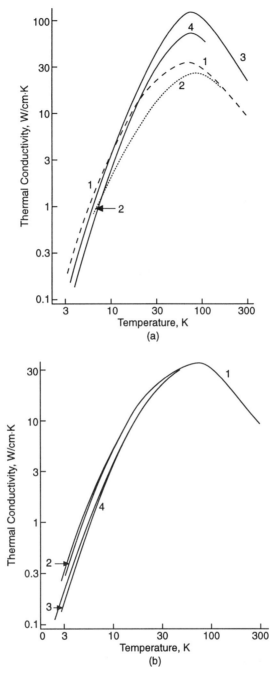

Figure 3.10 The temperature dependence of the thermal conductivity of diamonds of different dimensions and types. (a) Sample 1 — Type I, 10.9·1.1·1.1 mm^3; sample 2 — Type I, 10.65·1.1·1.1 mm^3 (nitrogen atom content 1.7·10^{20} cm^{-3}); sample 3 — Type IIa, 10·0.7·1.25 mm^3; sample 4 — Type IIb, 7·1.1·1.2 mm^3. (b) Four Type I samples of different square cross section: sample 1 — 3.9^2 mm^2; sample 2 — 3.2^2 mm^2; sample 3 — 1.7^2 mm^2; sample 4 — 1.1^2 mm^2. (After R. Berman, *Physical Properties of Diamond*, R. Berman, Ed., Clarendon Press, Oxford, 1965, 371. With permission.)

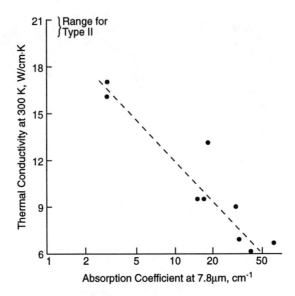

Figure 3.11 Correlation between the thermal conductivity and the absorption coefficient of Type Ia diamonds at 7.8 μm and 300 K. (After R. Berman, *The Properties of Diamond*, J.E. Field, Ed., Academic Press, New York, 1979, 4. With permission.)

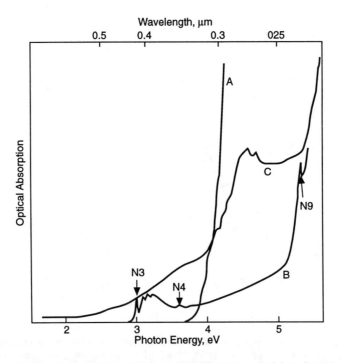

Figure 3.12 The visible and ultraviolet spectra typical for diamonds containing nitrogen. Curve A is for diamonds that contain nitrogen in the A form; curve B is for specimens containing nitrogen in the B form; curve C relates to the specimens containing singly-substituted nitrogen. (After C.D. Clark, E.W.J. Mitchell, and B.J. Parsons, *The Properties of Diamond*, J.E. Field, Ed., Academic Press, New York, 1979, 23. With permission.)

is equal to 1.2, which is close to its magnitude for the elastically isotropic cubic crystals (A = 1.0). Consequently, Young's modulus in diamond does not vary greatly with direction, but there is a noticeable difference in the speed of propagation of the longitudinal and transversal

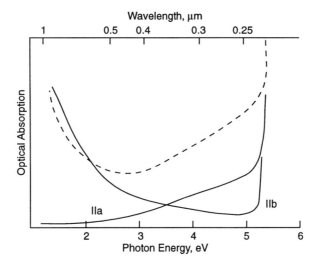

Figure 3.13 Absorption spectra of Type II diamonds. The curve labeled IIb is characteristic of semiconducting diamonds containing boron. The broken curve is for a Type IIb diamond and shows that the IIa absorption may also be present. (After C.D. Clark, E.W.J. Mitchell, and B.J. Parsons, *The Properties of Diamond*, J.E. Field, Ed., Academic Press, New York, 1979, 23. With permission.)

Table 3.1 Elastic Properties of Diamond

Elastic modulus	Magnitude (TPa)	Pressure dependence (dc/dP)	Temperature dependence $10^5 \alpha$; $[\alpha_c = (1/c)\frac{dc}{dT}, K^{-1}]$
c_{11}	1.079	0.60 ± 0.07	−1.4 ± 0.2
c_{12}	0.124	0.31 ± 0.07	−5.7 ± 1.5
c_{44}	0.578	0.30 ± 0.03	−1.25 ± 0.1

From H.J. McSkimin and P. Andreath, *J. Appl. Phys.*, 43, 985, 1972; 43, 2944, 1972.

Table 3.2 Knoop Hardness, GPa (kgp/mm²), of Various Diamonds

Plane	Direction	Type I	Type II	Australian
(001)	[110]	81.3 (8,300)	89.2 (9,100)	86.2 (8,800)
(001)	[100]	96.0 (9,800)	100.9 (10,300)	100.9 (10,300)
(110)	[1$\bar{1}$0]	86.2 (8,800)	92.1 (9,400)	95.1 (9,700)
(110)	[001]	105.8 (10,800)	112.7 (11,500)	107.8 (11,000)
(111)	[1$\bar{1}$0]	54.9 (5,600)	74.5 (7,600)	75.5 (7,700)
(111)	[11$\bar{2}$]	61.7 (6,300)	107.8 (11,000)	99.0 (10,100)

From C.A. Brooks, *The Properties of Diamond*, J.E. Field, Ed., Academic Press, New York, 1979, 383.

elastic waves. Young's modulus of diamond has a magnitude close to E = 1.054 TPa[3.152] (E = 1.050 TPa, from Reference 3.153). The diamond bulk modulus, B = $(c_{11} + 2c_{12})/3$, calculations based on Grimsditch and Ramdas'[3.153] results give the magnitude of 0.442 TPa. The highest magnitudes of elastic moduli are usually associated with Type II diamonds.

Mechanical properties of diamond are particularly interesting because their anisotropy in this monoatomic material is not complicated by the ionic components of the interatomic forces. Microhardness (by Mott's[3.154] definition) of diamonds was measured in different crystallographic directions with a Knoop indentor by Brookes[3.155,3.156] and by Bakul et al.[3.157] Table 3.2 presents the results of measurements[3.156] on various diamonds. The data on dependence of hardness of diamond on the Knoop indentor orientation regarding the crystal

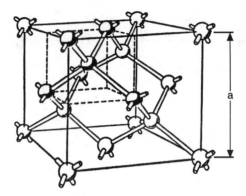

Figure 3.14 The unit cell of the diamond crystal structure.

correlate with the data of Brasen[3.158] and Watts and Willoughby[3.159] on indium phosphide, whereas, for example, some ternary and quaternary solid solutions of indium phosphide and gallium arsenide show a fundamentally different character of the hardness anisotropy.[3.159]

Optical properties of diamonds were used to ascribe them to one of two groups: Types I and II.[3.86] By now, as was mentioned earlier, we classify diamonds in four subgroups (Types Ia, Ib, IIa, and IIb). The diamonds which possess semiconducting properties belong to Type IIb. Both Type-II-subgroup diamonds contain little or no nitrogen as an impurity in contrast with the Type I diamonds. There are certain similarities in the optical absorption spectra of all insulating diamonds which are shown in Figure 3.12, after Dyer et al.[3.160] and Clark et al.[3.161] A typical absorption spectrum of Type IIa diamond is taken from the review by Clark et al.[3.162] The common feature of the spectra (with a few features related to the presence of impurities, mainly nitrogen) is an almost smooth increase of optical absorption with the photon energy higher than approximately 1.7 eV through the region of the fundamental absorption edge near 5.5 eV. The Type IIb diamonds' spectra are similar to one shown in Figure 3.13 (after Clark et al.[3.162]). This shape of the absorption spectrum is associated with the presence of boron in the Type IIb natural diamonds, as was shown by Chrenko,[3.91] Lightowler and Collins,[3.103] and reviewed by Sellschop.[3.90] In the infrared spectrum of a semiconducting diamond (see, e.g., review by Clark et al.,[3.162] there are three strong peaks present at 0.305, 0.348, and 0.364 eV, related to the first, second, and third excited states of boron acceptors.

Diamond has the fcc unit cell (Figure 3.14) with the centers of four nonadjacent octets also occupied by carbon atoms (space group Fd3m).

The band structure of diamond was calculated first by Herman[3.164] who used the method of orthogonalized plane waves. Kleinman and Phillips,[3.165] using a better model, essentially confirmed the results of Herman's calculations. The maximum of the electron energy in the valence band is in the center of the Brillouin zone ($\vec{k} = 0$, symmetry $\Gamma_{25'}$). The conducting electrons have the energy minimum in <100> direction (Δ-axis) of the reciprocal space (see Figure 3.15). This minimum, which represents the indirect energy gap, is equal to 5.50(5) eV (from the data of Himpsel et al.[3.167]). The indirect energy gap extrapolated to T = 0 K is equal to 5.49 eV, and its dependence on temperature between 100 K and 700 K may be expressed as E_g (eV) = $(0.182 + 8.28 \times 10^{-9} T^2)^{-1}$ in accordance with the data presented by Clark et al.[3.161] The form of the diamond band structure was reviewed by Pryce,[3.168] Painter et al.,[3.169] and Stoneham.[3.163] The diamond band structure calculated by Chelikowsky and Louie[3.170] is presented in Figure 3.16.

Lattice dynamics of a crystal is particularly interesting because the information regarding the character of interatomic interactions and magnitude and orientation of interatomic forces, together with the data on chemical composition of the crystal, determines a majority of its properties: thermal, mechanical, optical, electrical, etc. The experimental methods of the phonon dispersion study are, most commonly, infrared and Raman spectroscopy and neutron

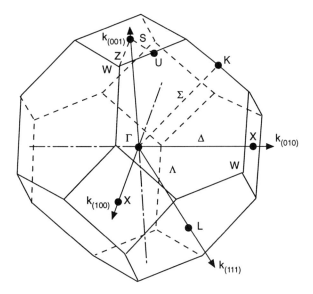

Figure 3.15 The (first) Brillouin zone for diamond. (After D. Long, *Energy Bands in Semiconductors*, John Wiley & Sons, New York, 1968. With permission.)

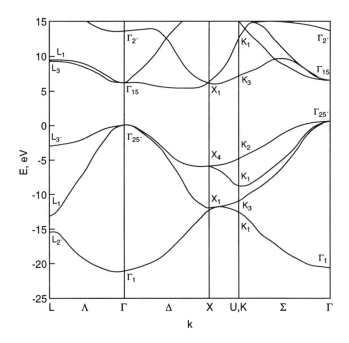

Figure 3.16 The diamond band structure calculated by an *ab initio* LCAO (linear combination of atomic orbitals) method. (After J.R. Chelikowsky and S.G. Louie, *Phys. Rev.*, B29, 3470, 1984. With permission.)

scattering. Stoneham[3.163] reviewed the empirical models of interatomic forces which include the bond polarizabilities model introduced by Smith,[3.171] the valence force potentials model (see, e.g., Musgrave and Pople[3.172] and Bell[3.173]), the shell model (Dolling and Cowley[3.174] and Peckman[3.175]), and the bond charge models (see, e.g., Phillips[3.176] and Go et al.[3.177]). The inelastic scattering of slow neutron experiments on diamonds (Warren et al.[3.179,3.180]) was used to determine the shape of the phonon dispersion curves. Their results, together with the

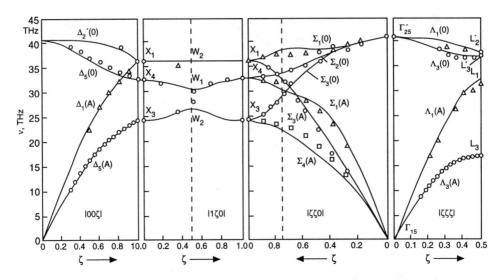

Figure 3.17 Phonon dispersion relations in diamond. Circles, triangles, and squares — experimental data from neutron scattering; the curves resulted from the shell model calculation. (After J.L. Warren, J.L. Yarnell, G. Dolling, and R.A. Cowley, *Phys. Rev.*, 158, 865, 1967. With permission.)

Table 3.3 The Phonon Frequency Data for Diamond

Symmetry	Branch	Frequency (THz)	Symmetry	Branch	Frequency (THz)
$\Gamma_{25'}$	TO and LO	39.9	X_4	TO	32.0
$L_{2'}$	LO	37.5	L_1	LA	30.2
$L_{3'}$	TO	36.2	X_3	TA	24.2
X_1	LA and LO	35.5	L_3	TA	16.9

From S.A. Solin and A.K. Ramdas, *Phys. Rev.*, B1, 1687, 1970.

magnitudes calculated on the basis of the shell model, are presented in Figure 3.17. The data are in satisfactory agreement with the magnitudes of phonon frequency at specific points of the Brillouin zone, presented by Solin and Ramdas[3.181] who used the Raman spectroscopy data (Table 3.3).

High mechanical strength of diamond and its consequently high surface energy (9.2 J/m^2 for (100), 6.5 J/m^2 for (110), and 5.3 J/m^2 for (111) plane[3.182]) are the natural obstacles in synthesis of diamonds in the shape most useful for electronic applications, namely, two-dimensional layers, both thin and thick. Nevertheless, by now we know several (low pressure) growth methods which have been successfully used for diamond growth. Among them are the hot filament method,[3.184] the microwave-enhanced chemical vapor deposition (CVD) method,[3.185] the RF plasma method,[3.186] the DC jet method,[3.187] and the welding torch method.[3.188,3.189]

The low-pressure synthesis of crystalline diamond from the gas phase is currently a subject of intensive research for the following reason. The traditional (metallurgical, stable) diamond synthesis methods in one form or another consist of precipitation of diamonds from a mixture of carbon, in the form of either graphite or diamond powder, with a suitable solvent-catalyst (Ni, Co, or Fe) under the pressure over 6 GPa and at the temperature above the eutectic point of the solvent (about 1500 K). Under these conditions, the known gem quality diamonds were grown to a size of not greater than 1 carat,[3.190,3.191] with production expenses much higher than their natural counterparts, and this expense picture has remained virtually unchanged during the last 20 years.

A metastable diamond growth is a process in which diamond is grown at the temperatures and pressures typical for the graphite stability form of carbon. One of the growth methods is based on the use of a carbon-containing solution that has low carbon solubility. Transfer of the solution periodically from the hot part of the growth apparatus to the cooler part and back may be followed by the supersaturation of the solution and deposition of carbon atoms on a properly chosen substrate (or seeds). The theory of the method and its requirements are described in References 3.192 to 3.195. In the patent by Brinkman et al.,[3.195] the authors describe, as an example, a growth process from the carbon solution in lead using temperatures 1870 K and 1520 K in the hot and cold parts of the growth system, respectively, with the massive growth rate up to 2.5 mm/day.

I assume, as was mentioned before, that the most promising growth method, from both technical and economical points of view, is diamond growth from the gas phase. Since 1962, when Eversole[3.196] patented a process of diamond growth from (1) a methyl-group containing gas and (2) a mixture of carbon dioxide and carbon monoxide, a substantial wealth of data has been published regarding diamond growth from carbon-containing gases. The most important feature of the growth process, as was noticed even by Eversole, is the role of atomic hydrogen in the carbon deposit area. The hydrogen suppresses the formation of graphite, so the growing layer has the diamond lattice. As was stated by Eversole, methane as well as ethane and propane may be successfully used for diamond growth, whereas carbon disulfide, carbon tetrachloride, and benzene do not work as the sources for the process.

Frenklach and Spear[3.197] suggested an elementary reaction mechanism of diamond growth on a diamond substrate by a vapor deposition process from some carbohydrides at low pressures. They believe that the growth is the result of interaction of the substrate surface with acetylene or "acetylenic" species (CH_3, C_2H_4, C_4H_2, etc.) where atomic hydrogen activates the substrate surface by abstracting hydrogen atoms from the surface and producing the radical sites. The sequence of propagation steps in the process of growth is followed by transformation of the initial sp-hybridization of the deposited acetylene carbons into sp^3-hybridization of diamond. Another point of view, supported in several publications,[3.198-3.200] suggests that atomic hydrogen from the gas phase reacts with graphite at a greater rate than with diamond and takes away gaseous products, allowing diamond nuclei to grow.

The role and selection of substrate for diamond growth, including the (100), (110), and (111) diamond substrates and solid and molten metal substrates, were analyzed by Machlin,[3.201] who came to the conclusion that stabilization of diamond for growth is possible on substrates such as (1) diamond surfaces, in ultrahigh vacuum (UHV) or atomic hydrogen, (2) solid metallic surfaces in UHV or atomic hydrogen, (3) the (111) surface of Cd(Se,Te) in atomic hydrogen, and (4) liquid metals in atomic hydrogen. The results of experimental investigation of diamond films grown on (100)-oriented silicon wafers using microwave plasma-assisted CVD from a mixture of methane and hydrogen[3.202] showed a strong tendency toward the formation of a large number of stalking faults and twins in the grown films, which may be considered as one of the main obstacles in growth of diamond structures for any particular application. High defect content was also reported by Williams and Glass,[3.203] who grew diamond films using the same method, on (111)-oriented silicon wafers. Another cycle of these experiments, performed by Williams et al.,[3.204] led to virtually the same results.

A microwave-induced plasma of methane and oxygen in hydrogen was used for the diamond polycrystalline film growth by Kamo et al.,[3.205] Badzian et al.,[3.206] and Harker and DeNatale[3.207] in the range of pressures from 400 Pa to 6.67 kPa (3 to 50 torr) and the substrate temperatures from 720 K to 1500 K. The authors of Reference 3.207 noted (see, also, Reference 3.209) that oxygen decreases defect density in the grown films and it works as a reactive etchant which abstracts the non-sp^3-bonded species forming on the growing film surface. They found that the fine-grained-oriented diamond films with Raman scattering spectra, amazingly similar to the spectra of natural Type IIa diamond, may be grown on silicon and epitaxially on diamond at 723 K from the mixture of hydrogen with 0.3 to 1.0 vol% of methane and 0.25 to 1.0% of oxygen.

Pickrell et al.[3.208] had successfully grown diamond films on fused silica (and on silicon) by microwave-plasma-enhanced CVD at 1 and 5% content of methane in hydrogen with a substrate temperature between 1250 K and 1270 K, a pressure of about 12 kPa (90 torr), and a gas flow rate of 100 sccm. They found that the Raman spectra of the films show the presence of nondiamond carbon when the growth is performed at a higher methane concentration. The next step forward was taken by Ramesham et al.,[3.210] and Ramesham and Roppel[3.211] who performed selective deposition of the polycrystalline diamond film at 873 K using a mixture of methane (4 vol%) and hydrogen (low-temperature deposition) by high-pressure microwave-plasma-assisted CVD on silicon and silicon dioxide[3.210] and also on silicone nitride, tantalum, molibdenum, alumina, and sapphire[3.211] substrates. Single-crystalline copper substrates were used by Ong et al.,[3.212] while Godbole and Narayan[3.13] deposited diamond films on Hastelloy, and Yang et al.[3.214] grew high quality diamond films on nickel polycrystals and single crystals.

One of the diamond and diamond-like carbon film growth methods, which utilizes laser energy (see, e.g., References 3.215 to 3.224), has apparently a promising future for growth of very pure diamond films.

To conclude this brief review of the current diamond synthesis methods, it makes sense to take another look at a technology particularly important for practical applications, namely, the combustion flame synthesis technique which permits the synthesis of diamond at rates above 12 nm/sec to the thickness greater than 0.3 mm with layer dimensions greater than 100 cm^2.[3.188,3.189,3.225-3.227] A detailed analysis of the diamond growth process is given by Ravi[3.226] and von Windheim and Glass.[3.227] An apparent additional advantage of this technology is the possibility to control the optical and semiconductor properties of the diamond layer by introduction of the proper dopants directly in the combustion zone.

Intensive research efforts in the fields of the properties of diamond, technology of its synthesis, and the ways of its practical application during the last 50 years show that this material has a prominent future both for engineering purposes and as a material whose properties may be very helpful in analysis of different theoretical models of solids.

3.3. SILICON

In 1824, the great chemist Jons J. Berzelius isolated silicon in the form of amorphous material by heating potassium with SiF$_4$ and painstakingly washing the reaction products. A second allotropic form, crystalline silicon was first prepared in 1854 by Jacque Deville. Silicon in its content in the earth's crust (25.7 mass%) is exceeded only by oxygen (46.7%).

Silicon consists of 3 stable isotopes, ^{28}Si (92.23%), ^{29}Si (4.67%), and ^{30}Si (3.10%), and 12 artificial radioactive isotopes with mass numbers from 22 to 39 and a half-time from 0.10 sec to 1.6×10^2 years. Valence electron configuration is $3s^23p^2$. Ionization energy for Si$^0 \rightarrow$ Si$^+ \rightarrow$ Si$^{2+} \rightarrow$ Si$^{3+} \rightarrow$ Si^{4+} is 8.15, 16.34, 33.46, and 45.13 eV, respectively. Electron affinity, Si^0 + e$^- \rightarrow$ Si, is 1.22 eV. Silicon atomic radius is 0.1175 nm; Si^{4+} radius is 0.039 nm.

Silicon is, by now, the most studied material among all known substances (with a possible competitor — germanium). It may be said that silicon is the substance most widely used as the active element of a majority of modern electronic devices utilizing its electrical, optical, photoelectrical, thermoelectrical, thermal, mechanical, and other properties.

The optical indirect energy gap of silicon at 300 K (transition $\Gamma_{25'v} - \Delta_{lc}$) of 1.1242 eV[3.228] (1.1700 eV at 0 K), together with the outstanding electrical and optical properties of SiO$_2$, makes silicon very attractive, in crystalline, thin-film, and amorphous forms, for application in rectifiers, transistors, thyristors, solar energy converters, field-effect devices, and so on. The energy band structure of silicon was calculated first by Herman[3.164b] and Jenkins[3.229] (see

ELEMENTAL SEMICONDUCTORS 57

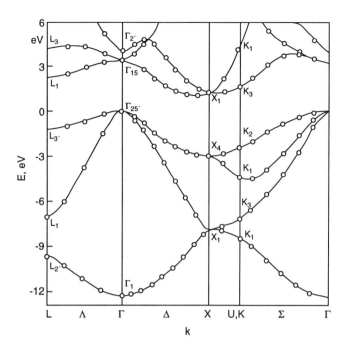

Figure 3.18 The silicon band structure. The curves are the result of nonlocal, energy-dependent pseudopotential calculations; the circles represent the results of calculation by the localized atomic orbital method. (After K.S. Shieh and P.V. Smith, *Phys. Stat. Sol.*, B129, 259, 1985. With permission.)

also Bell et al.[3.230] and reviews by Herman[3.231,3.232] and Lax[3.233]). The latest results of calculations by Sieh and Smith[3.234] are presented in Figure 3.18. The data are in satisfactory agreement with the known experimental results (see, e.g., References 3.228 and 3.235 to 3.240). The silicon energy gap calculated from the temperature dependence of electrical conductivity and Hall effect measurements is equal to 1.21 eV in the range of temperatures from 4 to 1000 K.[3.241] This is in a very good agreement with the Barber's[3.236] magnitude of 1.205 eV at 0 K.

Crystal structure of silicon under standard conditions is the same as diamond (space group O_h^7 – Fd3m), but under pressure between 11.2 and 12.5 GPa it acquires the body-centered tetragonal (white tin) structure (space group D_{4h}^{19} – I4$_1$/amd).[3.243,3.244] Olijnuk et al.[3.242] reported also about at least two hexagonal allotropes of silicon, namely, a simple hexagonal one (space group D_{6h}^1 – P6/mmm), stable above 16.4 GPa (see also Reference 3.244), and a hexagonal close-packed one, stable above 45 GPa. The allotropes of silicon and other group IVb elements are presented in Table 3.4.

The lattice parameter of the diamond-like silicon at 295.7 K, measured by Becker et al.[3.245] in vacuum on a high-purity single crystal, is equal to 0.543102018 nm. The temperature dependence of the silicon lattice parameter, as well as many other diamond-like substances, has an interesting feature, namely, that at low-enough temperatures the coefficient of linear thermal expansion becomes negative. This effect was reported first by Fizeau[3.246] in 1867 when he measured thermal expansion of cubic silver iodide. Negative thermal expansion of silicon was discovered first by Valentiner and Wallot[3.247] and later by Erfling,[3.248] who found that with temperature increase, the coefficient of thermal expansion passes through zero around 120 to 140 K and becomes negative at lower temperatures.

Gibbons,[3.249] Novikova and Strelkov,[3.250] Carr et al.,[3.251] Sparks and Swenson,[3.252] Lion et al.,[3.253] and Okada and Tokumaru[3.254] studied the thermal expansion of silicon from very low to high temperatures. Their data are presented in Table 3.5 and Figure 3.19. A brief review

Table 3.4 Allotropes of the Group IV Elements[a]

Element	T (°C)	Pressure (GPa)	Pearson symbol	Space group	SB	Prototype	Lattice Parameters (nm)		
							a	c	c/a
C(gr)	25	atm	hP4	P6$_3$/mmc	A9	C(gr)	0.24612	0.6709	2.7258
C(dia)	25	>60	cF8	Fd3m	A4	C(dia)	0.35669		
C(hd)	25	HP	hP4	P6$_3$/mmc		C(hd)	0.2522	0.4119	1.633
Si	25	atm	cF8	Fd3m	A4	C(dia)	0.54306		
	25	>9.5	tI4	I4$_1$/amd	A5	Sn	0.4686	0.2585	0.552
	25	>16.5	cI16	Im3m		Sn	0.6636		
	25	>16 atm	hP4	P6$_3$/mmc	A3'	La	0.380	0.628	1.653
Ge	25	atm	cF8	Fd3m	A4	C(dia)	0.56574		
	25	>12	tI4	I4$_1$/amd	A5	Sn	0.4884	0.2692	0.551
	25	>12 atm	tP12	P4$_1$2$_1$2		Ge	0.593	0.698	1.18
	LT	>12	cI16	Im3m		Si	0.692		
Sn	<13	atm	cF8	Fd3m	A4	C(dia)	0.64982		
	25	atm	tI4	I4$_1$/amd	A5	Sn	0.58318	0.31838	0.5456
	25	>9	tI2	?		Sn	0.370	0.337	0.91
Pb	25	atm	cF4	Fm3m	A1	Cu	0.49502		
	25	>10.3	hP2	P6$_3$/mmc	A3	Mg	0.3265	0.5387	1.650

[a] Abbreviations: SB — Structurbericht designation; gr — graphite; dia — diamond; hd — hexagonal close-packed; atm — atmospheric pressure; HP — high pressure; LT — low temperatures.

From H.W. King, *Handbook of Chemistry and Physics*, 73rd ed., D.R. Lide, Ed., CRC Press, Boca Raton, FL, 1992–1993, 12-10.

Table 3.5 Silicon Coefficient of Linear Thermal Expansion (in Units of 10^{-8} K^{-1})

T (K)	Ref. 3.252	Ref. 3.251	T (K)	Ref. 3.252	Ref. 3.251	Ref. 3.255	T (K)	Ref. 3.251	Ref. 3.255
4	0.006		22	−0.25	−0.45		80	−50	−48
5	0.011		24	−1.24	−1.45		90	−44.5	−44
6	0.019		26	−1.90	−2.6		100	−34	−33
8	0.045		28	−2.68	−3.8		120	−4	−5
10	0.088		30	−3.60	−5.2		150	52	
12	0.152		35		−9.0		200	144	143
14	0.24		40		−13.5		250	212	
16	0.32		50		−24		280	244	
18	0.18		60		−39	−38	283		235
20	−0.25	−0.45	70		−50				

of the theoretical analysis of this unusual (for cubic crystals) behavior of thermal expansion will be considered later.

Specific heat of silicon was studied by Pearlman and Keesom[3.257] and Flubacher et al.[3.258] The data on the atomic heat capacity at constant pressure, C_p, together with $C_p - C_V$ (C_V is the atomic heat capacity at constant volume) magnitudes calculated using the formula $C_p - C_V = \beta^2 TV/\eta$ and data on the volume coefficient of thermal expansion, β, from Reference 3.249, and the isothermal compressibility, η, from Reference 3.259 (V is the gram-atom volume), are presented in Table 3.6 for silicon and germanium. The results of calculations of the Debye temperature from Reference 3.258 are shown in Figure 3.20.

A detailed study of the silicon elastic properties was performed by McSkimin[3.259] who used the method of propagation and reflection of ultrasonic (10 to 30 MHz) longitudinal and transversal waves in a single crystal in a temperature range of 78 to 300 K. Using the density magnitude of 2.331 g/cm^3 at 298 K (Hennis[3.260] found density of a high purity crystal at this temperature equal to 2.329002 g/cm^3, while its magnitude calculated from the X-ray

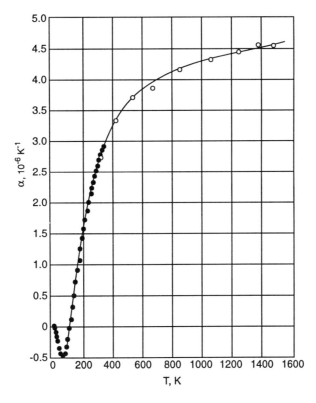

Figure 3.19 Temperature dependence of the silicon coefficient of linear thermal expansion. (After Y. Okada and Y. Tokumaru, *J. Appl. Phys.*, 56, 314, 1984. With permission.)

measurements at 295.7 K is 2.329042 g/cm³) and the thermal expansion data of Reference 3.247, McSkimin calculated the silicon elastic constants c_{11}, c_{12}, and c_{44} (Figure 3.21).

Thermal conductivity of high purity silicon crystals is controlled mainly by lattice vibrations; the free charge component is small, up to 1000 K.[3.260] Slack[3.261] measured silicon thermal conductivity and found that, in the range of 100 to 1000 K, it is reverse proportional to temperature. Fulkerson et al.[3.262] presented the results of the thermal conductivity measurements analytically (for the same temperature range) as

$$\kappa^{-1} \, (\text{cm} \cdot \text{K/W}) = 0.1598 + 1.532 \cdot 10^{-3} \, T + 1.583 \cdot 10^{-6} \, T^2$$

Their data on thermal conductivity are noticeably lower than those by Slack in the range between 100 and 500 K, but are in satisfactory agreement at higher temperatures. A detailed investigation of the silicon thermal conductivity between 2 and 150 K was performed by White and Woods[3.285] and, at higher temperatures, by McCarthy and Ballard.[3.286]

Intrinsic electrical conductivity of silicon at 300 K is close to 3.16 µS/cm.[3.241] Intrinsic charge carrier concentration of $1.02 \cdot 10^{10}$ cm⁻³ [3.265] is in satisfactory agreement with the Morin and Maita[3.241] magnitude of $1.38 \cdot 10^{10}$ cm⁻³. The temperature dependence of the intrinsic carrier concentration from Reference 3.241 is shown in Figure 3.22. The current carrier's mobility in silicon, measured by Prince,[3.266] was (at 290 K) 1350 ± 100 cm²/V·sec for electrons and 480 ± 15 cm²/V·sec for holes with a temperature dependence of $T^{-2.5}$ and $T^{-2.7}$ for the electron and hole mobility, respectively. Morin and Maita[3.241] found electron mobility at 300 K equal to 1450 cm²/V·sec. Jacobini et al.[3.267] suggested that the temperature dependence of the electron mobility near room temperature is $\mu_n = 1.43 \cdot 10^9 \, T^{-2.42}$ cm²/V·sec.

Table 3.6 Data on the Atomic Heat Capacity of Silicon and Germanium

T(K)	Silicon C_p [a]	$C_p - C_v$	Germanium C_p	$C_p - C_v$
2.5	(0.0000271)		0.0001413	
5	(0.000219)		0.001199	
10	0.001845		0.01394	
15	0.007300		0.07672	
20	0.02265		0.2174	
25	0.05702		0.4140	
30	0.1149		0.6316	
35	0.1962		0.8531	
40	0.2954		1.067	
50	0.5272		1.479	
60	0.7725		1.879	
70	1.019		2.269	
80	1.261		2.640	
90	1.502		2.989	0.001
100	1.739		3.302	0.001
120	2.205		3.838	0.003
140	2.650		4.261	0.005
160	3.057		4.587	0.007
180	3.420	0.001	4.837	0.009
200	3.735	0.001	5.033	0.011
220	4.009	0.001	5.190	0.013
240	4.246	0.002	5.318	0.015
260	4.455	0.002	5.423	0.017
280	4.638	0.003	5.511	0.020
300	4.796	0.004	5.590	0.023

[a] Units — cal/gram-atom·K.

From P. Flubacher, A.J. Leadbetter, and J.A. Morrison, *Philos. Magn.*, 4, 273, 1959.

The ohmic mobility, calculated by Morin and Maita[3.241] from the conductivity measurements, was equal to $1450(300/T)^{2.6}$ for electrons and $500(300/T)^{2.3}$ for holes. The Hall effect measurements on a high purity silicon sample gave the magnitude of the hole mobility of 370 cm²/V·sec at 300 K[3.268] and $2 \cdot 10^5$ cm²/V·sec at 20 K.[3.269] The conductivity measurements by Mitchel and Hemenger[3.270] gave the hole mobility 505 cm²/V·sec at 300 K; Ludwig and Watters[3.271] found it close to $1.6 \cdot 10^6$ cm²/V·sec at 20 K. The correlation between drift mobility of the current carriers and the conductivity of p-Si and n-Si samples is shown in Figure 3.23.[3.271] Ludwig and Watters came to the conclusion that this dependence may be explained by dissipation of the carriers by ionized impurity centers. The effective mass of electrons and holes in silicon was measured by Dexter et al.[3.264] Their data are in good agreement with those from References 3.236, 3.263, 3.264, 3.272, and 3.273 and are presented in Table 3.7.

Electrical conductivity of silicon at high temperatures and in liquid state was investigated by Regel et al.,[3.274,3.275] Glazov et al.,[3.276] and Baum et al.[3.277] Near the melting point (1685 K[3.278]), the conductivity of solid silicon is close to 600 S/cm and its magnitude for melt is close to 12 kS/cm. Liquid silicon behaves as a metal: its conductivity is changing antibatically with temperature (Figure 3.24, from Reference 3.276). Density of solid and liquid silicon near the melting point is close to 2.30 and 2.53 g/cm³, respectively. The contraction of the melt may be explained by the change of the solid silicon coordination number (4) to a greater magnitude (6 or 8), typical for metals, as was found, e.g., by Hendus[3.279] from X-ray analysis of liquid germanium.

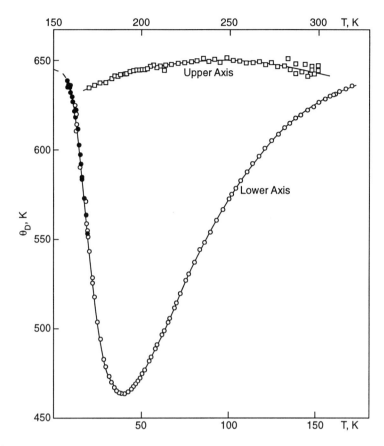

Figure 3.20 The Debye characteristic temperature of silicon vs. absolute temperature. Solid circles — Calorimeter I (12 K to 20 K). Open circles — Calorimeter II (12 K to 170 K). Squares — Calorimeter II (170 K to 300 K). (After P. Flubacher, A.J. Leadbettter, and J.A. Morrison, *Philos. Mag.*, 4, 273, 1959. With permission.)

The optical transition parameters of silicon may be evaluated from the band structure (Figure 3.18) calculated by Sieh and Smith.[3.234] The experimental magnitudes of the optical transition energies are compiled in Table 3.8.

Dependence of the silicon refractive index on photon energy was studied by Briggs,[3.282] who found that in the range of 0.47 to 1.13 eV, the refractive index, n_r, is changing from 3.443 to 3.553. Extrapolation of these data to the long-wave part of the spectrum, using the expression $(n_r^2 - 1)^{-1} = A - B\lambda^{-2}$, gives the long-wave refractive index of 3.43, in good agreement with the data of Carlson[3.283] and Philip and Taft,[3.284] which, in its turn, gives the silicon dielectric constant, $\varepsilon/\varepsilon_0$, equal to 11.8. Experimental magnitudes of the silicon optical constants from References 3.237, 3.283, and 3.284 as a function of the photon energy are shown in Figure 3.25.

Silicon is a diamagnetic material with the molar magnetic susceptibility of $-4.9 \cdot 10^{-5}$ SI units ($-3.9 \cdot 10^{-6}$ cgs units).[3.287] It is virtually independent of temperature up to the silicon melting point.

The semiconductor silicon technology consists of a few basic steps. The initial material is silica, SiO_2, which is mixed with coal or wood chips and heated in a furnace at the temperature of 1800 to 2300 K to reach a reduction of silica in reaction

$$SiO_2 + 2C \rightarrow Si(\text{melt}) + 2CO$$

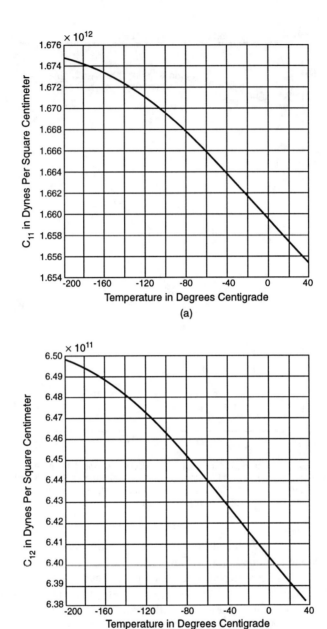

Figure 3.21 The silicon elastic moduli vs. temperature. (a) c_{11}, (b) c_{12}, (c) c_{44}. (After H.J. McSkimin, *J. Appl. Phys.*, 24, 988, 1953. With permission.)

The silicon lumps produced have a purity of 98 to 99% and are called metallurgical grade silicon, MG-Si.[3.288,3.289] The next step is a formation of various (liquid) chlorosilanes by reaction, e.g.,

$$Si + 3HCl \rightarrow SiHCl_3 + H_2$$

of MG-Si with anhydrous HCl in a fluidized-bed reactor at 500 to 700 K. This step is followed by purification of trichlorosilane by distillation (boiling point 304.95 K). The purified

ELEMENTAL SEMICONDUCTORS

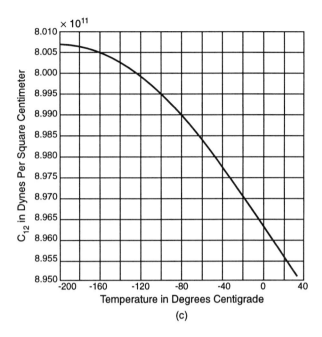

(c)

Figure 3.21 (continued)

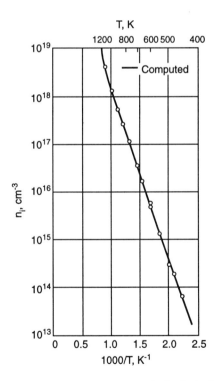

Figure 3.22 Carrier concentration in silicon vs. temperature in the intrinsic range. (After F.J. Morin and J.P. Maita, *Phys. Rev.*, 96, 28, 1954. With permission.)

trichlorosilane is evaporized, mixed with high-purity hydrogen, and directed into the deposition reactor. In the reactor, at a temperature of about 1400 K, trichlorosilane reacts with hydrogen and releases silicon:

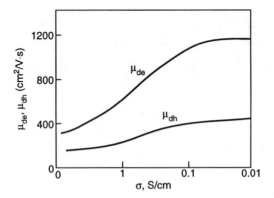

Figure 3.23 The room temperature carrier drift mobility in silicon vs. electrical conductivity. (After G.W. Ludwig and R.L. Watters, *Phys. Rev.*, 101, 1699, 1956. With permission.)

Table 3.7 Effective Mass of the Current Carriers in Silicon

Carrier	Temperature (K)	Effective mass characteristics	Magnitude[a]	Ref.
Electron	300	Longitudinal, m_{eL}	0.98 ± 0.04	3.264
	1.26	Longitudinal, m_{eL}	0.9136	3.272
	300	Transversal, m_{eT}	0.19 ± 0.01	3.264
	1.26	Transversal, m_{eT}	0.1905	3.272
	300	Anisotropy coefficient, $K = m_{eL}/m_{eT}$	5.1	3.264
	300	Density-of-states, $m_{eds} = (m_{eL}m_{eT}^2)^{1/3}$ [b]	0.33	3.264
	300	Density-of-states, $m_{eds} = (m_{eL}m_{eT}^2)^{1/3}$	1.18	3.273
	600	Density-of-states, $m_{eds} = (m_{eL}m_{eT}^2)^{1/3}$	1.28	3.273
	77	Density-of-states, $m_{eds} = (m_{eL}m_{eT}^2)^{1/3}$	1.08	3.273
	4.2	Density-of-states, $m_{eds} = (m_{eL}m_{eT}^2)^{1/3}$	1.026	3.236
	300	Ohmic (conductive), $m_{eC} = 3m_{eL}(2K+1)$	0.26	3.264
Hole	300	Heavy, m_{ph}	0.49	3.264
	4.2	Heavy, m_{ph}	0.537	3.236
	300	Light, m_{pl}	0.16	3.264
	4.2	Light, m_{pl}	0.153	3.236
	300	Density-of-states, $m_{pds} = (m_{pn}^{3/2} + m_{pl}^{3/2})^{2/3}$	0.55	3.264
	4.2	Density-of-states, $m_{pds} = (m_{pn}^{3/2} + m_{pl}^{3/2})^{2/3}$	0.591	3.236

[a] Effective mass is in the units of the free electron mass, m_o.
[b] The appropriate expression for the conductive band density-of-states effective mass is $m_{ds} = Z^{2/3}(m_L m_T^2)^{1/3}$, where Z is the number of the equivalent minima in the conductive band. For silicon, Z = 6. With this correction, the magnitude of m_{ds} in Reference 3.264 would be 1.09.

$$SiHCl_3 + H_2 \rightarrow Si + 3HCl$$

The silicon from the vapor phase deposits on the thin high purity silicon rods (slim rods), usually having a U-shape, heated by electric current to about 700 K at the initiation of the silicon deposition process. The growing polycrystalline silicon is called semiconductor-grade silicon, SG-Si, and may be used for single crystal growth. There are some other methods of the SG-Si production, such as the monosilane process and granular polysilicon deposition, but their description is beyond the limits of this book.

The next step is the growth of silicon single crystals from high-purity polycrystals. There are two methods which are used most commonly. One of them is the float-zone method[3.290] and the other is a pulling method designed initially by Czochralski[3.291] and developed for Si crystal growth by Teal and Little.[3.292,3.293] A schematic view of a typical Czochralski crystal growth system is shown in Figure 3.26 (after Abe[3.294]).

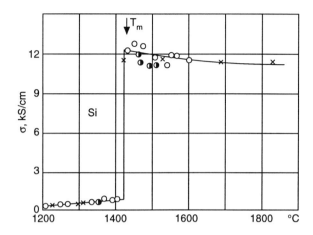

Figure 3.24 Temperature dependence of the electrical conductivity of solid and liquid silicon. (After V.M. Glazov, S.N. Chizhevskaya, and N.N. Glagoleva, *Liquid Semiconductors*, Nauka Publishing House, Moscow, 1967. With permission.)

Table 3.8 Optical Transitions in Silicon

Transition		Energy (eV)	Ref.	Comments
From	To			
$\Gamma_{25'v}$	Γ_{15c}	3.4	3.280	
$\Gamma_{25'v}$	$\Gamma_{2'c}$	4.185	3.237	4.2 K
$\Gamma_{25'v}$	$\Gamma_{2'c}$	4.135	3.237	190 K
$L_{3'v}$	L_{1c}	3.45	3.280	
$L_{3'v}$	L_{3c}	5.50	3.280	
$\Gamma_{25'v}$	L_{1c}	2.4 ± 0.15	3.281	
$\Gamma_{25'v}$	L_{1c}	2.04	3.239	
$\Gamma_{25'v}$	L_{3c}	4.15 ± 0.1	3.281	
$\Gamma_{25'c}$	Δ_{1c}	1.1242	3.228	300 K

The grown single crystals are trimmed, oriented in a desirable crystallographic direction, and sliced into wafers. The wafers are polished and etched, and, after a detailed inspection, may be used for semiconductor device manufacturing. Some details of the device technology will be described later.

Physical, chemical, and technological properties of silicon are, by now, investigated wider and deeper than the rest of the elements of Mendeleev's Periodic table (with germanium as the only exception). Nevertheless, there is one area where silicon still has not been sufficiently studied. This area is silicon superconductivity. The recent results of investigations show that, at certain conditions, the superconductivity onset temperature of silicon is close to 12 K at the pressure of 1 GPa.[3.295] There is reason to believe that a proper technology can be developed to make electronic devices utilizing this effect.

3.4. GERMANIUM

The existence of germanium was predicted by D.I. Mendeleev (1871), who named the still unknown element as eka-silicon and described the chemical properties it must have. It was first isolated (1887) by Clemens A. Winkler from a silver ore (argyrodite, Ag_8GeS_6) in the form of the sulfide GeS_2.[3.297] The atomic number of germanium is 32 and its atomic weight is 72.61. Germanium consists of five stable isotopes, Ge^{70}, Ge^{72}, Ge^{74}, Ge^{76}, and Ge^{78}.

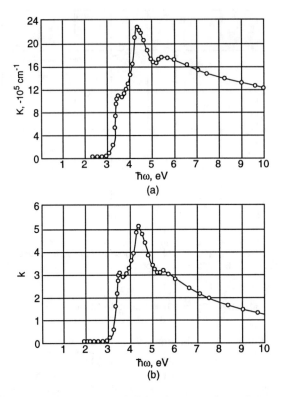

Figure 3.25 The silicon optical parameters and dielectric constant vs. photon energy. (a) Absorption coefficient, (b) extinction coefficient, (c) index of refraction, (d) real, ε_1, and imaginary, ε_2, parts of the dielectric constant. (After H.R. Phillipp and E.A. Taft, *Phys. Rev.*, 120, 37, 1960. With permission.)

Electron configuration (ground state) is $4s^2 4p^2$. Ionization energies for $Ge^0 \rightarrow Ge^+ \rightarrow Ge^{2+} \rightarrow Ge^{3+}$ are 8.13, 15.95, 34.22, and 45.70 eV, respectively.

Germanium has valency 2 and 4. It is stable in air, water, and oxygen at room temperature. At 1000 K, germanium reacts energetically in air and in the oxygen stream, forming GeO and GeO_2. Germanium reacts intensively with halogens, forming tetrahalogenides. Powder of germanium ignites in the atmosphere of chlorine and fluorine even at room temperature. In the electrochemical row, Ge is located between Cu and Ag. It does not react with HCl and diluted H_2SO_4, but concentrated sulfuric acid slowly dissolves it. Aqua regia and HNO_3 etch germanium, especially at elevated temperatures. It reacts slowly with alkali, but the addition of hydrogen peroxide to alkali sharply accelerates the reaction. Germanium dissolves in 3% hydrogen peroxide with precipitation of GeO_2. Germanium does not form carbides and may be melted in graphite crucibles.

The main application of germanium is in the semiconductor industry, but there are some other important uses, such as a catalyst, a phosphor in fluorescent lamps, and an infrared filter. GeO_2 is useful as a component of glass for microscope objectives and wide-angled camera lenses. Since the successful development in the silicon industry, the field of organogermanium chemistry has extensively grown. Some interesting applications of organogermanium compounds are connected with their chemotherapeutic effects.

Elemental Ge is a brittle (up to 800 K) substance of silverish color with pycnometric density of 5.3234 g/cm³ at 298 K[3.296] and X-ray density of 5.32426 g/cm³; melting point is 1211.5 K, boiling point is 3107 K. Germanium heat of fusion is 34.71 kJ/g-atom; its heat of vaporization is 333.9 kJ/g-atom (3107 K).

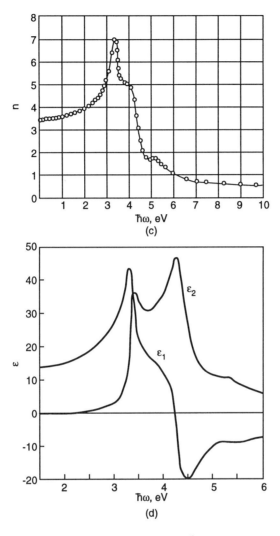

Figure 3.25 (continued)

Primary germanium may be extracted from the minerals argyrodite; germanite, $Cu_3(Fe,Ge)S_4$; ultrabasite, $Ag_{22}Ge_3Pb_{28}Sb_4S_{51}$; and some zinc ores. The main way to extract germanium is through the processing of combustion products of certain coals (yield about 30 g/t).[3.298-3.301]

Interest in the properties of germanium, after the discovery of the amplification effect in semiconductors[3.302] and the invention of the transistor,[3.303,3.304] was so strong that by now these material properties are known better than those of any other element of the Periodic table.

The lattice constant of germanium at 298.15 K, measured by Baker and Hart,[3.305] is equal to 0.56579060 nm in a satisfactory agreement with a = 0.565754 ± 0.000001 nm reported by Smakula and Kalnajs[3.306] and with a magnitude calculated from the density measurements.[3.296] The results of measurements of the lattice constant by Singh[3.307] are shown in Figure 3.27.

The thermal expansion of the single-crystalline germanium was studied by Gibbons,[3.249] Novikova,[3.308] McCammon and White,[3.309] and Zhdanova.[3.310] It was found that the germanium coefficient of linear thermal expansion is negative below 40 K[3.249] or below 48 K.[3.308] The data of Gibbons and Novikova are presented in Figure 3.28. The authors of Reference 3.309 reported that germanium has a small positive expansion while warming up to 14 K. From

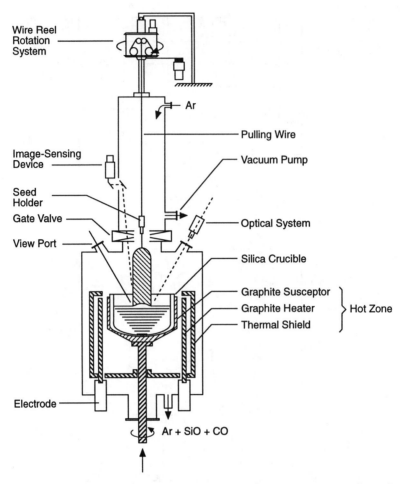

Figure 3.26 Schematics of a typical Czochralski silicon crystal growing system. (After T. Abe, *VLSI Electronics Microstructure Science,* N.G. Einspruch and H.R. Huff, Eds., Vol. 12, Academic Press, New York, 1985. With permission.)

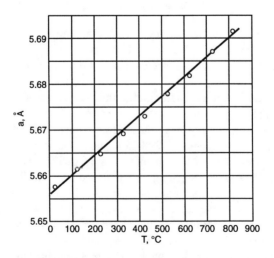

Figure 3.27 Temperature dependence of the germanium lattice parameter. (After H.P. Singh, *Acta Crystall.,* 24a, 469, 1968. With permission.)

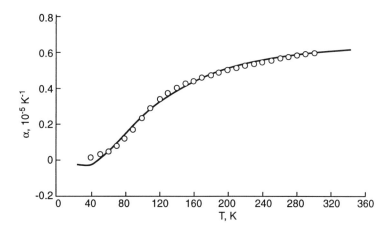

Figure 3.28 Temperature dependence of the Ge coefficient of linear thermal expansion. The curve represents the experimental data of Reference 3.308; the circles are the data of Reference 3.249. (After S.I. Novikova, *Sov. Phys. Solid State*, 2, 37, 1960. With permission.)

14 K, the coefficient of thermal expansion becomes negative, and at 42 K it passes again through zero and becomes positive. Zhdanova studied the dependence of thermal expansion of germanium on the gallium content in it on several single crystals in different crystallographic directions at temperatures from 77 to 340 K. She found that the samples with higher hole concentration (from $1.15 \cdot 10^{15}$ to $5.2 \cdot 10^{19}$ cm^{-3}) have a higher coefficient of thermal expansion.

The heat capacity of germanium was studied by Esterman and Weertman,[3.311] Hill and Parkinson,[3.312] Flubaker et al.,[3.258] and by Piesbergen.[3.313] The experimental data on the atomic heat capacity, C_p, (in cal/g-atom·K) from Reference 3.313 are presented in Table 3.9 together with the results of calculations of the $C_p - C_V$ and C_V. The data on the Debye temperature are presented in Table 3.9 as well. The data on C_p in the table are in satisfactory agreement with the results reported in Reference 3.258 (see Table 3.6). The Debye temperature of germanium as a function of temperature is shown in Figure 3.29.

Elastic properties of germanium were studied by Bridgman,[3.312A] Bond et al.,[3.313A] McSkimin,[3.259] Fine,[3.314] and Burenkov et al.[3.315] The temperature dependence of the second-order elastic constants is shown in Figure 3.30 (from Reference 3.259). The magnitudes of elastic constants in Reference 3.315 ($c_{11} = 124.0$ GPa, $c_{12} = 41.3$ GPa, $c_{44} = 68.3$ GPa at 298 K) are only a few percent different from the data of McSkimin. Fine[3.314] measured the Young's and torsion moduli of germanium between 1.7 and 80 K in the <100> and <111> directions, and calculated the germanium elastic constants (Figure 3.31). One can see that the elastic moduli approach constant magnitudes are very low temperatures, as required by the third law of thermodynamics. The data of Reference 3.314 on elastic moduli of germanium at 77 K ($E_{100} = 104.2$ GPa, $E_{111} = 158.4$ GPa, and $S_{100} = c_{44} = 68.7$ GPa) are in good agreement with the values from Reference 3.259.

The band structure of germanium was investigated experimentally using cyclotron resonance (see review by Herman[3.231,3.232] and Lax[3.233]), electrical resistivity measurements (see, e.g., Reference 3.241), optical[3.316] and magnetic[3.317] transmission measurements, and photoemission measurements.[3.318-3.320] A theoretical calculation of the germanium band structure was first performed by Herman,[3.321] who concluded that the maximum of the valence band is located at the center of the Brillouin zone, and the conduction band has four equivalent minima located along directions <111> near the border of the Brillouin zone. Later, Chelikowsky and Cohen[3.322] came to the conclusion that the germanium conduction band has eight equivalent minima at the end points with symmetry L_6 (end points of the <111> direction). The band structure of germanium is shown in Figure 3.32. Germanium is an

Table 3.9 Germanium Heat Capacity and Debye Temperature

T (K)	C_p	$C_p - C_v$	C_v	θ_D (K)
12	0.0337	—	0.0337	288
15	0.0827	—	0.0827	267
20	0.218	—	0.218	257
25	0.419	—	0.419	257
30	0.640	—	0.640	264
35	0.857	—	0.857	276
40	1.074	—	1.074	287
45	1.284	—	1.284	298
50	1.488	—	1.488	309
60	1.884	—	1.884	329
70	2.265	0.001	2.264	343
80	2.635	0.001	2.634	354
90	2.985	0.002	2.983	361
100	3.303	0.002	3.301	367
110	3.584	0.003	3.581	371
120	3.835	0.004	3.831	374
130	4.056	0.005	4.051	376
140	4.267	0.006	4.261	377
150	4.439	0.007	4.432	377
160	4.589	0.008	4.581	377
170	4.720	0.009	4.711	378
180	4.840	0.010	4.830	377
190	4.954	0.011	4.943	374
200	5.061	0.012	5.049	370
210	5.160	0.013	5.147	365
220	5.241	0.014	5.227	360
230	5.312	0.015	5.297	357
240	5.373	0.016	5.357	354
250	5.425	0.017	5.408	351
260	5.467	0.018	5.449	350
270	5.506	0.019	5.487	349
273.2	5.518	0.019	5.499	348

From V. Piesbergen, *Z. Naturforsch.*, 18a, 141, 1963.

indirect semiconductor: its most narrow energy gap at 291 K (transition $\Gamma_{8v} - L_{6c}$) is equal to 0.664 cV,[3.316] whereas the direct energy gap at 293 K (transition $\Gamma_{8v} - \Gamma_{7c}$) is close to 0.805 eV.[3.317] Both indirect and direct energy gaps decrease when the temperature increases (Figure 3.33, a and b).

The effective mass tensor of electrons and holes in germanium was studied using the cyclotron[3.264,3.325] and magnetophonon[3.326] resonance and piezomagnetoreflectance[3.327] measurements. The anisotropy coefficient (the ratio of the longitudinal and transversal electron effective mass, $K = m_{eL}/m_{eT}$) was studied by Pergola and Sette[3.328] and Furth and Waniek[3.329] using the magnetoresistance measurements. Willardson et al.[3.330] were, apparently, the first to assume that some (about 2.5%) of the holes in germanium have a smaller effective mass as an explanation for the temperature dependence of the Hall constant in p-Ge in the weak magnetic fields. The data on the effective mass of electrons and holes in germanium are presented in Table 3.10.

The electrical transport properties of the intrinsic germanium are characterized by the electron mobility of 3900 cm^2/V·sec at 300 K[3.331] and its temperature dependence $\mu_n = 4.9 \cdot 10^7 \, T^{-1.655}$ between 77 and 300 K,[3.241] and current carrier concentration equal to $n_i = 2.33 \cdot 10^{13}$ cm^{-3} at 300 K and its temperature dependence of

$$n_i = 1.76 \cdot 10^{16} \, T^{3/2} \exp(-0.785/2k_B T)$$

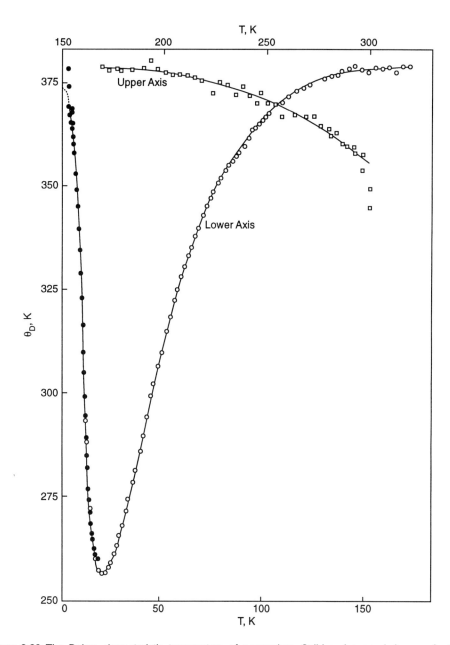

Figure 3.29 The Debye characteristic temperature of germanium. Solid and open circles are for low temperatures (up to 170 K); the squares are for temperatures between 170 K and 300 K. (After P. Flubacher, A.J. Leadbetter, and J.A. Morrison, *Philos. Mag.*, 4, 273, 1959. With permission.)

where $k_B T$ is in electron-volts.[3.241] In accordance with Morin and Maita,[3.241] the hole mobility between 100 and 300 K fits the formula

$$\mu_p = 1.05 \cdot 10^9 \, T^{-2.33}$$

and its magnitude at 300 K is 1800 cm²/V·sec.

Near melting point, electrical conductivity of the solid germanium is close to 1200 S/cm, while the liquid germanium at this temperature (1210.4 K[3.278]) has conductivity about ten

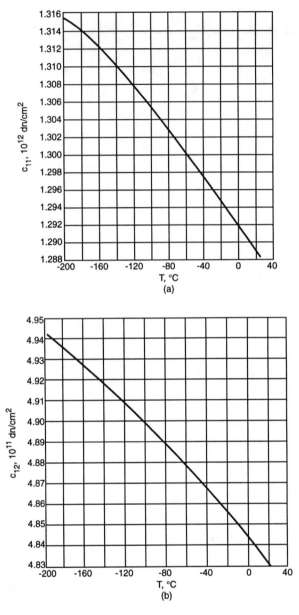

Figure 3.30 The germanium second-order elastic moduli. (a) c_{11}, (b) c_{12}, and (c) c_{44}. (After H.J. McSkimin, J. Appl. Phys., 24, 988, 1953. With permission.)

times greater.[3.276,3.332] Conductivity of the liquid germanium decreases with temperature increase, i.e., germanium melt has metallic conductivity as is also true with silicon.

The thermal conductivity of germanium at room temperature is close to 0.6 W/cm·K and strongly depends on the impurity content and crystal imperfection.[3.343] The temperature dependence of the thermal conductivity was reported by Rosenberg,[3.336] White and Woods,[3.337] Slack,[3.338] Carruthers et al.,[3.339] Geballe and Hull,[3.340] Glassbrenner and Slack,[3.341] and Bakhchieva et al.[3.342] The temperature dependence of the thermal conductivity is shown in Figure 3.34, a and b (Reference 3.254, p. 36). Dependence of the thermal conductivity of germanium on the impurity content was reported by Fagen et al.,[3.344] Goff and Pearlman,[3.345] Mathur and Pearlman,[3.346] and Sota et al.[3.347] Slack[3.338] studied dissipation of phonons by isotopes in the germanium crystals.

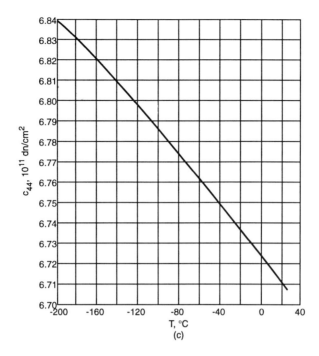

Figure 3.30 (continued)

The molar magnetic susceptibility of germanium at 293 K is $-9.65 \cdot 10^{-4}$ SI units ($-7.68 \cdot 10^{-5}$ cgs units)[3.333,3.334] and virtually independent of temperature ($-1.1 \cdot 10^{-3}$ SI units near melting point[3.335]).

Index of refraction of germanium in the photon energy range from 0.477 eV to 0.690 eV was investigated by Briggs[3.282] who, using a prism made from pure germanium, found that the index of refraction was changing from 4.068 at 0.477 eV to 4.143 at 0.689 eV. Extrapolation of these data to zero photon energy, E, using the formula

$$\left(n^2 - 1\right)^{-1} = A + BE^2,$$

gives $n_{\lambda = \infty} = 4.005$ and the low-frequency dielectric permeability $\varepsilon/\varepsilon_o = 16.04$ in good agreement with the results of Avery and Clegg,[3.348] who evaluated the index of refraction of germanium in the wavelength range from 0.35 µm to 1.3 µm. These results agree well with the results of calculations reported by van Camp et al.[3.350] ($\varepsilon/\varepsilon_o = 16.23$). Lark-Horowitz and Meissner[3.349] found that the reflectance, R, of germanium is virtually independent of wavelength in the range from 2 µm to 150 µm with $R \simeq 0.36$. The formula

$$n = \frac{1 + \left|R^{1/2}\right|}{1 - \left|R^{1/2}\right|}$$

gives n = 4.0 for this wavelength range in conformance with the results of Edwin et al.[3.351] The spectral distribution of the index of refraction, extinction coefficient, and absorption coefficient for germanium in the range from 0.7 eV to 10 eV is presented in Figure 3.35 (from Philipp and Taft[3.352]).

Aspnes and Studna[3.353] measured the dielectric constant of germanium by spectroscopical ellipsometry in the wavelength range from 200 to 830 nm. Their experimental data, together

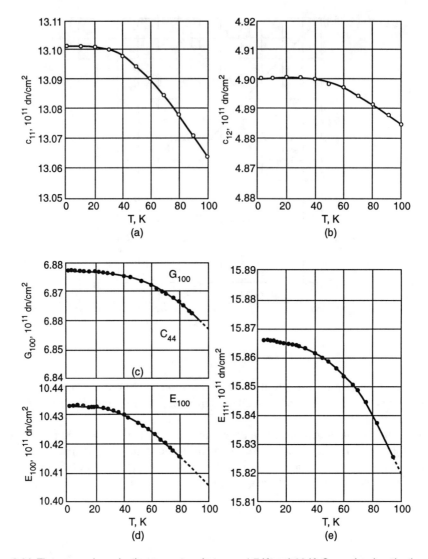

Figure 3.31 The germanium elastic parameters between 1.7 K and 80 K. Second-order elastic moduli: (a) c_{11}, (b) c_{12}, (c) c_{44}, (d) Young's modulus, (e) shear modulus, (f) compressibility, and (g) Poisson's ratio. (After M.E. Fine, *J. Appl. Phys.*, 26, 862, 1955. With permission.)

with the results of calculations of the refractive index, extinction and absorption coefficients, and reflectivity, are collected in Table 3.11.

There is a large number of reports on the behavior of the impurities in germanium, relevant to their solubility in it (see, e.g., References 3.354 and 3.355), the diffusion and self-diffusion coefficients (see, e.g., References 3.355 and 3.356), and the energy levels of impurities and other defects. A comprehensive compilation of these data is presented in Reference 3.256 (p. 37).

3.5. GRAY TIN

The history of gray tin (α-Sn) as a semiconductor starts, apparently, from 1949 when Russian Academician A.F. Ioffe suggested to one of his associates, N.A. Goryunova, to investigate whether gray tin is a semiconductor or not. The crystal structure of gray tin (space

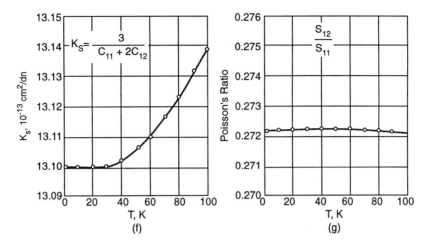

Figure 3.31 (continued)

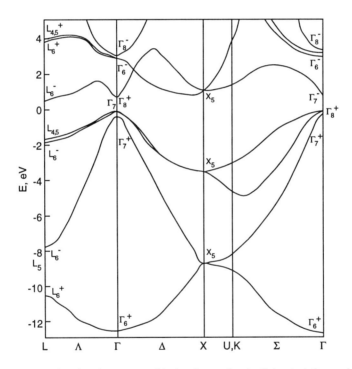

Figure 3.32 The germanium band structure. (Nonlocal pseudopotential calculation, spin-orbit splitting included.) (After J.R. Chelikowsky and M.L. Cohen, *Phys. Rev.*, B30, 556, 1976. With permission.)

group Fd3m – O_{h7}, structure type A4) is similar to that of silicon or germanium, which implies that there is an analogy between the properties of the aforementioned elements and those of the gray tin.

The results of the investigation[3.357,3.358] showed that gray tin is a semiconductor with the (direct) energy gap close or equal to zero (see also Reference 3.371) and with a very high (up to 10^5 cm^2/V·s) electron and hole mobility.

The equilibrium transition temperature from white tin (β-Sn) to the low-temperature phase of gray tin is 286.35 ± 0.01 K,[3.359] but the process has a substantial hysteresis. The highest

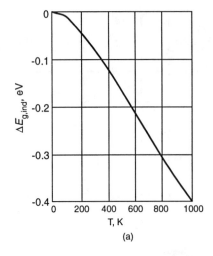
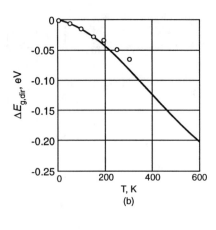

Figure 3.33 Deviation of the germanium energy gap from its magnitude at 0 K with temperature. Dots and circles — experimental data, curves — results of calculation. (a) Indirect gap. (After P. Lautenschlager, P.B. Allen, and M. Cardona, *Phys. Rev.*, B31, 2163, 1985. With permission.) (b) Direct gap. (After P.B. Allen and M. Cardona, *Phys. Rev.*, B27, 4760, 1983. With permission.)

Table 3.10 Current Carrier Effective Mass Data — Germanium

Carrier	Temperature (K)	Effective mass characteristics[a]	Magnitude	Ref.
Electron	300	Longitudinal, m_{eL}	1.64 ± 0.03	3.264
	120	Longitudinal, m_{eL}	1.59	3.326
	300	Transversal, m_{eT}	0.0819 ± 0.003	3.264
	120	Transversal, m_{eT}	0.0823	3.326
Hole	4	Heavy, m_{ph}	0.284	3.264
	4	Light, m_{pl}	0.0438	3.264
	4	Spin-orbit splitting, m_{pso}	0.077	3.264
	30	Spin-orbit splitting, m_{pso}	0.095	3.327
	4	Density-of-states, m_{pds}[b]	0.39	3.264

[a] See comments to Table 3.7.
[b] Density-of-state effective mass for holes can be calculated using expression $m_{pds} = (m_{pl}^{3/2} + m_{ph}^{3/2})^{2/3}$, if the spin-orbit contribution and influence of warping are excluded.

rate of transition relates to 240 K. A description of the gray tin preparation methods and its stabilization at temperatures above 286 K was published by Ewald.[3.360,3.361] The methods of growth of the large "almost perfect" crystals of gray tin were developed by Ewald and Tufte[3.362] and the crystals were investigated by Blumberg and Eisinger.[3.363] Ewald and Tufte used a mercury-tin amalgam for growth of the gray tin single crystals from liquid phase at a temperature between 240 and 250 K with a rate of about 1 cm/month. The crystals grown under these conditions contained about 10^{17} n-type impurities per cubic centimeter. Growth above 250 K produced p-type crystals, probably as a result of the excessive content of mercury atoms in their crystal lattice. Ewald and Kohnke[3.360,3.364,3.365] developed the growth technology for the gray tin whiskers and measured the Hall constant and electrical conductivity on the grown specimens.

Gray tin's lowest conduction band and highest valence band degenerate at Γ (symmetry Γ_8).[3.366,3.367] The second conduction band has L_6 minimum at 0.094 eV (at 0 K).[3.365] The band structure of gray tin calculated by Chelikowsky and Cohen[3.366] is presented in Figure 3.36

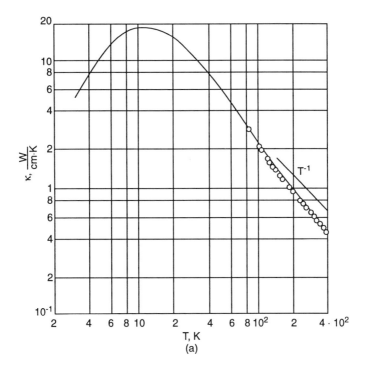

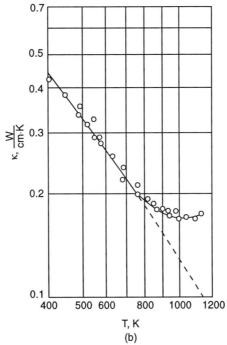

Figure 3.34 Thermal conductivity of germanium at low (a) and high (b) temperatures with broken line representing the extrapolated lattice component. (After C.J. Glassbrenner and G.A. Slack, *Phys. Rev.*, 134A, 1058, 1964 and S.R. Bakhchieva, N.P. Kekelidze, and M.G. Kekua, *Phys. Stat. Solidi*, A83, 139, 1984. With permission.)

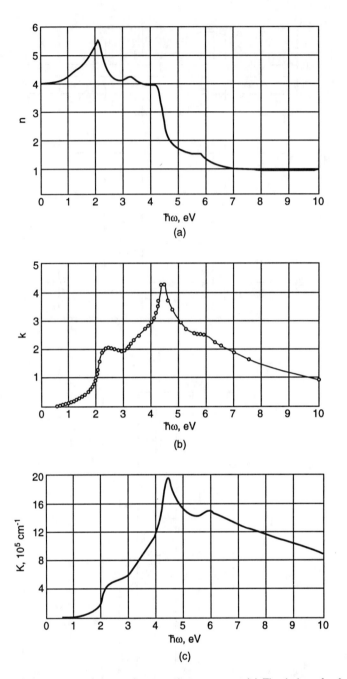

Figure 3.35 Optical parameters of germanium vs. photon energy. (a) The index of refraction, (b) the extinction coefficient, and (c) the absorption coefficient. (After H.R. Philipp and E.A. Taft, *Phys. Rev.*, 113, 1002, 1959. With permission.)

together with the data from the angular resolved photoemission experiments.[3.367] The data agree well with the energy band parameters reported by Pollak et al.[3.388] The specifics of the band structure is that the density of states of the L_6 conduction band is from one to two orders of magnitude greater than that of the Γ_{8c} band.[3.368] This makes the Γ_{8c} conduction negligible, at least in the first approximation, if the L_{6c} band is populated. The data of Reference 3.367 are in satisfactory agreement with the results of semiempirical band structure calculations by Harrison.[3.369]

ELEMENTAL SEMICONDUCTORS

Table 3.11 Germanium Optical Constants

Photon energy, (eV)	Dielectric constant ($\varepsilon/\varepsilon_o$)		Refractive index	Extinction coefficient	Reflectance	Absorption coeff (10^3 cm^{-1})
	Real	Imaginary				
1.5	21.560	2.772	4.653	0.298	0.419	45.30
2.0	30.361	10.427	5.588	0.933	0.495	189.12
2.5	13.153	20.695	4.340	2.384	0.492	604.15
3.0	12.065	17.514	4.082	2.145	0.463	652.25
3.5	9.052	21.442	4.020	2.667	0.502	946.01
4.0	4.123	26.056	3.905	3.336	0.556	1352.55
4.5	−14.655	16.782	1.953	4.297	0.713	1960.14
5.0	−8.277	8.911	1.394	3.197	0.650	1620.15
5.5	−6.176	7.842	1.380	2.842	0.598	1584.57
6.0	−6.648	5.672	1.023	2.774	0.653	1686.84

From D.E. Aspnes and A.A. Studna, *Phys. Rev.*, B27, 985, 1983.

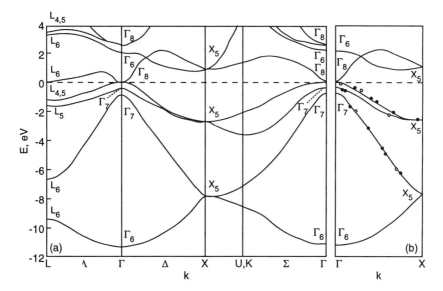

Figure 3.36 Band structure of gray tin. (a) Results of a nonlocal pseudopotential calculation; (b) comparison of these results with the data acquired from the angular-resolved photoemission experiments. (After (a) J.R. Chelikowsky and M.L. Cohen, *Phys. Rev.*, B14, 556, 1976 and (b) H. Höchst and I. Hernández-Calderón, *Surf. Sci.*, 126, 25, 1983. With permission.)

The lattice parameter of gray tin (0.64892 nm at 293 K) and its coefficient of linear thermal expansion in the range of 140 to 298 K (4.8·10^{-6} K^{-1}) were determined by Thewlis and Davey.[3.370] The temperature dependence of the coefficient of thermal expansion of the gray tin samples, prepared by pressing its powder, was investigated by Novikova[3.372] (see Figure 3.37). At 45 K, the coefficient of thermal expansion is equal to zero and is negative at lower temperatures, similar to silicon and germanium. The pycnometric density of gray tin at 291 K of 5.765 g/cm^3 is in satisfactory agreement with the X-ray density of 5.7702 g/cm^3.

The mechanical properties of gray tin are typical for a semiconductor: the crystals are brittle in sharp contrast with the soft and easily deformed crystals of the metallic white tin. The second-order elastic moduli of gray tin are c_{11} = 69 GPa, c_{12} = 29 GPa, and c_{44} = S = 36 GPa[3.389] (S is the shear modulus), and therefore, Young's modulus is equal to E = (c_{11} + c_{12} + 2c_{44})/2 = 85 GPa. Young's and shear moduli of (polycrystalline) white tin are close to 41 GPa and 16 GPa, respectively.

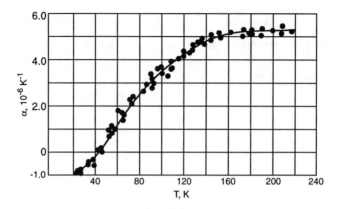

Figure 3.37 Temperature dependence of the coefficient of thermal expansion of gray tin. (After S.I. Novikova, *Sov. Phys. Solid State*, 2, 2087, 1961. With permission.)

Electrical properties of gray tin were investigated by Busch et al.,[3.373-3.376] Kendal[3.377,3.378] (pressed pellets), Kohnke and Ewald[3.365] (filaments), Lavine and Ewald,[3.368] and Li-Wei Tu et al.[3.379] (epilayers). The energy gap at room temperature, in accordance with Busch and Mooser,[3.374] is 0.08 eV. The magnitude of 0.094 eV was reported in Reference 3.365 together with the temperature dependence of $-5 \cdot 10^{-5}$ eV/K. (Extrapolation to 300 K gives the energy gap of 0.079 eV.) The intrinsic photoconductivity measurements at low temperatures, reported by Becker,[3.380] give the energy gap of 0.075 eV.

Because of the narrow energy gap, gray tin possesses intrinsic conductivity even at room temperature, if the specimens do not have a high impurity content. Its conductivity at 300 K is between 2 and 5 kS/cm in both bulk and filament[3.364] and thin film[3.379] specimens. The temperature dependence of the Hall constant and electrical conductivity measured on the gray tin filament samples[3.365] is shown in Figure 3.38. The temperature dependence of the concentration and mobility of electrons and holes calculated from the data is presented in Figure 3.39. The results of Li-Wei Tu et al.[3.379] (Figure 3.40) show that the intrinsic conductivity of a gray tin epitaxial film with a thickness of 37.5 nm grown on a (001) CdTe wafer possesses the intrinsic conductivity at T > 200 K. The authors of Reference 3.379 found that the gray tin epilayer's energy gap is virtually independent of thickness of the films (if thickness is greater than 20 nm) and is close to 0.1 eV.

Gray tin samples dopped with the Group II (zinc) or Group III (aluminum, indium) elements have p-type conductivity, whereas the Group V elements (arsenic or antimony) are used to prepare samples of the n-type conductivity.

Effective mass of the electrons and holes in gray tin was estimated first by Kohnke and Ewald[3.365] and by Blumberg and Eisinger,[3.363] who received the magnitudes for electrons from 0.3 to 0.8 m_o. Later, Booth and Ewald[3.381] and Lavine and Ewald[3.368] found the light-electron density-of-states effective mass equal to 0.0236 m_o at 1.3 K, and the heavy electron mass of 0.21 m_o at 4.2 K. Groves et al.,[3.381] using the results of magnetoreflection experiments, found that the Γ_{8v}-zone hole effective mass is equal to 0.195 m_o and the Γ_{7v}-zone hole effective mass is 0.058 m_o.

The molar magnetic susceptibility of gray tin at 273 K is close to $-4.65 \cdot 10^{-4}$ SI units ($-37 \cdot 10^{-6}$ cgs units);[3.367] its dielectric constant is 24 at 300 K.[3.382]

Optical properties of gray tin in the range of the photon energy from 1.25 to 5.0 eV were reported by Hanyu[3.383] (see Figure 3.41).

The ability of gray tin to conserve its crystal structure at temperatures above the transition point if the material is dopped with (about 0.75 wt%) germanium[3.361] or bismuth,[3.385] or if a thin gray tin epilayer is grown on a substrate with diamond-like crystal structure, presents a

ELEMENTAL SEMICONDUCTORS

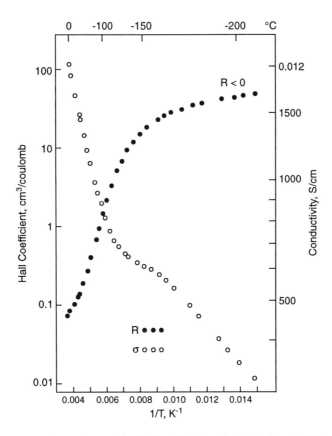

Figure 3.38 Temperature dependence of the Hall coefficient and electrical conductivity of gray tin. (After E.E. Kohnke and A.W. Ewald, *Phys. Rev.*, 102, 1481, 1956. With permission.)

substantial practical interest. It was noted by Sandomirskii[3.386] that the thermodynamic properties and kinetic coefficients of a solid depend on its characteristic geometric dimensions, if the dimensions are comparable with the de Broglie wavelength of the charge carriers (quantum size effect). Then, if a dimension (e.g., thickness) of a gray tin epitaxial layer is of the mentioned order of magnitude, its energy gap may be tailored from zero to a desirable magnitude by a proper choice of the layer thickness and the substrate material and orientation.[3.387] Li-Wei Tu et al.[3.379] evaluated the energy gap of the gray tin epilayers, grown on the cadmium telluride substrates, from the temperature dependence of electrical resistivity. They found a pronounced dependence of the energy gap on thickness (Figure 3.42). These results, together with the data on the charge carrier mobility, show that gray tin ought to be considered as a very promising material for, e.g., far-infrared radiation detectors. In spite of its promising properties for particular applications, gray tin has not been substantially investigated.

3.6. GROUP V — PHOSPHORUS

Among the elements of Group V, only phosphorus in its orthorhombic and monoclinic modifications has semiconductor properties. The other pnictogens — arsenic, antimony, and bismuth — are semimetals (see, e.g., Dresselhaus[3.390]). Moss,[3.1] in the process of measurements of electrical, optical, and photoelectrical properties of thin films, found, notwithstanding, that gray arsenic behaves as a semiconductor with the energy gap of 1.14 eV (from temperature dependence of electrical conductivity). The photoconductivity of gray arsenic is

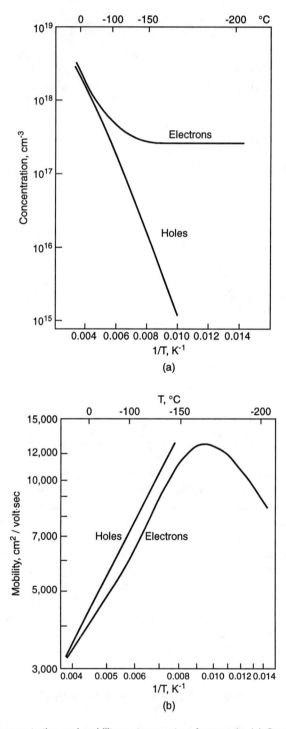

Figure 3.39 Carrier concentration and mobility vs. temperature for gray tin. (a) Carrier concentration, (b) mobility. (After E.E. Kohnke and A.W. Ewald, *Phys. Rev.*, 102, 1481, 1956. With permission.)

characterized by the long-wave limit between 1.19 eV and 1.23 eV in agreement with the data on the conductivity vs. temperature measurements. Moss reported also on photoconductivity in thin layers of antimony with activation energy close to 0.1 eV. It is possible that these results prove the existence of the quantum dimensional effect mentioned before.[3.386]

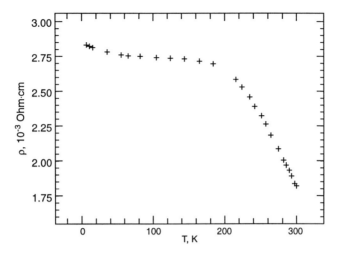

Figure 3.40 Temperature dependence of the electrical resistivity of an epitaxially grown 37.5-nm-thick gray tin film. (After L.-W. Tu, G.K. Wong, and J.B. Ketterson, *Appl. Phys. Lett.,* 54, 1010, 1989. With permission.)

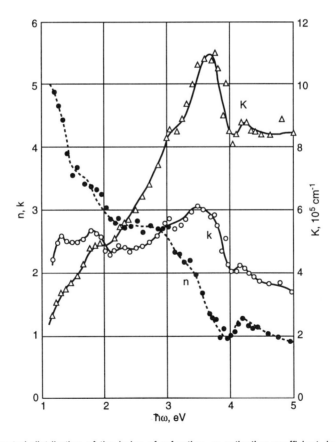

Figure 3.41 Spectral distribution of the index of refraction, n; extinction coefficient, k; and absorption coefficient, K; in gray tin. (After T. Hanyu, *J. Phys. Soc. Jpn.,* 31, 1738, 1971. With permission.)

Because phosphorus is the only semiconducting element in Group V, let us discuss its properties in more detail. The element was discovered between 1669 and 1675 by Hennig Brand (often called the last of the alchemists) in the search for the philosopher's stone. Robert

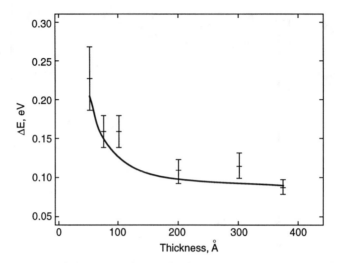

Figure 3.42 Correlation between the energy gap and the thickness of the gray tin epilayers. (After L.-W. Tu, G.K. Wong, and J.B. Ketterson, *Appl. Phys. Lett.*, 54, 1010, 1989. With permission.)

Boyle independently prepared phosphorus in 1680, but, in contrast with Brand, he published the results without delay.

The atomic number of phosphorus is 15, atomic weight is 30.973762. It consists of only one stable isotope, ^{31}P, with natural abundance of almost 100%. Among 16 known radioactive isotopes, beta-minus isotope ^{32}P ($T_{1/2}$ = 14.28 day) may be considered as the only one important for practical use. Electron configuration of phosphorus is $3s^2 3p^3$. The ionization energy of phosphorus in the sequence $P^0 \rightarrow P^+ \rightarrow P^{2+} \rightarrow P^{3+} \rightarrow P^{4+} \rightarrow P^{5+}$ is 10.48669, 19.7694, 30.2027, 50.4439, and 65.0251 eV (Reference 3.75, p. 10-205).

The most common allotrope is yellow (or white) phosphorus that can be produced by liquifaction of phosphorus vapor (which consists of tetrahedral molecules P_4) with following crystallization of the liquid. Yellow phosphorus exists in two forms: (1) high-temperature cubic alpha-form with lattice constant a = 1.85 nm and 56 P_4 molecules in unit cell, density of 1.828 g/cm^3 at 293 K, melting point of 317.3 K, molar heat of sublimation of 56.0 kJ/mol, specific heat of 768 J/kg K, molar magnetic susceptibility of $-3.35 \cdot 10^{-4}$ SI units ($2.66 \cdot 10^{-5}$ cgs units), volume resistivity at 300 K close to 10^{15} Ohm·cm, and the energy gap near 2.1 eV;[3.1] and (2) low-temperature rhombohedral beta-phosphorus (transition temperature close to 196 K) with density of 1.88 g/cm^3 at 190 K. Yellow phosphorus has a very high chemical activity (self-ignition in the air at 317 K).

Two other modifications of phosphorus, namely, black and red, are semiconductors with the thermal energy gap at room temperature of 0.34 and 1.50 eV, respectively. Black crystalline phosphorus is, probably, the most thermodynamically stable allotrope of the element. It crystallizes in the orthorhombic lattice D_{2h}^{18}-Cmca (Pearson symbol oC8) with parameters a = 0.33136 nm, b = 1.0478 nm, and c = 0.43763 nm (Reference 3.75, p. 12-10). The melting point is 1300 K and density 2.69 g/cm^3 at 300 K. Black phosphorus was produced first by Bridgman[3.391] by heating yellow phosphorus to 473 K under a pressure of 1.3 GPa.

The Brillouin zone of black phosphorus and its band structure are presented in Figure 3.43 (Reference 3.43). The current carrier effective mass in black phosphorus is close to 0.34 m_0 (electrons) and 0.26 m_0 (holes). At room temperature, the carrier mobility is about 220 cm^2/V·s (electrons) and about 350 cm^2/V·s (holes) with electrical conductivity between 0.5 S/cm and 1.0 S/cm.

Thermal conductivity of polycrystalline black phosphorus is 13.2 W/m·K at 273 K and 12.1 W/m·K at 298 K.[3.343] The dielectric constant is 6.1, index of refraction is near 2.4.

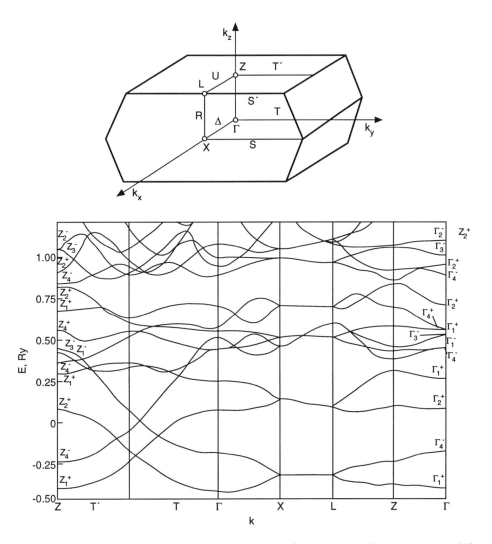

Figure 3.43 Brillouin zone (a) and the band structure (b) of orthorhombic black phosphorus. (After H. Asahina, A. Morita, and K. Shindo, *Proc. 15th Int. Conf. Phys. Semicond.*, Kioto, 1980 and S. Tanaka and Y. Toyozawa, Eds., *Phys. Soc. Jpn.*, Tokyo, 1980, p. 85. With permission.)

Red (violet) phosphorus exists in several modifications, among them amorphous red-I, monoclinic red-II, hexagonal or tetragonal red-IV, and triclinic red-V. Depending on the synthesis method, red phosphorus density magnitude lies in the limits 2.0 g/cm³ and 2.4 g/cm³ and melting point between 858 K and 875 K (863 K under phosphorus vapor pressure of 4.3 MPa, sublimates at 690 K under vapor pressure of 0.1 MPa).

Moss[3.1] found that the volume resistivity, ρ, of several red phosphorus samples was changing linearly in coordinates $\log \rho - T^{-1}$ with magnitudes close to $2 \cdot 10^9$ Ohm·cm at 325 K and $4 \cdot 10^6$ Ohm·cm at 420 K.

In spite of the relative simplicity of red phosphorus production — it can be made from yellow phosphorus by baking the latter in its vapor at 525 K — and its interesting electrical ($E_g \simeq 1.5$ eV), optical ($\varepsilon/\varepsilon_o = 6.7$; $n = 2.6$), and photoelectrical properties, this material has not found any application as an elemental semiconductor up to now.

Black phosphorus, apparently, has intense light absorption in the visible part of the spectrum, which in combination with high carrier mobility may be of interest for solar/electrical energy conversion. Its electrical and optical properties have been investigated only superficially; there are no data on electron work function and secondary electron emission parameters.

Electronic and optical properties of single-crystalline and epitaxially grown phosphorus are virtually unknown; all available information was obtained using polycrystalline specimens. The technology of purification of phosphorus has been successfully developed: red phosphorus with the purity of 99.99995% is available[3.74] under $7 per gram. The logical and urgent next step ought to be the development of the single crystal growth technology and investigation of the properties of the crystals.

3.7. SELENIUM

Selenium and tellurium are the only chalcogens that possess semiconductor properties. Selenium was discovered by Berzelius in 1818; its atomic number is 34, atomic weight is 78.96(3). It consists of six stable isotopes: ^{74}Se (0.89%), ^{76}Se (9.36%), ^{77}Se (7.63%), ^{78}Se (23.77%), ^{80}Se (49.61%), and ^{82}Se (8.74%). The most important among 15 radioactive isotopes is ^{75}Se ($T_{1/2}$ = 119.78 days). Electron configuration of selenium is $4s^2 4p^4$, ionization energy in the sequence $Se^0 \rightarrow Se^+ \rightarrow Se^{2+} \rightarrow Se^{3+}$ is 9.75238, 21.19, 30.8204 eV. Ionic radii in crystals are[3.392] 0.198 nm (Se^{-2}) and 0.050 nm (Se^{+6}) (both magnitudes relate to a coordination number of six). Atomic radius of selenium is 0.117 nm.[3.393] The melting point of selenium under atmospheric pressure is 490 to 493 K. Liquid selenium has a very strong tendency to overcooling, thus forming amorphous or vitreous selenium (a-Se). The glass transformation temperature, T_g, strongly depends on the temperature change rate. Kasap and Yannacopoulos[3.394] showed that T_g varies from 303 K at $8.33 \cdot 10^{-4}$ K/sec to 338 K at 1.67 K/sec. The same authors[3.395] found that T_g has a different dependence on the cooling rate, varying from about 308 K at $-6.3 \cdot 10^{-3}$ K/sec to 320 K at -0.67 K/sec. Thus, T_g at a given magnitude of the temperature change rate is about 10 to 15 K higher at heating in comparison with cooling. Viscosity of the selenium glass at room temperature is close to 10^{12} Pa·s[3.396] and decreases exponentially with the temperature increase. The Vickers microhardness of the a-Se glass at room temperature is close to 350 N/mm^2;[3.397] its shear modulus is 3.7 GPa.[3.398] Amorphous selenium is widely used as a xerographic photoreceptor material, and optimization of its properties for this application attracts unabated attention of the material scientists. Amorphous selenium is a semiconductor with the resistivity at room temperature of 10^{12} to 10^{16} Ohm·cm[3.399,3.400] and has an energy gap of 1.7 eV.[3.1] Being heated, a-Se transforms into gray trigonal selenium.

Another allotrope modification, monoclinic selenium, occurs in three phases with different lattice parameters. This allotrope may be prepared by crystallization from solutions in carbon disulfide. The lattice is formed by puckered S_8 rings. Monoclinic α-Se (space group C_{2h}^5–$P2_1/n$) has lattice parameters a = 0.905 nm, b = 0.908 nm, c = 1.161 nm, β = 90°46'; its density is 4.44 g/cm^3. When slowly heated to 383 to 393 K, it transforms into beta-Se (a = 0.807 nm, b = 0.931 nm, c = 1.285 nm, β = 93°48') with a density of 4.50 g/cm^3. The energy gap of α-Se at room temperature is between 2.1 and 2.5 eV.[3.401] Moss[3.1] suggests E_g = 1.6 eV, which, apparently, is too low. The resistivity of α-Se is of the order of 10^{15} Ohm·cm; the carrier mobility ranges from 1.5 to 3.6 cm^2/V·sec (electrons) and from 0.2 to 1.3 cm^2/V·sec (holes). Electrical properties of beta-Se are virtually unknown. Comparison of the available physicochemical properties of amorphous and monoclinic selenium (Table 3.12) shows that the monoclinic allotrope still has yet to be investigated.

The most studied allotrope of selenium is trigonal gray Se. It is the only modification which is thermodynamically stable below melting point, at atmospheric pressure. The crystal structure of gray Se is formed by the spiral chains of selenium atoms with the chain axes

Table 3.12 Some Physicochemical Properties of Amorphous and Monoclinic Selenium

Parameter	Allotrope		
	Amorphous, a-Se	Alpha-Se	Beta-Se
Density, g/cm³	4.28	4.44	4.50
Transformation temperature, K	360 (to gray Se)	388 (to beta-Se)	395 (to gray Se)
Specific heat, J/g·K (between 293 and 323 K)	0.426	—	—
Coefficient of thermal expansion, K⁻¹, at 291 K	$37 \cdot 10^{-6}$	—	—
Thermal conductivity at 300 K, mW/cm·K	5.2	—	—
Molar magnetic susceptibility at 300 K, SI units	$3.23 \cdot 10^{-4}$	—	—

parallel to the c-axis (Figure 3.44). The chains may have a clockwise (symmetry group D_3^4 – $P3_121$) or a counterclockwise (D_3^6 – $P3_221$) direction. Three atoms form one turn of the spiral. Lattice parameters (300 K) are a = 0.43642 ± 0.00008 nm, c = 0.49588 ± 0.00008 nm, and the chain radius r = 0.09819 ± 0.00002 nm. Pycnometric density of gray Se is 4.79 g/cm³, with X-ray density 4.8088 ± 0.0019 g/cm³. The melting point of gray Se is 494.33 K, molar heat capacity C_p = 25.4 J/mol·K (300 K), and the heat of fusion at melting point is 5.09 kJ/mol. The temperature dependence of the thermal conductivity of gray Se and a-Se is compiled in Table 3.13 (Reference 3.447, p. 4-146).

The coefficient of the linear thermal expansion of gray Se at 291 K is $17.9 \cdot 10^{-6}$ K⁻¹ along the c-axis and $74.1 \cdot 10^{-6}$ K⁻¹ perpendicular to the c-axis. The mechanical properties of gray Se, measured on polycrystals, are: Brinell hardness close to 750 N/mm², Young modulus 55 GPa, and shear modulus close to 6.6 GPa.

The extensive use of selenium in electrical rectifiers, photocells, photoexponometers, and other semiconductor devices instigated a great number of works devoted to investigation of its properties. Gray Se crystals always possess the p-type conductivity.[3.402,3.403] The nature of this phenomenon has not been completely explained, but it seems reasonable to assume that the cause of it lies in the specifics of the gray Se crystal structure. The chains forming the crystal lattice have finite lengths and the free bonds form the acceptor levels. Many attempts to change the conductivity type by doping selenium with controlled amounts of different elements were, in general, unsuccessful (see, e.g., Reference 3.404). The addition of iodine and some other group VII elements increases the p-component of conductivity (and total conductivity), whereas doping selenium with mercury or zinc decreases the hole conductivity. The only known element which produces n-type conductivity in gray Se is gold.[3.406] Berger and Petrov[3.406] found that the solid solution of more than 0.03 wt% tantalum in pure selenium is an n-type semiconductor with the electron concentration of $4 \cdot 10^{15}$ cm⁻³ and the mobility of 0.2 cm²/V·sec at room temperature in the purest samples. The change in the conductivity type was accompanied by transition from "metallic" temperature dependence of the conductivity, typical for undopped selenium, to semiconductor dependence.

The energy gap of gray Se evaluated from the optical absorption experiments by Choyke and Patrick[3.407] is near 1.8 eV at 300 K; its temperature coefficient is about $-9 \cdot 10^{-4}$ eV/K. Optical properties of gray Se in the range of 400 to 23,000 nm and its band structure were studied by Gobrecht and Tausend.[3.408] The dielectric constant of gray Se at low frequencies is 12.2 parallel to the c-axis and 7.4 perpendicular to the c-axis, whereas at high frequencies, it is close to 11.6 and 7.0, respectively.

Electrical conductivity of (gray) selenium in a wide range of temperatures, including liquid state, was analyzed by Glazov et al.[3.276] Their data show (Figure 3.45) that in the liquid state selenium retains its semiconductor properties. The energy gap of liquid selenium is between 1.6 and 1.7 eV for the as-received samples, with p-type conductivity, while the specimens from which oxygen has been removed have the energy gap in the liquid state of about 0.5 eV and the n-type conductivity.[3.409]

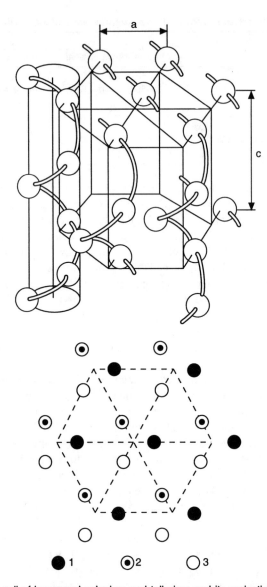

Figure 3.44 The unit cell of hexagonal selenium and tellurium and its projection on a horizontal plane. 1, 2, and 3 — sequence of atoms in the spiral chains along the c axis.

3.8. TELLURIUM

Tellurium was extracted first in 1782 from a gold ore (probably muthmannite, [Ag,Au]Te) by Austrian mineralogist F.J. Müller, but his discovery was forgotten until German chemist M.H. Klaproth recognized it as an element (1798), giving, however, a proper credit to the original discoverer. Native tellurium, in the shape of prismatic crystals, is found in Rumania, West Australia, and Colorado, USA.

The atomic number of tellurium is 52; its atomic weight is 127.60. It consists of five stable isotopes: ^{120}Te (0.095%), ^{122}Te (2.59%), ^{124}Te (4.79%), ^{125}Te (7.12%), ^{126}Te (18.93%), and ^{128}Te (31.70%). Among the radioactive isotopes, ^{123}Te (0.905%, $T_{1/2} = 1.3 \cdot 10^{13}$ years) and ^{130}Te (33.87%, $T_{1/2} = 2.5 \cdot 10^{21}$ years) have such a long half-life that they are frequently considered as a part of the stable chemical substance.[3.75]

Table 3.13 Thermal Conductivity Vs. Temperature

Allotrope	Thermal conductivity, mW/cm·K, at temperature (K)		
	273	298	373
Gray selenium			
Parallel to c-axis	48.1	45.2	48.3
Perpendicular to c-axis	13.7	13.1	13.9
Amorphous selenium	4.28	5.19	8.18 (323 K)

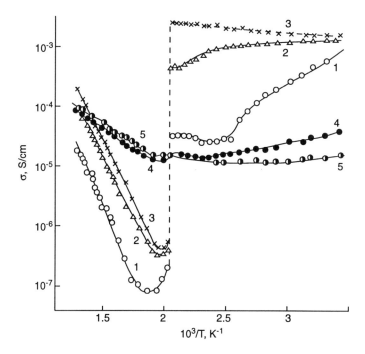

Figure 3.45 The temperature dependence of electrical conductivity of solid and liquid selenium. (After V.M. Glazov, S.N. Chizhevskaya, and N.N. Glagoleva, *Liquid Semiconductors*, Nauka Publishing House, Moscow, 1967. With permission.)

The electron configuration of tellurium is $4d^{10}5s^25p^4$. The ionization energy (in eV) in the sequence $Te^0 \rightarrow Te^+ \rightarrow Te^{2+} \rightarrow Te^{3+} \rightarrow Te^{4+} \rightarrow Te^{5+} \rightarrow Te^{6+} \rightarrow Te^{7+}$ is 9.0096, 18.6, 27.96, 37.41, 58.75, 70.7, and 137.0 (Reference 3.75, p. 10-205).

Tellurium crystallizes in the trigonal lattice similar to that of gray selenium (Figure 3.44) with lattice constants (300 K) a = 0.44570 ± 0.00008 nm, c = 0.59290 ± 0.00010 nm, and the chain radius r = 0.11945 nm (Reference 3.447, p. 4-156). The X-ray density at 300 K is 6.2316 g/cm^3; the pycnometric density 6.20 to 6.25 g/cm^3. The tellurium melting point is 722.72 K and the boiling point is 1261.15 K. Enthalpy of fusion is 13.53 kJ/mol; the enthalpy of vaporization at the boiling point is 114.1 kJ/mol. The specific heat of tellurium at 298 K is 0.202 J/g·K (Reference 3.447, p. 5-79); its thermal expansion coefficient (polycrystal, at 293 K) is 16.8·10^{-6} K^{-1}. The temperature dependence of thermal conductivity of single-crystalline tellurium is presented in Table 3.14 (Reference 3.447, p. 4.149).

Young's moduli of tellurium at room temperature are 43.5 and 20.9 GPa along and across the c-axis, respectively; compressibility at 310 K is 15 1/TPa. Shear modulus (polycrystal, at 293 K) is 15.5 GPa.

The electrical properties of tellurium are much more understood than those of selenium, first of all, because the former has only one allotropic modification. As does selenium, tellurium

Table 3.14 Thermal Conductivity of Tellurium

Crystallographic direction	Thermal conductivity, mW/cm·K, at temperature (K)		
	273	298	373
Parallel to the c-axis	36.0	33.8	29.2
Perpendicular to the c-axis	20.8	19.7	17.3

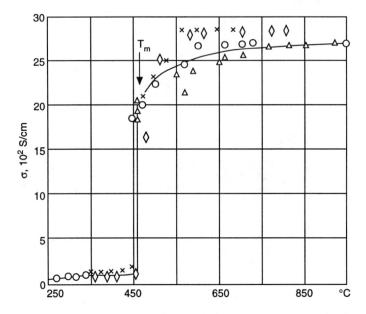

Figure 3.46 Temperature dependence of the electrical conductivity of solid and liquid tellurium. Data of different authors. (After V.M. Glazov, S.N. Chizhevskaya, and N.N. Glagoleva, *Liquid Semiconductors*, Nauka Publishing House, Moscow, 1991. With permission.)

has substantial anisotropy of its electrical parameters. The mobility of both electrons and holes at room temperature is about twice greater along the c-axis than in a perpendicular direction ($\mu_{n\|c} = 2380$, $\mu_{n\perp c} = 1150$, $\mu_{p\|c} = 1260$, and $\mu_{p\perp c} = 650$ cm²/V·sec), while the effective mass of both electrons and holes in the c-direction is smaller ($m^*_{n\|c} = 0.050\ m_o$, $m^*_{n\perp c} = 0.060\ m_o$, $m^*_{p\|c} = 0.109\ m_o$, and $m^*_{p\perp c} = 0.114\ m_o$) with the intrinsic electron concentration $n_i = 5.6 \cdot 10^{15}$ cm^{-3} (Reference 3.13, p. 7). At room temperature, purified single-crystalline tellurium is an intrinsic semiconductor with the thermal energy gap of 0.32 to 0.38 eV.[3.412,3.413] The temperature coefficient of the energy gap is positive in contrast with the majority of semiconductors, including selenium, and is close to $1.9 \cdot 10^{-4}$ eV/K.[3.412] The mobility of both electrons and holes is proportional to $T^{-3/2}$.[3.412] A steady increase in purity of tellurium (currently, the purest commercially available tellurium has 6N+ grade) is still followed by a continuous increase of the carrier mobility. An interesting feature of the temperature dependence of the electrical transport properties of tellurium consists of triple change of the sign of the Hall coefficient, R. At low temperatures, R = constant > 0. When temperature reaches 230 K to 235 K, changes sign as is common for many p-type semiconductors, but at about 500 K, R becomes positive again. Melting of tellurium is accompanied by a sharp decrease of the R magnitude. When the temperature of the tellurium melt reaches 850 K, R becomes negative again.[3.413] During the melting process, tellurium increases its conductivity by an order of magnitude;[3.276] this is followed by a gradual growth of conductivity with temperature in the liquid state (Figure 3.46). Thus, tellurium remains a semiconductor in the liquid state.

Optical properties of tellurium were measured by Moss,[3.1] Hartig and Loferski,[3.415] Loferski,[3.416] and Caldwell and Fan.[3.417] Hodgson[3.418] measured the optical properties of melted tellurium. Photoconductivity of thin tellurium films was studied by Moss.[3.1] Tellurium index of refraction, n_r, for polycrystals, extrapolated to $1/\lambda = 0$, is 4.8.[3.1] Tellurium single crystals show a strong dependence of n_r on the angle between the electric vector of the polarized incident radiation $\vec{E}$ and the orientation of the c-axis of the crystal.[3.415] In the range 4 to 8 μm, $n_r \simeq 4.90$ for $\vec{E} \perp c$ and $n_r \simeq 6.25$ for $\vec{E} \| c$. The optical absorption of tellurium single crystals also depends on the angle:[3.416] the absorption edge is at 0.324 eV for $\vec{E} \| c$ and at 0.374 eV for $\vec{E} \perp c$. From the investigation of the photoconductivity of (polycrystalline) tellurium films, Moss[3.1] received for the energy gap a magnitude of 0.35 eV. The optical data are in satisfactory agreement with the magnitude of the thermal energy gap.[3.412,3.413]

Currently, tellurium is utilized mainly as a component of a very large group of semiconducting chalcogenides used in electro-optical and electronic devices (see, e.g., References 3.194 and 3.419), but its low energy gap and especially its dependence on the tellurium single crystal orientation relative to the direction of the plane of polarization of the incident light, in combination with high carrier mobility and simplicity of the purification and crystal growth technology, contain a substantial potential for the use of elemental tellurium in the instruments of night vision, and, in particular, the ones which are based on the analysis of reflected infrared radiation.

3.9. CHEMICAL INTERACTION BETWEEN THE ELEMENTAL SEMICONDUCTORS: SOLID SOLUTIONS AND COMPOUNDS

The only binary alloys and compounds which are included in this section are ones with constituents that belong to the semiconductors considered in this chapter.

3.9.1. Boron-Carbon

Solubility of carbon in boron at room temperature does not exceed 0.2 at%.[3.420] Several boron-carbon phases, B_4C, $B_{13}C_2$, $B_{15}C_2$, and B_9C, were investigated in the search for high-temperature semiconductor materials for thermoelectric applications. The (mainly hypothetical) phase diagram of the B-C system was suggested by Samsonov et al.[3.421] The boron-rich part of it (100 – 60 at%) was reported by Elliott.[3.422] B_4C and $B_{13}C_2$ crystallize in rhombohedral structure with lattice parameters a = 0.56 nm and c = 1.207 nm (Lipp and Röder[3.423]) for B_4C, and a = 0.567 nm and c = 1.219 nm (Allen[3.424]) for $B_{13}C_2$.

The most investigated of boron carbides, B_4C has a melting point of 2490 K,[3.425] a boiling point of >3800 K, a density of 2.52 g/cm^3; its energy gap (300 K) is 0.48 eV; the hole mobility, measured on the pressed (300 K) samples, is within the limits of 0.3 to 1.0 cm^2/V·sec; the thermal conductivity varies from 0.3 W/cm·K at 300 K to 0.15 W/cm·K at 1000 K.[3.426] The Knoop hardness is 29,700 N/mm^2 [3.427] and the coefficient of linear thermal expansion 5.65 to 5.87 1/MK.[3.428] The specific heat of several boron carbides (B_4C, $B_{15}C_2$, and B_9C) was investigated by Türkes et al.[3.55] Their data are in satisfactory agreement with the results of Gilchrist and Preston[3.429] and Kelley.[3.430] The electronic and lattice properties of $B_{13}C_2$, $B_{15}C_2$, and B_9C have not been studied as extensively as B_4C. Existence of a wide range of solid solutions based on B_4C (in the limits from $B_{12}C$ to B_4C[3.422]) gives reason to assume that the alloys with a relatively low thermal conductivity and high thermal emf can be found in this composition range. Among the boron-carbon compounds, B_9C has the highest Seebeck coefficient and electrical conductivity,[3.435,3.436] with the thermal conductivity of the order of 50 mW/cm·K at room temperature, which makes this material especially interesting for thermoelectric applications.

The sequence of the boron-carbon solid solutions is formed by the alloys in which carbon atoms fill the sites between the boron icosahedra in the boron rhombohedral lattice (see Section 3.1 of this chapter and Reference 3.10).

3.9.2. Boron-Silicon

The phase diagram of the system B-Si has not been investigated in detail. Two known compounds of boron with silicon are B_6Si and B_4Si. Both phases are semiconductors with the energy gap close to 0.5 eV, melting points about 1860 K (B_6Si) and 1270 K (B_4Si);[3.431] the coefficient of linear thermal expansion of $18.2 \cdot 10^{-6}$ K^{-1} [3.432] for B_6Si and $6.0 \cdot 10^{-6}$ K^{-1} [3.433] for B_4Si; the hardness (Knoop) of 28,700 and 22,900 N/mm^2 for B_6Si and B_4Si, respectively.[3.434]

3.9.3. Boron-Phosphorus

There are two known compounds in the B-P system: B_6P and BP. The latter is one of the III-V diamond-like compounds and will be considered in Chapter 4. B_6P has a rhombohedral structure, space group $D_{3d}^5 - R\bar{3}m$, formed by B_{12} icosahedra. The lattice parameters are a = 0.5984 nm and c = 1.1850 nm.[3.437,3.438] The optical energy gap of B_6P at 300 K is 3.3 eV and melting point >2300 K.[3.439] Electrical, optical, and physicochemical properties of B_6P have been studied only superficially.

3.9.4. Carbon-Silicon

The binary system C-Si contains only one compounds, carborundum, SiC,[3.440] the properties of which will be considered in Chapter 4. The solubility either of carbon or silicon in SiC is apparently very small.

3.9.5. Silicon-Germanium

The binary system silicon-germanium represents a continuous series of substitutional solid solutions in the total range of concentrations (Figure 3.47). This is the only case for both elements where they possess not only a wide range of solid solutions, but the solutions are balanced. The lattice constant of the solutions follows the Vegard law: the deviation from the linear dependence of the lattice constant on the concentration (contraction) does not exceed 0.9 pm (about 0.2%).[3.441]

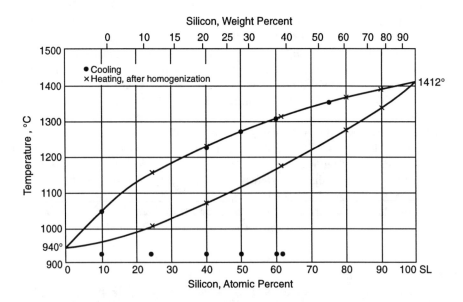

Figure 3.47 Germanium-silicon phase diagram. (After M. Hansen and K. Anderko, *Constitution of Binary Alloys*, McGraw-Hill, New York, 1958. With permission.)

Electrical conductivity of the Si-Ge alloys was reported by Levitas[3.442] and Busch and Vogt.[3.443] The composition dependence of the alloy's intrinsic conductivity is shown in Figure 3.48 (after Reference 3.443). At $x \simeq 0.15$, there is a change of the slope of the curve. A bend at about the same composition of the curve of compositional dependence of the indirect energy gap (Figure 3.49, after Reference 3.446) gives reason to assume that the origin of this bend is connected to the specifics of the band structures of silicon and germanium. Herman[3.444] and Braunstein et al.[3.445] explained this effect by comparing the conductive band minima of Si and Ge. At the low silicon concentration, electrons occupy the minima of germanium which are oriented in the [111] direction; at $x \simeq 0.15$, the germanium conduction band minima and the silicon [100] minima intersect, and at $x > 0.15$ the silicon conduction band minima determine the electron generation-recombination process and, consequently, the electrical properties of the alloys

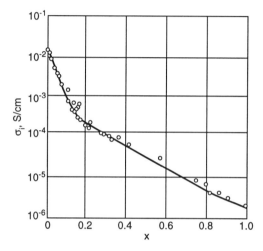

Figure 3.48 Composition dependence of the intrinsic electrical conductivity of silicon-germanium alloys. (After G. Busch and O. Vogt, *Helv. Phys. Acta*, 33, 437, 1960. With permission.)

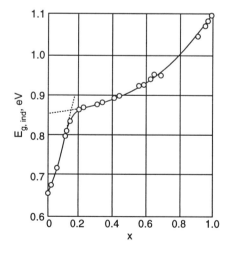

Figure 3.49 Compositional dependence of the indirect energy gap of Si_xGe_{1-x} alloys at 296 K (based on one-phonon analysis of the optical absorption edge). Near $x = 0.15$, a crossover occurs of the Ge-like [111] and the Si-like [100] conduction band minima. (After R. Braunstein, A.R. Moore, and F. Herman, *Phys. Rev.*, 109, 695, 1958. With permission.)

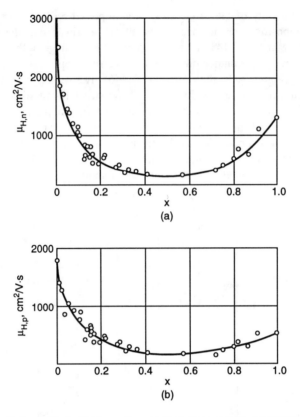

Figure 3.50 Carrier Hall mobility in Si_xGe_{1-x} alloys at room temperature. (a) Electron mobility, (b) hole mobility. (After G. Busch and O. Vogt, *Helv. Phys. Acta*, 33, 437, 1960. With permission.)

The electron and hole mobility in the Si-Ge alloys, as can be expected, are the smooth functions of the alloy composition and pass through a minimum near x = 0.5 (Figure 3.50).

3.9.6. Silicon-Tin and Germanium-Tin

Silicon and tin as well as germanium and tin form eutectic binary alloy systems. The only application of the Ge-Sn alloys was in the experiments in which Ewald used the diluted (under 1%) solid solutions of germanium in tin for stabilization of the gray tin crystal structure not only at ambient, but also at relatively high temperatures.[3.446]

For additional information regarding the binary semiconducting solid solutions see Section 7.8 of this book.

REFERENCES

3.1. T. S. Moss, *Photoconductivity in the Elements*, Butterworth & Co. Ltd., London, 1952.
3.2. V. I. Matkovich and J. Economy, *Boron and Refractory Borides*, V. I. Matkovich, Ed., Springer-Verlag, New York, 1977, 78.
3.3. L. V. McCarty, J. S. Kasper, F. N. Horn, B. F. Decker, and A. E. Newkirk, *J. Am. Chem. Soc.*, 80, 2592, 1958.
3.4. B. F. Decker and J. S. Kasper, *Acta Crystallogr.*, 12, 503, 1959.
3.5. J. L. Hoard, R. E. Hughes, and D. E. Sands, *J. Am. Chem. Soc.*, 80, 4507, 1958.
3.6. C. P. Talley, S. LaPlaca, and B. Post, *Acta Crystallogr.*, 13, 271, 1960.

3.7. J. L. Hoard, D. B. Sullenger, C. H. L. Kennard, and R. E. Hughes, *J. Solid State Chem.*, 1, 268, 1970.
3.8. W. Weber and M. F. Thorpe, *J. Phys. Chem. Solids*, 36, 967, 1975.
3.9. H. Werheit, *Boron*, Landolt-Börnstein New Series, Vol. III/17e, O. Madelung, M. Schulz, and H. Weiss, Eds., Springer-Verlag, New York, 1983, 9.
3.10. H. L. Yakel, Boron-rich solids, in *AIP Conf. Proc. 140*, AIP, 1986, 97.
3.11. B. Morosin, A. W. Mullendore, D. Emin, and G. A. Slack, *Mater. Res. Soc. Symp. Proc.*, 97, 151, 1987.
3.12. A. C. Switendick and B. Morosin, *AIP Conf. Proc. 231*, AIP, 1991, 205.
3.13. O. Madelung, Ed., *Semiconductors Other Than Group IV Elements and III-V Compounds*, Springer-Verlag, New York, 1992.
3.14. D. Geist, R. Kloss, and H. Follner, *Acta Crystallogr.*, B26, 1800, 1970.
3.15. B. Calmer, *Acta Crystallogr.*, B33, 1951, 1977.
3.16. R. E. Hughes, C. H. L. Kennard, D. B. Sullinger, H. A. Weakleim, D. E. Sands, and J. L. Hoard, *J. Am. Chem. Soc.*, 85, 361, 1963.
3.17. L. Pauling, *The Nature of the Chemical Bond*, 3rd ed., Cornell University Press, Ithaca, NY, 1960, 364.
3.18. L. Pauling and Z. S. Herman, *Advances in Boron and the Boranes*, Vol. 5, J. F. Liebman, A. Greenberg, and R. E. Williams, Eds., AIP, 1988, 517.
3.19. K. Ploog and E. Amberger, *J. Less Common Met.*, 23, 33, 1971.
3.20. G. Will and K. H. Kossobutzki, *J. Less Common Met.*, 47, 33, 1976.
3.21. R. Naslain, *Boron and Refractory Borides*, V. I. Matkovich, Ed., Springer-Verlag, New York, 1977, 166.
3.22. R. A. Assink, J. E. Schirber, D. A. Loy, B. Morosin, and G. A. Carlson, *J. Mater. Res.*, 7, 2136, 1992.
3.23. M. Vlasse, R. Naslain, J. S. Kasper, and K. Ploog, *J. Less Common Met.*, 67, 1, 1979.
3.24. J. G. Bower, *Progress in Boron Chemistry*, Vol. 2, R. J. Brotherton and H. Steinberg, Eds., Pergamon Press, Elmsford, NY, 1970, 231.
3.25. JANAF Thermochemical Tables, Suppl. Dow Chemical Midland, Michigan, 1965.
3.26. C. E. Holcombe, Jr., D. D. Smith, J. D. Lore, W. K. Duerlesen, and D. A. Carpenter, *High Temp. Sci.*, 5, 349, 1973.
3.27. P. Runow, *J. Mater. Sci.*, 7, 499, 1972.
3.28. L. Gorski, *Sov. Phys. Solid State*, 9, K169, 1965.
3.29. C. P. Talley, *J. Appl. Phys.*, 32, 1787, 1961.
3.30. J. L. Hoard and A. E. Newkirk, *J. Am. Chem. Soc.*, 82, 70, 1960.
3.31. Z. Olemska, A. Badzian, K., Pietrzak, and T. Niemyski, *J. Less Common Met.*, 11, 351, 1966.
3.32. E. Weintraub, *Trans. Am. Electrochem. Soc.*, 16, 165, 1909.
3.33. E. Weintraub, *J. Ind. Eng. Chem.*, 5, 106, 1913.
3.34. A. H. Warth, *Trans. Am. Electrochem. Soc.*, 47, 62, 1925.
3.35. R. Freimann and A. Stieber, *Comp. Rend.*, 199, 1109, 1934.
3.36. J. Lagrenaudie, *J. Phys. Radium*, 13, 554, 1952.
3.37. J. Lagrenaudie, *J. Phys. Radium*, 14, 14, 1953.
3.38. W. C. Shaw, D. E. Hudson, and G. C. Danielson, *Phys. Rev.*, 89, 900A, 1953; 91, 208A, 1953.
3.39. W. C. Shaw, D. E. Hudson, and G. C. Danielson, *Phys. Rev.*, 107, 419, 1957.
3.40. J. Jaumam and J. Schnell, *Z. Naturforsch.*, 20A, 1639, 1965.
3.41. T. N. Anderson, O. H. Berzijian, and H. Eyring, *J. Electrochem. Soc.*, 114, 8, 1967.
3.42. R. A. Brungs, *Boron*, Vol. 2, G. K. Gaulé, Ed., Plenum Press, New York, 1965, 119.
3.43. R. A. Brungs and V. P. Jacobsmeyer, *J. Phys. Chem. Solids*, 25, 701, 1964.
3.44. M. M. Usmanova, *Radiatsionnyie Effekty v Tverdykh Telakh* (Radiation Effects in Solids), Inst. Yadernoi Fiziki Akad. Nauk Uz. SSR, 1963, 75 (see also p. 160).
3.45. W. Dietz and H. Hermann, *Boron*, Vol. 2, G. K. Gaulé, Ed., Plenum Press, New York, 1965, 107.
3.46. I. R. King, F. Wawner, G. Taylor, and C. Talley, *Boron*, Vol. 2, G. K. Gaulé, Ed., Plenum Press, New York, 1965, 45.
3.47. E. I. Adirovich and L. M. Goldshtein, *Sov. Phys. Solid State*, 8, 2467, 1966.

3.48. H. Werheit, *Festkörperprobleme X*, O. Madelung, Ed., Pergamon Press, Elmsford, NY, 1970, 189.
3.49. R. Franz, H. Werheit, and A. J. Nadolny, Boron-rich solids, in *AIP Conf. Proc. 140*, AIP, 1986, 340.
3.50. R. A. Brungs, Thesis, St. Louis University, St. Louis, MO, 1962.
3.51. G. A. Slack, *Phys. Rev.*, 139A, 507, 1965.
3.52. G. A. Slack, D. W. Oliver, and E. H. Horn, *Phys. Rev.*, B4, 1714, 1971.
3.53. O. A. Golikova, V. K. Zaitsev, V. M. Orlov, A. V. Petrov, L. S. Stilbans, and E. N. Tkalenko, *Phys. Status Solidi*, A21, 405, 1974.
3.54. J. C. Thompson and W. J. McDonald, *Phys. Rev.*, 132, 82, 1963.
3.55. P. R. H. Türkes, E. T. Swartz, and R. O. Pohl, Boron-rich solids, in *AIP Conf. Proc. 140*, AIP, 1986, 346.
3.56. H. L. Johnston, H. N. Hersh, and E. C. Kerr, *J. Am. Chem. Soc.*, 73, 1112, 1951.
3.57. N. Bilir, W. A. Phillips, and T. H. Geballe, *Low Temperature Physics*, LT14, Vol. 3, M. Krusius and M. Vuorio, Eds., North-Holland, Amsterdam, 1975, 9.
3.58. I. M. Silvestrova, L. M. Belyayev, Y. V. Pisarevski, and T. Niemyski, *Mater. Res. Bull.*, 9, 1101, 1974.
3.59. L. G. Wisnyi and S. W. Pijanowski, U.S. Atomic Energy Commission, TID-7530 (Pt. 1), 46 (1957).
3.60. W. Neft and K. Seiger, *Boron*, Vol. 2, G. K. Gaulé, Ed., Plenum Press, New York, 1965, 143.
3.61. G. K. Gaulé, J. T. Breslin, and R. R. Patty, *Boron*, Vol. 2, G. K. Gaulé, Ed., Plenum Press, New York, 1965, 169.
3.62. A. J. Nadolny, J. W. Ostrowski, and M. Z. Przegalinska-Mierszkowska, *Phys. Status Solidi*, 16, K133, 1966.
3.63. E. Amberger and W. Dietze, *Z. Anorg. Allg. Chem.*, 3–4, 131, 1964.
3.64. A. E. Newkirk, *Boron, Metallo-Boron Compounds and Boranes*, R. M. Adams, Ed., Interscience, New York, 1964, 301.
3.65. F. L. Gebhart and V. P. Jacobsmeyer, *Boron*, Vol. 2, G. K. Gaulé, Ed., Plenum Press, New York, 1965, 133.
3.66. B. Kierzek-Pecold, J. Kolodziejczak, and I. Pracka, *Phys. Status Solidi*, 22, K147, 1967.
3.67. H. W. Herring, NASA Technical Note TN D3202, January, 1966.
3.68. F. E. Wawner, Jr., *Modern Composite Materials*, L. J. Broutman and R. H. Krock, Eds., Addison-Wesley, Reading, MA, 1967, 244.
3.69. C. J. Speerschneider and J. A. Sartel, *Boron*, Vol. 2, G. K. Gaulé, Ed., Plenum Press, New York, 1965, 269.
3.70. A. A. Giardini, J. A. Kohn, L. Toman, and D. W. Eckhart, *Boron. Synthesis, Structure and Properties*, J. A. Kohn, W. F. Nue, and G. K. Gaulé, Eds., Plenum Press, New York, 1960, 140.
3.71. A. E. Newkirk, *From Borax to Boranes* (Advances in Chemistry Series), American Chemical Society, Washington, D.C., 1961, 27.
3.72. R. J. Starks, J. T. Buford, and P. E. Grayson, Boron-rich solids, in *AIP Conf. Proc. 140*, AIP, 1986, 373.
3.73. H. J. Becher and R. Mattes, *Boron and Refractory Borides*, V. I. Matkovich, Ed., Springer-Verlag, New York, 1977, 107.
3.74. Alfa Catalog, Research Chemicals, Johnson Matthey, 1993–1994.
3.75. N. E. Holden, *Handbook of Chemistry and Physics*, 74th ed., D. R. Lide, Ed., CRC Press, Boca Raton, FL, 1993–1994, 11-35.
3.76. G. K. Gaulé, R. L. Ross, and J. L. Bloom, *Boron*, Vol. 2, G. K. Gaulé, Ed., Plenum Press, New York, 1965, 317.
3.77. Belgian Patent 667,939 (16 November 1965) to Consortium Industrie GmbH.
3.78. S. Ruben, U.S. Patent 3,216,955 (9 November 1965).
3.79. Y. Tzeng, M. Yoshikawa, M. Merakawa, and A. Feldman, Eds., *Applications of Diamond Films and Related Materials*, Elsevier, New York, 1991.
3.80. J. C. Angus and C. C. Hayman, *Science*, 241, 913, 1988.
3.81. K. E. Spear, *J. Am. Ceram. Soc.*, 72, 171, 1989.
3.82. F. J. Himpsel, J. A. Knapp, J. A. Van Vechten, and D. E. Eastman, *Phys. Rev.*, B20, 624, 1979.
3.83. H. Nakazava, Y. Nakazava, M. Kamo, and K. Osumi, *Thin Solid Films*, 151, 199, 1987.

3.84. H. Werheit, *Boron,* Landolt-Börnstein New Series, Vol. III/17c, O. Madelung, M. Schulz, and H. Weiss, Eds., Springer-Verlag, New York, 1983.
3.85. F. P. Bundy, *Science,* 137, 1055, 1962.
3.86. R. Robertson, J. J. Fox, and A. E. Martin, *Philos. Trans. R. Soc.,* A232, 463, 1934.
3.87. J. F. H. Custers, *Physica,* 18, 489, 1952.
3.88. S. Tolansky and H. Komatsu, *Science,* 157, 1173, 1967.
3.89. T. Evans and P. Rainey, *Proc. R. Soc. London,* A334, 111, 1975.
3.90. J. P. F. Sellschop, *The Properties of Diamond,* J. E. Field, Ed., Academic Press, New York, 1979, 107.
3.91. R. M. Chrenko, *Phys. Rev.,* B7, 4560, 1973.
3.92. H. Wartenberg, *Phys. Z.,* 13, 1123, 1912.
3.93. T. Mendelsson and H. Dember, *Rev. Fac. Sci. Univ. Istanbul,* A6, 18, 1940.
3.94. I. G. Austin and R. Wolfe, *Proc. Phys. Soc. London,* B69, 329, 1956.
3.95. P. T. Wedepohl, Thesis, Reading University, 1957.
3.96. P. T. Wedepohl, *Proc. Phys. Soc. London,* B70, 177, 1957.
3.97. P. J. Kemmey, Thesis, Reading University, 1960.
3.98. A. W. S. Williams, Thesis, University of London, London, 1970.
3.99. A. T. Collins and A. W. S. Williams, *J. Phys.,* C4, 1789, 1971.
3.100. A. T. Collins and E. C. Lightowler, *The Properties of Diamond,* J. E. Field, Ed., Academic Press, New York, 1979, 79.
3.101. A. W. S. Williams, E. C. Lightowler, and A. T. Collins, *J. Phys. C. Solid State,* 3, 1727, 1970.
3.102. R. M. Chrenko, *Nature Phys. Sci.,* 229, 165, 1971.
3.103. E. C. Lightowler and A. T. Collins, Diamond Research (Suppl. Industr. Diam. Rev.), 1976, 14.
3.104. V. S. Vavilov, Proc. 12th Int. Conf. Physics of Semiconductors, M. H. Pilkuhn, Ed., Teubner, Stuttgart, 1974, 277.
3.105. R. Berman and F. Simon, *Z. Elektrochem.,* 59, 333, 1955.
3.106. R. Berman, *Physical Properties of Diamond,* R. Berman, Ed., Clarendon Press, Oxford, 1965, 371.
3.107. R. Berman, *The Properties of Diamond,* J. E. Field, Ed., Academic Press, New York, 1979, 4.
3.108. E. J. Prosen, R. S. Jessup, and F. D. Rossini, *J. Res. NBS,* 33, 447, 1944.
3.109. P. Hawtin, J. B. Lewis, N. Moul, and R. H. Phillips, *Phil. Trans. R. Soc.,* A261, 67, 1966.
3.110. W. DeSorbo and W. W. Tyler, *J. Chem. Phys.,* 21, 1660, 1953.
3.111. U. Bergenlid, R. W. Hill, F. J. Webb, and J. Wilks, *Philos. Mag.,* 45, 851, 1954.
3.112. P. H. Keesom and N. Pearlman, *Phys. Rev.,* 99, 1119, 1955.
3.113. J. C. Bowman and J. A. Krumhansl, *J. Phys. Chem. Solids,* 6, 367, 1958.
3.114. C. F. Lucks, H. W. Deem, and W. D. Wood, *Bull. Am. Ceram. Soc.,* 39, 313, 1960.
3.115. N. S. Rasor and J. D. McClelland, *J. Phys. Chem. Solids,* 15, 17, 1960.
3.116. W. DeSorbo, *J. Chem. Phys.,* 21, 876, 1953.
3.117. D. L. Burk and S. A. Friedburg, *Phys. Rev.,* 111, 1275, 1958.
3.118. J. E. Desnoyers and J. A. Morrison, *Philos. Mag.,* 3, 42, 1958.
3.119. A. C. Victor, *J. Chem. Phys.,* 36, 1903, 1962.
3.120. J. McDonald, *J. Chem. Eng. Data,* 10, 243, 1965.
3.121. E. D. West and S. Ishihara, *Advances in Thermophysical Properties at Extreme Temperatures and Pressures,* ASME, New York, 1965, 146.
3.122. J. B. Nelson and D. P. Riley, *Proc. Phys. Soc. London,* 57, 477, 1945.
3.123. E. G. Steward and B. P. Cook, *Nature,* 185, 78, 1960.
3.124. E. G. Steward, B. P. Cook, and E. A. Kellett, *Nature,* 187, 1015, 1960.
3.125. E. A. Kellett and B. P. Richards, *J. Nucl. Mats.,* 12, 184, 1964.
3.126. M. E. Straumanis and E. Z. Aka, *J. Am. Chem. Soc.,* 73, 5643, 1951.
3.127. J. Thewlis and A. R. Davey, *Philos. Mag.,* 40, 409, 1956.
3.128. B. J. Skinner, *Am. Min.,* 42, 39, 1957.
3.129. W. Kaiser and W. L. Bond, *Phys. Rev.,* 115, 857, 1959.
3.130. S. I. Novikova, *Fizika Tverdogo Tela (Solid State Phys.),* 2, 1617, 1960.
3.131. R. Mykolajewycz, J. Kalnajs, and A. Smakula, *J. Appl. Phys.,* 35, 1773, 1964.
3.132. A. C. J. Wright, Thesis, Reading University, 1966.
3.133. G. A. Slack and S. F. Bartram, *J. Appl. Phys.,* 46, 89, 1975.

3.134. H. G. Drickamer, R. W. Lynch, R. L. Clenenden, and E. A. Perez-Albuerne, *Solid State Physics,* Vol. 19, F. Seitz and D. Turbull, Eds., 1966, 135.
3.135. R. W. Powell, C. Y. Ho, and P. E. Liley, *J. Phys. Chem. Ref. Data,* 1, 279, 1972; N. E. Holden, *Handbook of Chemistry and Physics,* 74th ed., D. R. Lide, Ed., CRC Press, Boca Raton, FL, 1993–1994, 12-136.
3.136. A. Eucken, *Verh. Deutsche Phys. Ges.,* 13, 829, 1911.
3.137. W. J. DeHaas and T. Biermasz, *Physica,* 5, 47, 1938.
3.138. P. G. Klemens, *Phys. Rev.,* 86, 1055, 1952.
3.139. R. Berman, F. E. Simon, and J. M. Ziman, *Proc. R. Soc.,* A220, 171, 1953.
3.140. R. Berman, E. L. Foster, and J. M. Ziman, *Proc. R. Soc. London,* A237, 344, 1956.
3.141. R. Berman, P. R. W. Hudson, and M. Martinez, *J. Phys. C. Solid State,* 8, L430, 1975.
3.142. G. A. Slack, *J. Phys. Chem. Solids,* 34, 321, 1973.
3.143. E. A. Burgemeister and M. Seal, Diamond Conf., Cambridge, 1976; (unpublished from R. Berman, *The Properties of Diamond,* J. E. Field, Ed., Academic Press, New York, 1979.
3.144. R. Berman and M. Martinez, Diamond Research Suppl. to Ind. Diamond Rev., 1976, 7.
3.145. W. Kaiser and W. L. Bond, *Phys. Rev.,* 115, 857, 1959.
3.146. M. Martinez, Thesis, Oxford University, New York, 1976.
3.147. L. H. Adams, *J. Wash. Acad. Sci.,* 11, 45, 1921.
3.148. E. D. Williamson, *J. Franklin Inst.,* 193, 491, 1922.
3.149. S. Bhagavantam and J. Bhimasenachar, *Proc. R. Soc.,* A187, 381, 1946.
3.150. E. Prince and W. A. Wooster, *Acta Crystallogr.,* 6, 450, 1953.
3.151. H. J. McSkimin and W. L. Bond, *Phys. Rev.,* 105, 116, 1957.
3.152. H. J. McSkimin and P. Andreath, *J. Appl. Phys.,* 43, 985, 1972; 43, 2944, 1972.
3.153. M. H. Grimsditch and A. K. Ramdas, *Phys. Rev.,* B11, 3139, 1975.
3.154. B. W. Mott, *Microindentation Hardness Testing,* Butterworth & Co. Ltd., London, 1956.
3.155. C. A. Brooks, *Nature (London),* 228, 660, 1970.
3.156. C. A. Brooks, *The Properties of Diamond,* J. E. Field, Ed., Academic Press, New York, 1979, 383.
3.157. V. N. Bakul, M. G. Loshak, and V. I. Malnev, *Sinteticheskiie Almazy (Synthetic Diamonds),* 6, 16, 1973.
3.158. D. Brasen, *J. Mater. Sci.,* 11, 791, 1976; 13, 1776, 1978.
3.159. D. Y. Watts and A. F. W. Willoughby, *J. Appl. Phys.,* 56, 1869, 1984; *Mater. Lett.,* 2, 355, 1984.
3.160. H. B. Dyer, F. A. Raal, L. du Preez, and J. H. N. Loubser, *Philos. Mag.,* 11, 763, 1965.
3.161. C. D. Clark, P. J. Dean, and P. V. Harris, *Proc. R. Soc. London,* A277, 312, 1964.
3.162. C. D. Clark, E. W. J. Mitchell, and B. J. Parsons, *The Properties of Diamond,* J. E. Field, Ed., Academic Press, New York, 1979, 23.
3.163. A. M. Stoneham, *The Properties of Diamond,* J. E. Field, Ed., Academic Press, New York, 1979, 185.
3.164. F. Herman, (a) *Phys. Rev.,* 88, 1210, 1952; (b) *Phys. Rev.,* 93, 1214, 1954.
3.165. L. Kleinman and J. C. Phillips, *Phys. Rev.,* 116, 880, 1959.
3.166. D. Long, *Energy Bands in Semiconductors,* Wiley, New York, 1968.
3.167. F. J. Himpsel, J. A. Knapp, J. A. van Vechten, and D. E. Eastman, *Phys. Rev.,* B20, 624, 1979.
3.168. M. H. L. Price, *Physical Properties of Diamond,* R. Berman, Ed., Clarendon Press, Oxford, 1965, 201.
3.169. G. S. Painter, D. E. Ellis, and A. R. Lubinsky, *Phys. Rev.,* B4, 3610, 1971.
3.170. J. R. Chelikowsky and S. G. Louie, *Phys. Rev.,* B29, 3470, 1984.
3.171. H. M. J. Smith, *Philos. Trans. R. Soc.,* 241, 105, 1948.
3.172. M. J. P. Musgrave and J. A. Pople, *Proc. R. Soc. London,* A268, 474, 1962.
3.173. M. I. Bell, *J. Chem. Phys.,* 62, 3357, 1975.
3.174. G. Dolling and R. A. Cowley, *Proc. Phys. Soc.,* 88, 463, 1965.
3.175. G. Peckham, *Solid State Comm.,* 5, 311, 1967.
3.176. J. C. Phillips, *Phys. Rev.,* 166, 832, 1968.
3.177. S. Go, H. Bilz, and M. Cardona, *Phys. Rev. Lett.,* 34, 580, 1975.
3.178. C. J. Rauch, *Phys. Rev. Lett.,* 7, 83, 1961.
3.179. J. L. Warren, R. G. Wenzel, and J. L. Yarnell, *Inelastic Neutron Scattering,* Vol. 1, IAEA, Vienna, 1965, 361.

3.180. J. L. Warren, J. L. Yarnell, G. Dolling, and R. A. Cowley, *Phys. Rev.,* 158, 805, 1967.
3.181. S. A. Solin and A. K. Ramdas, *Phys Rev.,* B1, 1687, 1970.
3.182. J. E. Field, *The Properties of Diamond,* J. E. Field, Ed., Academic Press, New York, 1979, 284.
3.183. P. K. Bachmann, *Adv. Mater.,* 2, 195, 1990.
3.184. S. Matsumoto, Y. Sato, M. Kamo, and N. Setaka, *Jpn. J. Appl. Phys., Part 2,* 21, L183, 1982.
3.185. P. K. Bachmann, W. Drawl, D. Knight, R. Weimer, and R. F. Messier, Diamond and Diamond-Like Materials, Extended Abstracts No. 15, G. H. Johnson, A. R. Badzian, and M. W. Geis, Eds., Material Research Society, Pittsburgh, 1988.
3.186. D. E. Meyer, N. J. Ianno, J. A. Woolman, A. B. Swartzlander, and A. J. Nelson, *J. Mater. Res.,* 3, 1397, 1988.
3.187. N. Ohtake and M. Yoshikawa, *J. Electrochem. Soc.,* 137, 717, 1990.
3.188. Y. Hirose, S. Amanuma, N. Okada, and K. Komaki, Proc. 1st Int. Symp. on Diamond and Diamond-Like Films, Electrochemical Society, 1989, 80.
3.189. Y. Matsui, A. Yuuki, M. Sahara, and Y. Hirose, *Jpn. J. Appl. Phys.,* 28, 1718, 1989.
3.190. R. H. Wentorf, *J. Phys. Chem.,* 75, 1833, 1971.
3.191. G. S. Woods and A. R. Lang, *J. Cryst. Growth,* 28, 215, 1975.
3.192. Y. Mita, *J. Phys. Soc. Jpn.,* 17, 784, 1962.
3.193. Y. V. Shmartsev, Y. A. Valov, and A. S. Borshchevskii, *Refractory Semiconductor Materials,* Plenum Press, New York, 1966.
3.194. L. I. Berger and V. D. Prochukhan, *Ternary Diamond-Like Semiconductors,* Plenum Press, New York, 1969.
3.195. J. A. Brinkman, C. J. Meecham, and H. M. Dieckamp, U.S. Patent No. 3,142,539, 1964; No. 3,175,885, 1965.
3.196. W. G. Eversole, U.S. Patent No. 3,030,187; No. 3,030,188, 1962.
3.197. M. Frenklach and K. E. Spear, *J. Mater. Res.,* 3, 133, 1988.
3.198. B. V. Derjaguin and D. V. Fedoseev, *Growth of Diamond and Graphite from Gas Phase,* Nauka, Moscow, 1977.
3.199. B. V. Spitsyn, L. L. Bouilov, and B. V. Derjaguin, *J. Cryst. Growth,* 52, 219, 1981.
3.200. A. Badzian, B. Simonton, T. Badzian, R. Messier, K. E. Spear, and R. Roy, *SPIE,* 683, 127, 1986.
3.201. E. S. Machlin, *J. Mater. Res.,* 3, 958, 1988.
3.202. W. Zhu, A. R. Badzian, and R. Messier, *J. Mater. Res.,* 4, 659, 1989.
3.203. B. E. Williams and J. T. Glass, *J. Mater. Res.,* 4, 373, 1989.
3.204. B. E. Williams, H. S. Kong, and J. T. Glass, *J. Mater. Res.,* 5, 801, 1990.
3.205. M. Kamo, Y. Saito, S. Matsumoto, and N. Setaka, *J. Cryst. Growth,* 62, 642, 1983.
3.206. A. R. Badzian, T. Badzian, R. Roy, R. Messier, and K. E. Spear, *Mater. Res. Bull.,* 23, 531, 1988.
3.207. A. B. Harker and J. F. DeNatale, *J. Mater. Res.,* 5, 818, 1990.
3.208. D. J. Pickrell, W. Zhu, A. R. Badzian, R. E. Newnham, and R. Messier, *J. Mater. Res.,* 6, 1264, 1991.
3.209. W. A. Weimer, F. M. Cerio, and C. E. Johnson, *J. Mater. Res.,* 6, 2134, 1991.
3.210. R. Ramesham, T. Roppel, C. Ellis, D. A. Jaworske, and W. Baugh, *J. Mater. Res.,* 6, 1278, 1991.
3.211. R. Ramesham and T. Roppel, *J. Mater. Res.,* 7, 1144, 1992.
3.212. T. P. Ong, Fulin Xiong, R. P. H. Chang, and C. W. White, *J. Mater. Res.,* 7, 2429, 1992.
3.213. V. P. Godbole and J. Narayan, *J. Mater. Res.,* 7, 2785, 1992.
3.214. P. C. Yang, W. Zhu, and J. T. Glass, *J. Mater. Res.,* 8, 1773, 1993.
3.215. S. Fujimori, T. Kasai, and T. Inamura, *Thin Solid Films,* 92, 71, 1982.
3.216. C. L. Marquardt, R. T. Williams, and D. J. Nagel, Plasma Synthesis and Etching of Electronic Materials, R. P. H. Chang and B. Abeles, Eds., MRS Symp. Proc. 38, Pittsburgh, PA, 1985, 325.
3.217. T. Sato, S. Furuno, S. Iguchi, and M. Hanabusa, *Appl. Phys.,* A45, 355, 1988.
3.218. J. Krishnaswamy, A. Rengan, J. Narayan, K. Vedam, and C. J. McHargue, *Appl. Phys. Lett.,* 54, 2455, 1989.
3.219. A. P. Malshe, S. M. Chaudhari, S. M. Kanetkar, S. B. Ogale, S. V. Rajarshi, and S. T. Kshirsagar, *J. Mater. Res.,* 4, 1238, 1989.

3.220. R. T. Demers and D. G. Harris, *Diamond Optics II,* Vol. 1146, A. Feldman and S. Holly, Eds., SPIE, Bellingham, WA, 1990, 68.
3.221. F. Davanloo, E. M. Juengerman, D. R. Jander, T. J. Lee, and C. B. Collins, *J. Appl. Phys.,* 67, 2081, 1990.
3.222. C. B. Collins, F. Davanloo, D. R. Jander, T. J. Lee, H. Park, and J. H. You, *J. Appl. Phys.,* 69, 7862, 1991.
3.223. D. L. Pappas, K. L. Saenger, J. Bruley, W. Krakow, J. J. Cuomo, T. Gu, and R. W. Collins, *J. Appl. Phys.,* 71, 5675, 1992.
3.224. Fulin Xiong, Y. Y. Wang, V. Leppert, and R. P. H. Chang, *J. Mater. Res.,* 8, 2265, 1993.
3.225. K. Okada, S. Komatsu, S. Matsumoto, and Y. Moriyoshi, *J. Cryst. Growth,* 108, 416, 1991.
3.226. K. V. Ravi, *J. Mater. Res.,* 7, 384, 1991.
3.227. J. A. von Windheim and J. T. Glass, *J. Mater. Res.,* 7, 2144, 1991.
3.228. W. Bludau, A. Onton, and W. Heinke, *J. Appl. Phys.,* 45, 1846, 1974.
3.229. D. P. Jenkins, *Physica,* 20, 967, 1954.
3.330. D. G. Bell, R. Heusman, D. P. Jenkins, and L. Pincherle, *Proc. Phys. Soc.,* A67, 562, 1954.
3.231. F. Herman, *Proc. IRE,* 43, 1703, 1955.
3.232. F. Herman, *Rev. Mod. Phys.,* 30, 102, 1958.
3.233. B. Lax, *Rev. Mod. Phys.,* 30, 122, 1958.
3.234. K. S. Sieh and P. V. Smith, *Phys. Status Solidi,* B129, 259, 1985.
3.235. G. Dresselhaus, A. F. Kip, and C. Kittel, *Phys. Rev.,* 92, 827, 1953; 95, 568, 1954; 98, 368, 1955.
3.236. H. D. Barber, *Solid State Electron.,* 10, 1039, 1964.
3.237. D. E. Aspnes and A. S. Studna, *Solid State Comm.,* 11, 1375, 1972.
3.238. R. A. Foreman and D. E. Aspnes, *Solid State Comm.,* 14, 100, 1974.
3.239. D. R. Masovic, F. R. Vukajlovic, and S. Zekovic, *J. Phys.,* C16, 6731, 1983.
3.240. P. Lautenschlager, P. B. Allen, and M. Cardona, *Phys. Rev.,* B31, 2163, 1985.
3.241. F. J. Morin and J. P. Maita, *Phys. Rev.,* 94, 1525, 1954; 96, 28, 1954.
3.242. H. Olijnuk, S. K. Sikka, and W. B. Golzapfel, *Phys. Lett.,* A103, 137, 1984.
3.243. J. Z. Hu and I. L. Spain, *Solid State Comm.,* 44, 263, 1984.
3.244. C. S. Mononi, J. Z. Hu, and I. L. Spain, *Phys. Rev.,* B34, 362, 1986.
3.245. P. Becker, P. Seyfried, and H. Siegert, *Z. Phys.,* B48, 17, 1982.
3.246. H. Fizeau, *Ann. Phys.,* 208, 2, Folge 132, 292, 1867.
3.247. S. Valentiner and J. Wallot, *Ann. Phys.,* 46, 858, 1915.
3.248. H. D. Erfling, *Ann. Phys.,* 41, 467, 1942.
3.249. D. F. Gibbons, *Phys. Rev.,* 112, 136, 1958.
3.250. S. I. Novikova and P. G. Strelkov, *Sov. Phys. Solid State,* 1, 1841, 1959.
3.251. R. H. Carr, R. D. McCammon, and G. K. White, *Philos. Mag.,* 12, 157, 1965.
3.252. P. W. Sparks and C. A. Swenson, *Phys. Rev.,* 163, 779, 1967.
3.253. K. G. Lyon, G. L. Salinger, C. A. Swenson, and G. K. White, *J. Appl. Phys.,* 48, 865, 1977.
3.254. Y. Okada and Y. Tokumaru, *J. Appl. Phys.,* 56, 314, 1958.
3.255. G. K. White, AIP Conf. Proc. 17, Int. Symp. Thermal Expansion of Solids 1973, Session 1, R. E. Taylor and G. L. Denman, Eds., AIP, New York, 1974, 1.
3.256. O. Madelung, Ed., *Semiconductors. Group IV Elements and III-V Compounds,* Springer-Verlag, New York, 1991.
3.257. N. Pearlman and P. H. Keesom, *Phys. Rev.,* 88, 398, 1952.
3.258. P. Flubacher, A. J. Leadbetter, and J. A. Morrison, *Philos. Mag.,* 4, 273, 1959.
3.259. H. J. McSkimin, *J. Appl. Phys.,* 24, 988, 1953.
3.260. P. Leroux-Hugon, N. X. Xinh, and M. Rodot, *J. Phys.,* 28, Cl-73, 1967.
3.261. G. A. Slack, *J. Appl. Phys.,* 35, 3460, 1964.
3.262. W. Fulkerson, J. P. Moore, R. K. Williams, R. S. Graves, and D. L. McElroy, *Phys. Rev.,* 167, 765, 1968.
3.263. R. A. Smith, *Semiconductors,* Cambridge University Press, New York, 1959, 310.
3.264. R. N. Dexter, H. J. Zeiger, and B. Lax, *Phys. Rev.,* 104, 637, 1956.
3.265. T. Wasserrab, *Z. Naturforsch.,* 32A, 746, 1977.
3.266. M. B. Prince, *Phys. Rev.,* 93, 1204, 1954.
3.267. C. Jacobini, C. Canali, G. Ottaviani, and A. Quaranta, *Solid State Electron.,* 20, 77, 1977.

3.268. A. Hang and W. Schmid, *Solid State Electron.,* 25, 665, 1982.
3.269. F. Szmulowicz, *Appl. Phys. Lett.,* 43, 485, 1983; *Phys. Rev.,* B28, 5943, 1983.
3.270. W. C. Mitchel and P. M. Hemenger, *J. Appl. Phys.,* 53, 6880, 1982.
3.271. G. W. Ludwig and R. L. Watters, *Phys. Rev.,* 101, 1699, 1956.
3.272. J. C. Hensel, H. Nasegawa, and M. Nakayama, *Phys. Rev.,* 138, A225, 1965.
3.273. M. Shur, *Physics of Semiconductor Devices,* Prentice-Hall, Englewood Cliffs, NJ, 1990, 623.
3.274. A. I. Blum, H. P. Mokrovskii, and A. R. Regel, *Sov. Phys. Tech. Phys.,* 21, 237, 1951; *Izv. USSR Acad Sci. Phys. Ser.,* 16, 139, 1952.
3.275. M. S. Ablova, O. D. Evpat'evskaya, and A. R. Regel, *Sov. Phys. Tech. Phys.,* 26, 1366, 1956.
3.276. V. M. Glazov, S. N. Chizhevskaya, and N. N. Glagoleva, *Liquid Semiconductors,* Nauka, Moscow, 1967.
3.277. B. A. Baum, P. V. Geld, and S. I. Suchil'nikov, *Izv. USSR Acad. Sci. Met. Mining Ser.,* No. 2, 149, 1964.
3.278. R. Hultgren, P. D. Desai, D. T. Hawkins, M. Gleiser, K. K. Kelley, and D. D. Wagman, *The Thermodynamic Properties of the Elements,* ASM, 1973.
3.279. H. Hendus, *Z. Naturforsch.,* 2A, 505, 1947.
3.280. R. R. L. Zucca and Y. R. Shen, *Phys. Rev.,* B1, 2668, 1970.
3.281. D. Straub, L. Ley, and F. J. Himpsel, *Phys. Rev.,* B33, 2607, 1986.
3.282. H. B. Briggs, *Phys. Rev.,* 77, 287, 1950; 77, 727, 1950.
3.283. R. O. Carlson, *Phys. Rev.,* 108, 1390, 1957.
3.284. H. R. Phillip and E. A. Taft, *Phys. Rev.,* 120, 37, 1960.
3.285. G. K. White and S. B. Woods, *Phys. Rev.,* 103, 569, 1956.
3.286. K. A. McCarthy and S. S. Ballard, *Phys. Rev.,* 99, 1104, 1955.
3.287. D. R. Lide, Ed., *Handbook of Chemistry and Physics,* 74th ed., CRC Press, Boca Raton, FL, 1993–1994, 11-35.
3.288. L. D. Crossman and J. A. Baker, *Semiconductor Silicon 1977,* H. R. Huff and E. Sirtle, Eds., Electrochemical Society, Princeton, NJ, 1981, 18.
3.289. F. Shimura, *Semiconductor Silicon Crystal Technology,* Academic Press, New York, 1989, 114.
3.290. P. H. Keck, W. Van Horn, J. Soled, and A. MacDonald, *Rev. Sci. Instrum.,* 25, 331, 1954.
3.291. J. Czochralski, *Z. Phys. Chem.,* 92, 219, 1918.
3.292. G. K. Teal and J. B. Little, *Phys. Rev.,* 78, 647, 1950.
3.293. G. K. Teal and E. Buehler, *Phys. Rev.,* 87, 190, 1952.
3.294. T. Abe, *VLSI Electronics Microstructure Science,* Vol. 12, N. G. Einspruch and H. R. Huff, Eds., Academic Press, New York, 1985, 3.
3.295. T. V. Valyanskaya and G. N. Stepanov, *Solid State Comm.,* 86, 723, 1993.
3.296. M. E. Straumanis and A. Z. Aka, *J. Appl. Phys.,* 23, 330, 1952.
3.297. C. A. Winkler, *Zs. Pract. Chem.,* 36, 177, 1887.
3.298. G. Morgan and G. R. Davies, *Chem Ind.,* 56, 717, 1937.
3.299. J. L. Andrieux and M. J. Andrieux, *Compt. Rend.,* 240, 2104, 1955.
3.300. L. Steygers, *Mining World Seattle,* 22, 34, 1960.
3.301. R. Monnier and P. Tissot, *Helv. Chim. Acta,* 47, 2203, 1964.
3.302. J. Bardeen and W. H. Brattain, *Phys. Rev.,* 75, 1208, 1949.
3.303. W. Shockley, G. L. Pearson, and J. R. Haynes, *Bell System Tech. J.,* 28, 344, 1949.
3.304. W. Shockley, *Electrons and Holes in Semiconductors,* Academic Press, New York, 1950.
3.305. J. F. C. Baker and M. Hart, *Acta Crystallogr.,* 31a, 2297, 1975.
3.306. A. Smacula and J. Kalnajs, *Phys. Rev.,* 99, 1737, 1955.
3.307. H. P. Singh, *Acta Crystallogr.,* 24a, 469, 1968.
3.308. S. N. Novikova, *Sov. Phys. Solid State,* 2, 37, 1960.
3.309. R. D. McCammon and G. K. White, *Phys. Rev. Lett.,* 10, 234, 1963.
3.310. V. V. Zhdanova, *Sov. Phys. Solid State,* 5, 2450, 1963.
3.311. I. Esterman and J. R. Weertman, *J. Chem. Phys.,* 20, 972, 1952.
3.312. R. W. Hill and D. H. Parkinson, *Philos. Mag.,* 43, 309, 1952.
3.312A. P. W. Bridgman, *Proc. Natl. Acad. Sci.,* 8, 362, 1922.
3.313. U. Piesbergen, *Z. Naturforsch.,* 18a, 141, 1963.
3.313A. W. L. Bond, W. P. Mason, H. J. McSkimin, K. M. Olson, and G. K. Teal, *Phys. Rev.,* 78, 176, 1950.

3.314. M. E. Fine, *J. Appl. Phys.*, 26, 862, 1955.
3.315. Y. A. Burenkov, S. P. Nikanorov, and A. V. Stepanov, *Sov. Phys. Solid State*, 12, 1940, 1971.
3.316. G. G. Macfarlane, T. P. McLean, J. E. Quarrington, and V. Roberts, *Phys. Rev.*, 108, 1377, 1957; 111, 1245, 1958.
3.317. S. Zwerdling, B. Lax, L. M. Roth, and K. J. Button, *Phys. Rev.*, 114, 80, 1959.
3.318. T. C. Hsieh, T. Miller, and T. C. Chiang, *Phys. Rev.*, B30, 7005, 1984.
3.319. A. L. Wachs, T. Miller, T. C. Hsieh, A. P. Shapiro, and T. C. Chiang, *Phys. Rev.*, B32, 2326, 1985.
3.320. J. M. Nichols, G. V. Hansson, O. U. Karlsson, P. E. S. Persson, R. I. G. Uhrberg, R. Engelhard, S. A. Flodström, and E. E. Koch, *Phys. Rev.*, B32, 6663, 1985.
3.321. F. Herman, *Phys. Rev.*, 93, 1214, 1954; *Phys. Grav.*, 20, 801, 1954.
3.322. J. R. Chelikowsky and M. L. Cohen, *Phys. Rev.*, B30, 556, 1976.
3.323. See Ref. 3.240.
3.324. P. B. Allen and M. Cardona, *Phys. Rev.*, B27, 4760, 1983.
3.325. D. Fink and R. Braunstein, *Phys. Status Solidi (b)*, 73, 361, 1976.
3.326. Y. Hirose, K. Shimonae, and C. Hamaguchi, *J. Phys. Soc. Jpn.*, 51, 2226, 1982.
3.327. R. L. Aggarwal, *Phys. Rev.*, B2, 446, 1970.
3.328. G. C. D. Pergola and D. Sette, *Phys. Rev.*, 104, 598, 1956.
3.329. H. P. Furth and R. W. Waniek, *Phys. Rev.*, 104, 343, 1956.
3.330. R. K. Willardson, T. C. Harman, and A. C. Beer, *Phys. Rev.*, 96, 1512, 1954.
3.331. M. B. Prince, *Phys. Rev.*, 92, 681, 1953.
3.332. D. K. Hamilton and K. G. Seidensticker, *J. Appl. Phys.*, 34, 2697, 1963.
3.333. G. Busch and N. Helfer, *Helv. Phys. Acta*, 27, 201, 1954.
3.334. G. Foex, Ed., *Tables de Constantes et Données Numériques*, Vol. 7, Relaxation Paramagnetique, Masson, Paris, 1957.
3.335. V. M. Glazov and S. N. Chizhevskaya, *Sov. Phys. Solid State*, 6, 1684, 1964.
3.336. H. M. Rosenberg, *Proc. Phys. Soc. London*, A67, 837, 1954.
3.337. Ref. 3.285
3.338. G. A. Slack, *Phys. Rev.*, 105, 829, 1957.
3.339. P. Carruthers, T. H. Geballe, H. M. Rosenberg, and J. M. Ziman, *Proc. R. Soc. London*, A238, 502, 1957.
3.340. T. H. Geballe and G. W. Hull, *Phys. Rev.*, 110, 773, 1958.
3.341. C. J. Glassbrenner and G. A. Slack, *Phys. Rev.*, 134A, 1058, 1964.
3.342. S. R. Bakhchieva, N. P. Kekelidze, and M. G. Kekua, *Phys. Status Solidi (a)*, 83, 139, 1984.
3.343. C. Y. Ho, R. W. Powell, and P. E. Liley, *J. Phys. Chem. Ref. Data*, 1, 279, 1972.
3.344. E. Fagen, J. F. Goff, and N. Pearlman, *Phys. Rev.*, 94, 1415, 1954.
3.345. J. F. Goff and N. Pearlman, *Phys. Rev.*, 140, A2151, 1965.
3.346. M. P. Mathur and N. Pearlman, *Phys. Rev.*, 180, 833, 1969.
3.347. T. Sota, K. Suzuki, and D. Fortier, *Phys. Rev.*, B31, 7947, 1985.
3.348. D. G. Avery and P. L. Clegg, *Proc. Phys. Soc.*, B66, 512, 1953.
3.349. K. Lark-Horowitz and K. W. Meissner, *Phys. Rev.*, 76, 1530, 1949.
3.350. P. E. van Camp, V. E. van Doren, and J. T. Devreese, *Bull. Am. Phys. Soc.*, 27, 249, 1982.
3.351. R. P. Edwin, M. T. Dudermel, and M. Lamare, *Appl. Optics*, 21, 878, 1982.
3.352. H. P. Philipp and E. A. Taft, *Phys. Rev.*, 113, 1002, 1959.
3.353. D. E. Aspnes and A. A. Studna, *Phys. Rev.*, B27, 985, 1983.
3.354. W. L. Hansen, E. E. Haller, and P. N. Luke, *IEEE Trans. Nucl. Sci.*, 29, 738, 1982.
3.355. N. A. Stolwijk, W. Frank, J. Holzl, S. J. Pearton, and E. E. Haller, *J. Appl. Phys.*, 57, 5211, 1985.
3.356. M. Werner, H. Mehrer, and H. D. Hochheimer, *Phys. Rev.*, B32, 3930, 1985.
3.357. A. I. Blum and N. A. Goryunova, *Akad. Nauk SSSR Doklady*, 75, 367, 1950.
3.358. N. A. Goryunova, Gray Tin, Ph.D. thesis, Leningrad State University, 1950.
3.359. F. Vnuk, A. DeMonte, and R. W. Smith, *J. Appl. Phys.*, 55, 4147, 1984.
3.360. A. W. Ewald, *Phys. Rev.*, 91, 244, 1953.
3.361. A. W. Ewald, *J. Appl. Phys.*, 25, 1436, 1954.
3.362. A. W. Ewald and O. N. Tufte, *J. Phys. Chem. Solids*, 8, 523, 1959; see also O. N. Tufte and A. W. Ewald, *Bull. Am. Phys. Soc. Ser. II*, 3, 128, 1958.

3.363. W. E. Blumberg and J. Eisinger, *Phys. Rev.,* 120, 1965, 1960.
3.364. A. W. Ewald and E. E. Kohnke, *Phys. Rev.,* 97, 607, 1955.
3.365. E. E. Kohnke and A. W. Ewald, *Phys. Rev.,* 102, 1481, 1956.
3.366. J. R. Chelikowsky and M. L. Cohen, *Phys. Rev.,* B14, 556, 1976.
3.367. H. Hoechst and I. Hernández-Calderón, *Surf. Sci.,* 126, 25, 1983.
3.368. C. F. Lavine and A. W. Ewald, *J. Phys. Chem. Solids,* 32, 1121, 1970.
3.369. W. A. Harrison, *J. Vac. Sci. Technol.,* 14, 1016, 1977.
3.370. J. Tewlis and A. R. Davey, *Nature (London),* 174, 1011, 1959.
3.371. S. Groves and W. Paul, *Phys. Rev. Lett.,* 11, 194, 1963.
3.372. S. I. Novikova, *Sov. Phys. Solid State,* 2, 2087, 1961.
3.373. G. Busch, J. Wieland, and H. Zoller, *Helv. Phys. Acta,* 23, 528, 1950; 24, 49, 1951.
3.374. G. A. Busch and E. Mooser, *Helv. Phys. Acta,* 26, 611, 1953.
3.375. G. A. Busch and J. Wieland, *Helv. Phys. Acta,* 26, 697, 1953.
3.376. G. A. Busch and R. Kern, *Solid State Phys.,* 11, 1, 1960.
3.377. J. T. Kendal, *Proc. Phys. Soc. (London),* B63, 821, 1950.
3.378. J. T. Kendal, *Philos. Mag.,* 45, 141, 1954.
3.379. Li-Wei Tu, G. K. Wong, and J. B. Ketterson, *Appl. Phys. Lett.,* 54, 1010, 1989.
3.380. J. H. Becker, *Bull. Am. Phys. Soc.,* 30, 1, 1955.
3.381. S. H. Groves, C. R. Pidgeon, A. W. Ewald, and R. J. Wagner, *J. Phys. Chem. Solids,* 31, 2031, 1970.
3.382. R. E. Lindquist and A. W. Ewald, *Phys. Rev.,* 135, A191, 1964.
3.383. T. Hanyu, *J. Phys. Soc. Jpn.,* 31, 1738, 1971.
3.384. L. I. Berger and B. R. Pamplin, *Handbook of Chemistry and Physics,* 74th ed., D. R. Lide, Ed., CRC Press, Boca Raton, FL, 1993–1994, 12-79.
3.385. E. Cohen and A. K. W. A. van Lieshout, *Z. Phys. Chem.,* 117A, 331, 1936.
3.386. V. B. Sandomirskii, *Sov. Phys. JETP,* 25, 101, 1967.
3.387. S. Takatani and Y. W. Chung, *Phys. Rev.,* B31, 2290, 1985.
3.388. F. H. Pollak, M. Cardona, C. W. Higginbotham, F. Herman, and J. P. Van Dike, *Phys. Rev.,* B2, 352, 1970.
3.389. D. L. Price, J. M. Rowe, and R. M. Nicklow, *Phys. Rev.,* B3, 1268, 1971.
3.390. M. S. Dresselhaus, *The Physics of Semimetals and Narrow-Gap Semiconductors,* Pergamon Press, Elmsford, NJ, 1971, 3.
3.391. P. W. Bridgman, *J. Am. Chem. Soc.,* 36, 144, 1914; 38, 609, 1916.
3.392. R. D. Shannon, *Acta Crystallogr.,* A32, 751, 1974.
3.393. D. M. Chizhikov and V. P. Shchastlivyi, *Selen i Selenidy (Selenium and Selenides),* Nauka, Moscow, 1964.
3.394. S. O. Kasap and S. Yannacopoulos, *J. Mater. Res.,* 4, 893, 1989.
3.395. S. Yannacopoulos and S. O. Kasap, *J. Mater. Res.,* 5, 789, 1990.
3.396. M. Cukierman and D. R. Uhlmann, *J. Non-Cryst. Solids,* 12, 199, 1973.
3.397. S. O. Kasap, M. Winnicka, and S. Yannacopoulos, *J. Mater. Res.,* 3, 609, 1988.
3.398. K. Vedam, D. L. Miller, and R. Roy, *J. Appl. Phys.,* 37, 3432, 1966.
3.399. D. N. Nasledov and E. K. Malyshev, *Sov. Phys. Tech. Phys.,* 16, 1127, 1946.
3.400. P. K. Weimer, *Phys. Rev.,* 79, 171, 1950.
3.401. B. Gudden and R. W. Pohl, *Z. Phys.,* 48, 384, 1928.
3.402. K. W. Plessner, *Proc. Phys. Soc.,* B64, 671, 1951.
3.403. H. W. Henkels, *J. Appl. Phys.,* 22, 1265, 1951.
3.404. H. K. Henish, *Rectifying Semiconductors Contacts,* Oxford, 1957.
3.405. V. G. Sidyakin, *Sov. Phys. Solid State,* 1, 8, 1959.
3.406. L. I. Berger and V. M. Petrov, *Inorg. Mater.,* 11, 959, 1975.
3.407. W. J. Choyke and L. Patrick, *Phys. Rev.,* 108, 25, 1957.
3.408. A. Gobrecht and A. Tausend, *Z. Phys.,* 161, 205, 1961.
3.409. G. B. Abdullaev and G. M. Aliev, *Sov. Phys. Solid State,* 6, 1018, 1964.
3.410. W. W. Scanlon and K. Lark-Horowitz, *Phys. Rev.,* 72, 530, 1947.
3.411. A. R. von Hippel, *J. Chem. Physl.,* 16, 372, 1948.
3.412. S. Tanuma, *Sci. Rep. Res. Inst. Tohoko Univ. Ser.,* A4, 353, 1952.
3.413. A. S. Epstein, H. Fritzsche, and K. Lark-Horotwitz, *Phys. Rev.,* 107, 412, 1957.

3.414. J. C. Perron, *Compt. Rend.*, 258, 4698, 1964.
3.415. P. A. Hartig and J. J. Loferski, *J. Opt. Soc. Am.*, 44, 17, 1954.
3.416. J. J. Loferski, *Phys. Rev.*, 93, 707, 1954.
3.417. R. S. Caldwell and H. Y. Fan, *Phys. Rev.*, 94, 1427, 1954.
3.418. J. N. Hodgson, *Philos. Mag.*, 8, 735, 1963.
3.419. J. L. Shay and J. H. Wernick, *Ternary Chalcopyrite Semiconductors: Growth, Electronic Properties and Applications,* Pergamon Press, Elmsford, NY, 1975.
3.420. R. P. Elliott, *Constitution of Binary Alloys,* First Supplement, McGraw-Hill, New York, 1965, 112.
3.421. G. V. Samsonov, N. I. Zhuravlev, and I. G. Emmanuel, *Fiz. Metal. Metalloved. (Phys. Metals Metallogr.),* 3, 309, 1956.
3.422. R. P. Elliott, IIT Res. Inst. ARF-2200-12, Final Rep., Contract AT(11-1)-578, 1961.
3.423. A. Lipp and M. Röder, *Z. Anorg. Allgem. Chem.*, 344, 225, 1966.
3.424. R. D. Allen, *J. Am. Chem. Soc.*, 75, 3582, 1953.
3.425. R. Kieffer, E. Gugel, G. Leimer, and P. Ettmayer, *Ber. Keram. Ges.*, 48, 385, 1971.
3.426. H. W. Deem and C. F. Lucks, *Thermophysical Properties of Matter,* Vol. 2, Thermal Conductivity, Non-Metallic Solids, Y. S. Touloukian and E. H. Buyco, Eds., Plenum Press, New York, 1970, 572.
3.427. A. Lipp and K. Schwetz, *Ber. Dtsch. Keram. Ges.*, 52, 335, 1975.
3.428. H. L. Yakel, *J. Appl. Crystallogr.*, 6, 471, 1973.
3.429. K. E. Gilchrist and S. D. Preston, *High Temp. High Pressure,* 11, 643, 1979.
3.430. K. K. Kelley, *J. Am. Chem. Soc.*, 63, 1137, 1941.
3.431. G. N. Makarenko, *Boron and Refractory Borides,* V. I. Matkovich, Ed., Springer-Verlag, New York, 1977, 310.
3.432. E. Lugschedier, H. Reinmann, and W. J. Quadakkers, *Ber. Dtsch. Keram. Ges.*, 56, 301, 1979.
3.433. H. L. Longquet-Higgins and M. V. Roberts, *Proc. R. Soc. London,* 230A, 110, 1955.
3.434. M. F. Rizzo and L. R. Bidwell, *J. Am. Ceram. Soc.*, 43, 550, 1960.
3.435. C. Wood and D. Emin, *Phys. Rev.*, B29, 4582, 1984.
3.436. M. Bouchacourt and F. Trevenot, *J. Mater. Sci.*, 20, 1237, 1985.
3.437. S. LaPlaca and B. Post, *Plannseeber Pulvermet,* 9, 109, 1961.
3.438. V. I. Matkovich, *J. Am. Chem. Soc.*, 83, 1804, 1961.
3.439. J. L. Peret, *J. Am. Ceram. Soc.*, 47, 44, 1964.
3.440. M. Hansen and K. Anderco, *Constitution of Binary Alloys,* McGraw-Hill, New York, 1958.
3.441. E. R. Johnson and S. M. Christian, *Phys. Rev.*, 95, 560, 1954.
3.442. A. Levitas, *Phys. Rev.*, 99, 1810, 1955.
3.443. G. Busch and O. Vogt, *Helv. Phys. Acta,* 33, 437, 1960.
3.444. F. Herman, *Phys. Rev.*, 95, 847, 1954.
3.445. R. Braunstein, A. R. Moore, and F. Herman, *Phys. Rev.*, 109, 695, 1958.
3.446. A. W. Ewald, *J. Appl. Phys.*, 25, 11, 1954; 25, 436, 1954.
3.447. D. R. Lide, Ed., *Handbook of Chemistry and Physics,* 73rd ed., CRC Press, Boca Raton, FL, 1992–1993.

CHAPTER 4

Binary IV-IV and III-V Semiconductors

Goryunova and Parthe[4.1] suggested a "spiral" (or conical) form of the Periodic table of the elements (Figure 4.1) in which two groups may be considered as two "axes" (shaded area). The anions on the right from the axes are common for all compounds. The cations which are located left from VIIIb group (A-cations) form, mainly, compounds with the octahedral coordination, i.e., each atom in their crystal lattice is surrounded by six nearest neighbors, forming an octahedron; the cations located on the left from IVb group (B-cations) form compounds with the tetrahedral coordination of atoms in which any atom is surrounded by four nearest neighbors, forming a tetrahedron. The authors of Reference 4.1 indicate that the A- and B-cations form completed electron shells similar to the shells of the VIIIb-group elements by two different ways: the former, by giving their valence electrons to the anions, while the latter fill their valence electron shells and, simultaneously, the shells of the anions by sharing the valence electrons between the neighboring atoms. Whereas many of the compounds with A-cations have ionic conductivity, the compounds with B-cations represent a vast group of semiconductor materials. The most prominent group among the binary semiconductors is a family of diamond-like materials, the crystallochemical analogs of the Group IV elements.

4.1. IVB-IVB COMPOUNDS — SILICON CARBIDE

Silicon carbide, SiC, is the only tetrahedral compound formed by the Group IVb elements. Its optical, electrical, mechanical, and thermal properties make SiC an interesting material for a wide range of applications. SiC has, apparently, a substantial advantage over many other binary compounds because it can be oxidized to form on its surface a stable layer of insulating SiO_2. Its disadvantage, for some applications, lies in its indirect energy gap.

Silicon carbide is a unique compound from the crystallographic point of view: it crystallizes in more than 200 polymorphic forms.[4.37] Essentially, all SiC polytypes may be divided into three groups: cubic (β-SiC, B3 type), denoted 3C-SiC, with crystal lattice similar to one of sphalerite (space group T_d^2-F$\bar{4}$3m); hexagonal polytype (denoted 2H-SiC) with the wurtzite-type crystal lattice, space group C_{6v}^4-P6$_3$mc; and the third group[4.2a] that includes all other known polytypes. The authors of Reference 4.2b suggest that all polymorphs of the third group may be considered as a natural superlattice in which the atomic layers with cubic and hexagonal symmetry follow in a certain alteration (that determines a particular polytype) along the c-axis. The most studied representatives of the third group are SiC polymorphs with the rhombohedral unit cell (denoted as 15R and 21R).

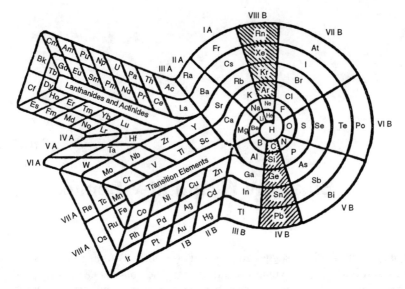

Figure 4.1 "Spiral" variant of the Periodic table. (After N. Goryunova and E. Parthe, *Mater. Sci. Eng.*, 2, 1, 1967. With permission.)

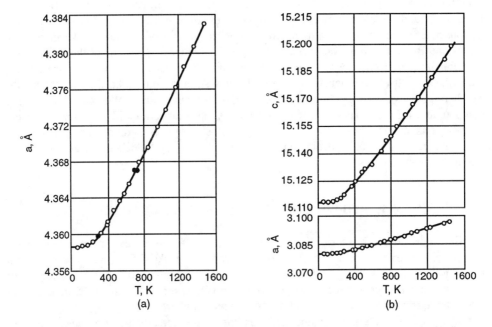

Figure 4.2 Temperature dependence of the SiC lattice parameters. (a) 3C-allotrope, (b) 6H-allotrope. (After A. Taylor and R.M. Jones, *Silicon Carbide — A High Temperature Semiconductor*, J.R. O'Connor and J. Smiltiens, Eds., Pergamon, 1960. With permission.)

The lattice parameters of the cubic 3C-SiC (0.43596 nm at 297 K) and the hexagonal 6H-SiC (a = 0.30806 nm and c = 1.51173 nm at 297 K) and their temperature dependence in the range 80 to 1500 K were measured by Taylor and Jones[4.3] (Figure 4.2). The coefficient of linear thermal expansion of 3C-SiC at 300 K is $2.77 \cdot 10^{-6}$ 1/K;[4.4,4.5] its magnitude at 1400 K is $5.45 \cdot 10^{-6}$ 1/K.[4.5] The temperature dependence of the coefficient of thermal expansion of the hexagonal α-SiC was reported by Touloukian et al.;[4.35] Berry et al.[4.36] studied the thermal expansion of β-SiC. Figure 4.3 (a and b) shows the temperature dependence of the thermal strain of α- and β-SiC in comparison with that of Si (from Reference 4.36).

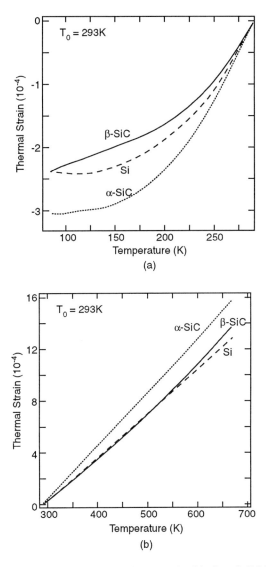

Figure 4.3 The thermal expansion of hexagonal (alpha-) and cubic (beta-) SiC in comparison with that of silicon. (a) Low temperatures, (b) high temperatures. (After B.S. Berry, W.C. Pritchet, R.I. Fuentes, and I. Babich, *J. Mater. Res.*, 6, 1061, 1991. With permission.)

The density of cubic SiC at 293 K is 3.166 g/cm^3;[4.6] the magnitude calculated from X-ray measurements is 3.214 g/cm^3; the density of 6H-SiC polytype at 300 K is 3.211 g/cm^3 [4.7] (the calculated magnitude is 3.214 g/cm^3). The specific heat at standard conditions is c_p = 0.671 J/g·K (3C-SiC) and c_p = 0.666 J/g·K (6H-SiC) (Reference 4.8, p. 5-25). Until now, there has been a substantial discrepancy in the phase transition temperatures of SiC. Ruff[4.9] and Nowotny et al.[4.10] suggest that the (peritectic) decomposition temperature of SiC is close to 2970 K at atmospheric pressure, whereas Scace and Slack[4.11] reported the magnitude of 3103 ± 40 K (3.5 MPa) and Dolloff[4.12] found that the peritectic temperature is 2813 ± 40 K. The decomposition temperature is, apparently, higher than 2870 K, because this is the temperature of the graphite substrates on which the SiC crystals are grown from the gas phase by the technology suggested by Lely,[4.13] which has been widely used up to now for SiC single crystal growth at the research and manufacturing facilities (see, e.g., Reference 4.14), especially after a substantial improvement of the method by Barrett et al.,[4.44] who grow high quality SiC rods greater than 25 mm in diameter. (Another method of the SiC synthesis, which certainly has a promising future, is the self-propagating high-temperature combustion synthesis.[4.80-4.82] Using this method, Narayan et al.[4.80] synthesized conglomerates of SiC crystals of relatively high quality.)

Both cubic and hexagonal SiC are indirect semiconductors. The minimum energy gap of 3C-SiC (Figure 4.4) is related to Γ_{15v}-X_{1c} transition. Its magnitude at 2 K is 2.416 eV.[4.16] The exitonic energy gap at 1.4 K is 2.38807 eV;[4.17] at room temperature its magnitude is close to 2.36 eV.[4.18] The direct energy gap of 3C-SiC at 300 K (transition $\Gamma_{15v} - \Gamma_{1c}$) is 6.0 eV.[4.19] The exitonic energy gap of several SiC polytypes vs. temperature is presented in Figure 4.5.[4.18] The 300-K minimum indirect energy gap of the 6H-SiC (α-SiC) is 2.86 eV.[4.20]

The electron effective mass in 3C-SiC measured by Anikin et al.[4.21] at 45 K is (1) longitudinal $m_l^* = 0.677\ m_o$; (2) transversal $m_t^* = 0.247\ m_o$. As is shown in Figure 4.4, the spin-orbit splitting is virtually absent; therefore, the valence bands in the cubic SiC are highly nonparabolic and the hole effective mass is strongly -dependent.[4.21,4.22]

The index of refraction of cubic SiC in the range 467 to 691 nm was measured by Shaffer and Naum,[4.23] who suggested an empirical dispersion formula:

$$n\ (467 \leq \lambda \leq 691\ nm) = 2.55378 + 34170\lambda^{-2}$$

This result is in satisfactory agreement with dispersion data of Powell,[4.24] who reported the index of refraction (Figure 4.6) for cubic and several birefringent hexagonal and rhombohedral SiC crystals.

The static dielectric constant of 3C-SiC at 300 K is $\varepsilon(0) = 9.72$; the high frequency magnitude $\varepsilon(\infty) = 6.52$.[4.25] The data for 6H-SiC — $\varepsilon_\perp(0) = 9.66$, $\varepsilon_\perp(\infty) = 6.52$ and $\varepsilon_\parallel(0) = 10.03$, $\varepsilon_\parallel(\infty)$ 6.70 [4.25] — are in agreement with earlier results of Ketelaar.[4.26]

SiC is a diamagnetic with the magnetic susceptibility of $-80.4 \cdot 10^{-6}$ SI units/gram-mol (Reference 4.8, p. 9-54).

Silicon carbide is one of the hardest materials. Its Mohs hardness is 9.5; the Knoop microhardness is 28.8 ± 3.5 GPa.[4.27] The elastic constants of 6H-SiC at 300 K are $c_{11} = 500$, $c_{12} = 92$, $c_{33} = 564$, and $c_{44} = 168$ GPa.[4.28] The room temperature magnitude of the cubic SiC Young's modulus is 410 GPa. It should be mentioned that the amorphous SiC films, also having very high mechanical properties (the hardness of 30 GPa and Young's modulus of 240 GPa[4.45]), present substantial interest in using the films as the X-ray masks.

SiC electrical transport properties strongly depend on the impurity content as is common for the wide-energy-gap semiconductors. In the early measurements, Busch and Labhart[4.29] found that the electron mobility in 6H-SiC passes through a maximum at about 200 K. The authors of Reference 4.14 showed that this effect is common for crystals of 6H-SiC dopped with nitrogen (Figure 4.7). The electron mobility at room temperature in 3C-SiC varies from 300 to 900 cm^2/V·sec; the hole mobility varies from 10 to 30 cm^2/V·sec.[4.30,4.31]

Thermal conductivity, κ, of hexagonal SiC was reported by Slack,[4.32] who found (Figure 4.8) that κ passes through a maximum close to 30 to 40 W/cm·K near 50 to 60 K and decreases as T^{-1} at the higher temperatures up to 400 K.

One of the early applications of SiC was related to the ability of the granular material to decrease sharply its electrical resistance in the strong electric field.[4.33] The devices based on this effect (varistors) are currently used for protection of electric lines from the damage by lightning. Another application was based on discovery of p-n transitions on some as-grown SiC crystals. These crystals emit visible light when current passes through them like the light-emitting diodes.[4.34] The properties of SiC and the information relevant to the applications of the material in semiconductor devices are presented in detail in the reviews by Van Vliet et al.,[4.38] and Edgar[4.39] (see also Reference 4.40) and in the conference proceedings.[4.2b,4.41,4.42] The high quality field effect transistor based on the 3C-SiC epitaxial film deposited on the

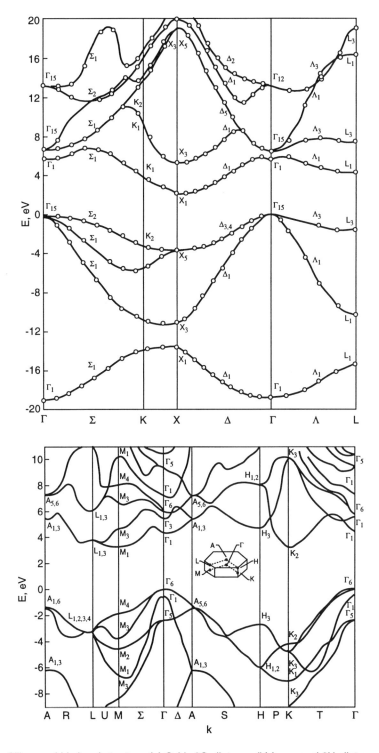

Figure 4.4 Silicon carbide band structure. (a) Cubic 3C-allotrope, (b) hexagonal 2H-allotrope. (After L.A. Hemstreet and C.Y. Fong, *Silicon Carbide — 1973*, R.C. Marshall, J.W. Faust, and C.E. Ryan, Eds., Univ. S. Carolina Press, Columbia, 1974. With permission.)

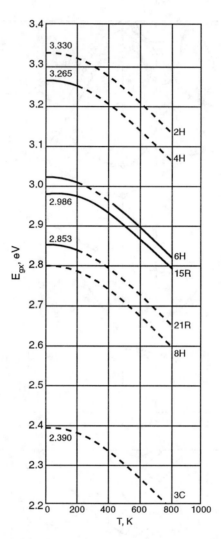

Figure 4.5 Temperature dependence of the excitonic energy gap of SiC allotropes. (After W.J. Choyke, in "Silicon Carbide - 1968". Proc. Intent. Conf. on Silicon Carbide, University Park, PA, October 1968. *Mater. Res. Bull. Suppl.,* Pergamon, 1969, p. S141. With permission.)

Si substrate was prepared by Davis et al.[4.43] and may be considered as the beginning of the period of wide use of SiC in the semiconductor industry.

The analytical purity silicon carbide is available commercially: α-SiC, 99.9% — $84.00/50 g; β-SiC, 99.8% — $31.50/100 g (1994).

4.2. IIIB-VB COMPOUNDS

The IIIb-Vb (further: III-V) compounds are substances with the predominantly covalent interatomic bonds; nevertheless, they will be grouped here in accordance with the common chemical classification, i.e., nitrides, phosphides, etc.

4.2.1. Crystal Structure

All Group III and V elements which are located in the second to fifth periods of the Periodic table form binary compounds with the structure similar to that of the minerals

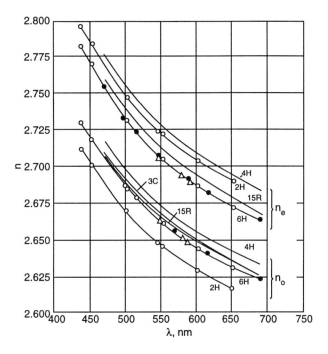

Figure 4.6 Dispersion of the index of refraction in some SiC polytypes. (After J.A. Powell, *J. Opt. Soc. Amer.*, 62, 341, 1972. With permission.)

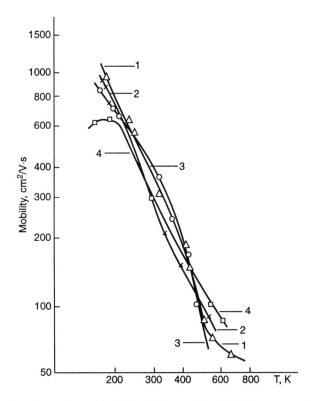

Figure 4.7 Temperature dependence of mobility in SiC crystals with electron concentration (at 300 K) of $4.2 \cdot 10^{16}$ cm^{-3} (sample 1), $7.8 \cdot 10^{16}$ cm^{-3} (sample 2), $8.2 \cdot 10^{16}$ cm^{-3} (sample 3), and $2.2 \cdot 10^{16}$ cm^{-3} (sample 4). (After L.I. Berger, V.M. Efimov, and L.A. Stroganova, *Proc. IREA*, 33, 247, 1971. With permission.)

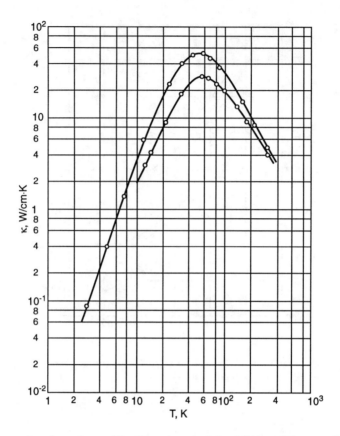

Figure 4.8 Temperature dependence of the thermal conductivity of 6H-type SiC across the c axis for two different samples. (After G.A. Slack, *J. Appl. Phys.*, 35, 3460, 1964. With permission.)

sphalerite or wurtzite. The sphalerite mineral group includes metacinnabarite, HgS; tiemannite, HgSe; coloradoite, HgTe; and sphalerite, ZnS. Wurtzite, ZnS, belongs to the group of hexagonal or rhombohedral minerals which also includes cinnabar, HgS, and greenockite, CdS. The crystal structure of sphalerite and wurtzite is presented in Figure 4.9.

4.2.2. Boron Nitride

Boron nitride, BN, has the molecular weight of 24.82, the average atomic weight, $A_{av} = (A_1 + A_2)/2$, of 12.41, the average atomic number, $N_{av} = (N_1 + N_2)/2$, equal to 6, and the standard enthalpy of formation of -254.4 kJ/mol (Reference 4.8, p. 5-4).

The melting point of BN is close to 3200 K (under N_2 pressure of about 7 GPa),[4.47,4.48] the enthalpy of melting 228.3 kJ/mol,[4.53] and the heat capacity at 298.15 K $C_p = 19.7$ J/mol·K.[4.49] The thermal conductivity of BN is very high: for cubic BN its magnitude of 8.0 W/cm·K is almost twice greater than that of silver.[4.50]

Boron nitride crystallizes in three allotrope modifications:[4.59] (1) cubic, c-BN, space group $F\bar{4}3m$, with the sphalerite-type lattice, the lattice parameter at 293 K of a = 0.36157 nm and density 3.4870 g/cm³;[4.51] (2) hexagonal, h-BN, with 12 atoms in a unit cell, the lattice parameters a = 0.25040 nm and c = 0.66612 nm[4.52] and the density of 2.18 g/cm³ (pycnometric) and 2.286 g/cm³ (X-ray); and (3) high-pressure (transition at 20 GPa[4.48] wurtzite-type phase, w-BN, corresponding to mineral lonsdaleite, with lattice parameters a = 0.255 nm and c = 0.420 nm[4.54] and the density of 3.454 g/cm³.

At standard conditions, the only stable allotrope is h-BN (space group $C6m2$-D_3^{12}). BN cubic allotrope is metastable at standard conditions and converts to the h-BN structure when

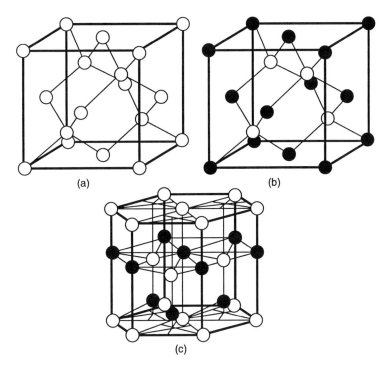

Figure 4.9 Unit cell of (a) diamond, (b) sphalerite, and (c) wurtzite.

heated in vacuum to 1700 K, whereas w-BN is metastable at any conditions in solid state. Hexagonal h-BN may be synthesized in the form of graphite-like white flakes either by the process described by Podszus[4.57] and improved by Meerson et al.[4.58] that consists of heating B_2O_3 in the NH_3 atmosphere at 1300 to 1500 K, or by roasting a mixture of borax, $Na_2B_4O_7 \cdot 10H_2O$, with ammonium chloride, NH_4Cl, at 500 to 600 K. Another synthesis method which is widely used in the semiconductor industry is the chemical vapor deposition (CVD) of h-BN using BCl_3 and NH_3. Sano and Aoki[4.60] reported the results of h-BN film deposition on the Al_2O_3 and SiO_2 flat substrates, Motojima et al.[4.61] deposited h-BN on the flat copper substrates, and Hurd et al.[4.56] successfully deposited h-BN films on SiO_2 crucibles used for GaAs synthesis. The disadvantage of this process consists of difficulties of uniform mixing of the boron trihalides (BCl_3 or BF_3) and ammonia to produce a stoichiometric deposit. So, attempts were undertaken to use volatile species which *ab initio* have a proper ratio of boron and nitrogen in them, such as borazine, $B_3N_3H_3$, trichloroborazine, $B_3N_3H_3Cl_3$,[4.62] or hexachloroborazine, $B_3N_3Cl_6$. Pavlović et al.[4.63] combined the two processes in one deposition unit: BCl_3 was premixed with NH_3 at 390 to 490 K forming trichloroborazine (or iminochloroborazine, ClBNH, or aminodichloroborane, Cl_2BNH_2) that was transported into the deposition zone by the stream of nitrogen carrier gas.

Hexagonal BN is, apparently, a direct transition semiconductor with the optical (direct) energy gap of 5.8 eV at room temperature.[4.55,4.56] The E_g magnitude received by Hoffman et al.[4.64] from reflectance measurements is 5.2 eV (300 K). There is a substantial variation of the data on E_g from different sources (from 3.3 to 5.8 eV) which may be referred to the differences in the synthesis methods. The band structure of h-BN calculated by Robertson[4.65] is shown in Figure 4.10. Catellani et al.[4.66] suggest an indirect energy gap of 3.9 eV (transition H → M, see Figure 4.11). Carpenter and Kirby[4.67] reported a thermal energy gap above 1000 K of 7.1 eV from the results of the resistivity measurements, and the resistivity magnitude close to $3 \cdot 10^{14}$ Ohm·cm at 1000 K. The discrepancy between this magnitude and $1.7 \cdot 10^{13}$ Ohm·cm (at 298 K) reported earlier by Taylor[4.68] may be referred to the difference in the sample purity.

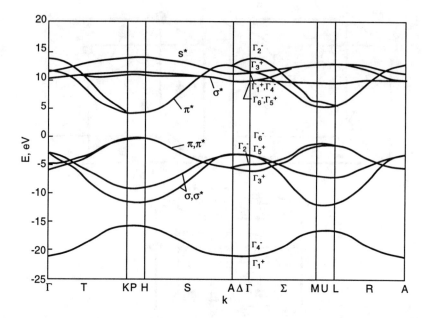

Figure 4.10 Band structure of hexagonal BN calculated by the tight-binding method. (After J. Robertson, *Phys. Rev.*, B29, 2131, 1984. With permission.)

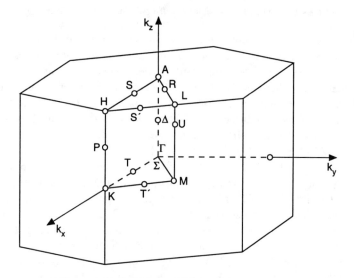

Figure 4.11 Brillouin zone of the hexagonal lattice.

The h-BN thermal diffusivity estimate in Reference 4.54 (about $1 \cdot 10^{-2}$ cm^2/sec) correlates with the thermal conductivity of about 18 mW/cm·K. This magnitude is in sharp contrast with the one presented in Reference 4.70, p. 90 (150 to 289 mW/cm·K at 573 K and 121 to 268 mW/cm·K at 1273 K) and with the data reported by Powell and Tye[4.71] (0.2 W/cm·K). Until now, thermal conductivity measurements have been performed mainly on polycrystalline specimens or on small single crystals; therefore, our information on thermal transport in h-BN in different crystallographic directions is still insufficient.

The thermal expansion of h-BN was investigated by Pease[4.72] who found from X-ray measurements that the average coefficient of thermal expansion near room temperature is $-2.9 \cdot 10^{-6}$ 1/K along c axis and $40.9 \cdot 10^{-6}$ 1/K across c axis. The average linear expansion

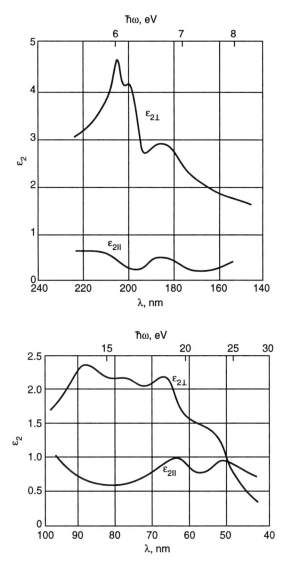

Figure 4.12 Ordinary and extraordinary dielectric functions ε_2 of hexagonal BN in the photon energy range (a) 5 to 9 eV and (b) 13 to 30 eV. (After R. Mamy, J. Thomas, G. Jezequel, and J.C. Lemonnier, *J. Phys (Paris) Lett.*, 42, L473, 1981. With permission.)

coefficient measured on polycrystalline specimens is $10.15 \cdot 10^{-6}$ 1/K in the range of 298 K to 623 K (Reference 4.68, p. 90) in satisfactory agreement with the average magnitude of $11.7 \cdot 10^{-6}$ 1/K from Reference 4.72.

Hexagonal boron nitride is a soft material with Mohs' hardness about two units and may be used as a vacuum and refractory lubricating material.

The dielectric properties of h-BN are typical for an anisotropic insulator: the 300-K static dielectric constant is 5.06 along and 6.85 across the c axis; the high frequency dielectric constant is 4.10 along and 4.95 across the c axis.[4.73] Mamy et al.[4.74] have reported the dependence of the ordinary and extraordinary dielectric functions on the UV photon wavelength (Figure 4.12). Their data agree with the magnitudes of the index of refraction (2.13 along and 1.65 across the c axis) reported by Ishii and Sato.[4.75] In the infrared part of the spectrum, there are two absorption maxima, at 7.28 μm and 12.3 μm (wave numbers 1374 cm^{-1} and 813 cm^{-1}, respectively), which may be related to directions along and across the c axis in the h-BN crystal.[4.76] These magnitudes agree with the photon wave numbers reported by Hoffman et al.[4.64]

Cubic boron nitride, c-BN, often called borazone,[4.77] has outstanding properties. Its hardness is virtually equal to that of diamond (10 units on Mohs' scale, with microhardness of 66 to 75 GPa[4.90]), and, in contrast with diamond which ignites in the air at about 1200 K, c-BN may be only slightly oxidized in the air at temperatures as high as 2800 to 2900 K. It is stable in vacuum up to 3000 K and resistant to acids.

The coefficient of thermal expansion of c-BN is $1.8 \cdot 10^{-6}$ 1/K.[4.35] The c-BN room temperature thermal conductivity is close to 13 W/cm·K,[4.79] second only to diamond. This magnitude makes c-BN a very attractive material for the heat sinks as well as for the high-current semiconductor devices. The second-order elastic modulus c_{11} estimated by Steigmeier[4.89] has a magnitude close to 0.71 TPa, which is only 30% smaller than that of diamond. The ultimate compressive strength of c-BN is between 4.15 and 5.33 GPa.[4.90]

Cubic boron nitride has an indirect energy gap of 6.4 ± 0.5 eV at room temperature;[4.80] its direct energy gap evaluated from reflectivity measurements[4.20] is equal to 14.5 eV. The calculations performed by Huang and Ching[4.84] and Prasad and Dubey[4.85] (see Figure 4.13) give for the indirect energy gap (transition $\Gamma_{15v} \to X_{1c}$) magnitudes of 6.99 eV and 8.6 eV, respectively, and for a direct energy gap (transition $\Gamma_{15v} \to \Gamma_{1c}$) magnitudes of 9.94 eV and 10.86 eV, respectively.

Cubic BN has the largest (indirect) energy gap among the diamond-like semiconductors. Its advantage consists of its ability to acquire both n- and p-type conductivity by doping, e.g., with silicon or berillium, respectively,[4.86] which was used to make light-emitting p-n junctions.[4.87] The authors of Reference 4.86 have shown that the donor and acceptor activation energy is about 0.4 to 0.5 eV (Figure 4.14); therefore, the noticeable extrinsic conductivity may occur even at room temperature. A similar dependence of electrical conductivity on temperature was reported earlier by Bam et al.[4.88] (Figure 4.15) after their measurements between 300 K and 900 K on polycrystalline samples of c-BN (the maximum magnitude of the conductivity activation energy of about 1.5 eV). They found that the charge carrier mobility (the sign was not determined) is increasing with temperature from 0.2 to 4 cm^2/V·sec between 500 K and 900 K, while the carrier concentration is decreasing from 10^{15} to 10^{14} cm^{-3}. A review of the c-BN electronic properties was compiled by Golikova.[4.91]

The c-BN optical and dielectric properties were investigated by Gielisse et al.,[4.95] who reported the index of refraction of 2.117 at 300 K and 589 nm, and the static and optical frequency dielectric constant of 7.1 and 4.5, respectively (all magnitudes at 300 K).

Recent progress reached in growth and application of diamond films triggered interest in the growth and investigation of c-BN thin films. Bulk pyrolytic c-BN may be obtained by heating h-BN at 1630 K and 6.2 GPa[4.77] or by nitrogenating the boron phosphide powder[4.92] (hot-pressed rods, 9.5 mm diameter, 150 mm long are available commercially, $60.00 per rod). There are numerous reports (see, e.g., a review[4.39] and the results of Ballal et al.[4.93]) on the growth of c-BN thin films, in particular, on silicon substrates. Doll et al.[4.96] have reported deposition of the high quality epitaxial films of c-BN on silicon. Okamoto et al.[4.97] deposited c-BN films with good adhesion on diamond.

Boron nitride in the form of powder, 99.5% purity, is available commercially ($34.70/50 g — 1994).

4.2.3. Boron Phosphide

Boron phosphide, BP, has molecular weight of 41.78, average atomic weight of 20.89, and the average atomic number of 10 (equal to that of SiC). BP decomposes at 1400 K at atmospheric pressure, but can be melted under the phosphorus pressure close to 100 GPa at about 3300 K.[4.98] BP has only one known allotrope at this time with the sphalerite structure and lattice parameter of 0.45383 nm at 297 K[4.4] and the density of 2.970 g/cm^3. The precision

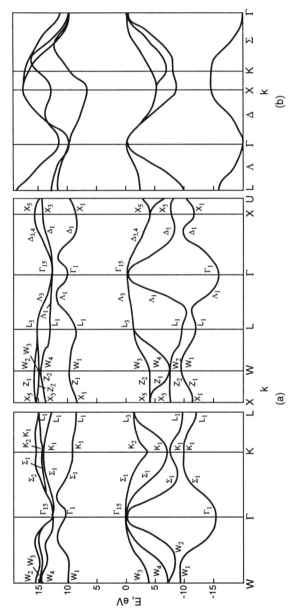

Figure 4.13 The band structure of cubic BN. (a) The results of the LCAO method calculation, including ionicity and fitting of APW results at high-symmetry points; (b) the results of *ab initio* calculations. (After C. Prasad and J.D. Dubey, *Phys. Stat. Sol.*, B125, 625, 1984. With permission.)

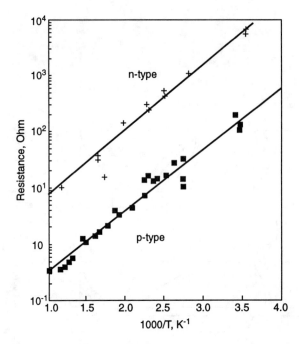

Figure 4.14 The temperature dependence of electrical resistance of the n- and p-type samples of β-BN. (After O. Mishima, J. Tanaka, S. Yamaoka, and O. Fukunada, *Science*, 238, 181, 1987. With permission.)

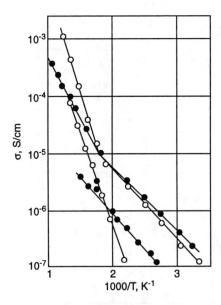

Figure 4.15 Temperature dependence of the electrical resistivity of different polycrystalline samples of cubic BN. (After I.S. Bam, V.M. Davidenko, V.G. Sidorov, L.I. Feldun, M.D. Skagalov, and Y.K. Shalabutov, *Sov. Phys. Semicond.*, 16, 331, 1976. With permission.)

X-ray measurements performed by Nishinaga[4.99] on several BN epitaxial wafers grown by Kumashiro et al.[4.100] on silicon substrates and having an excess of boron (p-type) or phosphorus (n-type) showed that the lattice parameter of n-type wafers was slightly greater (about 0.4538571 nm) than that of the p-type wafers (about 0.4538094 nm). Kim and Shono[4.101]

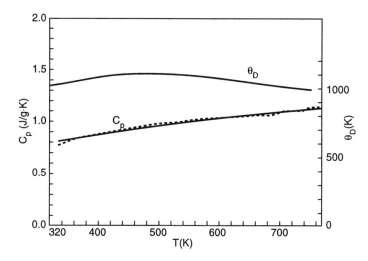

Figure 4.16 Temperature dependence of the specific heat and Debye temperature of boron phosphide. (After Y. Kumashiro, *J. Mater. Res.*, 5, 2933, 1991. With permission.)

assumed that in a nonstoichiometric BP, the excess phosphorus atoms occupy the boron sites, while the excess boron atoms occupy the phosphorus sites, extending or shrinking the crystal lattice, respectively. Slack and Bartram[4.102] reported the magnitude of the coefficient of thermal expansion of BP at 400 K of $3.65 \cdot 10^{-6}$ 1/K, while in Reference 4.35, its magnitude is $2.9 \cdot 10^{-6}$ 1/K. The pycnometric density of BP at 300 K is 2.89 g/cm³.[4.103]

BP has a relatively high microhardness of 32 GPa (3200 kgp/cm²) at the 100-g load.[4.104] The elastic moduli of BP at room temperature evaluated by Wettling and Windschleif[4.105] from the Brillouin scattering experiments are $c_{11} = 0.515$ TPa, $c_{12} = 0.10$ TPa, and $c_{44} = 0.16$ TPa.

The pure boron phosphide crystals are almost transparent. The n-type crystals are colored orange-red, while the p-type samples are dark red.[4.106]

The temperature dependence of the BP heat capacity was reported by Kumashiro et al.[4.109] together with the results of the Debye temperature calculations. Their data are presented in Figure 4.16.

In accordance with the results of calculations by Hemstreet and Fong[4.107] and Huang and Ching,[4.108] BP is an indirect gap semiconductor with the minimum (indirect) energy gaps of 2.2 eV (transition $\Gamma_{15v} \rightarrow X_{1c}$)[4.107] and 1.98 eV (transition $\Gamma_{15v} \rightarrow L_{1c}$). The BP band structure (after Reference 4.107) is shown in Figure 4.17. The direct energy gap at the center of the Brillouin zone (transition $\Gamma_{15v} \rightarrow \Gamma_{1c}$) is close to 3.25 eV at 300 K. The estimation of the electron and hole-effective mass based on the results of thermoelectrical and electrical measurements[4.110,4.111] gives magnitudes $m_n^* = 0.67\ m_o$ and $m_p^* = 0.042\ m_o$.

The electric transport properties of BP were reported, in particular, by Stone and Hill,[4.104] Kato et al.,[4.111] Yugo and Kimura,[4.112] and Kumashiro.[4.106] In contrast with the data of Reference 4.110, presented in Figure 4.18, the authors of References 4.104 and 4.111 found that the mobility of both electrons and holes increases with temperature increase up to about 400 to 500 K approximately as $T^{3/2}$, reaching magnitudes up to 200 cm²/V·sec at maximum, and decreases with the further growth of temperature. Wang et al.[4.114] and Sono et al.[4.115] reported the magnitudes of the hole mobility at room temperature between 285 and 500 cm²/V·sec. These mobility data make boron phosphide a promising material for the high-temperature semiconductor devices, taking into consideration its chemical inertness (BP cannot be dissolved in acids or in boiling aqueous alkaline solutions; none of the known etchants attacks it; it can be dissolved only in the alkali melts.[4.70]

Thermal conductivity of BP was reported by Slack,[4.79] Yugo et al.,[4.113] and Kumashiro et al.[4.109] Slack found that it passes through a maximum of 14.0 W/cm·K at 80 K and is equal

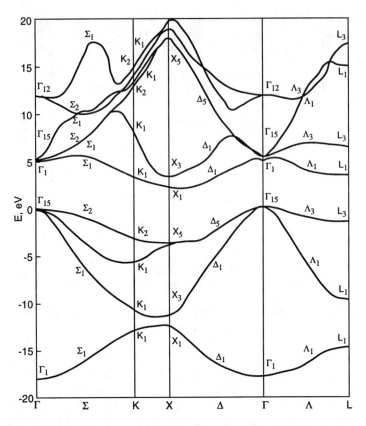

Figure 4.17 Band structure of boron phosphide. Nonlocal empirical potential method calculation. (After L.A. Hemstreet and C.Y. Fong, *Phys. Rev.*, B6, 1464, 1972. With permission.)

to 3.6 W/cm·K at 300 K. The data of Reference 4.109 (Figure 4.19) are in good agreement with the results of Reference 4.79, while the measurements on thin films, grown by chemical vapor deposition (CVD) on Si (100) substrate,[4.113] give values of thermal conductivity two orders of magnitude smaller. In accordance with References 4.79, 4.109, and 4.113, at the temperatures above the ambient, thermal conductivity is proportional (with slight deviations) to T^{-1}.

Boron phosphide has interesting thermoelectric properties. In the range of 300 to 700 K, both n-type and p-type BN single crystal wafers investigated by Kumashiro et al.[4.110] have Seebeck coefficient of about 300 to 500 μV/K, decreasing slowly with the temperature growth.

The static dielectric constant of BP is equal to 11.0 at room temperature,[4.116] in very good agreement with the magnitudes of the index of refraction (from 3.34 at 454 nm to 3.00 at 632.8 nm[4.105]).

The first successful attempts to synthesize BP were reported by Besson[4.117] and Moissan.[4.118] Popper and Ingles'[4.119] two-temperature synthesis method was successfully used by Rundquist[4.103] and Perri et al.[4.120] for growth of BP crystals. Wickery[4.92] was, apparently, the first who used a variant of CVD of BP films by thermal decomposition of $BCl_3 \cdot PCl_5$. Williams and Ruherwein,[4.121] Chu et al.,[4.122] and Nishinage et al.[4.123] used a range of boron and phosphorus hydrides and halides for CVD of BP films. Stone and Hill[4.104] synthesized needle-shaped crystals and granules of BP using reaction in the vapor phase and crystallization from the melt. Kobayashi et al.[4.124] grew BP crystals using a high-pressure/high-temperature method. A bibliography of the BP synthesis methods was compiled by Kumashiro.[4.106] In his own experiments, Kumashiro received BP single crystals up to $5 \times 5 \times 3$ mm³ in size and

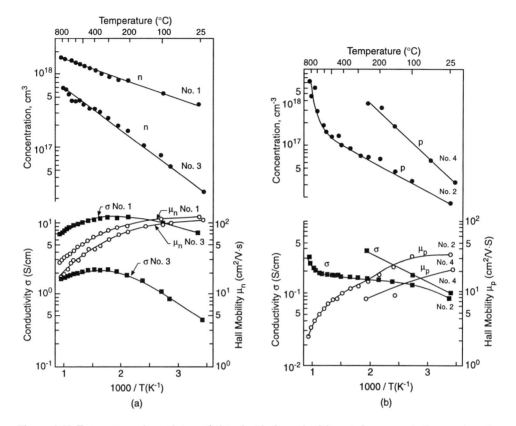

Figure 4.18 Temperature dependence of the electrical conductivity, carrier concentration, and carrier mobility in several BP samples. (a) n-type samples, (b) p-type samples. (After Y. Kumashiro, *J. Mater. Res.*, 5, 2933, 1990. With permission.)

epitaxial wafers with an area of 10 × 20 mm² and up to 0.3 mm thick, which were deposited on the silicon substrates of various orientation.

Technical-grade boron phosphide (powder, 97.5% purity) is available commercially ($84.00/5 g — 1994).

4.2.4. Boron Arsenide

Boron arsenide, BAs, has a molecular weight of 85.73, an average atomic weight of 42.87, and an average atomic number of 19. BAs crystals are of metallic-gray color and have a sphalerite structure with the lattice parameter at 300 K of 0.4777 nm and a density of 5.22 g/cm³.[4.125,4.126] The melting point of BAs is close to 2300 K.[4.120,4.121] It has the microhardness close to 19 GPa and possesses high abrasive ability.

The band structure of BAs was calculated by Stuckel[4.127] (see Figure 4.20). Per Stuckel, BAs has an indirect energy gap of about 1.8 to 1.9 eV (transition $\Gamma \rightarrow X$), whereas a direct energy gap at the center of the Brillouin zone is about 3.5 eV. The optical absorption experiments by Chrenko[4.80] on thin BAs layers led him to the assumption that BAs has an indirect gap of 0.67 eV and a direct gap of 1.46 eV (see, also, Reference 4.120).

Boron arsenide is stable when heated in arsenic vapors; it is resistive to corrosion[4.68] and can be synthesized from the mixture of boron and arsenic in evacuated and sealed fused silica ampules held at 1073 K during periods of time from 12 to 50 hr.

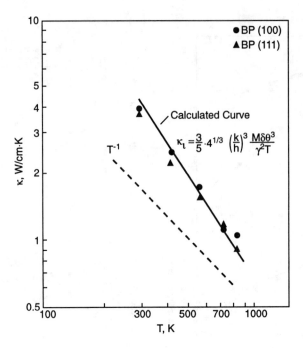

Figure 4.19 Temperature dependence of the thermal conductivity of the BP single-crystalline wafers grown on (110)Si and (111)Si substrates. The solid line represents the results of the lattice thermal conductivity calculation based on three-phonon processes; the dashed line represents the T^{-1} dependence. (After Y. Kumashiro, *J. Mater. Res.*, 5, 2933, 1990. With permission.)

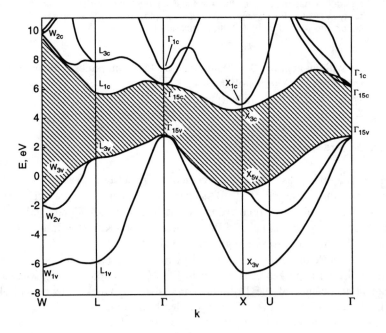

Figure 4.20 Band structure of boron arsenide calculated by a combined self-consistent OPW and pseudopotential method. (After D.J. Stackel, *Phys. Rev.*, B1, 3458, 1970. With permission.)

4.2.5. Aluminum Nitride

Aluminum nitride, AlN, has the molecular weight of 40.99, the average atomic weight of 20.49, the average atomic number of 10, and the enthalpy of formation of –240.2 kJ/mol. The melting point of AlN under a moderate nitrogen pressure (from uncorrected pyrometer data of Class[4.128]) is close to 3100 K. When being heated in vacuum, AlN dissociates at about 2200 K.[4.129,4.130] In water, at room temperature, AlN slowly decomposes evolving ammonia. In the air, AlN cannot be oxidized up to 1000 K. At T > 1000 K, a protective layer of Al_2O_3 or AlO_xN_y grows on the surface of AlN specimens and prevents oxidation of the bulk material until the temperature rises to 1350 K.[4.131,4.132]

AlN has two allotropic forms: (1) a low-pressure wurtzite type (space group C_{6v}^4 – $P6_3mc$), which may be presented as two interpenetrated hexagonal close-packed sublattices of aluminum and nitrogen displaced from each other along [0001] by 0.385c, with parameters (300 K) a = 0.311 nm and c = 0.498 nm[4.133-4.135] (Loughin et al.[4.139] and Ching and Harmon[4.140] reported a = 0.311 nm and c = 0.4978 nm); and (2) an allotrope with the rock salt structure and lattice parameter 0.3982 nm.[4.136] The transition from the former to the latter allotrope happens at 16 GPa (at 1800 K);[4.138] Ueno et al.[4.141] found from the *in situ* high-pressure X-ray experiments at room temperature that the transition pressure is 22.9 GPa.

The X-ray density of the low-pressure AlN allotrope is 3.255 g/cm³ at room temperature, hardness (of the ceramic samples) is close to 20 GPa,[4.142] and the thermal conductivity and coefficient of linear thermal expansion measured on polycrystals are equal to 2.60 W/cm·K and 4.6·10^{-6} 1/K, respectively.[4.143] A detailed investigation of thermal conductivity of single-crystalline, hot-pressed and cold-pressed AlN samples was performed by Slack,[4.80] who found that the thermal conductivity of AlN single crystals passes through a maximum of about 4 W/cm·K at 150 to 200 K, and the purest AlN single crystals have the room temperature magnitude close to 3.2 W/cm·K. Hot-pressed AlN samples have thermal conductivity of about 2.0 W/cm·K[4.144] and thermal expansion coefficient of 4.3·10^{-6} 1/K,[4.145] close to that of silicon between 293 and 473 K.[4.146] The Debye temperature of AlN is 1150 K,[4.155] its room temperature specific heat is equal to 0.732 J/g·K.

AlN has a relatively high mechanical strength — up to 500 MPa — even at temperatures as high as 1700 K.[4.147,4.148] An estimation of the AlN elastic characteristics by Chin et al.[4.155] gave 265 and 44.2 GPa for the longitudinal and transversal elastic constants, respectively.

Aluminum nitride is a direct semiconductor with the experimentally estimated minimum energy gap of 6.20 to 6.28 eV.[4.149,4.150] The AlN band structure calculated by Kobayashi et al.[4.151] is shown in Figure 4.21. The electron effective mass in AlN in accordance with Reference 4.155 is $(0.48 \pm 0.05)m_0$. Some parameters of the band structure of the AlN rock salt allotrope, the indirect minimum energy gap of 8.9 eV (transition Γ – X) and the direct energy gap of 11.6 eV (transition Γ – Γ), were estimated by Pandley et al.[4.160]

Almost all reports on the measurements of the AlN electrical conductivity vs. temperature relate to the extrinsic conductivity because of its wide energy gap. Being doped by selenium, AlN has n-type conductivity, and doping with zinc forms the p-type samples.[4.152] The substitutional oxygen at nitrogen sites produces a donor level at 2.2 eV below the conduction band bottom.[4.153] Oxygen plays, possibly, the key role in the transport properties of AlN.[4.79,4.154]

The high electrical resistivity of undopped AlN makes it difficult (if not impossible) to measure the Hall effect and estimate the carrier mobility. Chin et al.[4.155] estimated the phonon limited electron drift mobility in AlN to be about 2000 and 300 cm²/V·sec at 77 K and 300 K, respectively. Edwards et al.[4.156] found the hole mobility in a doped AlN single crystal equal to 14 cm²/V·sec at 290 K, but they pointed out that this magnitude has a substantial uncertainty.

The static dielectric constant of AlN is relatively small, which makes this material interesting for the microwave applications. Thorp et al.[4.157] reported $\varepsilon(0)$ = 9.2 which is close

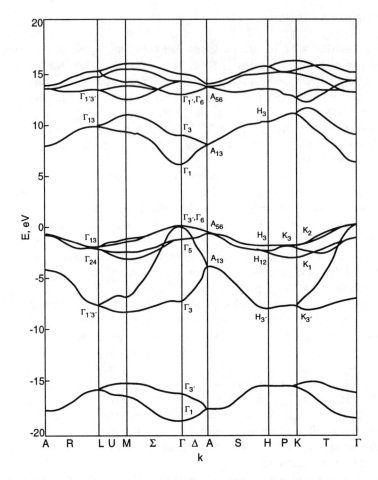

Figure 4.21 Band structure of aluminum nitride. Semi-empirical tight-binding method calculation. (After A. Kobayashi, O.F. Sankey, S.M. Volz, and J.D. Dow, *Phys. Rev.*, B28, 935, 1983. With permission.)

to the magnitude of 9.14 (300 K) found by Collins et al.[4.158] The high-frequency dielectric constant is 4.84,[4.158] in agreement with the magnitude of 4.77 in Reference 4.155. The AlN index of refraction close to 2.0 was reported by Ohuchi and French.[4.159] AlN is transparent in the visible and near-infrared regions.[4.164]

Until now, the only reported device fabricated with AlN as a semiconductor is an UV electroluminescence source — a two-terminal device made with a relatively low resistivity material.[4.161] The device emitted a broad spectrum from UV (215 nm) to greater than 500 nm. The other applications include surface acoustic wave devices, based on the AlN films, which utilize the AlN piezoelectric properties and the high surface acoustic wave velocity in AlN.[4.162,4.163] The other AlN applications are in integrating circuit packaging, ignition modules, RF and microwave packages, heat sinks, laser diode heat spreaders, and cutting tools.[4.143] A brief review of the AlN applications and the methods of its synthesis was compiled by Edgar.[4.39]

4.2.6. Aluminum Phosphide

Aluminum phosphide, AlP, has the molecular weight of 57.955, average atomic weight of 28.98, average atomic number of 14, and the enthalpy of formation of –166.6 kJ/mol.

At the normal pressure, a stable allotrope of AlP is one with the crystal structure of sphalerite, space group T_d^2 – $F\bar{4}3m$ (AlP-I phase). A high-pressure phase, AlP-II, has the rock salt structure. The transition from AlP-I to AlP-II at room temperature requires pressure from 14 to 17 GPa.[4.165]

The melting point of AlP is close to 2820 K, under equilibrium phosphorus vapor pressure.[4.166] The Mohs hardness of AlP is 5.5; the second-order elastic moduli estimated by Kagaya and Soma[4.167] are close to 160 GPa (c_{11}), 75 GPa (c_{12}), and 40 GPa (c_{44}).

The lattice constant of AlP under standard conditions reported by different authors varies from 0.5460 to 0.54635 nm. The latter result was reported by Bessolov et al.,[4.168] who performed X-ray diffraction measurements of the (pulverized) epitaxial film of AlP grown on a GaP substrate (mismatch 0.2%). The X-ray density of AlP is close to 2.36 g/cm^3; the magnitude reported by Caveney[4.169] is 2.40 g/cm^3. The specific heat of AlP at 400 K and constant pressure is 0.477 J/g·K.[4.170] An estimated magnitude of the AlP thermal conductivity[4.79] is 1.3 W/cm·K (Steigmeier[4.171] [see, also, Reference 4.172] suggests a room temperature magnitude of 0.9 W/cm·K).

Aluminum phosphide is an indirect semiconductor. A theoretical estimation of its minimum energy gap by Herman[4.173] has given a magnitude of 3.0 ± 0.3 eV. The results of measurements at 4K are 2.505 eV for the indirect and 3.63 eV for the direct energy gap; the direct energy gap at 300 K is 3.62 eV.[4.174] The band structure of AlP calculated by Huang and Ching is shown in Figure 4.22. Their data on the carrier effective mass are the following (in the units of m_o):

Electrons	Longitudinal		3.67
	Transversal		0.212
Holes	Heavy	Along [100]	0.513
	Heavy	Along [111]	1.372
	Light	Along [100]	0.211
	Light	Along [111]	0.145

The electron and hole mobility in AlP has been studied on thin films and on the single crystals with a relatively high carrier concentration. In accordance with Reid et al.,[4.175] the electron mobility at room temperature has a magnitude of 60 cm^2/V·sec. Wiley[4.176] reports the estimated hole mobility of 450 cm^2/V·sec. The available results of the AlP electrical conductivity and Hall constant measurements (see, e.g., References 4.175 and 4.177) show that the samples used have high impurity/defect level, so the data relate essentially to the extrinsic conductivity. Grimmeis et al.[4.177] found that the indirect energy gap at 293 K is equal to 2.42 eV; they report also that rectification, photo-emf, and electroluminescence were observed.

The static dielectric constant of AlP is 9.8; its high-frequency limit is close to 7.5. These magnitudes were derived from the index of refraction measurements (Reference 4.178, Figure 4.23).

Aluminum phosphide is readily decomposed by water (similar to boron phosphide); it actively reacts with acids and alkalis with the evolution of phosphine. The AlP powder of technical purity (99.5%) is available commercially ($10.80/gram — 1994).

4.2.7. Aluminum Arsenide

Aluminum arsenide, AlAs, has a molecular weight of 101.90, an average atomic weight of 50.95, the average atomic number of 23, and the enthalpy of formation of –148 kJ/mol.

AlAs crystallizes in the sphalerite structure with the lattice parameter of 0.5660 nm at 293 K.[4.179] The X-ray density of AlAs at 293 K is equal to 3.732 g/cm^3 (Adachi[4.182] found a magnitude of 3.760 g/cm^3). The average coefficient of linear thermal expansion in the range

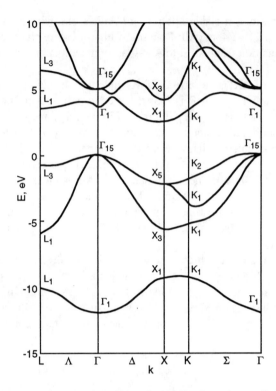

Figure 4.22 Band structure of aluminum phosphide calculated by the orthogonalized LCAO method. (After M. Huang and W.Y. Ching, *J. Phys. Chem. Solids*, 46, 977, 1985. With permission.)

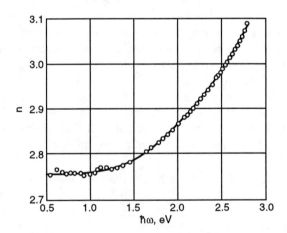

Figure 4.23 Room temperature index of refraction of AlP vs. photon energy. (After B. Monemar, *Solid State Comm.*, 8, 1295, 1970. With permission.)

from 300 to 1100 K, calculated from the data of Reference 4.179, is close to $5.2 \cdot 10^{-6}$ 1/K, which is substantially lower than the magnitude reported by Pashintsev and Sirota[4.180] (about $8 \cdot 10^{-6}$ 1/K).

The melting point of AlAs is 2013 ± 20 K,[4.181] its specific heat at constant pressure is close to 0.45 J/g·K, and its Debye temperature, calculated by Steigmeier,[4.171] is 417 K.

The microhardness of AlAs (50-g load) is close to 5.0 GPa (Reference 4.70, p. 94). The elastic moduli of AlAs estimated by Adachi[4.182] are $c_{11} = 120.2$ GPa, $c_{12} = 57.0$ GPa, and $c_{44} =$

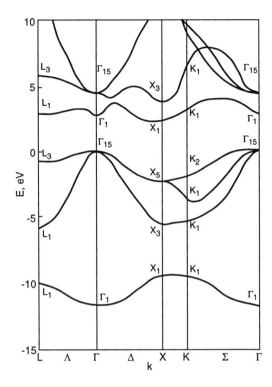

Figure 4.24 Band structure of aluminum arsenide. The orthogonalized LCAO method calculation. (After M. Huang and W.Y. Ching, *J. Phys. Chem. Solids*, 46, 977, 1985. With permission.)

58.9 GPa. The magnitude of c_{11} is in very good agreement with the estimation by Steigmeier[4.171] (122.0 GPa).

AlAs is an indirect semiconductor. Its minimum energy gap (transition $\Gamma_{15v} - X_{1c}$) is 2.229 eV and 2.153 eV at 4 K and 300 K, respectively.[4.174] Herman's[4.173] estimation suggests $E_{g,ind.} = 2.16 \pm 0.10$ eV, while Lee et al.[4.183] found $E_{g,ind.} = 2.363$ eV (295 K). The AlAs band structure calculated by Huang and Ching is shown in Figure 4.24. The effective mass of electrons and holes in AlAs is 0.11 m_o and 0.22 m_o, respectively.[4.170] Adachi[4.182] analyzed the available literature data and suggested $m_n^*(\Gamma) = 0.150$ m_o, the heavy hole effective mass, m_{ph}^*, from 0.409 m_o to 1.022 m_o, and the light hole effective mass, m_{pl}^*, from 0.109 m_o to 0.153 m_o depending on the crystallographic direction.

The electron mobility in AlAs cited in Reference 4.170 is 180 cm^2/V·sec at 300 K (see, also, Reference 4.184). This magnitude agrees with the results of Ettenberg et al.,[4.179] who studied the electric transport properties of the AlAs epitaxial layers and found the electron mobility varying from 75 to 290 cm^2/V·sec at 300 K, while at 77 K it is equal to 18 cm^2/V·sec.

The AlAs thermal conductivity at 300 K as cited by Leroux-Hugon et al.[4.185] is close to 0.84 W/cm·K. The magnitude presented in Reference 4.172 is almost ten times smaller.

The static dielectric constant of AlAs is 10.06, while extrapolation to the very high frequencies gives 8.16.[4.186] The index of refraction as reported in Reference 4.186 varies from 3.0 at 820 nm to 3.2 at 540 nm, in good agreement with the results of calculations by Campi and Papuzza.[4.187]

Although AlAs of the technical purity (99.5%) is available commercially ($15.00/gram — 1994), the process of polycrystal and especially single crystal growth of the semiconductor purity material still remains a very difficult task, first of all, because of the high arsenic vapor pressure at elevated temperatures and the low rate of the compound formation. In early experiments, AlAs was synthesized by melting the elemental components (with sublimation

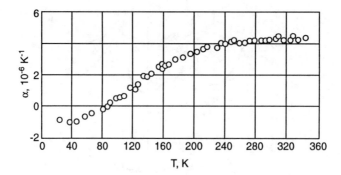

Figure 4.25 Temperature dependence of the AlSb coefficient of linear thermal expansion. (After Yu.I. Pashintsev and N.N. Sirota, *Dokl. Akad. Nauk SSSR*, 3, 38, 1959. With permission.)

of arsenic) in evacuated and sealed fused silica ampules. This was followed by grinding the product of synthesis and repeating the procedure of heating. The synthesized material was porous and inconvenient for the investigation of its properties. The two-temperature method used by Folberth and Oswald[4.188] permitted the experimenters to keep the vapor pressure (of arsenic) inside of the sealed ampule at a chosen level, while having the aluminum melted. Borshchevskii and Tretyakov[4.189] improved the method by application of vibrational stirring of the reaction components in the ampule (the applied frequency was 100 Hz).

Work with AlAs requires certain precautiousness as it reacts with water and with moist air. Covering of the inner walls of the fused silica ampules with a protective layer, e.g., of graphite, prevents formation of alumosilicates which substantially decreases the mechanical strength of the ampules.

4.2.8. Aluminum Antimonide

Aluminum antimonide, AlSb, has the molecular weight of 148.73, average atomic weight of 74.37, the average atomic number of 32, and the enthalpy of formation of –96.3 kJ/mol.

AlSb crystallizes into sphalerite structure with lattice parameter of 0.61355 nm[4.191] (Goryunova[4.125] recommends the magnitude of 0.61361 nm). Under the pressure of greater than 7.7 GPa, AlSb undergoes a transition into an orthorhombic[4.191] (or tetragonal, similar to white tin) structure with lattice parameters a = 0.5327 nm and c = 0.2892 nm (Reference 14.125, p. 97).

The X-ray density of the ambient pressure AlSb allotrope at room temperature is 4.277 g/cm^3; its pycnometric magnitude is 4.15 g/cm^3 (Reference 4.125, p. 105). In accordance with Reference 4.185, the coefficient of thermal expansion of AlSb is equal to $3.5 \cdot 10^{-6}$ 1/K at room temperature. Novikova and Abrikosov[4.194] found its magnitude slightly greater than $4 \cdot 10^{-6}$ 1/K (Figure 4.25). Osamura and Murakami[4.192] reported that AlSb melts congruently at 1338 K; this is in good agreement with the early data of Urazov,[4.193] who investigated the Al-Sb phase diagram and found only one compound — aluminum antimonide — with the melting point of 1333 K.

The heat capacity of AlSb was measured by Piesbergen.[4.195] The results of his measurements, together with the calculated Debye temperature, are presented in Table 4.1.

The AlSb microhardness is 4 GPa (Reference 4.70, p. 95). The elastic constants at 300 K, calculated by Bolef and Menes[4.196] from the results of the ultrasonic wave propagation experiments using the density magnitude of 4.36 g/cm^3, are c_{11} = 89.39, c_{12} = 44.29, and c_{44} = 41.55 GPa. The bulk modulus, $B = (c_{11} + 2c_{12})/3$, is 59.3 GPa. These figures are in satisfactory agreement with the results reported by McSkimin et al.[4.197] and Weil.[4.198] The AlSb elastic Debye temperature at 0 K, calculated by Steigmeier and Kudman,[4.171] is 292 K.

Table 4.1 The AlSb Atomic Heat Capacity (cal/g-atom·K) and Debye Temperature

T (K)	C_p	$C_p - C_v$	C_v	θ (K)
20	0.350	—	0.350	219
25	0.640	—	0.640	220
30	0.920	—	0.920	230
35	1.180	—	1.180	241
40	1.410	—	1.410	254
45	1.610	—	1.610	268
50	1.788	—	1.788	282
60	2.115	—	2.115	307
70	2.415	—	2.415	329
80	2.710	0.001	2.709	347
90	2.995	0.002	2.993	360
100	3.265	0.002	3.263	371
110	3.510	0.002	3.508	380
120	3.730	0.003	3.727	387
130	3.944	0.004	3.940	391
140	4.145	0.004	4.141	393
150	4.312	0.005	4.307	396
160	4.463	0.006	4.457	398
170	4.593	0.006	4.587	400
180	4.709	0.007	4.702	402
190	4.816	0.008	4.808	403
200	4.920	0.009	4.911	402
210	5.012	0.010	5.002	400
220	5.100	0.010	5.090	397
230	5.183	0.011	5.172	394
240	5.262	0.012	5.250	387
250	5.340	0.013	5.327	379
260	5.408	0.013	5.395	371
270	5.450	0.014	5.436	370
273.2	5.460	0.014	5.446	370

Adapted from U. Piesbergen, *Z. Naturforsch.*, 18A, 141, 1963.

Aluminum antimonide is an indirect semiconductor with a minimum energy gap (transition $\Gamma_{15v} - \Delta_{1c}$) from 1.615 eV at 300 K to 1.686 eV at 27 K.[4.199] The direct energy gap at the center of the Brillouin zone (transition $\Gamma_{15v} - \Gamma_{1c}$) varies from 2.300 eV at 295 K to 2.384 eV at 25 K.[4.200] The band structure of AlSb calculated by Huang and Ching[4.108] is shown in Figure 4.26. The electron effective mass parallel to [100] and corrected for nonparabolicity[4.201] is 1.8 m_o; the electron effective mass perpendicular to [100] is equal to 0.259 m_o. The estimated[4.107] hole effective mass is (1) heavy holes: $m_{ph}^*(\|[100]) = 0.336\ m_o$, $m_{ph}^*(\|[111]) = 0.872\ m_o$; (2) light holes: $m_{pl}^*(\|[100]) = 0.123\ m_o$ and $m_{pl}^*(\|[111]) = 0.091\ m_o$.

The electrical transport properties of AlSb were reported by Welker,[4.202] Reid and Willardson,[4.203] Nasledov and Slobodchikov,[4.204] Stirn and Becker,[4.205] and Glazov et al.[4.206] Because the hole mobility in AlSb at all temperatures is greater than that of electrons, the intrinsic material always possesses p-type conductivity. The AlSb thermal energy gap estimated from the intrinsic conductivity measurements above 1000 K is equal to 1.57 eV.[4.204] The n-type crystals may be prepared by doping AlSb with tellurium. The electron mobility in these crystals varies from 700 cm²/V·sec at 77 K to 200 cm²/V·sec at 295 K.[4.205] The hole mobility in a nondoped single crystal varies from 2000 cm²/V·sec at 77 K[4.205] to about 400 cm²/V·sec at 300 K.[4.203]

The electrical conductivity of AlSb in solid and liquid states[4.206] is increasing with temperature, in contrast with Ge, Si, and the other investigated diamond-like III-V compounds, which have a negative temperature coefficient of electrical conductivity in the liquid state. During the melting process, AlSb increases its electrical conductivity by a factor of about 60.

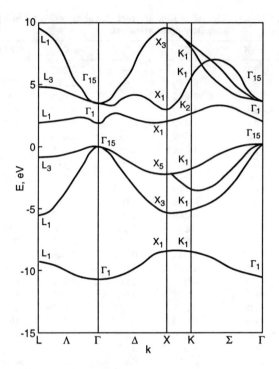

Figure 4.26 The AlSb band structure. The orthogonalized LCAO method calculation. (After M. Huang and W.Y. Ching, *J. Phys. Chem. Solids*, 46, 977, 1985. With permission.)

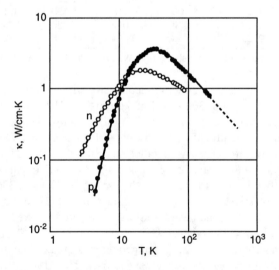

Figure 4.27 Temperature dependence of the thermal conductivity of an n-type and a p-type sample of aluminum antimonide. (After V.M. Mushdaba, A.Ya. Nashelskii, P.V. Tamarin, and S.S. Shalyt, *Fiz. Tverd. Tela*, 10, 2866, 1968; *Sov. Phys. Solid State*, 10, 2265, 1968. With permission.)

The thermal conductivity of AlSb at room temperature is 0.59 W/cm·K as cited in Reference 4.185, in satisfactory agreement with the data of Muzhdaba et al.[4.207] presented in Figure 4.27.

At room temperature, the index of refraction of AlSb changes from 3.445 at $\lambda = 1100$ nm to 2.995 at 20,000 nm, then passes through the region of anomalous dispersion between

30,000 and 31,300 nm pertaining to lattice absorption (the absorption coefficient is equal to 11.73 at 31,300 nm), and decreases again from about 12.5 (!) at 31,300 nm to 3.652 at 40,000 nm.[4.208,4.209] The static dielectric constant of AlSb at 300 K is 12.04; its high frequency limit is 10.24.[4.210]

AlSb is a diamagnetic with the room temperature molar magnetic susceptibility of about $-24 \cdot 10^{-5}$ SI units/g-mol (Reference 4.206, p. 87), which is virtually independent of temperature up to the AlSb melting point. During the melting process, the total magnetic susceptibility increases to the magnitude of about $-21 \cdot 10^{-5}$ SI units/g-mol, and in the liquid state, it slowly increases with temperature. Increase of magnetic susceptibility during the melting process and with the following temperature growth may be explained (in agreement with the results of the electrical conductivity measurements) by the increase of the free electron concentration and, consequently, increase of the electron paramagnetism. Among many investigated III-V diamond-like compounds, AlSb is the only one material with the semiconductor character of conductivity in the liquid state. An explanation for this specific of AlSb (and possibly some other Al-V compounds) has not been found thus far.

Being a diamond-like compound, AlSb decreases its specific volume during the transition from solid to liquid state.[4.206,4.211] But, while Si and Ge decrease their volume by about 10% and 4.75%, respectively, the AlSb volume decrease is almost 13%, greater than any other investigated tetrahedral substance. The relative change of volume of AlSb correlates with the change of electrical conductivity during the melting process.

The vapor pressure of antimony at the melting temperature of AlSb is relatively low (about 4 kPa); therefore, the compound can be synthesized in evacuated and sealed ampules. As with other aluminum-contained compounds, precautionary measures have to be undertaken to prevent the reaction of the ampule material with the AlSb components.

The AlSb single crystals can be grown by the Czochralski technique in an inert gas (argon) atmosphere (see, e.g., Reference 4.212). Aluminum antimonide shot with the purity of 99.999% is available commercially for $197.00/100 g (1994).

4.2.9. Gallium Nitride

Gallium nitride, GaN, has the molecular weight of 83.727, average atomic weight of 41.86, the average atomic number of 19, and the enthalpy of formation of -104.2 kJ/mol.

GaN crystallizes in the wurtzite structure (space group $P6_3mc$) with the lattice parameters a = 0.3189 nm and c = 0.5182 nm[4.213] (Lagerstedt and Monemar[4.214] found that the lattice parameters of the GaN crystals vary in the limits 0.3160 to 0.3190 nm [a] and 0.5125 to 0.5190 nm [c]). The GaN pycnometric density is 6.1 g/cm^3. The GaN average coefficient of thermal expansion between 273 and 900 K, calculated using the lattice parameter measurements by Maruska and Tietjen,[4.215] is close to $5.4 \cdot 10^{-6}$ 1/K and $7.2 \cdot 10^{-6}$ 1/K along the a and c axes, respectively. These magnitudes are noticeably higher in comparison with the data of Sheleg and Savastenko[4.216] (Figure 4.28). Ross et al.[4.213] cite $\alpha(a) = 5.6 \cdot 10^{-6}$ 1/K and $\alpha(c) = 3.2 \cdot 10^{-6}$ 1/K.

In the atmosphere of nitrogen or ammonia, GaN melts at about 1770 K. In vacuum, it decomposes when temperature approaches 1270 K[4.217] (or 1320 K[4.218]). GaN dissolves very slowly in acids and alkalis (Reference 4.70, p. 98).

Until now, the bulk single crystals of GaN have not been grown, so its microhardness is still unknown. The GaN elastic moduli calculated from the X-ray diffraction measurements are $c_{11} = 296$, $c_{12} = 130$, $c_{13} = 158$, $c_{33} = 267$, and $c_{44} = 24.1$ GPa (Reference 4.219, p. 89).

GaN is a direct semiconductor with the energy gap at room temperature of 3.4 eV.[4.215] Monemar[4.220] found the magnitude of 3.503 eV at 1.6 K (3.504 eV being extrapolated to 0 K) and 3.44 eV at 300 K. The data of Reference 4.220 in the range of temperatures below 295 K may be presented analytically by the equation

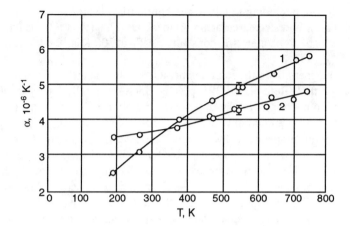

Figure 4.28 Temperature dependence of the GaN coefficient of linear thermal expansion across (curve 1) and along (curve 2) of the c axis. (After A.V. Sheleg and V.A. Savastenko, *Vesti Akud. Nauk BSSR, Ser. Fiz. Materm. Nauk*, 3, 126, 1976. With permission.)

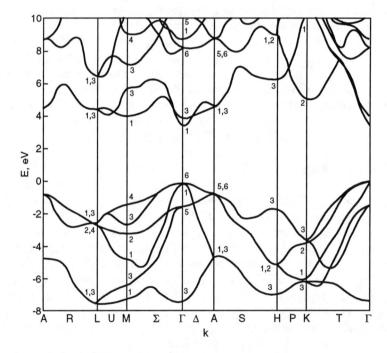

Figure 4.29 The gallium nitride band structure. The empirical pseudopotential method calculation. (After S. Bloom, G. Harbeke, E. Meier, and I.B. Ortenburger, *Phys. Stat. Sol.*, B66, 161, 1974. With permission.)

$$E_g(T, K) = 3.504 - 5.08 \cdot 10^{-4} T^2/(996 - T) \; (eV)$$

The earlier experiments by Kauer and Rabenau[4.221] gave the GaN energy gap of 3.25 cV at 300 K. The GaN band structure calculated by Bloom et al.[4.222] is shown in Figure 4.29. At room temperature, the electron effective mass of GaN is $(0.27 \pm 0.06)m_o$[4.223] and that of holes is $(0.8 \pm 0.2)m_o$.[4.224]

The electric transport properties of GaN have been a subject of numerous research reports. The majority of these reports contain information on the n-type GaN. This conductivity type

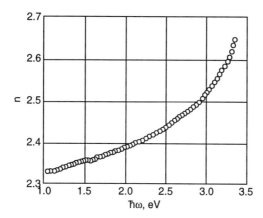

Figure 4.30 The GaN index of refraction vs. photon energy at 300 K and with $\vec{E} \perp c$. (After E. Ejder, *Phys. Stat. Sol.*, A6, K39, 1971. With permission.)

is a consequence of the nitrogen vacancies presented in the undoped material.[4.215] Seifert et al.[4.226] believe that the oxygen impurity may be an important factor in the electron conductivity domination.

Ilegems and Montgomery[4.227] found that the electron Hall mobility in GaN passes through a maximum of about 800 cm^2/V·sec near 150 K and its room temperature magnitude is close to 220 cm^2/V·sec. The highest electron Hall mobility reached on the epitaxial GaN films is 900 cm^2/V·sec at 300 K in agreement with the results of the mobility calculations by Gelmont et al.[4.229] and Chin et al.[4.230]

Only relatively recently, the results of successful experiments directed to production of the p-type GaN have been reported. Amano et al.[4.231] prepared Mg-doped epitaxial layers of p-GaN with the room temperature mobility of about 10 cm^2/V·sec and volume resistivity of 30 to 40 Ohm·cm. Doping permits variation of conductivity of GaN in wide limits.[4.39]

The thermal conductivity of GaN was reported by Sichel and Pankove.[4.232] With the temperature increase, the thermal conductivity (along the c axis) increases, passes through a smooth maximum of about 1.3 W/cm·K near 200 K, and decreases to a magnitude of 1.1 W/cm·K at room temperature. The thermal conductivity of large GaN single crystals may be expected to be much higher.

The molar magnetic susceptibility of GaN is $-34.9 \cdot 10^{-5}$ SI units/g-mol (Reference 4.125, p. 106). The static dielectric constant (room temperature) is 10.4 and 9.5 along and across the c axis, respectively.[4.233] The high-frequency magnitude is 5.8 and 5.35 along and across the c axis, respectively.[4.233,4.234]

The index of refraction, n, of GaN across the c axis was measured by Ejder[4.235] (Figure 4.30) using the interference method. Ejder found the magnitude of n along the c axis slightly (1 to 2%) lower, which contradicts the results of the dielectric constant measurements.[4.233,4.234]

Gallium nitride forms continuous solid solutions with AlN and InN. It, thus, creates an opportunity to tailor a material with a desirable energy gap between 2 and 6 eV. GaN is a promising material for the gigahertz power devices due to the high magnitude of the saturated electron drift velocity.[4.236] As a direct band gap semiconductor, GaN can be used to make optoelectronic devices with higher efficiency in comparison with those from materials such as diamond or silicon carbide. Availability of the p-type GaN makes it possible to produce p-n junctions and light-emitting diodes for the violet, blue, green, and yellow parts of the spectrum.[4.231] The electrical, surface, and optical properties of the high quality thin films of GaN are analyzed in References 4.231 and 4.238 to 4.242. A comprehensive review of the GaN properties is presented in Reference 4.243.

Gallium nitride powder may be synthesized by passing a stream of NH_4 above melted gallium at 1470 K,[4.244] or by the decomposition of $(NH_4)GaF_6$.[4.245] The bulk crystal growth technique is up to now unknown. The GaN films may be grown by the reactive sputtering,[4.246] chemical vapor deposition (CVD),[4.247] metal-organic CVD,[4.248] plasma-enhanced CVD,[4.249,4.250] reactive ion beam deposition,[4.251] and molecular beam epitaxy.[4.225,4.252]

4.2.10. Gallium Phosphide

Gallium phosphide, GaP, has a molecular weight of 100.69, the average atomic weight of 50.35, the average atomic number of 23, and the enthalpy of formation of −72.0 kJ/mol.

At the pressure below 21 GPa to 25 GPa, the crystal structure of GaP is one of sphalerite (space group T_d^2-$F\bar{4}3m$) with the lattice parameter at room temperature of 0.5447 ± 0.00002 nm. At higher pressures, GaP transforms into an allotrope with the rock salt structure.[4.268-4.270] Bessolov et al.[4.260] found a = 0.54505 ± 0.00002 nm at normal pressure and room temperature in the tetrahedral GaP allotrope. The pycnometric density of GaP at 300 K is 4.10 g/cm³ (Reference 4.125, p. 105), and the X-ray density is between 4.129 and 4.136 g/cm³. The lattice parameter and thermal expansion of GaP were reported in a large number of works (see, e.g., References 4.4 and 4.253 to 4.263). A theoretical consideration of the low-temperature thermal expansion was done by Soma,[4.264] who came to the conclusion that in contrast with many a material with tetrahedral coordination of atoms in the crystal lattice, GaP does not have a low-temperature range in which its coefficient of thermal expansion is negative. Soma and Deus et al.[4.263] assume that this is the consequence of the dominating covalent component of the interatomic bonds in GaP. But silicon, e.g., has virtually only the covalent interatomic bonds, and it has a wide (from about 4 K to 120 K) range of temperatures where its thermal expansion is negative, while diamond, another covalent material, does not possess a negative expansion at any temperature. On the other hand, the AgI sphalerite allotrope has the negative coefficient of thermal expansion from the lowest investigated temperatures up to 330 K. The only common feature for all diamond-like crystals is their crystal structure and formation of the sp^3-hybrid bonds; therefore, it is reasonable to assume that at low-enough temperatures, all diamond-like crystals pass through a temperature range with the negative coefficient of thermal expansion. Development of sufficiently sensitive dilatometers for very low temperatures (see, e.g., Ackerman and Anderson's[4.265] description of a high-sensitivity dilatometer based on a SQUID detector) may shed some light on this question. The temperature dependence of the GaP coefficient of thermal expansion is shown in Figure 4.31a (after Reference 4.263), together with some other literature data. Pierron et al.[4.349] recommend the magnitude of $(5.81 \pm 0.13) \cdot 10^{-6}$ 1/K for the GaP coefficient of linear thermal expansion at room temperature.

The melting point of GaP is 1743 K[4.266,4.271] under the phosphorus vapor pressure of about 30 MPa[4.276] (Imar et al.[4.267] report T_m = 1730 ± 5 K).

At elevated temperatures, solid stoichiometric GaP can dissolve up to 0.04 at% of phosphorus and up to 0.05 at% of gallium. The highest solidus temperature of GaP relates to the composition $Ga_{1.0002}P_{0.9998}$.[4.272-4.274]

The room-temperature heat capacity of GaP at constant pressure is 21.9 J/mol·K;[4.275] the Debye temperature (at 300 K) is 445 K[4.277] (or 446 K[4.89]).

The hardness of GaP is 5 units of Mohs' scale, the Knoop microhardness with 25-g load is 9450 ± 1550 MPa, depending on the indentor orientation, and the Vickers microhardness at 50-g load is 9.40 ± 0.35 GPa (Reference 4.70, p. 98). The second-order elastic moduli of GaP at 300 K are c_{11} = 140.5 GPa[4.278] (138.2 GPa in Reference 4.89), c_{12} = 62.03 GPa, and C_{44} = 70.33 GPa.[4.278] The elastic moduli vs. temperature are shown in Figure 4.31b (after Reference 4.279). The Debye temperature, calculated from the elastic data, is 320.5 K.[4.89]

Gallium phosphide is an indirect semiconductor. The minimum indirect energy gap at 300 K is 2.272 eV (transition $\Gamma_{15v} - \Delta_{1c}$), and extrapolation to 0 K gives a magnitude of

BINARY IV-IV AND III-V SEMICONDUCTORS

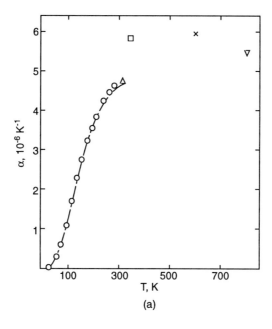

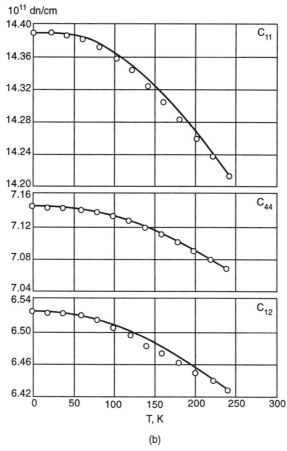

Figure 4.31 (a) Temperature dependence of the coefficient of linear thermal expansion of GaP from the data of different authors. (After P. Deus, U. Voland, and H. Oettel, *Exp. Tech. Phys.*, 32, 433, 1984. With permission.) (b) Temperature dependence of the GaP second-order elastic moduli. (After W.F. Boyle and R.J. Sladek, *Phys. Rev.*, B11, 2933, 1975. With permission.)

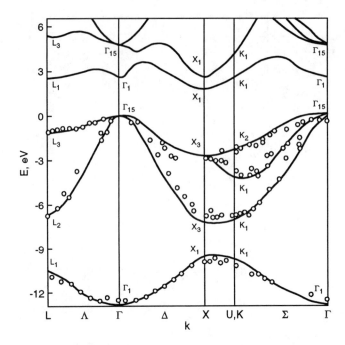

Figure 4.32 Band structure of GaP. Solid curves: the results of calculation by the pseudopotential method (spin-orbit interaction is excluded). (After J.R. Chelikowsky and M.L. Cohen, *Phys. Rev.*, B14, 556, 1976. With permission.) The circles represent data from angle-resolved photoemission. (See also F. Solal, G. Jezequel, F. Houzay, A. Barski, and R. Pinchaux, *Solid State Comm.*, 52, 37, 1984 and A.C. Sharma, N.M. Ravindra, S. Auluck, and V.K. Srivastava, *Phys. Stat. Sol.*, B120, 715, 1983.)

2.350 eV.[4.280] These figures, received from the wavelength modulation transmission measurements, are in very good agreement with the earlier data by Panish and Casey.[4.281] A direct energy gap (transition $\Gamma_{15v} - \Gamma_{1c}$) evaluated by Nelson et al.[4.282] from the photoconductivity measurements is 2.780 eV at 300 K; its extrapolation to 0 K gives 2.895 eV.

The GaP band structure was calculated by Chelikowsky and Cohen[4.283] and was estimated from the photoemission measurements by Williams et al.[4.284] The data of these two reports agree with each other with discrepancies in the limits 5% to 8%. The band structure after Reference 4.283 is presented in Figure 4.32. The effective mass of electrons in GaP parallel and perpendicular to [100] is equal to 0.254 m_o and 4.8 m_o, respectively.[4.285] The effective mass of heavy holes parallel to [111] is 0.67 m_o, and the light holes in the same direction have the effective mass of 0.17 m_o; the spin-orbit split-off band effective mass is 0.4649m_o.[4.287]

Up to almost 800 K, high-purity GaP possesses extrinsic conductivity. At room temperature, the Hall mobility of electrons and holes is 150 cm^2/V·sec to 160 cm^2/V·sec and 120 cm^2/V·sec to 135 cm^2/V·sec, respectively (References 4.170 and 4.219, p. 96). The hole drift mobility in GaP was calculated by Takeda et al.,[4.288] who found its magnitude close to $4 \cdot 10^3$ cm^2/V·sec at 77 K and to 70 cm^2/V·sec at 300 K. Doping high-purity GaP with shallow donors and acceptors forms so-called semi-insulating gallium phosphide with the volume resistivity of the order of 10^8 to 10^{11} Ohm·cm (Reference 4.219, p. 96).

The thermal conductivity of GaP was measured by Muzhdaba et al.[4.207] (Figure 4.33) and analyzed by Slack.[4.79] It passes through a maximum of about 7 W/cm·K near 30 K and decreases to 1 W/cm·K at 300 K. Slack includes GaP in the group of the adamantine materials with high thermal conductivity (the smallest in this group).

GaP is a diamagnetic with the molar magnetic susceptibility of $-33.4 \cdot 10^{-5}$ SI units per gram-mol (Reference 4.125, p. 105). The static dielectric constant of GaP at 300 K is equal

BINARY IV-IV AND III-V SEMICONDUCTORS

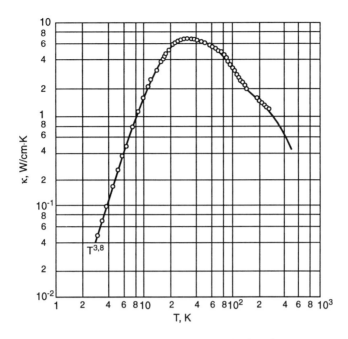

Figure 4.33 Temperature dependence of the GaP thermal conductivity for a p-type sample. (After V.M. Muzhdaba, A.Ya. Nashelskii, P.V. Tamarin, and S.S. Shalyt, *Fiz. Tverd. Tela*, 10, 2866, 1968; *Soviet Phys. Solid State*, 10, 2265, 1968. With permission.)

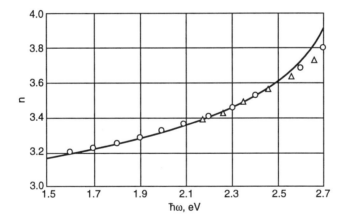

Figure 4.34 Spectral distribution of the GaP index of refraction. (After D. Campi and C. Papuzza, *J. Appl. Phys.*, 57, 1305, 1985. With permission.) (See also H. Burkhard, H.W. Dinges, and E. Kuphal, *J. Appl. Phys.*, 53, 655, 1982.)

to 11.11, decreasing to 10.86 at 75.7 K.[4.289] The high-frequency dielectric constant at room temperature is 9.11. The dependence of the GaP index of refraction on the photon energy is shown in Figure 4.34.

When an electric current passes through GaP crystals, electroluminescence is observed with the emitted light of an orange to dark-red color (Reference 4.70, p. 99).

GaP of 99.999% purity in the form of amber lumps is available commercially ($26.50/gram — 1994).

The most prospective applications of gallium phosphide are in the high-temperature solid-state electronic, optoelectronic, and magneto-optical devices.[4.290]

4.2.11. Gallium Arsenide

Gallium arsenide, GaAs, has a molecular weight of 144.64, the average atomic weight of 72.32, the average atomic number of 32, and the enthalpy of formation of –83.7 kJ/mol.

In many features, GaAs is similar to germanium, but in certain aspects its properties are superior to those of Ge, primarily in the direct energy gap transition. There are two GaAs allotropes known: the normal-pressure one with the sphalerite structure, stable to the pressure of 17.2 GPa; and the high-pressure allotrope, stable above this pressure, with a distorted rock salt structure (space group is, probably, D_{2h}^1-Pmmm).[4.268] The lattice parameter of the normal pressure GaAs at 300 K is 0.565325 nm[4.291] (cf. a = 0.56534 nm[4.292]). The lattice parameter of GaAs noticeably depends on the deviation from its stoichiometric composition (Figure 4.35, from Reference 4.293); this may be explained, in accordance with the T-x projection of the GaAs phase diagram (Figure 4.36), that solubility of the elemental components in GaAs is very small, and the excessive atoms occupy the interstitial positions. The thermodynamic calculations by Ivashchenko et al.[4.274] show that the solubility of As in GaAs below 1100 K is virtually equal to zero, while the solubility of gallium does not exceed 0.01 at%. The T-x, T-p, and p-x projections of the GaAs phase diagram are presented in Figure 4.36.[4.294]

GaAs thermal expansion follows the patterns of many other adamantine substances. At low temperatures (between about 10 and 55 K[4.295,4.296]), the coefficient of linear thermal expansion is negative, with the maximum absolute magnitude of $5.0 \cdot 10^{-7}$ 1/K at about 40 K;[4.295] it becomes zero at 55 K and remains positive at higher temperatures. Above 80 K, the data of References 4.95 and 4.96 are in satisfactory agreement with the results of Reference 4.297. Below 10 K, the oscillations of atoms in the crystal lattice of GaAs are essentially harmonic and therefore the coefficient of thermal expansion is very close to zero (see Figure 4.35b[4.296]). At 300 K its reported magnitude varies between $5.39 \cdot 10^{-6}$ 1/K,[4.185] $6.0 \cdot 10^{-6}$ 1/K,[4.293] and $6.86 \cdot 10^{-6}$ 1/K.[4.349]

At atmospheric pressure, GaAs melts congruently at 1510 ± 3 K. A relatively low (98 kPa) saturated vapor pressure upon GaAs at the melting point simplifies its synthesis in the evacuated fused silica ampules and single crystal growth from the GaAs melt encapsulated by liquid boron sesquioxide by the Czochralski method. The GaAs specific heat at room temperature is 0.326 J/g·K,[4.298] and the Debye temperature at 300 K is 344 K.[4.171] The temperature dependence of the GaAs heat capacity and Debye temperature is presented in Table 4.2 (from Reference 4.195).

Mechanical and elastic properties of GaAs attract particular interest in connection with certain applications, first of all in electronic devices and lasers. There are many indications that dislocations introduced in the process of fabrication of solid-state lasers decrease their lifetime and worsen their performance. Information regarding deformation of laser materials may be used in the fabrication process to decrease the defect formation. The Mohs hardness of GaAs is 4.5, the square pyramid microhardness (50-g load) is equal to 7.50 GPa (Reference 4.125, p. 105), and the Knoop microhardness (25-g load), reported by Wolff et al.,[4.344] is 7.50 ± 0.42 GPa depending on the crystal orientation regarding the indentor axes and loading direction. Roberts et al.[4.345] performed a detailed investigation of the GaAs Knoop hardness as a function of the indentor orientation regarding the crystal and temperature (from 293 to 713 K). Swaminathan et al.[4.303] studied the effect of doping on the microhardness of the active layers used in GaAs injection lasers. They found that the n-type doping (with Te) changes the Knoop microhardness to H = 6.40 ± 0.14 GPa, while p-type doping (with Ge) changes it to H = 5.98 ± 0.10 GPa. The elastic parameters of GaAs were reported in several publications and the data are collected in Table 4.3.

The stress-strain correlation on undoped and doped GaAs single crystals at elevated temperatures was studied by Djemel and Costaing,[4.3-4] who found that it strongly depends on the direction of compression; the impurity hardening was caused by Si, and Te doping softened

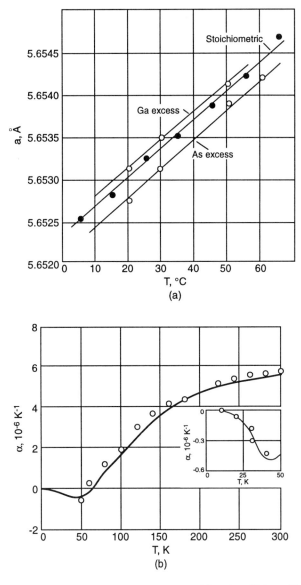

Figure 4.35 Temperature dependence of the lattice parameter (a) and coefficient of linear thermal expansion (b) of gallium arsenide. (After (a) M.E. Straumanis and C.D. Kim, *Acta Cryst.*, 19, 256, 1965 and (b) T. Soma, J. Sato, and H. Matsuo, *Solid State Commun.*, 42, 889, 1982. With permission.)

the GaAs samples. These results are in agreement with the reports[4.305-4.307] (see, also, Reference 4.308). In accordance with Grigor'yeva,[4.309] the GaAs Young's modulus is equal to 9.1 TPa and its shear modulus is 3.6 TPa. The Poisson ratio for GaAs is 0.29, the microbreaking strength is 1.89 GPa, and the brittleness criterion is equal to 3.0 (Reference 4.70, p. 99).

Gallium arsenide is a direct gap semiconductor. Its minimum energy gap is from 1.35 eV[4.310] to 1.45 eV (Reference 4.125, p. 105) at 300 K. Extrapolation of the electrical measurements data to 0 K gives a magnitude of 1.604 eV.[4.311] From optical measurements, the GaAs minimum energy gap (transition $\Gamma_{8v} - \Gamma_{6c}$) is 1.51914 eV at 0 K (extrapolation[4.312]) and 1.424 eV at 300 K.[4.313] The band structure of GaAs has been calculated by Chelikowsky and Cohen[4.283] (Figure 4.37) and investigated experimentally by Williams et al.[4.284] The data of these two reports are in reasonable agreement. The density of state electron effective mass

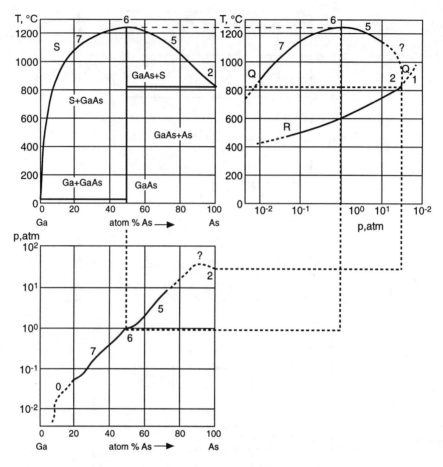

Figure 4.36 The temperature-composition, T-x, temperature-pressure, T-p, and pressure-composition, p-x, projections of the Ga-As phase diagram. (After J. van den Boomgaard and K. Schol, *Philips Res. Rep.*, 12, 127, 1957. With permission.)

in GaAs is 0.063 m_o and 0.067 m_o at 300 K and 0 K, respectively.[4.298] These magnitudes are in good agreement with the mass of 0.0650 m_o calculated by Lindemann et al.[4.414] from the cyclotron resonance data. The density-of-state hole effective mass at 300 K is 0.50 m_o for the heavy and 0.076 m_o for the light holes; its magnitude for the spin-orbit mass is 0.145 m_o, while the combined magnitude of the heavy and light hole bands density of state effective mass is 0.53 m_o.[4.298]

The intrinsic carrier concentration, n_i, in GaAs, in accordance with Reference 4.219 (p. 105), may be described as

$$n_i(T) = 1.05 \cdot 10^{16} \; T^{3/2} \exp(-1.604/2k_BT)$$

(T in K, k_BT in eV) for the range from 33 K to 475 K. At 1000 K, n_i is close to $3 \cdot 10^{16}$ cm^{-3}. At room temperature, $n_i = 2.1 \cdot 10^6$ cm^{-3} [4.317] with intrinsic resistivity equal to 10^8 Ohm·cm.

The data on the carrier mobility in GaAs vary in the wide limits depending on methods of synthesis, defect content, measurement methods, etc. Hilsum and Rose-Innes[4.290] suggested the magnitudes between 400 and 8500 cm^2/V·sec at room temperature for electrons. Blakemore[4.298] reported the magnitudes of 8000 cm^2/V·sec for electrons and 320 cm^2/V·sec for holes. Peng[4.315] and Bebney and Jay[4.316] studied correlation between mobility and the carrier concentration in the liquid-phase epitaxy-grown GaAs and have found that the n-type

Table 4.2 The GaAs Atomic Heat Capacity (cal/g-atom·K) and Debye Temperature

T (K)	C_p	$C_p - C_v$	C_v	θ (K)
12	0.045	—	0.045	261
15	0.094	—	0.094	255
20	0.239	—	0.239	249
25	0.445	—	0.445	251
30	0.670	—	0.670	260
35	0.906	—	0.906	270
40	1.130	—	1.130	281
45	1.350	—	1.350	292
50	1.562	—	1.562	302
60	1.975	—	1.975	320
70	2.380	0.001	2.379	333
80	2.763	0.001	2.762	342
90	3.122	0.002	3.120	348
100	3.448	0.002	3.446	351
110	3.730	0.003	3.727	355
120	3.978	0.004	3.974	357
130	4.195	0.005	4.190	358
140	4.397	0.006	4.391	358
150	4.558	0.007	4.551	359
160	4.697	0.008	4.689	360
170	4.815	0.009	4.806	361
180	4.920	0.010	4.910	361
190	5.020	0.011	5.009	360
200	5.112	0.012	5.100	358
210	5.193	0.013	5.180	356
220	5.267	0.014	5.253	353
230	5.327	0.015	5.312	352
240	5.377	0.016	5.361	352
250	5.420	0.017	5.403	352
260	5.450	0.018	5.432	356
270	5.478	0.019	5.459	359
273.2	5.481	0.019	5.462	362

Adapted from U. Piesbergen, *Z. Naturforsch.*, 18A, 141, 163.

Table 4.3 The GaAs Elastic Parameters

c_{11} (GPa)	c_{12} (GPa)	c_{44} (GPa)	Compressibility in 10^{-12} m²/N	Bulk modulus GPa	Ref.
119.0	53.8	59.5			4.344
118.8					4.89
			12.97		4.195
118.88	53.8	59.4			4.301
118.1[a]	53.2[a]	59.4[a]		74.8[a]	4.302
122.1[b]	56.6[b]	59.9[b]		78.4[b]	4.302
122.6[c]	57.1[c]	60.0[c]		78.9[c]	4.302

[a] At 300 K.
[b] At 77.4 K.
[c] At 0 K (extrapolated).

films have electron mobility between about 2000 and 7000 cm²/V·sec in the range of concentrations from 10^{18} to 10^{15} cm^{-3}, respectively. The electron mobilities in excess of $2 \cdot 10^5$ cm²/V·sec have been attained by Abrokwah et al.[4.398] on the high-purity GaAs thin films. The hole mobility lies in the limits from about 100 to 300 cm²/V·sec in the range of the hole concentration from $4 \cdot 10^{18}$ to $1 \cdot 10^{17}$ cm^{-3}, respectively. Even in GaAs polycrystals (see, e.g., Reference 4.318), the electron mobility reaches 4000 cm²/V·sec. The data on the GaAs carrier

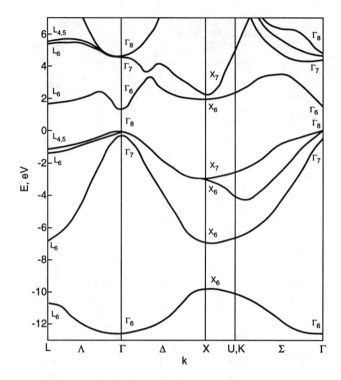

Figure 4.37 The GaAs band structure obtained by the nonlocal pseudopotential calculation method. (After J.R. Chelikowsky and M.L. Cohen, *Phys. Rev.*, B14, 556, 1976. With permission.)

mobility vs. temperature from References 4.319 and 4.320 are presented in Figures 4.38 and 4.39. The GaAs electrical conductivity in the liquid state has been studied by Glazov et al.[4.206] During the melting process, the conductivity increases by about 26 times; the GaAs melt behaves as a metal; namely, the conductivity decreases with the temperature increase (Figure 4.40).

Thermal conductivity, κ, of GaAs has been the subject of many reports. The room temperature magnitude of 0.35 W/cm·K was reported in Reference 4.297; the Ioffe[4.321] and Zaslavskii et al.[4.322] magnitude (0.52 W/cm·K) was higher than that of Leroux-Hugon et al.[4.185] (κ = 0.46 W/cm·K); Carlson et al.[4.323] found κ = 0.54 W/cm·K. Analysis of the data of various authors, based on description of the samples used and the measurement techniques, gives reason to consider 0.46 W/cm·K as the adequate room-temperature magnitude of κ for GaAs single crystals with the lowest defect content and highest electrical resistivity. Dependence of κ on impurity concentration at different temperatures is shown in Figure 4.41 (after Holland[4.324]).

GaAs is a diamagnetic with the molar magnetic susceptibility of $-40.7 \cdot 10^{-5}$ SI units/g-mol.[4.325] Glazov et al.[4.206] studied the temperature dependence of the GaAs magnetic susceptibility. The volume (per 1 cm³) GaAs magnetic susceptibility (in the units of Gauss system) is shown in Figure 4.42. In accordance with the Blakemore[4.298] recommendations, the static dielectric constant of GaAs may be described as $\varepsilon(0) = 12.40(1 + 1.2 \cdot 10^{-4}\ T)$; the high-frequency limit $\varepsilon(\infty) = 10.60(1 + 9.0 \cdot 10^{-5}\ T)$ (both at temperatures below 600 K). This gives the index of refraction of $n = 3.255(1 + 4.5 \cdot 10^{-5}\ T)$. From the analysis of several reports, Blakemore concludes that in the infrared part (0.3 to 1.4 eV), the index of refraction may be described as

$$n = \left[7.10 + 3.78/\left(1 - 0.18 E^2\right)\right]^{1/2}$$

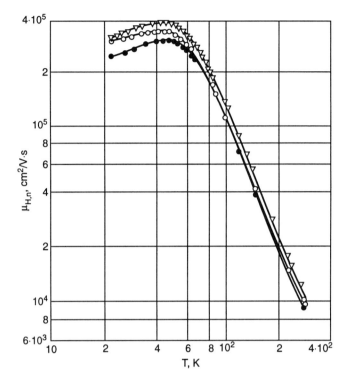

Figure 4.38 The Hall mobility of electrons in three high-purity GaAs epitaxial layers with the 300 K electron concentration close to $3 \cdot 10^{13}$ cm^{-3}. Triangles and circles — experimental data; solid curves — results of calculation. (After L. Lin, Y. Lin, X. Zhong, Y. Zhang, and H. Li, *J. Cryst. Growth*, 56, 344, 1982. With permission.)

where E is the photon energy in eV. The GaAs index of refraction in the range from 1 eV to 1.7 eV has been studied by Campi and Papuzza[4.326] (Figure 4.43). An estimation of n from the data of Aspnes and Studna[4.327] on the GaAs dielectric constant (Figure 4.44) shows a consistency of $n = n(\hbar\omega)$ from their results with the magnitudes for lower photon energies.

GaAs is now, probably, the most important material for electrooptical devices and consequently the most studied among the binary diamond-like semiconductors. Several international meetings have been devoted exclusively to reports and discussions of the GaAs properties, technology of the crystal and film preparation, and technology and parameters of the electrooptical devices (first of all, lasers built using GaAs).

The interaction of GaAs with metal films that may be used for the gate metallization and ohmic contacts has been reported by Tsai and Williams,[4.328] Lin et al.,[4.329] and Sands et al.[4.330] The zero-dimensional intrinsic defects in GaAs crystal lattice, such as gallium vacancies and arsenic antisites, serve as the medium for and enhance the diffusion of dopants and impurities in both bulk crystals and thin layers. A thermodynamic discussion of this process was presented by Iguchi,[4.331] who analyzed diffusion of Si, Zn, and Be through the above-mentioned defects, taking into consideration as well the crystal growth methods and surface-localized point defects. Formation of the linear defects, first of all dislocations, in the process of the GaAs crystal growth by the Czochralski method is still one of the unsolved problems in application of the method. Lambropoulos[4.332] has discussed the correlation between plastic and elastic stress in the growing crystal and the dislocation density in it. Control of the thermal stress anisotropy in the growing GaAs crystals is a way to decrease noticeably the dislocation density. Heteroepitaxial growth of GaAs on Si is also followed by the dislocation formation due to the difference in the thermal expansion between GaAs and Si and a resulting thermal stress generation. During the cooling process from the deposition to room temperature, density

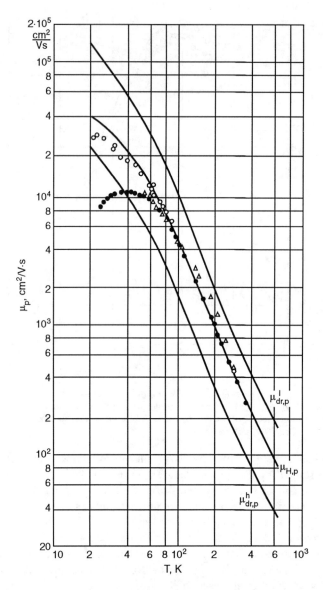

Figure 4.39 Temperature dependence of the hole mobility in GaAs. The upper curve — the light hole drift mobility; the middle curve — the Hall mobility; the lower curve — the heavy hole drift mobility. The triangles and circles are experimental data. (After P. Leyral, F. Litty, G. Bremond, A. Nonailhat, and G. Guillot, *Semi-Insulating III–V Materials,* Evian, 1982, S. Macram-Ebeid and B. Tuck, Eds., Nantwich-Shiva Publishing, 1982.)

of dislocation may have a 10^4-fold increase (see, e.g., Reference 4.336). Ohashi[4.337] has studied the ways of suppression of dislocation accumulation and found that selective growth of the film and introduction of dopant (Si) into a part of the growing film decrease the dislocation density to 10^{-3} of the magnitude for the undoped part. Earlier, Yamaguchi[4.338] reported a successful experiment in which the dislocation density in GaAs layers was controlled by the thermal cycle annealing in the process of the MOCVD heteroepitaxial growth of GaAs on Si substrates. By annealing at 1173 K, the dislocation density could be decreased up to 100 times after 10 to 15 cycles.

An important and, until now, unsolved problem relevant to the two-dimensional defects in GaAs is formation of the antiphase domains in the MBE-grown GaAs on Si.[4.333,4.334] The antiphase domain boundaries (APBs) are these defects. A comprehensive investigation of

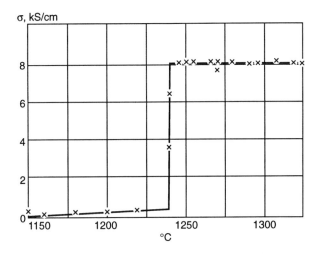

Figure 4.40 Temperature dependence of the GaAs electrical conductivity in the solid and liquid states. (After V.M. Glazov, S.N. Chizhevskaya, and N.N. Glagoleva, *Liquid Semiconductors*, Nauka Publishing House, Moscow, 1967. With permission.)

APBs was undertaken by Georgakilas et al.,[4.335] who investigated the formation of APBs, the scattering of electrons by APBs (the electron Hall mobility decreases from about 1700 cm^2/V·sec to about 500 cm^2/V·sec with APBs density growth from $1 \cdot 10^{-3}$ to $11 \cdot 10^{-3}$ APBs per cm). The authors of Reference 4.335 discussed and experimentally showed the ways to annihilate the APBs.

Among the three-dimensional defects in GaAs, the amorphous phase formed in the process of ion implantation (see, e.g., Reference 4.339) or as a result of solid-state reactions between thin films of GaAs and a metal.[4.340,4.341] Jones and Santana[4.342] analyzed parameters which characterize the amorphization by the Si$^+$ ion implantation in a large group of binary diamond-like compounds and in silicon. Solid-state amorphization reactions between GaAs and cobalt at temperatures between 530 and 570 K were observed and studied thermodynamically by Shiau et al.[4.343]

Polycrystalline GaAs of 99.999% purity and the carrier concentration of under $1 \cdot 10^{16}$ cm^{-3} is available commercially ($12.10 to $123.00 per gram — 1995).

4.2.12. Gallium Antimonide

Gallium antimonide, GaSb, has a molecular weight of 191.47, an average atomic weight of 95.74, the average atomic number of 41, and the enthalpy of formation of –43.9 kJ/mol.

At ambient pressure, GaSb has the sphalerite structure with the lattice constant at 300 K a = 0.6096120 ± 0.0000009 nm (Reference 4.125, p. 97) (Straumanis and Kim[4.293] found a = 0.6095882 nm at 273.15 K). At a pressure above 7.65 GPa (Reference 4.219, p. 116), GaSb crystal structure at room temperature is one of white tin (space group D_{4h}^{19}-I4$_1$/amd)[4.346] with lattice parameters a = 0.5348 nm and c = 0.2937 (Reference 4.125, p. 97).

The X-ray density of GaSb at 300 K is 5.614 g/cm^3; the pycnometric density is 5.61 g/cm^3. The GaSb coefficient of linear thermal expansion at room temperature is $6.1 \cdot 10^{-6}$ 1/K;[4.185] its dependence on temperature below 340 K was reported by Novikova and Abrikosov[4.194] (Figure 4.45), who found that the coefficient of thermal expansion is equal to zero at 52 K and is negative below this temperature.

GaSb melts without decomposition at 975.0 ± 0.3 K[4.346] and forms eutectics with gallium (degenerate, solidus line at 303 K) and with antimony (solidus at 856 K, eutectic point in the T-x projection of the Ga-Sb phase diagram is located at 13 at% Ga).[4.347] The solubility of both Ga and Sb in solid GaSb, if any, does not exceed a few hundredths of a percent.

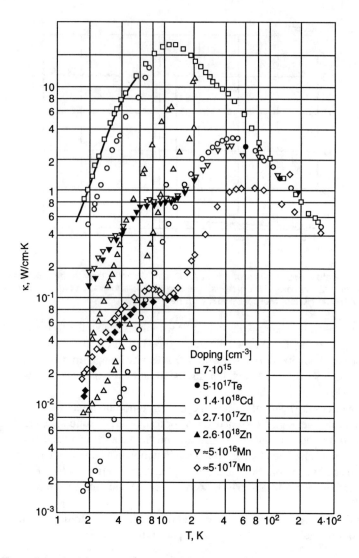

Figure 4.41 Thermal conductivity vs. temperature of GaAs samples with various impurity content. (After M.G. Holland, *Proc. 7th Int. Conf. Phys. Semicond.*, Paris, Dunod, 1964. With permission.)

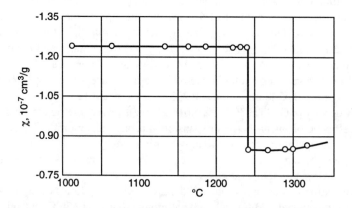

Figure 4.42 Specific magnetic susceptibility (in the Gauss system units) of GaAs in the solid and liquid states. (After V.M. Glazov, S.N. Chizhevskaya, and N.N. Glagoleva, *Liquid Semiconductors*, Nauka Publishing House, Moscow, 1967. With permission.)

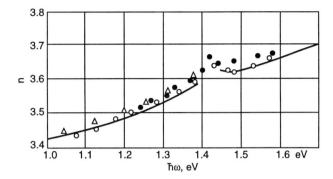

Figure 4.43 The GaAs index of refraction vs. photon energy. Solid lines — calculated; triangles and circles — experimental data. (After D. Campi and C. Papuzza, *J. Appl. Phys.*, 57, 1305, 1985. With permission.) (See also H. Burkhard, H.W. Dinges, and E. Kuphal, *J. Appl. Phys.*, 53, 655, 1982.)

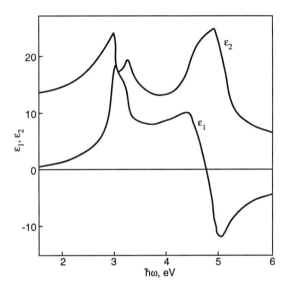

Figure 4.44 Real and imaginary parts of the GaAs dielectric constant vs. photon energy. (After D.E. Apness and A.A. Studna, *Phys. Rev.*, B27, 985, 1983. With permission.)

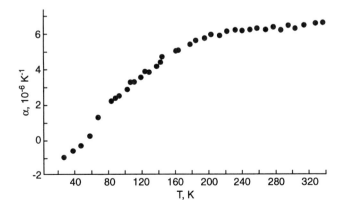

Figure 4.45 Temperature dependence of the GaSb coefficient of linear thermal expansion (After S.I. Novikova and N.Kh. Abrikosov, *Sov. Phys. Solid State*, 5, 1558, 1964. With permission.)

Table 4.4 The GaSb Atomic Heat Capacity (cal/g-atom·K) and Debye Temperature

T (K)	C_p	$C_p - C_v$	C_v	θ (K)
12	0.120	—	0.120	188
15	0.240	—	0.240	186
20	0.530	—	0.530	189
25	0.828	—	0.828	200
30	1.117	—	1.117	212
35	1.380	—	1.380	224
40	1.640	—	1.640	236
45	1.890	—	1.890	246
50	2.130	—	2.130	255
60	2.600	0.001	2.599	268
70	3.038	0.001	3.037	277
80	3.438	0.002	3.436	282
90	3.785	0.003	3.782	285
100	4.076	0.004	4.072	288
110	4.314	0.005	4.309	290
120	4.525	0.006	4.519	291
130	4.710	0.007	4.703	290
140	4.870	0.008	4.862	289
150	5.005	0.009	4.996	287
160	5.115	0.010	5.105	286
170	5.208	0.011	5.197	285
180	5.293	0.012	5.281	283
190	5.372	0.013	5.359	280
200	5.446	0.014	5.432	274
210	5.514	0.015	5.499	268
220	5.570	0.016	5.554	262
230	5.620	0.017	5.603	256
240	5.657	0.018	5.639	252
250	5.690	0.019	5.671	249
260	5.720	0.020	5.700	245
270	5.748	0.021	5.727	241
273.2	5.755	0.021	5.734	240

Adapted from U. Piesbergen, *Z. Naturforsch.*, 18A, 141, 1963.

The GaSb average atomic heat capacity and the Debye temperature calculated on its basis are presented in Table 4.4.[4.195] Steigmeier[4.89] used the Marcus and Kennedy[4.348] approach to calculate the GaSb Debye temperature (at 0 K) from the data on elastic moduli. The magnitude reported in Reference 4.89 (265.5 K) agrees with the Debye temperature extrapolation to T = 0 K of the $θ_D = f(C_v)$ curve from (Reference 4.195, Figure 3).

The GaSb Mohs hardness is 4.5; its Knoop microhardness varies in the limits 4.48 ± 0.27 GPa, depending on the indentor orientation relative to the crystal. The Vickers microhardness is 4.20 ± 0.10 GPa. The GaSb Young's modulus is 76.0 GPa; the Poisson ratio has a magnitude of 0.30. The GaSb microbreaking strength is 1.51 GPa; the brittleness criterion is equal to 1.8 (Reference 4.70, p. 103).

The GaSb elastic moduli at room temperature are (in GPa):

c_{11}	c_{12}	c_{44}	Reference
88.5	40.4	43.2	4.301
88.3	—	43.2	4.196
88.4	40.3	43.2	4.350
88.34	40.23	43.22	4.279

and their temperature dependence is shown in Figure 4.46.

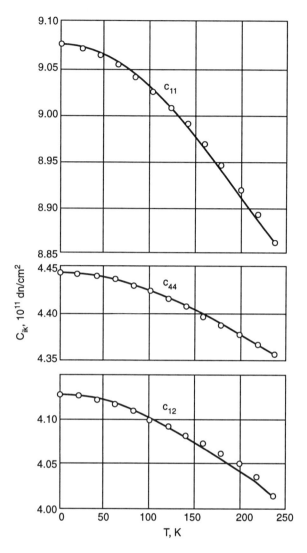

Figure 4.46 Temperature dependence of the GaSb second-order elastic moduli. (After W.F. Boyle and R.J. Sladek, *Phys. Rev.*, B11, 1587, 1975. With permission.)

Gallium antimonide is a direct semiconductor with the minimum energy gap (transition $\Gamma_{8v} - \Gamma_{6c}$) of 0.75 eV and 0.822 eV at 300 K and 0 K, respectively.[4.351] The GaSb band structure was calculated by Chelikowsky and Cohen[4.283] (Figure 4.47) and studied experimentally by Williams et al.[4.284] The electron effective mass at the conduction band minimum in the Brillouin zone center is equal to $0.0412\ m_0$ (between 1 K and 30 K) (Reference 4.219, p. 116). Pickering[4.352] cites a magnitude of $0.045\ m_0$. The hole effective mass is equal to $(0.28 \pm 0.13)m_0$ (heavy holes), $(0.050 \pm 0.005)m_0$ (light holes), and $0.82\ m_0$ (density-of-state)[4.353] (see, also, Reference 4.354). The large uncertainty in the heavy hole effective mass has been pointed out by Kranzer.[4.355]

The electric transport properties of GaSb have certain specifics related (see Figure 4.47) to its band structure. There is one maximum in the valence band (Γ_{8v}) and three minima (Γ_{8c}, L_{6c}, and X_6). The reviews of the results of the early electrical, galvanomagnetic, and thermoelectrical measurements of GaSb are compiled in References 4.310, 4.356, and 4.357. The GaSb electrical properties were extensively measured by Hrostowski and Tanenbaum,[4.358]

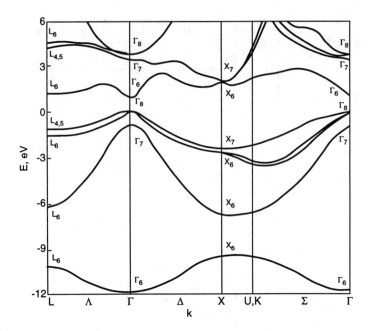

Figure 4.47 Band structure of gallium antimonide calculated by the nonlocal pseudopotential method. (After J.R. Chelikowsky and M.L. Cohen, *Phys. Rev.,* B14, 556, 1976. With permission.)

Leifer and Dunlap,[4.359] Ure (Reference 4.360, p. 1), Stirn (Reference 4.360, p. 67), Roberts et al.,[4.361] Kranzer,[4.355] Mathur and Jain,[4.363] Lee and Woolley,[4.364] and Heller and Hamerly.[4.354]

The undoped single crystals of GaSb grown by the Czochralski method usually have p-type conductivity. Doping with Se or Te produces the n-type material. Welker and Weiss[4.310] reported electron mobility of 4000 cm^2/V·sec and hole mobility of 700 cm^2/V·sec for room temperature. The authors of Reference 4.364 evaluated the contribution in the electron mobility in the Γ and L bands from various scattering mechanisms in the temperature range from about 20 to 340 K (Figure 4.48). The hole mobility vs. temperature has been estimated in References 4.363 and 4.365. There is a noticeable difference between their data, related, possibly, to the difference in the methods used for the measurements and in the calculation approach. The mobility measurements on the GaSb samples with the variable donor concentration were reported by Wiley (Reference 4.360, Vol. 10, 1975, p. 91) and Campos et al.[4.366] The temperature dependence of the Hall constant and volume resistivity of a GaSb crystal (p = 2·10^{16} cm^{-3} and μ_p = 2500 cm^2/V·sec at 77 K and p = 1.2·10^{17} cm^{-3} and μ_p = 680 cm^2/V·sec at 300 K) is shown in Figure 4.49.[4.354]

The electrical properties of GaSb in solid and liquid state were investigated by Glazov et al.[4.206] The transition from the solid to liquid state is accompanied by the 40-time volume resistivity decrease. The GaSb melt has the metal-type temperature dependence of resistivity.

The GaSb thermal conductivity at room temperature is 0.330 W/cm·K.[4.299] The magnitude of 0.439 W/cm·K mentioned by Ioffe[4.321] without a reference is, apparently, too high. The results of the measurements of thermal conductivity on several samples of n-GaSb and p-GaSb, reported by Holland,[4.324] are shown in Figure 4.50. The thermal conductivity vs. temperature curves pass through a maximum, the magnitude and temperature of which depend on the impurity content.

The molar magnetic susceptibility of GaSb at room temperature is –48.3·10^{-5} SI units/g-mol (Reference 4.125, p. 105). The temperature dependence of the volume magnetic susceptibility at elevated temperatures, including liquid state, is shown in Figure 4.51 (from Reference 4.206, p. 87, in the units of Gauss system). The increase of the magnetic susceptibility in the

BINARY IV-IV AND III-V SEMICONDUCTORS

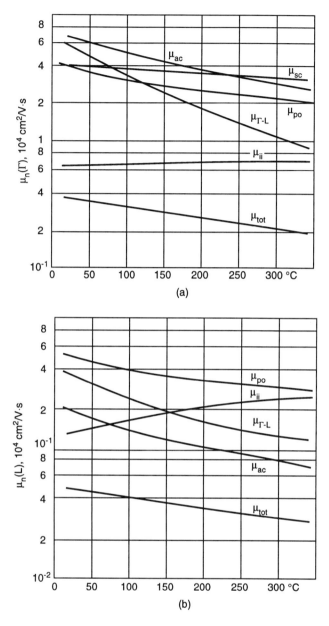

Figure 4.48 Electron mobility in a GaSb sample with n = 1.49·10^{18} cm^{-3}. (a) The Γ-band mobility, (b) the L-band mobility, total, μ_{tot}, and the contributions from various scattering mechanisms (ac — longitudinal acoustic phonon scattering; po — polar optical mode scattering, ii — ionized impurity scattering; sc — space charge scattering; Γ-L — intervalley scattering). (After H.J. Lee and J.C. Woolley, *Canad. J. Phys.*, 59, 1844, 1981. With permission.)

melting process may be explained by the sharp increase of the free electron content and the resulting increase of the paramagnetic component of susceptibility.

At 300 K, the static dielectric constant of GaSb is 15.69 and its high-frequency magnitude is equal to 14.44 (Reference 4.219, p. 118), while Mathur and Jain[4.363] recommend 15.0 and 13.8, respectively. Hence, the index of refraction magnitude has to be 3.7, in good agreement with the results of the spectroscopic ellipsometry measurements and calculations by Aspnes and Studna[4.327] presented in Table 4.5.

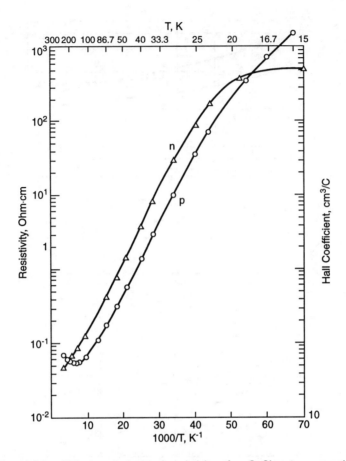

Figure 4.49 The Hall coefficient and electrical resistivity of a GaSb p-type sample with $p_{300\,K}$ = $1.2 \cdot 10^{17}$ cm^{-3} vs. temperature. (After M.W. Heller and R.G. Hamerly, *J. Appl. Phys.*, 57, 4626, 1985. With permission.)

Gallium antimonide has properties similar to those of Ge with the advantage of being a direct gap semiconductor. The disadvantage of GaSb is its relatively low melting point that decreases the temperature range in which the semiconductor devices based on GaSb can work efficiently for reasonably long periods of time. The ability of GaSb to form continuous solid solutions with other III-V compounds (AlSb, GaAs, and InSb) makes GaSb a prospective material for use in tailoring of semiconductor substances with desirable electronic and/or crystallographic properties.

4.2.13. Indium Nitride

Indium nitride, InN, has a molecular weight of 128.83, an average atomic weight of 64.41, an average atomic number of 28, and the enthalpy of formation of −19.3 kJ/mol.

InN crystallizes in the wurtzite structure with the lattice parameters at room temperature of a = 0.3540 ± 0.0004 nm and c = 0.5704 ± 0.0004 nm (Reference 4.125, p. 97). The X-ray experiments on the InN epitaxial layers have given a = 0.35446 nm and c = 0.57034 nm (Reference 4.219, p. 122). The pycnometric density at room temperature is 6.88 g/cm^3; the X-ray density is 6.916 g/cm^3.[4.367] Pearson Handbook's[4.368] magnitude (6.81 g/cm^3 at 298.15 K) may be considered too low.

The InN melting point is at 1473 K (Reference 4.125, p. 106) under the nitrogen pressure of about 3 MPa (the magnitude of 1373 K is presented in Reference 4.369). InN dissociates in vacuum at temperatures above 893 K.[4.218] The InN heat capacity at constant pressure is

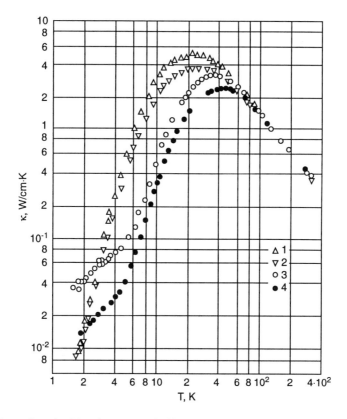

Figure 4.50 Thermal conductivity of two n-type GaSb samples with impurity content of $4 \cdot 10^{18}$ cm^{-3} (1) and $1.4 \cdot 10^{18}$ cm^{-3} (2) and of two p-type GaSb samples with impurity content of $1 \cdot 10^{17}$ cm^{-3} (3) and $2 \cdot 10^{17}$ cm^{-3} (4). (After M.G. Holland, *Proc. 7th Int. Conf. Phys. Semicond.*, Dunod, Paris, 1964. With permission.)

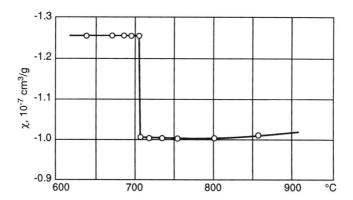

Figure 4.51 Specific magnetic susceptibility (in the Gauss system units) for GaSb in the solid and liquid states. (After V.M. Glazov, S.N. Chizhevskaya, and N.N. Glagoleva, *Liquid Semiconductors*, Nauka Publishing House, Moscow, 1967. With permission.)

32.3 J/g·K;[4.368] the Debye temperature is 1033 K.[4.230] The longitudinal and transversal elastic moduli of InN, estimated by Chin et al.,[4.230] are 265 and 442 GPa, respectively.

InN is a direct semiconductor. Its band structure calculated by Foley and Tansley[4.370] is shown in Figure 4.52. The InN optical energy gap measured by Tyagai et al.[4.371] is 2.05 eV at 300 K. Osamura et al.[4.372] have found the magnitude of 2.11 eV at 78 K. The electron

Table 4.5 The GaSb Optical Parameters at 300 K

Photon energy (eV)	Dielectric constant Real part	Dielectric constant Imaginary part	Index of refraction	Extinction coefficient	Reflectance	Absorption coefficient (10^3 cm^{-1})
1.5	19.135	3.023	4.388	0.344	0.398	52.37
2.0	25.545	14.442	5.239	1.378	0.487	279.43
2.5	13.367	19.705	4.312	2.285	0.484	579.07
3.0	9.479	15.738	3.832	2.109	0.444	641.20
3.5	7.852	19.267	3.785	2.545	0.485	902.86
4.0	−1.374	25.138	3.450	3.643	0.583	1477.21
4.5	−8.989	10.763	1.586	3.392	0.651	1547.17
5.0	−5.693	7.529	1.369	2.751	0.585	1394.02
5.5	−5.527	6.410	1.212	2.645	0.592	1474.51
6.0	−4.962	4.520	0.935	2.416	0.610	1469.28

From D.E. Aspnes and A.A. Studna, *Phys. Rev.*, B27, 985, 1983.

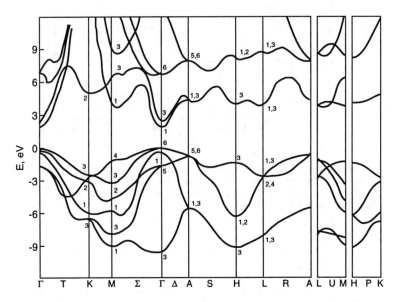

Figure 4.52 The InN band structure. Pseudopotential method calculation. (After C.P. Foley and T.L. Tansley, *Phys. Rev.*, B33, 1430, 1986. With permission.)

effective mass is close to 0.12 m_o,[4.230] in agreement with Reference 4.371. The heavy and light hole effective mass is 0.5 m_o and 0.17 m_o, respectively.[4.370]

The transport properties of InN have not been widely studied, primarily because the relatively pure, large, and stoichiometric single crystals have not been available for the measurements. From known reports (see, e.g., References 4.370 and 4.373 to 4.378), the InN pressed powder samples show a positive temperature coefficient of electrical resistivity, but the epitaxial layers have a semiconductor character of conductivity. The experimental data on the InN electron mobility[4.378] are in satisfactory agreement with the results of calculations by Chin et al.[4.230] (Figure 4.53). The authors of Reference 4.230 come to the conclusion that the theoretical maximum electron mobility in InN is equal or greater than 30,000 cm^2/V·sec at 77 K and 4400 cm^2/V·sec at 300 K. These magnitudes give reason to believe that InN presents a doubtless interest for electrooptical applications.

The molar magnetic susceptibility of InN is −51.2·10^{-5} SI units/g-mol (Reference 4.125, p. 105). The static dielectric constant is 15.3; the high-frequency magnitude is 8.4.[4.230] Tyagai et al.[4.371] found the high-frequency dielectric constant of 9.3 from the measurements on a

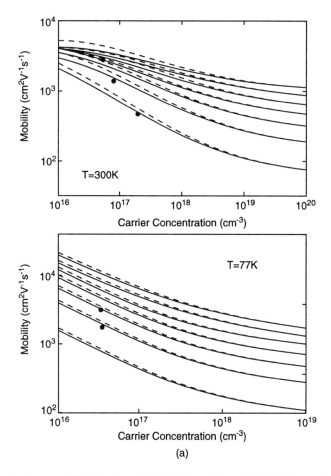

Figure 4.53 The electron mobility in indium nitride. (a) The electron drift (solid curves) and Hall mobility (dashed lines) vs. carrier concentration, calculated for the compensation ratio, N_A/N_D, of 0.00 (uppermost curves), 0.15, 0.30, 0.45, 0.60, 0.75, and 0.90 (lowest curves), and temperatures of 300 K and 77 K (the dots represent experimental data); (b) the temperature dependence of the electron drift mobility in InN calculated for carrier concentration (1) 10^{16} cm^{-3}, (2) 10^{17} cm^{-3}, and (3) 10^{18} cm^{-3} and different magnitudes of N_A/N_D. (After V.W. Chin, T.L. Tansley, and T. Osotchan, *J. Appl. Phys.*, 75, 7365, 1994. With permission.)

heavily doped film of InN. The index of refraction as a function of the photon energy, measured at 300 K by the interference method on an InN sample with electron concentration of $3 \cdot 10^{20}$ cm^{-3}, is 2.56 at 1.24 eV, 2.93 at 1.51 eV, and 3.12 at 1.88 eV.[4.371]

The information regarding technology of preparation of InN is limited. One of the common methods of its synthesis is by decomposition of $(NH_4)_3InF_6$ at 873 K. Kistenmacher et al.[4.380] have successfully grown thin InN films on mica and sapphire by the reactive rf-magnetron sputtering; the electron mobility in the grown films is up to 30 cm^2/V·sec at the carrier concentration of $2 \cdot 10^{20}$ cm^{-3} to $4 \cdot 10^{20}$ cm^{-3}. InN 100-mesh powder of 99.8% purity is available commercially ($56.80 to $76.00 per gram — 1995).

4.2.14. Indium Phosphide

Indium phosphide, InP, has a molecular weight of 145.79, an average atomic weight of 72.90, the average atomic number of 23, and the enthalpy of formation –92.1 kJ/mol.

At ambient pressure, InP crystallizes in the sphalerite structure with lattice parameter at 291.15 K of 0.586875 ± 0.000001 nm[4.190] and of 0.58871 nm at 912 K.[4.257] The coefficient of

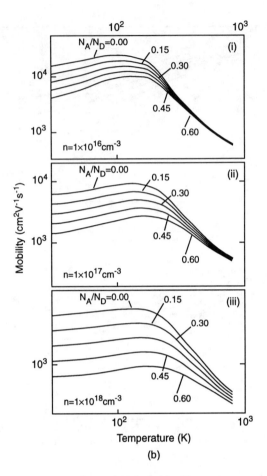

Figure 4.53 (continued)

thermal expansion at 298.15 K is $4.75 \cdot 10^{-6}$ 1/K (Reference 4.317, p. 627); Steigmeier and Kudman[4.171] suggest the magnitude of $4.60 \cdot 10^{-6}$ 1/K; a magnitude of $4.5 \cdot 10^{-6}$ 1/K is cited in Reference 4.310; Deus and Schneider[4.388] prefer a mean value of $4.3 \cdot 10^{-6}$ 1/K at room temperature. The temperature dependence of the InP coefficient of thermal expansion has been calculated by Soma et al.,[4.296] who used for this purpose the experimental data on the pressure dependence of the InP elastic moduli and the phonon wavenumbers. The InP Grüneisen constant is equal to zero at 0 K and at about 100 K; it is negative between these temperatures and, consequently, the coefficient of thermal expansion is also negative. The results of Reference 4.296 are presented in Figure 4.54.

At the pressure of 13.3 GPa and higher, InP possesses the face-centered cubic (NaCl-type) crystal lattice (space group O_h^5-Fm3m).[4.386]

InP melts congruently at 1331 K (Reference 4.125, p. 105) (Tmar et al.[4.382] believe that the magnitude of 1335 K better fits the available literature data). The projections of the InP phase diagram on the T-x, p-x, and T-p planes are shown in Figure 4.55 (after van den Boomgaard and Schol[4.294]).

The InP heat capacity was studied by Piesbergen.[4.195] His data on heat capacity and the Debye temperature are presented in Table 4.6. The InP specific heat at 298.2 K is 0.311 J/g·K.[4.195] Steigmeier and Kudman[4.171] reported the InP Debye temperature of 321 K at 0 K.

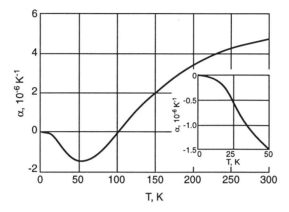

Figure 4.54 Temperature dependence of the InP coefficient of linear thermal expansion calculated using experimental data on the pressure derivatives of elastic moduli and phonon wavenumbers. (After T. Soma, J. Sato, and H. Matsuo, *Solid State Comm.*, 42, 889, 1982. With permission.)

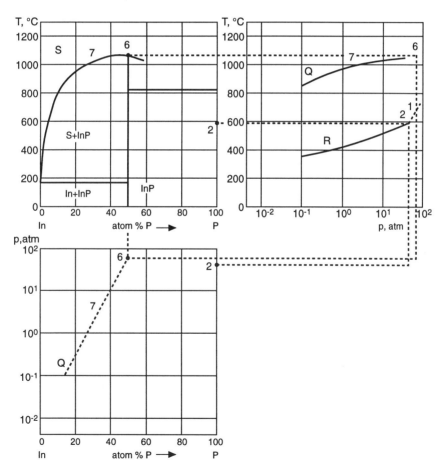

Figure 4.55 The temperature-composition, T-x, temperature-pressure, T-p, and pressure-composition, p-x, projections of the In-P phase diagram. (After J. van den Boomgaard and K. Schol, *Philips Res. Rep.*, 12, 127, 1957. With permission.)

Table 4.6 The InP Average Atomic Heat Capacity (cal/g-atom·K) and Debye Temperature

T (K)	C_p	$C_p - C_v$	C_v	θ (K)
12	0.063	—	0.063	234
15	0.162	—	0.162	213
20	0.388	—	0.388	212
25	0.648	—	0.648	219
30	0.892	—	0.892	232
35	1.128	—	1.128	246
40	1.345	—	1.345	260
45	1.540	—	1.540	274
50	1.710	—	1.710	288
60	2.023	—	2.023	315
70	2.318	—	2.318	338
80	2.600	0.001	2.599	357
90	2.872	0.001	2.871	373
100	3.132	0.002	3.130	385
110	3.370	0.002	3.368	396
120	3.595	0.003	3.592	404
130	3.810	0.003	3.807	408
140	4.012	0.004	4.008	412
150	4.190	0.004	4.186	414
160	4.342	0.005	4.337	417
170	4.474	0.006	4.468	421
180	4.595	0.006	4.589	423
190	4.711	0.007	4.704	424
200	4.818	0.008	4.810	424
210	4.917	0.008	4.909	422
220	5.000	0.009	4.991	422
230	5.070	0.010	5.060	423
240	5.135	0.010	5.125	423
250	5.193	0.011	5.182	424
260	5.250	0.012	5.238	423
270	5.307	0.012	5.295	420
273.2	5.320	0.013	5.307	420

Adapted from U. Piesbergen, *Z. Naturforsch.*, 18A, 141, 1963.

The square pyramid (Vickers) microhardness of InP is 4.35 ± 0.20 GPa (50-g load).[4.384] The InP Knoop hardness was studied by Wolff et al.,[4.344] Brasen,[4.384] and Watts and Willoughby.[4.385] Figure 4.56 represents the results of the Knoop hardness investigation (50-g load) on a pure InP sample (n = $1.5 \cdot 10^{16}$ cm^{-3}) and on a sulfur-doped sample (n = $1 \cdot 10^{19}$ cm^{-3}).[4.385] The measurements were done on the {100} surface with the indentor's long diagonal initially aligned along one <110> direction and finally along an adjacent <110> direction. It was found that doping increases the microhardness uniformly by about 12%.

The InP bulk modulus for the sphalerite allotrope is 71.0 GPa,[4.386] in satisfactory agreement with the results of calculations by Tsai et al.[4.387] (67.8 GPa). The sound velocity in InP is 5.13 km/sec (Reference 4.317, p. 627) and the magnitude of the Young modulus is 127 GPa. The InP elastic modulus c_{11} is equal to 102.8 GPa.[4.89]

InP is a direct semiconductor with the minimum energy gap of 1.34 eV (Reference 4.125, p. 104) (1.344 eV is found by Bugajski and Levandivski[4.389]). At 1.6 K, E_g = 1.4236 eV.[4.390] The InP band structure has been calculated by Chelikowsky and Cohen[4.283] (Figure 4.57) and studied experimentally by Williams et al.[4.284] The valence band has a maximum, typical for all adamantine-type semiconductors, at Γ. There are three minima in the conduction band; the lowest is at Γ and the other two at X and at L. The electron effective mass at Γ valley is 0.077 m_o and 0.068 m_o at 300 K and 500 K, respectively; at L valley, it is equal to 0.325 m_o (300 K), and at X valley its magnitude is 0.26 m_o (300 K) (Reference 4.317, p. 627). From

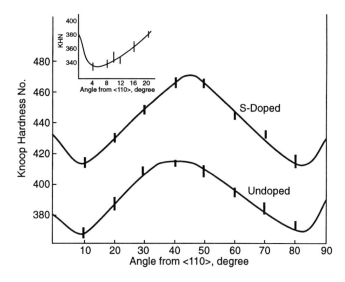

Figure 4.56 Knoop hardness anisotropy of undoped and sulfur-doped InP. (After D.Y. Watts and A.F.W. Willoughby, *J. Appl. Phys.*, 56, 1869, 1984. With permission.)

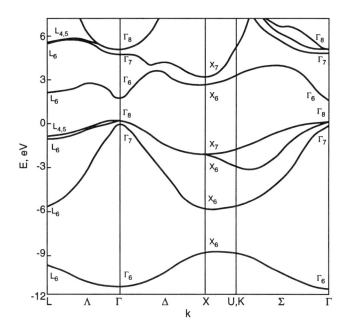

Figure 4.57 Band structure of InP calculated by the nonlocal pseudopotential method. (After J.R. Chelikowsky and M.L. Cohen, *Phys. Rev.*, B14, 556, 1976. With permission.)

the cyclotron resonance experiments, Leotin et al.[4.391] found that at 110 K heavy hole effective mass is 0.60 and 0.56 m_o in the magnetic field parallel to [111] and [100], respectively; the light holes have effective mass of 0.12 m_o, and the spin-orbit hole effective mass is 0.121 m_o.

The intrinsic carrier concentration, in accordance with Weiss,[4.392] is equal to $1.2 \cdot 10^8$ cm^{-3} at 293 K; its temperature dependence for the range 700 K to 920 K may be expressed as $n_i = 8.4 \cdot 10^{15} T^{3/2} \exp(-1.34/2k_B T)$,[4.393] where T is in K and $k_B T$ is in eV. The data on carrier mobility vary in wide limits in the available reports. Fairhurst et al.[4.394] find the electron mobility in the InP layers at 77 K of 131,600 cm^2/V·sec. Taylor and Anderson[4.396] reported a maximum electron mobility of 120,520 cm^2/V·sec in the thin InP layers at 77 K. Zhu et al.[4.395] conclude

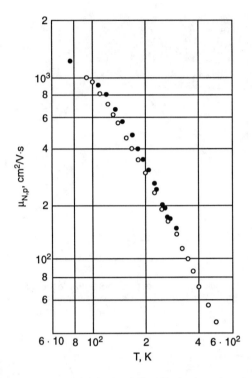

Figure 4.58 Temperature dependence of the Hall mobility of holes in undoped p-type InP samples. (After J.D. Wiley, *Semiconductors and Semimetals,* Vol. 10, R.K. Willardson and A.C. Beer, Eds., Academic Press, 1975. With permission.)

that the typical magnitude is 131,600 cm^2/V·sec and 5370 cm^2/V·sec for 77 and 300 K, respectively. Shur[4.317] cites the magnitudes of 40,000 to 60,000 cm^2/V·sec (at 77 K) and 4600 cm^2/V·sec (at 300 K). The hole mobility in InP is 150 cm^2/V·sec at 300 K.[4.317] The temperature dependence of the hole Hall mobility in InP is shown in Figure 4.58 (after Wiley[4.176]).

The InP thermal conductivity, κ, at room temperature is 0.68 W/cm·K.[4.171] The temperature dependence of κ is shown in Figure 4.59 (after Aliev et al.[4.397]).

InP is a diamagnetic with the molar magnetic susceptibility of $-18.2 \cdot 10^{-5}$ SI units per g-mol (Reference 4.125, p. 105). The InP static dielectric constant is 12.56 and 11.93 at 300 and 77 K, respectively.[4.390] The high-frequency dielectric constant is 9.61.[4.317] The dependence of dielectric and optical properties of InP on the photon energy at room temperature is presented in Table 4.7 (after Aspnes and Studna[4.327]).

Bulk ingots of InP may be synthesized by heating of the elemental components (with a slight excess of indium) in evacuated and sealed fused silica ampules. The following zone recrystallization purifies the ingot from the indium excess and produces relatively large InP single crystal.

Polycrystalline InP with the carrier concentration of $5 \cdot 10^{15}$ cm^{-3} and the purity of 99.9999% is available commercially at $91.40 to $106.00 per gram (1995).

4.2.15. Indium Arsenide

Indium arsenide, InAs, has a molecular weight of 189.74, the average atomic weight of 94.87, the average atomic number of 41, and the enthalpy of formation of −54.5 kJ/mol. InAs has three allotropes. At ambient pressure and up to 7.0 ± 0.2 GPa, a stable allotrope is one with the sphalerite structure and the lattice parameter at 1 MPa and 298.15 K of 0.605838 ±

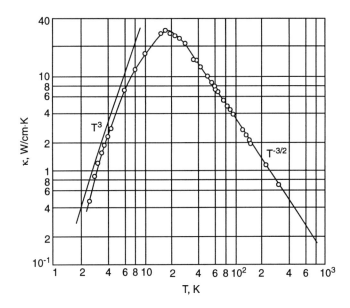

Figure 4.59 Temperature dependence of the InP thermal conductivity. (After S.A. Aliev, A.Ya. Nashelskii, and S.S. Shalyt, *Fiz. Tverd. Tela*, 7, 1590, 1965. With permission.)

Table 4.7 The InP Optical Parameters at 300 K

Photon energy (eV)	Dielectric constant Real part	Dielectric constant Imaginary part	Index of refraction	Extinction coefficient	Reflectance	Absorption coefficient (10^3 cm^{-1})
1.5	11.904	1.400	3.456	0.203	0.305	30.79
2.0	12.493	2.252	3.549	0.317	0.317	64.32
2.5	14.313	3.904	3.818	0.511	0.349	129.56
3.0	17.759	10.962	4.395	1.247	0.427	379.23
3.5	6.400	12.443	3.193	1.948	0.403	691.21
4.0	6.874	10.871	3.141	1.730	0.376	701.54
4.5	8.891	16.161	3.697	2.186	0.449	996.95
5.0	–7.678	14.896	2.131	3.495	0.613	1771.52
5.5	–4.528	7.308	1.426	2.562	0.542	1428.14
6.0	–2.681	5.644	1.336	2.113	0.461	1285.10

After D.E. Aspnes and A.A. Studna, *Phys. Rev.*, B27, 985, 1983.

0.000005 nm.[4.190] Between 7.0 and 17 GPa, InAs possesses the rock salt structure with the lattice parameter at 7.5 GPa and 300 K of 0.5514 nm (Reference 4.125, p. 99).

At 17.0 ± 0.4 GPa, the InAs structure transforms into one of white tin.[4.399] The X-ray density of InAs at 300 K and the atmospheric pressure are 5.668 g/cm^3 (calculated using a = 0.605838 nm). The mean value of the InAs coefficient of thermal expansion, α, at room temperature is 4.5·10^{-6} 1/K.[4.388] At temperatures below 60 K, α is negative.[4.388] The temperature dependence of α is shown in Figures 4.60[4.400] and 4.61 (the X-ray data are from Reference 4.401, the dilatometric magnitudes from Reference 4.402).

The InAs average atomic heat capacity and the Debye temperature between 11 and 273.2 K are presented in Table 4.8.[4.195] The InAs specific heat at 298.2 K is 0.252 J/g·K. The Debye temperature at 0 K is 249 K[4.171] (Steigmeier[4.89] found the 0-K magnitude of 262 K).

The InAs Vickers microhardness is 3.3 ± 0.1 GPa.[4.381] The Knoop microhardness with a 25-g load is 3.84 ± 0.26 GPa (Reference 4.70, p. 109). The InAs bulk modulus of 57.9 GPa and the elastic moduli c_{11} = 83.3 GPa, c_{12} = 45.3 GPa, and c_{44} = 39.6 GPa were reported by

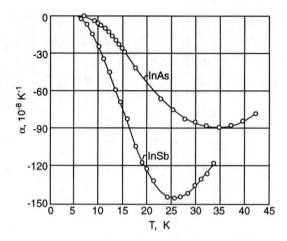

Figure 4.60 Low-temperature coefficient of thermal expansion of InAs and InSb. (After P.W. Sparks and C.A. Swenson, *Phys. Rev.*, 163, 779, 1967. With permission.)

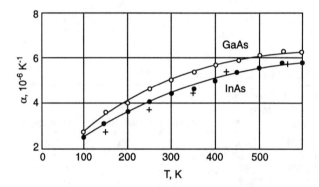

Figure 4.61 Temperature dependence of the coefficient of linear thermal expansion of GaAs and InAs (o and ● from X-ray data,[4.401] + representing the dilatometric data for InAs[4.402]).

Gerlich.[4.404] The elastic properties cited in References 4.309 and 4.70, p. 109 (Young's modulus equal to 81.1 GPa, and the shear modulus of 42.0 GPa), are in reasonable agreement with the previously cited second-order elastic constants.

InAs is a direct energy gap semiconductor with the minimum energy gap of 0.354 eV at 295 K[4.405] and 0.4180 eV at 4.2 K[4.406] (transition $\Gamma_{6v} - \Gamma_{8c}$). The InAs band structure has been calculated by Chelikowsky and Cohen[4.283] and studied experimentally by Williams et al.[4.284] The InAs band structure is shown in Figure 4.62 (after Reference 4.283). Zwerdling et al.[4.407] have performed a direct evaluation of the electron effective mass. From their measurements, $m_n^* = 0.03\ m_o$. Chasmar and Stratton[4.408] have found from the Hall and Seebeck effects measurements that $m_n^* = 0.035\ m_o$ at room temperature and $m_n^* = 0.0575\ m_o$ at 500 K. From the experimental data on magnetophonon resonance, Takayama et al.[4.409] found $m_n^*(150\ K) = 0.0231\ m_o$ and $m_n^*(250\ K) = 0.0219\ m_o$. For the hole effective mass, Neuberger[4.170] cites the magnitudes of 0.41 m_o (heavy holes) and 0.024 m_o (light holes). The calculations based on the valence band parameters[4.410] give the heavy holes effective mass of 0.43 m_o along [111] and 0.35 m_o along [100]. From the magneto-absorption measurements, Pidgeon et al.[4.411] found for light holes $m_p^* = 0.026\ m_o$ at 20 K.

Folberth et al.[4.412] showed that the intrinsic carrier concentration in the range of 350 K to 900 K may be expressed as $n_i(cm^{-3}) = 2.14 \cdot 10^{15}\ T^{3/2} \exp(-0.47/2k_B T)$ with T in K and $k_B T$

Table 4.8 The InAs Average Atomic Heat Capacity
(cal/g-atom·K) and Debye Temperature

T (K)	C_p	$C_p - C_v$	C_v	θ (K)
11	0.125	—	0.125	170
15	0.316	—	0.316	170
20	0.614	—	0.614	179
25	0.912	—	0.912	192
30	1.188	—	1.188	206
35	1.444	—	1.444	220
40	1.684	—	1.684	232
45	1.918	—	1.918	244
50	2.150	—	2.150	253
60	2.590	—	2.590	269
70	3.000	0.001	2.999	280
80	3.370	0.002	3.368	288
90	3.705	0.002	3.703	292
100	3.996	0.003	3.993	295
110	4.240	0.004	4.236	298
120	4.443	0.004	4.439	301
130	4.624	0.005	4.619	302
140	4.769	0.006	4.763	303
150	4.908	0.007	4.901	303
160	5.020	0.007	5.013	303
170	5.118	0.008	5.110	303
180	5.200	0.009	5.191	303
190	5.271	0.010	5.261	303
200	5.338	0.010	5.328	303
210	5.400	0.011	5.389	301
220	5.455	0.012	5.443	298
230	5.510	0.013	5.497	294
240	5.560	0.013	5.547	289
250	5.602	0.014	5.588	284
260	5.637	0.015	5.622	281
270	5.663	0.016	5.647	280
273.2	5.670	0.016	5.654	280

Adapted from U. Piesbergen, Z. Naturforsch., 18A, 141, 1963.

in eV. The extrapolation to 300 K gives $n_i = 1.3 \cdot 10^{15}$ cm^{-3}. From their results, the thermal energy gap magnitude lies between 0.45 and 0.49 eV.

The electron mobility in InAs is quite high: at 77 K, in the purest samples, it is up to $1 \cdot 10^5$ cm^2/V·sec; at 300 K, it is up to $3.3 \cdot 10^4$ cm^2/V·sec (D.L. Rode in Reference 4.176). The hole mobility in InAs at 300 K is 100 to 450 cm^2/V·sec,[4.412] depending on the sample purity.

The melting process increases the electrical conductivity of the InAs by about 90%.[4.206] In the liquid state, InAs behaves as a metal; its conductivity decreases noticeably with the temperature increase.

The thermal conductivity, κ, between 3 K and 200 K was measured by Tamarin and Shalyt[4.413] (Figure 4.63). Its magnitude reaches a maximum of 40 W/cm·K (sample with n = $1.6 \cdot 10^{16}$ cm^{-3}) at 10 K and is virtually independent of n at temperatures above 30 K. Extrapolation of the[4.413] data to higher temperatures gives the magnitudes of about 0.20 W/cm·K at 300 K. Abrahams et al.[4.318] found $\kappa \simeq 0.29$ W/cm·K at 300 K. The results of measurements of κ on InAs polycrystals[4.402] are somewhat lower (0.12 W/cm·K at 300 K).

InAs is a diamagnetic with the molar magnetic susceptibility of $-69.5 \cdot 10^{-5}$ SI units per g-mol (Reference 4.125, p. 105). The InAs static dielectric constant at 300 K is 15.15; at high-frequency limit, it is 12.25,[4.414] which gives for the optical index of refraction the magnitude of n = 3.5, in agreement with the average magnitude of n for the wavelength range of 276 to 827 nm ($\bar{n} = 3.54$) that can be derived from the results of the spectroscope

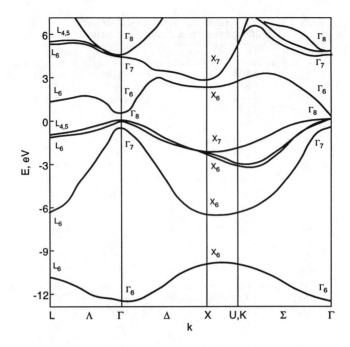

Figure 4.62 Band structure of InAs calculated by the nonlocal potential method. (After J.R. Chelikowsky and M.L. Cohen, *Phys. Rev.*, B14, 556, 1976. With permission.)

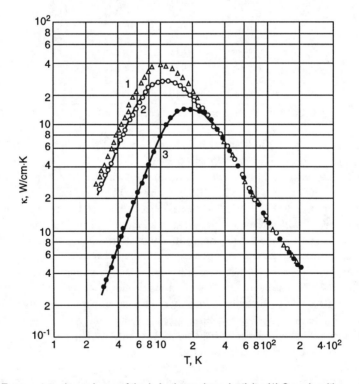

Figure 4.63 Temperature dependence of the InAs thermal conductivity. (1) Sample with $n = 1.6 \cdot 10^{16}$ cm^{-3}; (2) sample with $n = 2 \cdot 10^{17}$ cm^{-3}; (3) sample with $p = 2 \cdot 10^{17}$ cm^{-3}. (After P.V. Tamarin and S.S. Shalyt, *Fiz. Tekh. Poluprovodn.*, 5, 1245, 1971. With permission.)

Table 4.9 The InAs Optical Parameters at 300 K

Photon energy (eV)	Dielectric constant Real part	Dielectric constant Imaginary part	Index of refraction	Extinction coefficient	Reflectance	Absorption coefficient (10^3 cm^{-1})
1.5	13.605	3.209	3.714	0.432	0.337	65.69
2.0	15.558	5.062	3.995	0.634	0.370	128.43
2.5	15.856	15.592	4.364	1.786	0.454	452.64
3.0	6.083	13.003	3.197	2.034	0.412	618.46
3.5	5.973	10.550	3.008	1.754	0.371	622.13
4.0	7.744	11.919	3.313	1.799	0.393	729.23
4.5	−1.663	22.006	3.194	3.445	0.566	1571.19
5.0	−5.923	8.752	1.524	2.871	0.583	1455.26
5.5	−3.851	6.008	1.282	2.344	0.521	1306.62
6.0	−2.403	6.005	1.434	2.112	0.448	1284.15

From D.E. Aspnes and A.A. Studna, *Phys. Rev.*, B27, 985, 1983.

ellipsometry experiments and, based on them, calculations by Aspnes and Studna[4.327] (see Table 4.9).

InAs is widely used as the material for infrared detectors. Figure 4.64 shows the typical spectral response of the commercially available InAs photon detectors at 300 K, 195 K, and 77 K.[4.415]

InAs may be prepared by melting the mixture of the elemental components in evacuated and sealed fused silica ampules with vibrational stirring.[4.189] The zone refining is ineffective for some electrically active impurities that possess high distribution coefficients in InAs (e.g., 0.77, 1.0, and 0.93 for Zn, S, and Se, respectively[4.416]). T-x, T-p, and p-x projections of the In-As phase diagram are shown in Figure 4.65.[4.294] One can see that the equilibrium As pressure upon the InAs melt near its melting point is relatively low (about 30 kPa); this simplifies the technology of the synthesis and single crystal growth.

4.2.16. Indium Antimonide

Indium antimonide, InSb, has the molecular weight of 236.57, the average atomic weight of 118.29, the average atomic number of 50, and the enthalpy of formation equal to −33.5 kJ/mol.

At pressures below 3 GPa, InSb has a sphalerite crystal structure with the lattice parameter at normal pressure and 300 K a = 0.647962 ± 0.000012 nm (Reference 4.125, p. 98). Giesecke and Pfister[4.190] suggest a = 0.647877 ± 0.000005 nm, while the magnitude reported by Straumanis and Kim[4.417] (a = 0.647937 nm at 298.15 K) is close to an average between References 4.125 and 4.417 magnitudes. Yu et al.[4.165] noticed that the data on the InSb high-pressure phases are conflicting, but Yoder et al.[4.419] confirmed that at the pressures above 3 GPa, InSb has the white tin crystal structure.

The temperature dependence of the lattice parameter of the adamantine InSb above room temperature is shown in Figure 4.66 (after Reference 4.417). The temperature dependence of the InSb coefficient of thermal expansion, α, at the temperatures between 20 and 340 K was reported by Novikova[4.420] (see Figure 4.67). The data of Reference 4.420 are in general agreement with the results reported by Gibbons.[4.421] At 57.5 K, α of InSb becomes equal to zero and remains negative at lower temperatures, at least up to 6 K. The room temperature magnitude is close to $5.0 \cdot 10^{-6}$ to $5.3 \cdot 10^{-6}$ 1/K.[4.420,4.421] (Deus and Schneider[4.388] suggest $\bar{\alpha} = 5.1 \cdot 10^{-6}$ 1/K, while Leroux-Hugon et al.[4.185] prefer the magnitude of $4.67 \cdot 10^{-6}$ 1/K per Reference 4.171.) The room temperature density of InSb is 5.7747 g/cm^3 [4.417] (the X-ray magnitude for 298.15 K is 5.7768 g/cm^3).

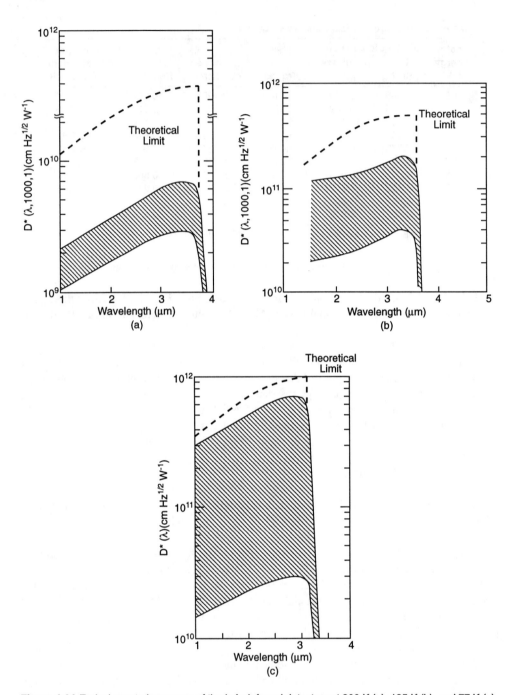

Figure 4.64 Typical spectral response of the InAs infrared detectors at 300 K (a), 195 K (b), and 77 K (c). (After W.L. Wolfe and G.J. Zissis, Eds., *Infrared Handbook,* IR Inform. and Analysis Center, Environ. Res. Inst. of Michigan, 1978. With permission.)

In accordance with the In-Sb phase diagram (Reference 4.70, p. 113), InSb is the only compound in the system. It melts congruently; the melting point magnitude from different reports[4.70,4.125,4.195,4.206] varies from 798.35 K to 809.15 K. From the available data, it is reasonable to assume that the best mean average magnitude is 802 ± 3 K. The InSb heat capacity at temperatures from 12 to 273.2 K was measured by Piesbergen[4.195] (see Table 4.10). He

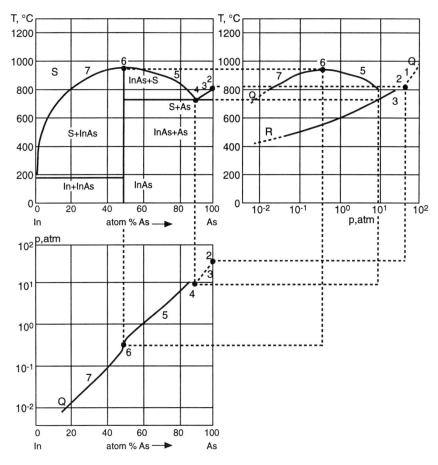

Figure 4.65 The temperature-composition, T-x; temperature-pressure, T-p; and pressure-composition, p-x, projections of the In-As phase diagram. (After J. van den Boomguaard and K. Schol, *Philips Res. Rep.*, 12, 127, 1957. With permission.)

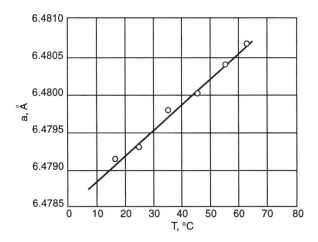

Figure 4.66 Temperature dependence of the InSb lattice parameter. (After M.E. Straumanis and C.D. Kim, *J. Appl. Phys.*, 36, 3822, 1965. With permission.)

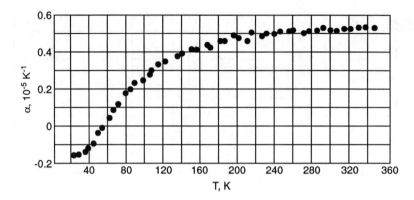

Figure 4.67 Temperature dependence of the InSb coefficient of linear thermal expansion. (After S.I. Novikova, *Sov. Phys. Solid State*, 2, 2087, 1961. With permission.)

Table 4.10 The InSb Average Atomic Heat Capacity (cal/g-atom·K) and Debye Temperature

T (K)	C_p	$C_p - C_v$	C_v	θ (K)
12	0.350	—	0.350	131
15	0.564	—	0.564	138
20	0.935	—	0.935	152
25	1.254	—	1.254	168
30	1.560	—	1.560	181
35	1.855	—	1.855	193
40	2.141	—	2.141	203
45	2.421	—	2.421	211
50	2.692	—	2.692	218
60	3.193	0.001	3.192	227
70	3.631	0.002	3.629	233
80	4.000	0.002	3.998	236
90	4.307	0.002	4.305	238
100	4.555	0.003	4.552	239
110	4.748	0.003	4.745	241
120	4.912	0.004	4.908	241
130	5.057	0.005	5.052	240
140	5.184	0.005	5.179	238
150	5.290	0.006	5.284	235
160	5.377	0.006	5.371	233
170	5.450	0.007	5.443	231
180	5.510	0.007	5.503	228
190	5.563	0.008	5.555	226
200	5.614	0.008	5.606	222
210	5.665	0.009	5.656	215
220	5.708	0.009	5.699	207
230	5.747	0.010	5.737	200
240	5.780	0.011	5.769	192
250	5.810	0.012	5.798	184
260	5.838	0.012	5.826	173
270	5.860	0.013	5.847	165
273.2	5.867	0.013	5.854	161

Adapted from U. Piesbergen, *Z. Naturforsch.*, 18A, 141, 1963.

found that the specific heat of InSb at 298.2 K is 0.209 J/g·K. The Steigmeier[4.89] evaluation of the Debye temperature from the InSb elastic moduli gave the magnitude of 202.5 K at 0 K.

The mechanical properties of InSb have been studied quite extensively. Its Vickers microhardness (50-g load) is 2.2 ± 0.1 GPa; the Knoop 25-g load microhardness is 2.2 ± 0.2 GPa

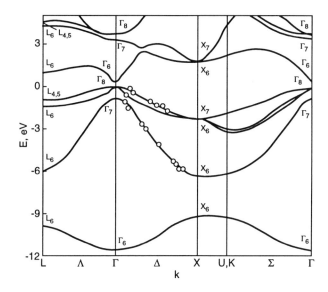

Figure 4.68 Band structure of InSb calculated by the nonlocal potential method. The open circles represent the experimental data from angular-resolved photoemission on (001) surface. (After J.R. Chelikowsky and M.L. Cohen, *Phys. Rev.*, B30, 4828, 1984. With permission.)

(Reference 4.70, p. 112). The InSb second-order elastic moduli at room temperature are c_{11} = 67.2 GPa, c_{12} = 36.7 GPa, and c_{44} = 30.2 GPa.[4.197] Slutsky and Garland[4.422] report c_{11} = 66.9 GPa, c_{12} = 36.45 GPa, and c_{44} = 30.2 GPa at 300 K, and c_{11} = 69.18 GPa, c_{12} = 37.88 GPa, and c_{44} = 31.32 GPa at 0 K (extrapolated). The InSb compressibility at room temperature is $15.09 \cdot 10^{-12}$ m²/N.[4.195]

InSb is a direct semiconductor. The first assumptions regarding its band structure were published by Herman,[4.423] who came to the conclusion, confirmed later, that both maximum of the valence band and minimum of the conduction band must be at the center of the Brillouin zone, and that there are two kinds of holes in the valence band. Later, the InSb band structure was calculated by Chelikowsky and Cohen[4.283,4.424] and investigated experimentally by Höchst and Hernandes-Calderon[4.425] and Williams et al.[4.284] The InSb band structure[4.283,4.424] is shown in Figure 4.68. One can see that the valence band is similar to that of other sphalerite semiconductors, with the only maximum at the Brillouin zone center. The conduction zone has the lowest minimum at the Brillouin zone center with the energy gap of 0.235 eV,[4.426] and two minima: one at L (1.03 eV above the valence band maximum (Reference 4.219, p. 142) and another at X (1.79 eV above the valence band maximum[4.427]). From the photogalvanomagnetic experiments by Goodwin et al.[4.428] and Littler and Seiler,[4.429] the InSb direct energy gap is 0.230 eV at 77 K and 0.2352 eV at 1.8 K.

The electron effective mass in InSb at the center of the Brillouin zone is 0.01359 m_o (at 4.2 K and n = $4.6 \cdot 10^{13}$ cm^{-3}).[4.430] This magnitude is in very good agreement with the early results received from the cyclotron resonance experiments by Dresselhaus et al.[4.431] (m_n^*/m_o = 0.013 ± 0.001) and Burstein et al.[4.432] (m_n^*/m_o = 0.015), and from the Seebeck effect measurements by Frederikse and Mielczarek[4.433] (m_n^*/m_o = 0.014). The electron effective mass at the L minimum is 0.09 m_o.[4.434] The heavy holes effective mass in InSb between 4 and 77 K is equal to 0.34 m_o along [100], 0.42 m_o along [110], and 0.45 m_o along [111] axis.[4.435] The light hole effective mass calculated from the magnetoplasma resonance experiments on p-InSb by Seiler et al.[4.436] is equal to 0.0147 m_o and 0.0158 m_o at 150 K and 77 K, respectively. Comparison of these data gives reason to assume that the heavy holes in InSb at 77 K present about 12% to 15% of the total number of holes.[4.433,4.437]

Even the earlier measurements of the InSb electrical properties[4.438,4.439] showed that the electron mobility is of the order of 10^4 to 10^5 cm²/V·sec and that it is about 80 to 100 times greater than the hole mobility. The greatest experimental magnitude of the electron Hall mobility at 300 K is close to $7 \cdot 10^4$ cm^2/V·sec (Madelung[4.440] cites $\mu_{n,H}(300\ K) = 7.7 \cdot 10^4$ cm^2/V·sec and $\mu_{n,H}(T) = 7.7 \cdot 10^4 (T/300)^{-5/3}$ cm^2/V·sec). Rode[4.441] showed from comparison of the data reported in literature that this temperature dependence is valid for both Hall and drift electron mobility in the limits 200 K and 800 K. The heavy hole mobility in InSb between 60 K and 500 K may be presented as $\mu_{ph}(T) \simeq 850(T/300)^{-1.95}$,[4.176] in satisfactory agreement with the earlier data by Goodwin[4.442] received from the photoconductivity and photomagnetic effect measurements. The mobility of light holes in InSb at 295 K is $3 \cdot 10^4$ cm^2/V·sec.[4.445]

The intrinsic carrier concentration in InSb was presented by Howarth et al.[4.443] in the form

$$n_i^2 = 1.8 \cdot 10^{32} (T/290)^3 \exp\left[-\frac{0.255}{k}\left(\frac{1}{T} - \frac{1}{290}\right)\right]\ (cm^{-3}) \qquad (4.1)$$

where k is a constant expressed in eV/K. Later, Cunningham and Gruber[4.444] suggested the expression

$$n_i = 5.76 \cdot 10^{14} T^{3/2} \exp(-0.26/2k_B T)\ cm^{-3} \qquad (4.2)$$

with $k_B = 8.617 \cdot 10^{-5}$ eV/K. From Equation 4.1, $n_i(290\ K) = 1.35 \cdot 10^{16}$ cm^{-3}; from Equation 4.2, $n_i(290\ K) = 1.56 \cdot 10^{16}$ cm^{-3} and $n_i(300\ K) = 1.96 \cdot 10^{16}$ cm^{-3}.

From the data on the mobility and concentration, the intrinsic conductivity at 300 K has a magnitude close to $2.3 \cdot 10^2$ S/cm. In the real InSb crystals, the measured conductivity at 300 K varies between 20 and 700 S/cm depending on the specimen purity and the chemical identity of the impurities.

The electrical conductivity of the melted InSb was reported by Ioffe and Regel,[4.446] Hamilton and Seidensticker,[4.447] and Glazov et al.[4.206] The melting process is accompanied by increase of electrical conductivity ($\sigma_{liquid}/\sigma_{solid} = 3.5$). The temperature dependence of σ of the liquid InSb is typical for a metal: σ (536°C) $\simeq$ 10 kS/cm and σ (700°C) $\simeq$ 9 kS/cm.[4.206]

The InSb thermal conductivity was studied by Busch and Schneider[4.453] and Stuckes.[4.454] Their data differ substantially at high temperatures. The low-temperature measurements were performed by Kosarev et al.;[4.455] their results are presented in Figure 4.69. The InSb room-temperature thermal conductivity is 0.17 W/cm·K[4.454] (Steigmeier and Kudman[4.171] cite a magnitude of 0.165 W/cm·K).

InSb is a diamagnetic with the molar magnetic susceptibility at room temperature of $-82.8 \cdot 10^{-5}$ SI units/g-mol (Reference 4.125, p. 105). The temperature dependence of the magnetic susceptibility was studied between 65 and 650 K by Stewens and Crawford[4.448] and between 700 and 930 K by Glazov et al.[4.206] The InSb dielectric constant at high frequencies is 15.68.[4.449] The static dielectric constant is equal to 16.8.[4.450] The InSb optical constants calculated by Aspnes and Studna[4.327] from the results of spectroscopic ellipsometry measurements are compiled in Table 4.11.

The technology of the InSb synthesis is one of the simplest among the III-V compounds. The ingots may be prepared by melting the stoichiometric mixture of the elemental components in evacuated and sealed fused silica ampules at a temperature that does not exceed 1100 to 1150 K. The zone recrystallization is very efficient for the InSb purification. This process is usually performed in the hydrogen atmosphere (see, e.g., References 4.451 and 4.452).

The single crystals of InSb can be grown by the Czochralski and Bridgman methods. The InSb thin films are very successfully used for highly sensitive magnetoresistance elements.[4.456]

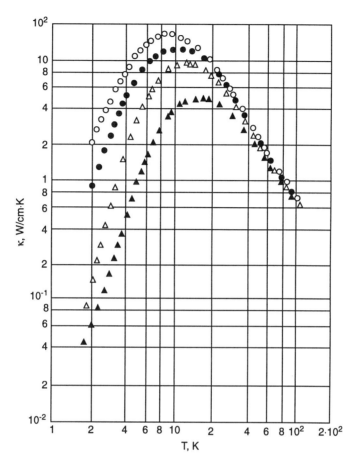

Figure 4.69 Temperature dependence of the thermal conductivity of several p-InSb samples. (After V.V. Kosarev, P.V. Tamarin, and S.S. Shalyt, *Phys. Stat. Sol.*, B44, 525, 1971. With permission.)

Table 4.11 The InSb Optical Parameters at 300 K

Photon energy (eV)	Dielectric constant Real part	Dielectric constant Imaginary part	Index of refraction	Extinction coefficient	Reflectance	Absorption coefficient (10^3 cm^{-1})
1.5	19.105	5.683	4.418	0.643	0.406	97.79
2.0	14.448	14.875	4.194	1.773	0.443	359.46
2.5	7.811	15.856	3.570	2.221	0.447	562.77
3.0	7.354	13.421	3.366	1.994	0.416	606.27
3.5	5.995	17.673	3.511	2.517	0.474	892.82
4.0	−6.722	19.443	2.632	3.694	0.608	1497.79
4.5	−6.297	8.351	1.443	2.894	0.598	1320.24
5.0	−4.250	6.378	1.307	2.441	0.537	1237.01
5.5	−4.325	4.931	1.057	2.333	0.563	1300.55
6.0	−3.835	3.681	0.861	2.139	0.572	1300.85

From D.E. Aspnes and A.A. Studna, *Phys. Rev.*, B27, 985, 1983.

Thin $In_{0.5-x}Sb_{0.5+x}$ ($0.07 \leq x \leq 0.10$) amorphous films present a certain interest for use as a recording media for phase change reversible optical storage disks.[4.457-4.459]

InSb is available commercially. The electronic grade (99.99%) shot is $6.30 to $8.20 per gram, the 99.999% pieces up to 6 mm in size — $14.70 to $19.85 per gram (1995 prices).

REFERENCES

4.1. N. Goryunova and E. Parthe, *Mater. Sci. Eng.*, 2, 1, 1967.
4.2. (a) P. J. Dean, W. J. Choyke, and L. Patrick, *J. Lumin.*, 15, 299, 1977; (b) *Amorphous and Crystalline Silicon Carbide II: Recent Developments,* Springer Proc. in Physics, Vol. 43, M. M. Rahman, C. Y.-W. Yang, and G. L. Harris, Eds., Springer-Verlag, New York, 1989.
4.3. A. Taylor and R. M. Jones, *Silicon Carbide — A High Temperature Semiconductor,* J. R. O'Connor and J. Smiltens, Eds., Pergamon Press, Elmsford, NY, 1960, 147.
4.4. G. A. Slack and S. F. Bartram, *J. Appl. Phys.*, 46, 89, 1975.
4.5. L. I. Berger, unpublished.
4.6. E. L. Kern, D. W. Hamill, H. W. Deem, and H. D. Sheets, *Mater. Res. Bull.*, 4, 25, 1969.
4.7. A. H. Gomes de Mesquita, *Acta Crystallogr.*, 23, 610, 1967.
4.8. D. R. Lide, Ed., *Handbook of Chemistry and Physics,* 74th ed., CRC Press, Boca Raton, FL, 1993–1994.
4.9. O. Ruff, *Trans. Electrochem. Soc.*, 68, 87, 1935.
4.10. H. Novotny, E. Parthe, R. Kieffer, and F. Benesovsky, *Monat. Chem.*, 85, 255, 1954.
4.11. R. I. Scace and G. A. Slack, *J. Chem. Phys.*, 30, 1551, 1959; see also in Reference 4.3, p. 24.
4.12. R. T. Dolloff, WADD Technical Rep. 60-143, Part 1, 1960.
4.13. I. A. Lely, *Ber. Dtsch. Keram. Ges.*, 32, 229, 1955.
4.14. L. I. Berger, V. M. Efimov, and L. A. Stroganova, *Trudy IREA (Proc. IREA),* 33, 247, 1971.
4.15. R. C. Marshall, J. W. Faust, and C. E. Ryan, Eds., *Silicon Carbide — 1973,* USC Press, Columbia, SC, 1974.
4.16. D. Bimberg, M. Altarelli, and N. O. Lipari, *Solid State Commun.*, 40, 437, 1981.
4.17. I. S. Gorban, V. A. Gubanov, V. G. Lysenko, A. A Pletyushkin, and V. B. Timofeev, *Sov. Phys. Solid State,* 26, 1385, 1984.
4.18. W. J. Choyke, *Mater. Res. Bull.*, 4, 141S, 1969.
4.19. R. Davlen, *J. Phys. Chem. Solids,* 26, 439, 1965.
4.20. H. R. Philipp and E. A. Taft, *Silicon Carbide — A High Temperature Semiconductor,* J. O'Connor and J. Smiltens, Eds., Pergamon Press, Elmsford, NY, 1960, 366.
4.21. M. M. Anikin, A. A. Lebedev, A. L. Syrkin, and A. V. Suvorov, *Sov. Phys. Semicond.*, 19, 69, 1985.
4.22. L. M. Hemstreet and C. Y. Fong, *Silicon Carbide — 1973,* R. C. Marshall, J. W. Faust, and C. E. Ryan, Eds., USC Press, Columbia, SC, 1974, 284.
4.23. P. T. B. Shaffer and R. G. Naum, *J. Opt. Soc. Am.*, 59, 1498, 1969.
4.24. J. A. Powell, *J. Opt. Soc. Am.*, 62, 341, 1972.
4.25. L. Patrick and W. J. Choyke, *Phys. Rev.*, B2, 2255, 1970.
4.26. Y. A. A. Ketelaar, *Z. Krist.*, 87, 436, 1934.
4.27. K. Zückler, *Halbleiterprobleme,* 3, 207, 1956.
4.28. G. Arlt and G. R. Schodder, *J. Acoust. Soc. Am.*, 37, 384, 1965.
4.29. G. Busch and H. Labhart, *Helv. Phys. Acta,* 19, 463, 1946.
4.30. W. E. Nelson, F. A. Holder, and A. Rosenblum, *J. Appl. Phys.*, 37, 333, 1966.
4.31. S. Nishino, J. A. Powell, and H. A. Will, *Appl. Phys. Lett.*, 42, 460, 1983.
4.32. G. A. Slack, *J. Appl. Phys.*, 35, 3460, 1964.
4.33. F. Ashworth, W. Needham, and W. W. Sillars, *Bell Syst. Tech. J.*, 26, 170, 1947.
4.34. L. Patrick, *J. Appl. Phys.*, 28, 765, 1957.
4.35. Y. S. Touloukian, R. K. Kirby, R. E. Taylor, and T. Y. R. Lee, Thermal expansion of nonmetallic solids, in *TPRC Data Series Vol. 13: Thermophysical Properties of Matter,* IFI/Plenum Press, New York, 1977.
4.36. B. S. Berry, W. C. Pritchet, R. I. Fuentes, and I. Babich, *J. Mater. Res.*, 6, 1061, 1991.
4.37. G. R. Fisher and P. Barnes, *Philos. Mag.*, B61, 217, 1990.
4.38. C. M. Van Vliet, G. Bosman, and L. L. Hench, *Annu. Rev. Mater. Sci.*, 18, 381, 1988.
4.39. J. H. Edgar, *J. Mater. Res.*, 7, 235, 1992.
4.40. Y. M. Tairov and Y. A. Vodakov, *Electroluminescence,* J. I. Pankove, Ed., Springer-Verlag, New York, 1989.
4.41. G. L. Harris and C. Y.-W. Yang, Eds., *Amorphous and Crystalline Silicon Carbide,* Springer-Verlag, New York, 1989.

4.42. J. T. Glass, R. Messier, and N. Fujimori, Eds., Diamond, silicon carbide and related wide band gap semiconductors, in Mater. Res. Soc. Symp. Proc. 162, Pittsburgh, PA, 1990.
4.43. R. F. Davis, J. W. Palmour, and J. A. Edmond, Diamond, silicon carbide and related wide band gap semicondcutors, in Mater. Res. Soc. Symp. Proc. 162, J. T. Glass, R. Messier, and N. Fujimori, Eds., Pittsburgh, PA, 1990, 463.
4.44. D. L. Barrett, R. G. Seidenstricker, W. Gaida, R. H. Hopkins, and W. J. Choyke, *J. Cryst. Growth*, 109, 17, 1991.
4.45. M. A. El Khakani, M. Chaker, A. Jean, S. Boily, J. C. Kieffer, M. E. O'Hern, M. F. Ravet, and F. Rousseaux, *J. Mater. Res.*, 9, 96, 1994.
4.46. R. A. Graham and A. B. Sawaoka, Eds., *High Pressure Explosive Processing of Ceramics*, Trans. Techn. Publ. Ltd., Aedermannsdorf, Switzerland, 1987.
4.47. R. H. Wentorff, *J. Chem. Phys.*, 26, 956, 1957; F. P. Bundy and R. H. Wentorff, *J. Chem. Phys.*, 38, 1144, 1963.
4.48. T. Akashi and A. B. Sawaoka, *High Pressure Explosive Processing of Ceramics*, R. A. Graham and A. B. Sawaoka, Eds., Trans. Techn. Publ. Ltd., Aedermannsdorf, Switzerland, 1987, 87.
4.49. D. R. Lide, Ed., *Handbook of Chemistry and Physics*, 74th ed., CRC Press, Boca Raton, FL, 1993–1994, 5-4.
4.50. F. R. Corrigan, High-pressure science and technology, in *Proc. 6th AIRAPT Conf.*, Vol. 1, K. D. Timmerhaus and M. S. Barber, Eds., Plenum Press, New York, 1977, 994.
4.51. T. Soma, S. Sawaoka, and S. Saito, *Mater. Res. Bull.*, 9, 755, 1974.
4.52. R. W. Linch and H. G. Drickamer, *J. Chem. Phys.*, 44, 181, 1966.
4.53. J. A. Van Vechten, *Phys. Rev.*, B7, 1479, 1973.
4.54. E. Parthe, *Crystal Chemistry of Tetrahedral Structures*, Gordon & Bleach, 1964.
4.55. A. Zunger, A. Katzin, and A. Halperin, *Phys. Rev.*, B13, 5560, 1976.
4.56. J. L. Hurd, D. L. Perry, B. T. Lee, K. M. Yu, E. D. Bourret, and E. E. Haller, *J. Mater. Res.*, 4, 350, 1989.
4.57. E. Podszus, *Z. Anorg. Chem.*, 30, 156, 1917.
4.58. G. A. Meerson, G. V. Samsonov, and N. Ya. Tseitlina, *Ogneupory (Refractories)*, 20, 72, 1955.
4.59. L. Vel, G. Demazeau, and J. Etourneau, *Mater. Sci. Eng.*, B10, 149, 1991.
4.60. M. Sano and M. Aoki, *Thin Solid Films*, 83, 247, 1981.
4.61. S. Motojima, Y. Tamura, and K. Sugiyama, *Thin Solid Films*, 88, 269, 1982.
4.62. R. N. Singh, Proc. 10th Int. Conf. on CVD, Electrochemical Society, Pennington, 1987, 543.
4.63. V. Pavlović, H.-R. Kötter, and C. Meixner, *J. Mater. Res.*, 6, 2393, 1991.
4.64. D. M. Hoffman, G. L. Doll, and P. C. Eklund, *Phys. Rev.*, B30, 6051, 1984.
4.65. J. Robertson, *Phys. Rev.*, B29, 2131, 1984.
4.66. A. Catellani, M. Pasternak, A. Baldereshi, H. J.-F. Jansen, and A. J. Freeman, *Phys. Rev.*, B32, 6997, 1985.
4.67. L. G. Carpenter and P. J. Kirby, *J. Phys.*, D15, 1143, 1982.
4.68. K. Taylor, *Ind. Eng. Chem.*, 47, 2506, 1955.
4.69. D. K. Potter and T. J. Ahrens, *J. Appl. Phys.*, 63, 910, 1987; H. Tan and T. J. Ahrens, *J. Mater. Res.*, 3, 1010, 1988.
4.70. N. A. Goryunova, *The Chemistry of Diamond-Like Semiconductors*, MIT Press, Cambridge, 1965.
4.71. R. W. Powell and R. P. Tye, *Proc. Brit. Ceram. Soc., Spec. Ceramics*, Academic Press, New York, 1963, 261.
4.72. R. S. Pease, *Acta Crystallogr.*, 5, 356, 1952.
4.73. R. Geick, C. H. Perry, and G. Rupprecht, *Phys. Rev.*, 146, 543, 1966.
4.74. R. Mamy, J. Thomas, G. Jezequel, and J. C. Lemonnier, *J. Phys. Lett.*, 42, L473, 1981.
4.75. T. Ishii and T. Sato, *J. Cryst. Growth*, 61, 689, 1983.
4.76. E. Brank, J. Margrave, and V. Meloche, *J. Inorg. Nucl. Chem.*, 5, 48, 1957.
4.77. R. H. Wentroff, *J. Chem. Phys.*, 26, 956, 1957.
4.78. A. G. Merzhanov and Borovinskaya, *U.S.S.R. Acad. Sci. Dokl. Chem.*, 204, 429, 1972.
4.79. G. A. Slack, *J. Phys. Chem. Solids*, 34, 321, 1973.
4.80. R. M. Chrenko, *Solid State Commun.*, 14, 511, 1974.
4.81. Z. A. Munir and U. Anselmi-Tamburini, *Mater. Sci. Rep.*, 3, 277, 1989.
4.82. J. B. Holt and S. D. Danmead, *Annu. Rev. Mater. Sci.*, 21, 305, 1991.

4.83. J. Nagayan, R. Raghunathan, R. Chowdhury, and K. Jagannadham, *J. Appl. Phys.,* 75, 7252, 1994.
4.84. M. Huang and W. Y. Ching, *J. Phys. Chem. Solids,* 46, 977, 1985.
4.85. C. Prasad and J. D. Dubey, *Phys. Status Solidi,* B125, 625, 1984.
4.86. O. Mishima, J. Tanaka, S. Yamaoka, and O. Fukunaga, *Science,* 238, 181, 1987.
4.87. O. Mishima, K. Era, J. Tanaka, and S. Yamaoka, *Appl. Phys. Lett.,* 53, 962, 1988.
4.88. I. S. Bam, V. M. Davidenko, V. G. Sidorov, L. I. Feldgun, M. D. Skagalov, and Yu. K. Shalabutov, *Sov. Phys. Semicond.,* 10, 331, 1976.
4.89. E. F. Steigmeier, *Appl. Phys. Lett.,* 3, 6, 1963.
4.90. K. Ichinose, M. Wakatsuki, T. Aoki, and Y. Mead, Proc. 4th Int. Conf. on High Pressure — 1974, J. Osugi, Ed.; Spec. Issue of the Rev. of Phys. Chem. of Jpn., Kawakita, Kioto, 1975, 436.
4.91. O. A. Golikova, *Phys. Status Solidi,* A51, 11, 1979.
4.92. R. Wickery, *Nature,* 184, 268, 1959.
4.93. A. K. Ballal, L. Salamanca-Riba, G. L. Doll, C. A. Taylor, and R. Clarke, *J. Mater. Res.,* 7, 1618, 1992.
4.94. D. J. Kester, K. S. Ailey, R. F. Davis, and K. L. More, *J. Mater. Res.,* 8, 1213, 1993.
4.95. P. J. Gielisse, S. S. Mitra, J. N. Plendl, R. D. Griffis, L. C. Mansur, R. Marshall, and E. A. Pascoe, *Phys. Rev.,* 155, 1039, 1967.
4.96. G. L. Doll, J. A. Sell, C. A. Taylor, and R. Clarke, *Phys. Rev.,* B43, 6816, 1991.
4.97. M. Okamoto, Y. Utsumi, and Y. Osaka, *Jpn. J. Appl. Phys.,* 29, L1004, 1990.
4.98. J. L. Peret, *J. Am. Ceram. Soc.,* 47, 44, 1964.
4.99. T. Nishinaga, *Oyo Butsuri,* 55, 1069, 1986.
4.100. Y. Kumashiro, Y. Okada, and S. Gonda, *J. Cryst. Growth,* 70, 507, 1985.
4.101. C. J. Kim and K. Shono, *J. Electrochem. Soc.,* 131, 120, 1984.
4.102. G. A. Slack and S. F. Bartram, *J. Appl. Phys.,* 46, 89, 1975.
4.103. S. Rundquist, Mineral Chemistry, 16th Congress IUPAC, Paris, 1958, 539.
4.104. B. Stone and D. Hill, *Phys. Rev. Lett.,* 4, 282, 1960.
4.105. W. Wettling and J. Windschleif, *Solid State Commun.,* 50, 33, 1984.
4.106. Y. Kumashiro, *J. Mater. Res.,* 5, 2933, 1990.
4.107. L. A. Hemstreet and C. Y. Fong, *Phys. Rev.,* B6, 1464, 1972.
4.108. M. Huang and W. Y. Ching, *Phys. Chem. Solids,* 46, 977, 1985.
4.109. Y. Kumashiro, T. Mitsuhashi, S. Okaya, F. Muta, T. Koshiro, Y. Takahashi, and M. Hirabayashi, *J. Appl. Phys.,* 65, 2147, 1989; Y. Kumashiro, T. Mitsuhashi, S. Okaya, F. Muta, T. Koshiro, Y. Takahashi, M. Hirabayashi, and Y. Okada, *High Temp. High Pressure,* 21, 105, 1989.
4.110. Y. Kumashiro, M. Hirabayashi, and T. Koshiro, *J. Less Common Met.,* 143, 159, 1988.
4.111. N. Kato, W. Kamura, M. Ivami, and K. Kawabe, *Jpn. J. Appl. Phys.,* 16, 1623, 1977.
4.112. S. Yugo and T. Kimura, *Phys. Status Solidi,* A59, 363, 1980.
4.113. S. Yugo, T. Sato, and T. Kimura, *Appl. Phys. Lett.,* 46, 842, 1985.
4.114. C. C. Wang, M. Cardona, and A. G. Fisher, *RCA Rev.,* 25, 159, 1964.
4.115. K. Sohno, M. Takigawa, and T. Nakada, *J. Cryst. Growth,* 24–25, 193, 1974.
4.116. T. Takenaka, M. Takigawa, and K. Sohno, *Jpn. J. Appl. Phys.,* 15, 2021, 1976.
4.117. M. A. Besson, *Compt. Rend. Paris,* 113, 78, 1891.
4.118. M. H. Moissan, *Compt. Rend. Paris,* 113, 624 and 726, 1891.
4.119. P. Popper and T. A. Ingles, *Nature,* 179, 1075, 1956.
4.120. J. A. Perri, S. LaPlaca, and B. Post, *Acta Crystallogr.,* 11, 310, 1958.
4.121. R. V. Williams and R. A. Ruherwein, *J. Am. Chem. Soc.,* 82, 1330, 1960.
4.122. T. L. Chu, J. M. Jackson, A. E. Hyslop, and S. C. Chu, *J. Appl. Phys.,* 42, 420, 1971.
4.123. T. Nishinaga, H. Ogawa, H. Watanabe, and T. Arizumi, *J. Cryst. Growth,* 13–14, 346, 1972.
4.124. T. Kobayashi, K. Susa, and S. Taniguchi, *Mater. Res. Bull.,* 9, 625, 1974.
4.125. N. A. Goryunova, Slozhnyye Almazopodobnyye Poluprovodniki (Multinary Diamond-Like Semiconductors), Soviet Radio, Moscow, 1968.
4.126. L. Merrill, *J. Phys. Chem. Ref. Data,* 6, 1205, 1977.
4.127. D. J. Stuckel, *Phys. Rev.,* B1, 3458, 1970.
4.128. W. Class, An Aluminum Nitride Melting Technique, NASA CR-1171, Fin. Rep. by Mater. Res. Corp. for Lewis Res. Cntr. Contract NAS3-10659.
4.129. T. Renner, *Z. Anorg. Chem.,* 298, 22, 1959.

4.130. D. N. Poluboyarinov, M. R. Gordova, I. G. Kuznetsova, M. D. Bershadskaya, and V. G. Avetikov, *Inorg. Mater.*, 15, 2055, 1979.
4.131. N. Azema, R. Durand, C. Dupuy, and L. Cot, *J. Eur. Ceram. Soc.*, 8, 291, 1991.
4.132. A. Bellosi, E. Landi, and A. Tampieri, *J. Mater. Res.*, 8, 565, 1993.
4.133. H. Scholz and K. H. Thiemann, *Solid State Commun.*, 23, 815, 1977.
4.134. S. Iwama, K. Hayakawa, and T. Arizumi, *J. Cryst. Growth*, 56, 265, 1982.
4.135. A. D. Westwood and M. R. Notis, *J. Am. Ceram. Soc.*, 74, 1226, 1991.
4.136. R. Pandey, A. Sutjiano, M. Seel, and J. E. Jaffe, *J. Mater. Res.*, 8, 1922, 1993.
4.137. I. Gorczyca, N. E. Christensen, P. Perlin, I. Grzegory, J. Jun, and M. Bockowski, *Solid State Commun.*, 79, 1033, 1991.
4.138. H. Vollstadt, E. Ito, M. Akishi, S. Akimoto, and O. Fukunaga, *Proc. Jpn. Acad.*, B66, 7, 1990.
4.139. S. Loughin, R. H. French, G. A. Slack, and J. B. Blum, Covalent Ceramics, G. S. Fischman, R. M. Spriggs, and T. S. Aselage, Eds., MRS Symp. Proc. EA-23, Pittsburgh, PA, 1990.
4.140. W. Ching and B. Harmon, *Phys. Rev.*, B34, 5305, 1986.
4.141. M. Ueno, A. Onodera, O. Shimomura, and K. Takemura, *Phys. Rev.*, B45, 10123, 1992.
4.142. A. A. Ivanko, *Handbook of Hardness Data*, G. V. Samsonov, Ed., Keter Press, Jerusalem, 1971.
4.143. L. M. Sheppard, *Ceram. Bull.*, 69, 1801, 1990.
4.144. R. Berman, *Thermal Conduction in Solids*, Clarendon Press, London, 1978.
4.145. R. R. Tummala, *Am. Ceram. Soc. Bull.*, 67, 752, 1992.
4.146. G. A. Slack, *J. Phys. Chem. Solids*, 34, 321, 1973.
4.147. P. Boch, J. C. Glandus, J. Jarrige, J. P. LeCompte, and J. Mexmin, *Ceram. Int.*, 8, 34, 1982.
4.148. G. DeWith and N. Hattu, *J. Mater. Sci.*, 18, 503, 1983.
4.149. H. Yamashita, K. Fukui, S. Misawa, and S. Yoshida, *J. Appl. Phys.*, 50, 896, 1979.
4.150. L. Roskovkova and J. Pastrnak, *Chech. J. Phys.*, B30, 586, 1980.
4.151. A. Kobayashi, O. F. Sankey, S. M. Volz, and J. D. Dow, *Phys. Rev.*, B28, 935, 1983.
4.152. T. L. Chu, D. W. Ing, and A. J. Noreika, *Solid State Electron.*, 10, 1023, 1967.
4.153. V. F. Veselov, A. V. Dobrynin, G. A. Naida, P. A. Pundur, E. A. Slotsenietse, and E. B. Sokolov, *Inorg. Mater.*, 25, 1250, 1989.
4.154. J. H. Harris and R. A. Youngman, *J. Mater. Res.*, 8, 154, 1993.
4.155. V. W. Chin, T. L. Tansley, and T. Osotchan, *J. Appl. Phys.*, 75, 7365, 1994.
4.156. J. Edwards, K. Kawabe, G. Stevens, and R. H. Tredgold, *Solid State Commun.*, 3, 99, 1965.
4.157. J. S. Thorp, D. Evans, M. Al-Naief, and M. Akhtaruzzman, *J. Mater. Sci.*, 25, 4965, 1990.
4.158. A. T. Collins, E. C. Lightowlers, and P. J. Dean, *Phys. Rev.*, 158, 833, 1967.
4.159. F. S. Ohuchi and R. H. French, *J. Vac. Sci. Technol.*, A6, 1695, 1988.
4.160. R. Pandey, A. Sutjianto, M. Seel, and J. E. Joffe, *J. Mater. Res.*, 8, 1922, 1993.
4.161. R. F. Rutz, *Appl. Phys. Lett.*, 28, 379, 1976.
4.162. T. Shiosaki, T. Yamamoto, T. Oda, and A. Kawabata, *Appl. Phys. Lett.*, 36, 643, 1980.
4.163. S. Yoshida, S. Misawa, Y. Fujii, S. Takada, H. Hayakawa, S. Gonda, and A. Itoh, *J. Vac. Sci. Technol.*, 16, 990, 1979.
4.164. W. M. Yin, E. J. Stofko, P. J. Zanzucchi, M. Ettenberg, and S. L. Gilbert, *J. Appl. Phys.*, 44, 292, 1973.
4.165. S. C. Yu, I. L. Spain, and E. F. Skelton, *Solid State Commun.*, 25, 49, 1978.
4.166. K. Osamura and Y. Murakami, *J. Phys. Chem. Solids*, 36, 1354, 1975.
4.167. H. M. Kagaya and T. Soma, *Phys. Status Solidi*, B127, 89, 1985.
4.168. V. N. Bessolov, S. G. Konnikov, V. I. Umanskii, and Y. P. Yakovlev, *Sov. Phys. Solid State*, 24, 875, 1982.
4.169. R. J. Caveney, *Philos. Mag.*, 17, 943, 1968.
4.170. M. Neuberger, *Handbook of Electronic Materials*, Vol. 2, IFI/Plenum Press, New York, 1971.
4.171. E. F. Steigmeier and I. Kudman, *Phys. Rev.*, 141, 767, 1966.
4.172. P. D. Maycock, *Solid State Electron.*, 10, 161, 1967.
4.173. F. Herman, *J. Electron.*, 1, 103, 1955.
4.174. B. Monemar, *Phys. Rev.*, B8, 5711, 1973.
4.175. F. G. Reid, S. E. Miller, and H. L. Goering, *J. Electrochem. Soc.*, 113, 467, 1966.
4.176. J. D. Wiley, *Semiconductors and Semimetals*, Vol. 10, R. K. Willardson and A. C. Beer, Eds., Academic Press, New York, 1975, 235.
4.177. H. G. Grimmeiss, W. Kishio, and A. Rabenau, *Phys. Chem. Solids*, 16, 302, 1960.

4.178. B. Monemar, *Solid State Commun.*, 8, 1295, 1970.
4.179. M. Ettenberg and R. J. Pfaff, *J. Appl. Phys.*, 41, 3926, 1970.
4.180. Yu. I. Pashintsev and N. N. Sirota, *Dokl. Akad. Nauk SSSR*, 3, 38, 1959.
4.181. W. Kishio, *Z. Anorg. Allg. Chem.*, 328, 187, 1964.
4.182. S. Adachi, *J. Appl. Phys.*, 58, R1, 1958.
4.183. H. J. Lee, L. Y. Juravel, J. C. Wooley, W. F. Spring, and A. J. Thorpe, *Phys. Rev.*, B21, 659, 1980.
4.184. J. Whitaker, *Solid State Electron.*, 8, 649, 1965.
4.185. P. Leroux-Hugon, N. X. Xinh, and M. Rodot, *J. Phys.*, 28, Cl-73, 1967.
4.186. R. E. Fern and A. Onton, *J. Appl. Phys.*, 42, 3499, 1971.
4.187. D. Campi and C. Paruzza, *J. Appl. Phys.*, 57, 1305, 1985.
4.188. O. G. Folbert and F. Oswald, *Z. Naturforsch.*, 9A, 1050, 1954.
4.189. A. S. Borshchevskii and D. N. Tretiyakov, *Sov. Phys. Solid State*, 1, 1483, 1959.
4.190. G. Giesecke and H. Pfister, *Acta Crystallogr.*, 11, 369, 1958.
4.191. M. Baublitz and A. L. Ruoff, *J. Appl. Phys.*, 54, 2109, 1982.
4.192. K. Osamura and Y. Murakami, *J. Phys. Chem. Solids*, 36, 931, 1975.
4.193. G. G. Urazov, *Izv. Inst. Fiz. Khim. Anal. (Proc. Inst. Phys. Chem. Anal.)* 1, 461, 1921.
4.194. S. I. Novikova and N. Kh. Abrikosov, *Sov. Phys. Solid State*, 5, 1558, 1964.
4.195. U. Piesbergen, *Z. Naturforsch.*, 18A, 141, 1963.
4.196. D. I. Bolef and M. Menes, *J. Appl. Phys.*, 31, 1426, 1960.
4.197. H. J. McSkimin, W. L. Bond, G. L. Pearson, and H. J. Hrostowski, *Bull. Am. Phys. Soc.*, Ser. 2, 1, 111, 1956.
4.198. R. Weil, *J. Appl. Phys.*, 43, 4271, 1972.
4.199. C. Alibert, A. Joullié, A. M. Joullié, and C. Ance, *Phys. Rev.*, B27, 4946, 1983.
4.200. A. Joullié, B. Girault, A. M. Joullié, and A. Zien-Eddine, *Phys. Rev.*, B25, 7830, 1982.
4.201. A. A. Kopylov, *Solid State Commun.*, 56, 1, 1985.
4.202. H. Welker, *Z. Naturforsch.*, 8A, 248, 1953; *Physica*, 20, 893, 1954.
4.203. F. J. Reid and R. K. Willardson, *J. Electron. Control*, 5, 54, 1958.
4.204. D. N. Nasledov and S. V. Slobodchikov, *Sov. Phys. Tech. Phys.*, 3, 669, 1958.
4.205. R. J. Stirn and W. M. Becker, *Phys. Rev.*, 141, 621, 1966; 148, 907, 1966.
4.206. V. M. Glazov, S. N. Chizhevskaya, and N. N. Glagoleva, *Zhidkiye Poluprovodniki (Liquid Semiconductors)*, Nauka, Moscow, 1967.
4.207. V. M. Muzhdaba, A. Ya. Nashel'skii, P. V. Tamarin, and S. S. Shalyt, *Sov. Phys. Solid State*, 10, 2265, 1969.
4.208. F. Oswald and R. Schade, *Z. Naturforsch.*, 9A, 611, 1954.
4.209. W. J. Turner and W. E. Reese, *Phys. Rev.*, 127, 126, 1962.
4.210. M. Hass and B. W. Henvis, *J. Phys. Chem. Solids*, 23, 1099, 1962.
4.211. F. Sauerwald, *Z. Metallkunde*, 14, 457, 1922.
4.212. H. A. Shell, *Z. Metallkunde*, 49, 140, 1958.
4.213. J. Ross, M. Rubin, and T. K. Gustafson, *J. Mater. Res.*, 8, 2613, 1993.
4.214. O. Lagerstedt and B. Monemar, *Phys. Rev.*, B19, 3064, 1979.
4.215. H. P. Maruska and J. J. Tietjen, *Appl. Phys. Lett.*, 15, 327, 1969.
4.216. A. U. Sheleg and V. A. Savastenko, *Vesti Akad. Nauk BSSR Ser. Fiz. Mat. Nauk*, 3, 126, 1976.
4.217. R. Groh, G. Gerey, L. Bartha, and J. I. Pankove, *Phys. Status Solidi*, A26, 353, 1971.
4.218. T. Renner, *Z. Anorg. Chem.*, 298, 22, 1959.
4.219. O. Madelung, Ed., Semiconductors. Group IV elements and III-V compounds, in Series *Data in Science and Technology*, R. Pörschke, Ed.-in-Chief, Springer-Verlag, Berlin, 1991.
4.220. B. Monemar, *Phys. Rev.*, B10, 676, 1974.
4.221. R. Kauer and A. Rabenau, *Z. Naturforsch.*, 12A, 942, 1957.
4.222. S. Bloom, G. Harbeke, E. Meier, and I. B. Ortenburger, *Phys. Status Solidi*, 66B, 161, 1974.
4.223. B. Rheinländer and H. Neumann, *Phys. Status Solidi*, 64B, K123, 1974.
4.224. J. I. Pankove, S. Bloom, and G. Harbeke, *RCA Rev.*, 36, 163, 1975.
4.225. M. J. Paisley, Z. Sitar, J. B. Posthill, and R. F. Davis, *J. Vac. Sci. Technol.*, A7, 701, 1989.
4.226. W. Seifert, R. Franzheld, E. Bitter, H. Sobotta, and V. Riede, *Cryst. Res. Technol.*, 18, 383, 1983.
4.227. M. Ilegems and H. C. Montgomery, *J. Phys. Chem. Solids*, 34, 885, 1973.
4.228. S. Nakamura, T. Mikai, and M. Senoh, *J. Appl. Phys.*, 71, 5543, 1992.
4.229. B. Gelmont, K. Kim, and M. Shur, *J. Appl. Phys.*, 74, 1818, 1993.

4.230. V. W. Chin, T. L. Tansley, and T. Osotchan, *J. Appl. Phys.*, 75, 7365, 1994.
4.231. H. Amano, M. Kito, K. Hiramatsu, and I. Akasaki, *Jpn. J. Appl. Phys.*, 28, L2112, 1989; see, also, *GaAs and Related Compounds, 198*, T. Ikoma and H. Watanabe, Eds., Inst. of Phys., Bristol, U.K., 1990, 725.
4.232. E. K. Sichel and J. I. Pankove, *J. Phys. Chem. Solids*, 38, 330, 1977.
4.233. A. S. Barker and M. Ilegems, *Phys. Rev.*, B7, 743, 1973.
4.234. D. D. Manchon, A. S. Barker, J. P. Dean, and R. B. Zetterstrom, *Solid State Commun.*, 8, 1227, 1970.
4.235. E. Ejder, *Phys. Status Solidi*, 6A, K39, 1971.
4.236. K. Das and D. K. Ferry, *Solid State Electron.*, 19, 851, 1976.
4.237. J. I. Pankove, *J. Lumin.*, 7, 114, 1973.
4.238. G. Jacob, R. Madar, and J. Hallis, *Mater. Res. Bull.*, 11, 445, 1976.
4.239. K. Naniwae, S. Itoh, H. Amano, K. Itoh, K. Hiramatsu, and I. Akasaki, *J. Cryst. Growth*, 99, 381, 1990.
4.240. I. Akasaki, H. Amano, Y. Koide, K. Haramatsu, and N. Savaki, *J. Cryst. Growth*, 98, 209, 1990.
4.241. H. Amano, I. Akasaki, K. Hiramatsu, and Y. Koide, *Thin Solid Films*, 163, 415, 1986.
4.242. S. Nakamura, T. Mukai, and M. Senoh, *J. Appl. Phys.*, 71, 5543, 1992.
4.243. D. Elwell and M. M. Elwell, *Prog. Cryst. Growth Character.*, 17, 53, 1988.
4.244. G. S. Zhdanov and G. V. Mirman, *Sov. Phys. ZETF*, 6, 1201, 1936.
4.245. H. Hahn and R. Juza, *Z. Anorg. Allg. Chem.*, B244, 111, 1940.
4.246. H. J. Hovel and J. J. Cuomo, *Appl. Phys. Lett.*, 20, 71, 1972.
4.247. D. Elwell, R. S. Feigelson, M. M. Simkins, and W. A. Tiller, *J. Cryst. Growth*, 66, 45, 1984.
4.248. M. Matloubian and M. Gershenson, *J. Electron. Mater.*, 14, 633, 1985.
4.249. H. Gotoh, T. Suga, H. Suzuki, and M. Kimata, *Jpn. J. Appl. Phys.*, 20, L545, 1981.
4.250. S. W. Choi, K. J. Bachmann, and G. Lukovsky, *J. Mater. Res.*, 8, 847, 1993.
4.251. K. Matsubara and T. Takagi, *Jpn. J. Appl. Phys.*, 22, 511, 1983.
4.252. S. Yoshida, S. Misawa, and S. Gonda, *J. Vac. Sci. Technol.*, B1, 250, 1983.
4.253. M. E. Straumanis and J.-P. Krumme, *J. Electrochem. Soc.*, 114, 640, 1967.
4.254. E. D. Pierron, D. I. Parker, and J. B. McNeely, *J. Appl. Phys.*, 18, 4669, 1967.
4.255. V. M. Glazov, A. Ya. Nashel'skii, and A. D. Lyutfalibekova, *Elektron. Tech.*, 14, 33, 1970.
4.256. H. L. Casey and F. A. Trumbore, *Mater. Sci. Eng.*, 6, 69, 1970.
4.257. I. Kudman and R. J. Pfaff, *J. Appl. Phys.*, 43, 3760, 1972.
4.258. S. I. Novikova, *Teplovoe Rasshireniye Tvyordykh Tyel (Thermal Expansion of Solids)*, Nauka, Moscow, 1974.
4.259. H. Oettel, G. Heide, and J. Sedivý, *Chech. J. Phys.*, B33, 175, 1983.
4.260. V. N. Bessolov, T. T. Dedgkaev, A. N. Efimov, N. F. Kartenko, and Yu. P. Yakovlev, *Sov. Phys. Solid State*, 22, 1652, 1980.
4.261. H. G. Brühl, K. Jacob, W. Seifert, and J. Mäge, *Krist. Tech.*, 14, K27, 1979.
4.262. P. Deus, U. Voland, and H. A. Schneider, *Phys. Status Solidi*, 80, K29, 1983.
4.263. P. Deus, U. Voland, and H. Oettel, *Exp. Tech. Phys.*, 32, 433, 1984.
4.264. T. Soma, *Solid State Commun.*, 34, 375, 1980.
4.265. D. D. Ackerman and A. C. Anderson, *Rev. Sci. Instrum.*, 53, 1657, 1982.
4.266. B. de Cremoux, *IEEE J. Quantum Electron.*, QE-17, 118, 1981.
4.267. M. Imar, A. Gabriel, C. Chatillon, and I. Ansara, *J. Cryst. Growth*, 68, 557, 1984.
4.268. M. Baublitz and A. L. Ruoff, *J. Appl. Phys.*, 53, 6179, 1982.
4.269. T. Soma and H.-Matsuo Kagaya, *Solid State Commun.*, 50, 261, 1984.
4.270. T. Soma, Y. Takahashi, and H.-Matsuo Kagaya, *Solid State Commun.*, 53, 801, 1985.
4.271. M. B. Panish, *J. Cryst. Growth*, 27, 6, 1974.
4.272. A. S. Jordan, A. R. Von Neida, R. Caruso, and C. K. Kim, *J. Electrochem. Soc.*, 121, 153, 1974.
4.273. J. A. Van Vechten, *J. Electrochem. Soc.*, 122, 423, 1975.
4.274. A. I. Ivashchenko, F. Ya. Kopanskaya, and G. S. Kuz'menko, *J. Phys. Chem. Sol.*, 45, 871, 1984.
4.275. V. V. Tarasov and A. F. Demidenko, *Phys. Status Solidi*, 30, 147, 1968.
4.276. C. J. Frosch and L. Derick, *J. Electrochem. Soc.*, 103, 251, 1961.
4.277. R. Weil and W. O. Groves, *J. Appl. Phys.*, 39, 4049, 1968.
4.278. Y. K. Yogurtcu, A. J. Miller, and G. A. Saunders, *J. Phys. Chem. Solids*, 42, 49, 1981.
4.279. W. F. Boyle and R. J. Sladek, *Phys. Rev.*, B11, 2933, 1975.

4.280. R. G. Humphreys, U. Rössler, and M. Cardona, *Phys. Rev.*, B18, 5590, 1978.
4.281. M. B. Panish and H. C. Casey, *J. Appl. Phys.*, 40, 163, 1969.
4.282. D. F. Nelson, L. F. Johnson, and M. Gershenson, *Phys. Rev.*, 135, A1399, 1964.
4.283. J. R. Chelikowsky and M. L. Cohen, *Phys. Rev.*, B14, 556, 1976.
4.284. G. P. Williams, F. Cerrina, G. J. Lapeyre, J. R. Anderson, R. J. Smith, and J. Hermanson, *Phys. Rev.*, B34, 5548, 1986.
4.285. N. Miura, G. Kido, M. Suekane, and S. Chukazumi, *J. Phys. Soc. Jpn.*, 52, 2838, 1983.
4.286. C. P. Schwerdtfeger, *Solid State Commun.*, 11, 779, 1972.
4.287. A. C. Sharma, N. M. Ravindra, S. Auluck, and V. K. Srivastava, *Phys. Status Solidi*, B120, 715, 1983.
4.288. K. Takeda, N. Matsumoto, A. Taguchi, H. Taki, E. Ohta, and M. Sakata, *Phys. Rev.*, B32, 1101, 1985.
4.289. G. A. Samara, *Phys. Rev.*, B27, 3494, 1983.
4.290. C. Hilsum and A. C. Rose-Innes, *Semiconducting III-V Compounds,* Pergamon Press, Elmsford, NY, 1961, 1.
4.291. J. B. Mullin, B. W. Straughan, C. M. H. Driscoll, and A. F. W. Willoughby, AIP Conf. Ser. 24, 1975, 275.
4.292. G. Giesecke and H. Pfister, *Acta Crystallogr.*, 11, 369, 1958.
4.293. M. E. Straumanis and C. D. Kim, *Acta Crystallogr.*, 19, 256, 1965.
4.294. J. van den Boomgaard and K. Schol, *Philips Res. Rep.*, 12, 127, 1957.
4.295. S. I. Novikova, *Sov. Phys. Solid State*, 3, 129, 1961.
4.296. T. Soma, J. Satoh, and H. Matsuo, *Solid State Commun.*, 42, 889, 1982.
4.297. N. N. Sirota and L. I. Berger, *Inzh. Fiz. Zh. Akad. Nauk BSSR*, 2, 104, 1959.
4.298. J. S. Blakemore, *J. Appl. Phys.*, 53, R123, 1982.
4.299. E. F. Steigmeier and I. Kudman, *Phys. Rev.*, 141, 767, 1966.
4.300. J. R. Drabble, *Solid State Commun.*, 4, 467, 1966.
4.301. T. B. Bateman, H. J. McSkimin, and J. M. Whelan, *J. Appl. Phys.*, 30, 544, 1959.
4.302. C. W. Garland and K. C. Park, *J. Appl. Phys.*, 33, 759, 1962.
4.303. V. Swaminathan, W. R. Wagner, and P. J. Anthony, *J. Electrochem. Soc.*, 130, 2468, 1983.
4.304. A. Djemel and J. Castaing, *Europhys. Lett.*, 2, 616, 1986.
4.305. D. Laister and G. M. Jenkins, *J. Mater. Sci.*, 8, 1218, 1973.
4.306. V. Swaminathan and S. M. Coplley, *J. Am. Ceram. Soc.*, 58, 482, 1975.
4.307. M. Steinhard and P. Haasen, *Phys. Status Solidi*, 93, 93, 1978.
4.308. I. Yonenaga, U. Onose, and K. Sumino, *J. Mater. Res.*, 2, 252, 1987.
4.309. L. F. Grigor'eva, Ph.D. thesis, Leningr. Inz. Stroit. Inst., 1960.
4.310. H. Welker and H. Weiss, *Solid State Physics,* Vol. 3, F. Seitz and D. Turnbull, Eds., Academic Press, New York, 1956, 51.
4.311. J. S. Blakemore, *J. Appl. Phys.*, 53, 520, 1983.
4.312. B. J. Scromme and G. E. Stillman, *Phys. Rev.*, B29, 1982, 1984.
4.313. D. D. Sell, H. C. Casey, and K. W. Wecht, *J. Appl. Phys.*, 45, 2650, 1974.
4.314. G. Lindemann, R. Lassnig, W. Seidenbusch, and E. Gornik, *Phys. Rev.*, B28, 4693, 1983.
4.315. Peng Rui-Wu, private communication; see, also, Peng Rui-Wu, *J. Cryst. Growth*, 56, 350, 1982.
4.316. B. T. Bebney and P. R. Jay, *Solid State Electron.*, 23, 773, 1980.
4.317. M. Shur, *Physics of Semiconductor Devices,* Prentice-Hall, Englewood Cliffs, NJ, 1990, 625.
4.318. M. S. Abrahams, R. Braunstein, and F. D. Rosi, *J. Phys. Chem. Solids*, 10, 204, 1959.
4.319. L. Lin, Y. Lin, X. Zhong, Y. Zhang, and H. Li, *J. Cryst. Growth*, 56, 344, 1982.
4.320. H. J. Lee and D. C. Look, *J. Appl. Phys.*, 54, 4446, 1983.
4.321. A. V. Ioffe, *Sov. Phys. Solid State*, 5, 2446, 1963.
4.322. A. I. Zaslavskii, V. M. Sergeeva, and I. A. Smirnov, *Sov. Phys. Solid State*, 2, 2884, 1960.
4.323. R. O. Carlson, G. A. Slack, and S. J. Silverman, *J. Appl. Phys.*, 36, 505, 1965.
4.324. M. G. Holland, Proc. 7th Int. Conf. Semicond. Physics, Dunod, Paris, 1964.
4.325. F. Bailly and P. Manca, *Proc. Symp. Chem. Bonds in Semiconductors,* N. N. Sirota, Ed., Nauka i Tekhnika, Minsk, U.S.S.R., 1967.
4.326. D. Campi and C. Papuzza, *J. Appl. Phys.*, 57, 1305, 1985.
4.327. D. E. Aspnes and A. A. Studna, *Phys. Rev.*, B27, 985, 1983.
4.328. C. T. Tsai and R. S. Williams, *J. Mater. Res.*, 1, 820, 1986.

4.329. J.-C. Lin, K.-C. Hsieh, K. J. Schulz, and Y. A. Chang, *J. Mater. Res.*, 3, 148, 1988.
4.330. T. Sand, E. D. Marshall, and L. C. Wang, *J. Mater. Res.*, 3, 914, 1988.
4.331. H. Iguchi, *J. Mater. Res.*, 6, 1542, 1991.
4.332. J. C. Lambropoulos, *J. Mater. Res.*, 3, 531, 1988.
4.333. H. Kroemer, Heteroepitaxy on Silicon, J. C. C. Fan and J. M. Poate, Eds., MRS Symp. Proc. 67, Pittsburgh, PA, 1986, 3.
4.334. N. Chand, F. Ren, A. T. Macrander, J. P. van der Ziel, A. M. Sergent, R. Hull, S. N. G. Chu, Y. K. Chen, and D. V. Lang, *J. Appl. Phys.*, 67, 2343, 1990.
4.335. A. Georgakilas, J. Stoemenos, K. Tsagaraki, P. Komninou, N. Flevaris, P. Panayotatos, and A. Christou, *J. Mater. Res.*, 8, 1908, 1993.
4.336. M. Tachikawa and H. Mori, *Appl. Phys. Lett.*, 56, 2225, 1990.
4.337. T. Ohashi, *J. Mater. Res.*, 7, 3032, 1992.
4.338. M. Yamaguchi, *J. Mater. Res.*, 6, 376, 1991.
4.339. D. K. Sadana, H. Choksi, J. Washburn, P. F. Byrne, and N. W. Cheung, *Appl. Phys. Lett.*, 44, 301, 1984.
4.340. V. A. Uskov, A. B. Fedotov, E. A. Erofeeva, A. I. Rodionov, and D. T. Dzhumakulov, *Inorg. Mater.*, 23, 186, 1987.
4.341. F.-Y. Shiau and Y. A. Chang, *Appl. Phys. Lett.*, 55, 1510, 1989.
4.342. K. S. Jones and C. J. Santana, *J. Mater. Res.*, 6, 1048, 1991.
4.343. F.-Y. Shiau, S.-L. Chen, M. Loomans, and Y. A. Chang, *J. Mater. Res.*, 6, 1532, 1991.
4.344. G. A. Wolff, L. Toman, N. J. Field, and J. C. Clark, Halbleiter und Phosphore, Braunschweig, 1958, 463.
4.345. S. G. Roberts, P. D. Warren, and P. B. Hirsch, *J. Mater. Res.*, 1, 162, 1986.
4.346. K. Berner and K. Smirous, *Chech. J. Phys.*, 5, 546, 1955.
4.347. W. Köster and B. Thoma, *Z. Metallkunde*, 46, 291, 1955.
4.348. P. M. Markus and A. J. Kennedy, *Phys. Rev.*, 114, 459, 1959.
4.349. E. D. Pierron, D. L. Parker, and J. B. McNeely, *J. Appl. Phys.*, 38, 4669, 1968.
4.350. H. J. McSkimin, *J. Appl. Phys.*, 39, 4127, 1968.
4.351. A. Joullié, A. Zein Eddin, and B. Girault, *Phys. Rev.*, B23, 928, 1981.
4.352. C. Pickering, *J. Electron. Mater.*, 15, 51, 1985.
4.353. R. A. Stradling, *Phys. Lett.*, 20, 217, 1966.
4.354. M. W. Heller and R. G. Hamerly, *J. Appl. Phys.*, 57, 4626, 1985.
4.355. D. Kranzer, *Phys. Status Solidi*, A26, 11, 1974.
4.356. E. W. Saker and F. A. Cunnel, *Research, London*, 7, 114, 1954.
4.357. L. Pincherle and J. M. Radcliffe, *Adv. Phys. (Philos. Mag. Suppl.)*, 5, 271, 1956.
4.358. H. J. Hrostowski and M. Tanenbaum, *Physica*, 20, 1065, 1954.
4.359. H. N. Leifer and W. C. Dunlap, *Phys. Rev.*, 95, 51, 1954.
4.360. R. K. Willardson and A. C. Beer, Eds., *Semiconductors and Semimetals*, Vol. 8, Academic Press, New York, 1972.
4.361. J. L. Roberts, B. Pistoulet, D. Barjon, and A. Raymond, *J. Phys. Chem. Solids*, 34, 2221, 1973.
4.362. L. I. Berger, *Trudy IREA (Proc. IREA)*, 26, 293, 1964.
4.363. P. C. Mathur and S. Jain, *Phys. Rev.*, B19, 3152, 1979.
4.364. H. J. Lee and J. C. Woolley, *Can. J. Phys.*, 59, 1844, 1981.
4.365. V. M. Glazov and S. N. Chizhevskaya, *Sov. Phys. Solid State*, 4, 1841, 1962.
4.366. M. D. Campos, A. Gouskov, L. Gouskov, and J. C. Pons, *J. Appl. Phys.*, 44, 2642, 1973.
4.367. R. Juza and H. Hahn, *Z. Anorg. Allg. Chem.*, 239, 282, 1938.
4.368. L. I. Marina and A. Ya. Nashelskii, *Zh. Fiz. Khim. (J. Phys. Chem.)*, 43, 963, 1969.
4.369. J. B. MacChesney, P. M. Bridenbauch, and P. B. O'Connor, *Mater. Res. Bull.*, 5, 783, 1970.
4.370. C. P. Foley and T. L. Tansley, *Phys. Rev.*, B33, 1430, 1986.
4.371. V. A. Tyagai, A. M. Evstigneev, A. N. Krasiko, A. F. Andreeva, and V. Ya. Malakhov, *Sov. Phys. Semicond.*, 11, 1257, 1977.
4.372. K. Osamura, S. Naka, and Y. Murakami, *J. Appl. Phys.*, 46, 3432, 1975.
4.373. R. Juza and A. Rabenau, *Z. Anorg. Allg. Chem.*, 285, 212, 1956.
4.374. R. Juza, A. Rabenau, and G. Pascher, *Z. Anorg. Allg. Chem.*, 285, 242, 1956.
4.375. T. L. Tansley, R. J. Egan, and E. C. Horrigan, *Thin Solid Films*, 164, 441, 1988.
4.376. H. J. Hovel and J. J. Cuomo, *Appl. Phys. Lett.*, 20, 71, 1972.

4.377. J. W. Trainor and K. Rose, *J. Electron. Mater.*, 3, 821, 1974.
4.378. T. L. Tansley and C. P. Foley, *Electron. Lett.*, 20, 1066, 1984.
4.379. T. L. Tansley and R. J. Egan, *Physica*, B185, 190, 1993.
4.380. T. J. Kistenmacher, W. A. Briden, J. S. Mriden, J. S. Morgan, D. Dayan, R. Fainchtein, and T. O. Poehler, *J. Mater. Res.*, 6, 1300, 1991.
4.381. G. A. Wolff and J. D. Broder, *Acta Crystallogr.*, 12, 313, 1959.
4.382. M. Tmar, A. Gabriel, C. Chatillion, and I. Ansara, *J. Cryst. Growth*, 68, 557, 1984.
4.383. A. S. Borshchevskii, N. A. Goryunova, and N. K. Takhtaryova, *Sov. Phys. Tech. Phys.*, 27, 1408, 1957.
4.384. D. Brasen, *J. Mater. Sci.*, 11, 791, 1976; 13, 1776, 1978.
4.385. D. Y. Watts and A. F. W. Willoughby, *J. Appl. Phys.*, 56, 1869, 1984.
4.386. S. B. Zhang and M. L. Cohen, *Phys. Rev.*, B35, 7604, 1987.
4.387. M.-H. Tsai, J. D. Dow, and R. V. Kasowski, *J. Mater. Res.*, 7, 2205, 1992.
4.388. P. Deus and H. A. Schneider, *Phys. Status Solidi*, A79, 411, 1983.
4.389. M. Bugajski and W. Lewandowski, *J. Appl. Phys.*, 57, 521, 1985.
4.390. H. Mathieu, Y. Chen, J. Camassel, J. Allegre, and D. S. Robertson, *Phys. Rev.*, B32, 4042, 1985.
4.391. J. Leotin, R. Barbaste, S. Askenazy, M. S. Skolnik, R. A. Stradling, and J. Tuchendler, *Solid State Commun.*, 15, 693, 1974.
4.392. H. Weiss, *Z. Naturforsch.*, 11A, 430, 1956.
4.393. O. G. Folberth and H. Weiss, *Z. Naturforsch.*, 10A, 615, 1955.
4.394. K. Fairhurst, D. Lee, D. S. Robertson, H. T. Parfitt, and W. H. E. Wilgoss, *J. Mater. Sci.*, 16, 1031, 1981.
4.395. L. D. Zhu, K. T. Chan, and J. M. Ballantyne, *Appl. Phys. Lett.*, 47, 47, 1985.
4.396. L. L. Taylor and D. A. Anderson, *J. Cryst. Growth*, 64, 55, 1983.
4.397. S. A. Aliev, A. Ya. Nashel'skii, and S. S. Shalyt, *Sov. Phys. Solid State*, 7, 1287, 1965.
4.398. J. K. Abrokwah, T. N. Peck, B. A. Waterson, G. E. Stillman, T. S. Low, and B. Skromme, *J. Electron. Mater.*, 12, 681, 1983.
4.399. Y. K. Vohra, S. T. Weir, and A. L. Ruoff, *Phys. Rev.*, B31, 7344, 1985.
4.400. P. W. Sparks and C. A. Swenson, *Phys. Rev.*, 163, 779, 1967.
4.401. N. N. Sirota and Yu. I. Pashintsev, *Inzh. Fiz. Zh. Akad. Nauk BSSR*, 1, 38, 1958.
4.402. L. I. Berger and N. N. Sirota, *Inzh. Fiz. Zh. Acad. Nauk BSSR*, 2, 102, 1959.
4.403. L. I. Berger and S. I. Radautsan, *Proc. 4th All-Union Conf. Semicond. Mater.*, Plenum Press, New York, 1963, 103.
4.404. D. Gerlich, *J. Appl. Phys.*, 34, 2915, 1963.
4.405. F. Lukes, *Phys. Status Solidi*, B84, K113, 1977.
4.406. A. V. Varfolomeev, R. P. Seisyan, and R. N. Yakimova, *Sov. Phys. Semicond.*, 9, 530, 1975.
4.407. S. Zwerdling, B. Lax, and L. M. Roth, *Phys. Rev.*, 108, 1402, 1957.
4.408. R. P. Chasmar and R. Stratton, *Phys. Rev.*, 102, 1682, 1956.
4.409. J. Takayama, K. Shimomae, and C. Hamaguchi, *Jpn. J. Appl. Phys.*, 20, 1265, 1981.
4.410. L. M. Kanskaya, S. I. Kokhanovskii, R. P. Seisyan, A. L. Efros, and V. A. Yukish, *Sov. Phys. Semicond.*, 17, 449, 1983.
4.411. C. R. Pidgeon, S. H. Groves, and J. Feinleib, *Solid State Commun.*, 5, 677, 1967.
4.412. O. G. Folberth, O. Madelung, and H. Weiss, *Z. Naturforsch.*, 9A, 954, 1954.
4.413. P. V. Tamarin and S. S. Shalyt, *Sov. Phys. Semicond.*, 5, 1097, 1971.
4.414. M. Hass and B. W. Henvis, *J. Phys. Chem. Solids*, 23, 1099, 1962.
4.415. W. L. Wolfe and G. J. Zissis, Eds., Infared Handbook, IR Inf. and Analysis Cntr., Environm. Res. Inst. of Michigan (for the ONR, DoN), 1978, 11-75.
4.416. E. Schillmann, *Z. Naturforsch.*, 11A, 6, 1956.
4.417. M. E. Straumanis and C. D. Kim, *J. Appl. Phys.*, 36, 3822, 1965.
4.418. D. G. Avery, D. W. Goodwin, W. D. Lawson and T. S. Moss, *Proc. Phys. Soc.*, B67, 761, 1954.
4.419. D. R. Yoder, R. Colella, and B. A. Weinstein, *Bull. Am. Phys. Soc.*, 27, 248, 1982.
4.420. S. I. Novikova, *Sov. Phys. Solid State*, 2, 2087, 1961.
4.421. D. F. Gibbons, *Phys. Rev.*, 112, 779, 1958.
4.422. L. J. Slutsky and C. W. Garland, *Phys. Rev.*, 113, 167, 1959.
4.423. F. Herman, *J. Electron.*, 1, 103, 1955.
4.424. J. R. Chelikowsky and M. L. Cohen, *Phys. Rev.*, B30, 4828, 1984.

4.425. H. Höchst and I. Hernandes-Calderon, *Surf. Sci.,* 126, 25, 1983.
4.426. C. L. Littler, D. G. Seiler, R. Kaplan, and R. J. Wagner, *Phys. Rev.,* B27, 7473, 1983.
4.427. S. Logothetidis, L. Vina, and M. Cardona, *Phys. Rev.,* B31, 947, 1985.
4.428. M. W. Goodwin, D. G. Seiler, and M. H. Weiler, *Phys. Rev.,* B25, 6300, 1982.
4.429. C. L. Littler and D. G. Seiler, *Appl. Phys. Lett.,* 46, 986, 1985.
4.430. D. M. Zengin, *J. Phys.,* D16, 653, 1983.
4.431. G. Dresselhaus, A. F. Kip, C. Kittel, and G. Wagoner, *Phys. Rev.,* 98, 556, 1955.
4.432. E. Burstein, G. S. Picus, and H. A. Gebbie, *Phys. Rev.,* 103, 825, 1956.
4.433. H. P. R. Frederikse and E. V. Mielczarek, *Phys. Rev.,* 99, 1889, 1955.
4.434. B. N. Zvonkov, N. N. Salashchenko, and O. N. Filatov, *Sov. Phys. Solid State,* 21, 777, 1979.
4.435. D. M. S. Bagguley, M. L. A. Robinson, and R. A. Stradling, *Phys. Lett.,* 6, 143, 1963.
4.436. D. G. Seiler, M. W. Goodwin, and A. Miller, *Phys. Rev. Lett.,* 44, 807, 1980.
4.437. H. J. Hrostovski, F. J. Morin, T. H. Geballe, and G. H. Wheatly, *Phys. Rev.,* 100, 1672, 1955.
4.438. M. Tanenbaum and J. P. Maita, *Phys. Rev.,* 91, 1009, 1953.
4.439. O. Madelung and H. Weiss, *Z. Naturforsch.,* 9A, 527, 1954.
4.440. O. Madelung, *Physics of III-V Compounds,* John Wiley & Sons, New York, 1964.
4.441. D. L. Rode, *Phys. Rev.,* B3, 3287, 1971.
4.442. D. W. Goodwin, Rep. Roy. Phys. Soc. London Meeting on Semiconds., April 1956, 137.
4.443. D. J. Howarth, R. H. Jones, and E. H. Putley, *Proc. Phys. Soc.,* B70, 124, 1957.
4.444. R. W. Cunningham and J. B. Gruber, *J. Appl. Phys.,* 41, 1804, 1970.
4.445. H. Schönwald, *Z. Naturforsch.,* 19A, 1276, 1964.
4.446. A. F. Ioffe and A. R. Regel, *Prog. Semicond. London,* 4, 237, 1960.
4.447. D. K. Hamilton and K. G. Seidensticker, *J. Appl. Phys.,* 34, 2697, 1963.
4.448. D. K. Stevens and J. H. Crawford, *Phys. Rev.,* 99, 487, 1955.
4.449. M. Hass and B. W. Henvis, *Phys. Chem. Solids,* 23, 1099, 1962.
4.450. J. R. Dixon and J. K. Furdyna, *Solid State Commun.,* 35, 195, 1980.
4.451. K. F. Hulme and J. B. Mullin, *J. Electron. Contr.,* 3, 160, 1956.
4.452. W. G. Pfann, *Solid State Phys.,* 4, 424, 1957.
4.453. G. Busch and M. Schneider, *Physica,* 20, 1084, 1954.
4.454. A. D. Stuckes, *Phys. Rev.,* 107, 427, 1957.
4.455. V. V. Kosarev, P. V. Tamarin, and S. S. Shalyt, *Phys. Status Solidi,* B44, 525, 1971.
4.456. M. Isai, T. Fukunaka, and M. Ohshita, *J. Mater. Res.,* 1, 547, 1986.
4.457. Y. Goto, K. Utsumi, A. Ishido, I. Tsugawa, and N. Koshino, MRS Symp. Proc. 74, 1987, 251.
4.458. J. Stuke, *J. Non Cryst. Solids,* 4, 1, 1970.
4.459. J. Solis, K. A. Rubin, and C. Ortiz, *J. Mater. Res.,* 5, 190, 1990.
4.460. M. I. Nathan, *Proc. IEEE,* 54, 10, 1966.

CHAPTER 5

Binary II-VI and I-VII Tetrahedral Semiconductors

5.1. BINARY II-VI COMPOUNDS

Binary II-VI compounds may be formed either by the group IIa or by the group IIb elements. Among the IIa-VIb compounds, only the beryllium chalcogenides, Be(S, Se, Te), and MgTe possess the adamantine structure (BeO is an ionic compound with the wurtzite structure); other compounds of this type crystallize either in the NaCl lattice or have a combined octahedral/tetrahedral coordination. In this chapter, only ten IIb-VIb compounds, eight semiconductors, and two semimetals with the tetrahedral coordination will be considered. The information pertinent to the properties of the IIa-VIb compounds is compiled, in particular, in Reference 5.1 (see, also, Reference 5.2).

5.1.1. Zinc Oxide

Zinc oxide, ZnO, has the molecular weight of 81.37, average atomic weight of 40.69, average atomic number of 19, and enthalpy of formation (298.15 K) of −350.5 kJ/mol (Reference 5.2, p. 5-20).

Under normal conditions, ZnO crystallizes into the wurtzite structure (space group $P6_3mc$). Natural zincite has lattice parameters a = 0.32495 ± 0.00005 nm and c = 0.52069 ± 0.00005 nm at 298 ± 5 K (Reference 5.2, p. 4-148). For synthetic ZnO, ASTM Data Card No. 5-0664 recommends a = 0.3249 nm and c = 0.5205 nm; Kumar et al.[5.3] have found that the epitaxial ZnO layers have a = 0.324 nm and c = 0.519 nm. There is a ZnO high-pressure allotrope with the rock salt structure and lattice parameter of 0.4280 nm (Reference 5.4, p. 98) (Sze [Reference 5.17, p. 848] cites a magnitude of 0.4580 nm). The ZnO coefficient of thermal expansion near room temperature is equal to $2.9 \cdot 10^{-6}$ and $4.8 \cdot 10^{-6}$ 1/K along and across the c axis, respectively.[5.5] The room temperature X-ray density of ZnO is 5.6750 ± 0.0018 g/cm^3 (Reference 5.2, p. 4-148), although from the above-cited ASTM data it is equal to 5.6803 g/cm^3 and the pycnometric magnitude is 5.606 g/cm^3.

ZnO has the melting point of 2248 K (Reference 5.2, p. 4-112), the enthalpy of fusion of 52.30 kJ/mol (Reference 5.2, p. 6-120), the specific heat at room temperature and constant pressure of 0.494 J/g·K (Reference 5.2, p. 5-20), and the Debye temperature extrapolated to 0 K is 416 K.[5.6] The ZnO hardness is 5.0 units in the Mohs scale. The second-order elastic moduli at room temperature are c_{11} = 207 GPa, c_{12} = 117.7 GPa, c_{13} = 106.1 GPa, c_{33} = 209.5 GPa, c_{55} = 44.8 GPa, and c_{66} = 4.46 GPa (Reference 5.7, p. 26).

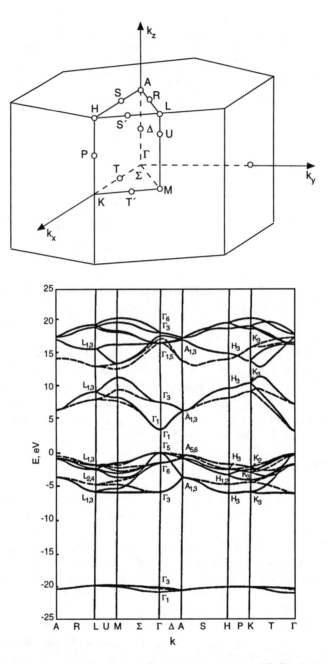

Figure 5.1 (a) Brillouin zone of the wurtzite lattice; (b) band structure of ZnO calculated using a semi-empirical tight-binding model. The energy bands corresponding to the Zn 3d-states are not included in the calculations. (After A. Kobayashi, O.F. Sankey, S.M. Volz, and J.D. Dow, *Phys. Rev.*, B28, 935, 1983. With permission.)

ZnO is a direct semiconductor. Its minimum energy gap is 3.2 eV at room temperature and 3.44 eV at 4 K.[5.8] The optical band gap measured on thin films[5.3,5.9] is close to 3.2 eV (300 K). The calculated ZnO band structure is shown in Figure 5.1 together with the Brillouin zone of the wurtzite lattice. The results of the experimental study of the ZnO band structure were reported by Kolb.[5.15] The carrier effective mass in ZnO is virtually scalar and equal to 0.24 m_o and 0.59 m_o for electrons and holes, respectively.[5.8]

BINARY II-VI AND I-VII TETRAHEDRAL SEMICONDUCTORS

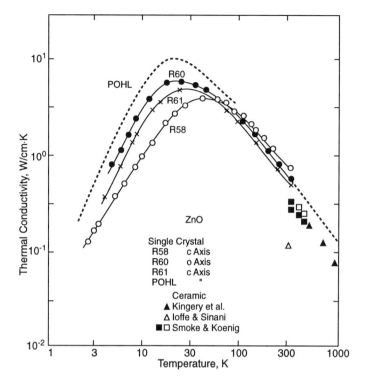

Figure 5.2 Thermal conductivity of ZnO single crystals from different sources. (After G.A. Slack, *Phys. Rev.*, B6, 3791, 1972. With permission.)

Like other wide energy gap semiconductors, ZnO has a potential for many optoelectronic and electronic devices. ZnO ceramics doped with some metal oxides possess nonlinear voltage-current characteristics[5.10-5.12] and are widely used in varistors for transient voltage surge protection in electronic circuits. A proper choice of dopants and the heat treatment make the ZnO ceramic varistors highly resistant to electrical, thermal, and mechanical stresses[5.13] (see, also, Reference 5.14).

The electron mobility in ZnO ceramics is between 100 and 200 cm²/V·sec (Reference 5.1, p. 26); its magnitude in single crystals (see, e.g., Reference 5.16) varies between 100 and 1000 cm²/V·sec. The hole mobility in ZnO is close to 180 cm²/V·sec. Qiu et al.[5.17] performed the resistivity and Hall effect measurements on the ZnO films doped with In and Al. They found that the electron mobility depends on the film heat treatment parameters and is close to 5 cm²/V·sec on as-prepared films. Electrical conductivity of ZnO:Zn polycrystals doped with Cu, Fe, Ni, and Mn was reported by Berger et al.[5.18]

The thermal conductivity, κ, of ZnO ceramics was reported by Ioffe and Sinani[5.19] (κ = 100 mW/cm·K at 300 K) and Kingery et al.[5.20] The temperature dependence of κ in the range of 3 to 300 K was investigated by Slack[5.6] on the ZnO single crystals along and across the c axis. These results are shown in Figure 5.2. At temperatures between 30 and 300 K, $\kappa_\perp$ is higher than $\kappa_\parallel$; it could be expected from the crystallographic consideration, although Martin and Wolf[5.21] did not notice any appreciable anistropy in their κ measurements on ZnO crystals. The magnitude of κ of the ZnO crystals at 300 K is close to 540 mW/cm·K.

ZnO is a diamagnetic with the gram-molar magnetic susceptibility of –57.8 SI units (Reference 5.2, p. 9-51). The ZnO static dielectric constant is 8.75 and 7.8 along and across the c axis, respectively; the optical frequency magnitude is 3.75 along and 3.70 across the c axis (Reference 5.7, p. 26). From these data, the index of refraction is expected to be about 1.92 to 1.94, close to the average experimental result of 2.01 (Reference 5.4, p. 105) and of

$n_o = 1.984$ and $n_e = 2.001$ for the ordinary and extraordinary indices, respectively, at the wavelength of 1400 nm.[5.8]

As was mentioned before, zinc oxide presents a substantial interest for practical applications. Transparent ZnO:(Al, Ga, In) thin films have a relatively high electrical conductivity and are less expensive and more stable than In-Sn-O films for use as transparent conductive electrodes in display devices and silicon solar cells. High-conductivity ZnO films have high reflectivity in the infrared part of the spectrum and are used as energy-efficient windows. The dependence of the luminescent properties of ZnO:Zn powder luminophors on the heavy metal dopants concentration and on the intensity of the low-energy electron beams was studied in Reference 5.43. The ZnO films with the c axis normal to the substrate possess high piezoelectric response and are used in surface and bulk acoustic wave devices and acousto-optical devices (see, e.g., in Reference 5.3).

ZnO is available commercially. The price of the 99.9995% purity powder is from $1.86 to $2.46 per gram (1995).

5.1.2. Zinc Sulfide

Zinc sulfide, ZnS, has the molecular weight of 97.44, average atomic weight of 42.72, average atomic number of 23, and the enthalpy of formation of –206.0 kJ/mol (sphalerite modification, eight atoms per unit cell) (Reference 5.2, p. 5-22) and –189.6 kJ/mol (wurtzite modification, twelve atoms per unit cell) (Reference 5.24, p. 122).

The ZnS sphalerite allotrope is stable below 1290 K; the wurtzite allotrope is stable above this temperature. The cubic sphalerite allotrope, c-ZnS, has a lattice constant at 300 K of 0.54145 ± 0.00009 nm (Reference 5.4, p. 98) (0.54093 ± 0.00005 nm in Reference 5.22, p. 4-144 and 0.5410 nm in Reference 5.7, p. 26). The density at room temperature varies in different reports from 4.075 g/cm^3 to 4.102 g/cm^3 (4.0885 ± 0.0011 g/cm^3 in Reference 5.22, p. 4-144). The c-ZnS coefficient of thermal expansion at room temperature is $6.36 \cdot 10^{-6}$ 1/K.[5.23] The hexagonal high-temperature allotrope, wurtzite (w-ZnS or 2H-ZnS) has lattice constants (at 300 K) $a = 0.38230 \pm 0.00010$ nm and $c = 0.62565 \pm 0.00010$ nm (Reference 5.22, p. 4-144) ($a = 0.3822$ nm and $c = 0.6260$ nm in Reference 5.7, p. 27). There are other known hexagonal ZnS polytypes, namely, 4H-ZnS ($a = 0.3814$ nm, $c = 1.246$ nm) and 6H-ZnS ($a = 0.3821$ nm, $c = 1.873$ nm);[5.25,5.26] 8H-ZnS ($a = 0.382$ nm, $c = 2.496$ nm) and 10H-ZnS ($a = 3.824$ nm and $c = 3.120$ nm).[5.27] A rhombohedral ZnS polytype was suggested by Buck and Strock.[5.28] A review of the ZnS crystallography was compiled by W.L. Roth (Reference 5.29, p. 117). The density of w-ZnS at room temperature is virtually equal to that of c-ZnS (4.087 to 4.105 g/cm^3; 4.0859 ± 0.0022 g/cm^3 [Reference 5.22, p. 4-144]). The w-ZnS average coefficient of thermal expansion at room temperature is $6.2 \cdot 10^{-6}$ 1/K.

Heating of either c-ZnS or w-ZnS is followed by the intensive sublimation of both zinc and sulfur. To prevent decomposition, the material is usually heated in the atmosphere of an inert gas, e.g., argon, under pressure of 10 to 20 MPa. Under pressure, the w-ZnS melting point is 2103 K[5.30] (see, also, Reference 5.31). The specific heat of c-ZnS at room temperature is $c_p = 0.47$ J/g·K (Reference 5.22, p. 5-22). The Debye temperature is 350 K (Reference 5.6 with reference to 5.32 and 5.33), while Ioffe[5.34] cites a magnitude of 530 K (unfortunately, she does not mention the source).

Cubic ZnS crystals have the Mohs hardness of 3.5 to 4.0. The c-ZnS elastic moduli at room temperature are $c_{11} = 104.6$ GPa, $c_{12} = 65.3$ GPa, and $c_{44} = 46.13$ GPa[5.35] (in Reference 5.7, p. 27, $c_{11} = 98.1$ GPa, $c_{12} = 62.7$ GPa, and $c_{44} = 44.83$ GPa). The Knoop hardness of w-ZnS (25-g load), normal to (011), is 1.78 ± 0.27 GPa depending on the indentor plane orientation (Reference 5.24, p. 122). The w-ZnS elastic moduli are $c_{11} = 123.4$ GPa, $c_{12} = 58.5$ GPa, $c_{13} = 45.5$ GPa, $c_{33} = 28.85$ GPa, $c_{44} = 32.45$, and $c_{66} = 139.6$ GPa (Reference 5.7, p. 27).

Both sphalerite and wurtzite ZnS allotropes are the direct gap semiconductors. The c-ZnS minimum energy gap at 300 K is 3.66 eV,[5.36] and its magnitude at 19 K is 3.78 eV (3.799 eV at 14 K[5.37]). The w-ZnS energy gap at room temperature is from 3.74 eV to 3.88 eV and from 3.78 eV to 3.87 eV along and across the c axis, respectively.[5.36] The band structures of c-ZnS and w-ZnS are presented in Figure 5.3. The effective mass of electrons in c-ZnS is 0.39 m_o;[5.38] the heavy and light holes have effective mass in c-ZnS of 1.76 and 0.23 m_o, respectively (Reference 5.7, p. 26). The effective mass of electrons in w-ZnS is $(0.27 \pm 0.03)m_o$ across the c axis.[5.39] The effective mass of holes is 1.4 m_o and 0.49 m_o along and across the c axis, respectively (Reference 5.7, p. 27).

The electric transport properties of c-ZnS are characterized by the maximum electron Hall mobility of 600 cm^2/V·sec at a carrier concentration of about 10^{12} cm^{-3} [5.40] (the concentration was controlled by the radiating light intensity). Aven and Mead[5.41] measured the electrical conductivity and Hall effect on the c-ZnS:I and w-ZnS:Al samples. Their data on the electron concentration and mobility are shown in Figure 5.4 (from S.S. Delwin [Reference 5.29, p. 595]). The Hall mobility of holes in c-ZnS:Cu has been reported to be 5 cm^2/V·sec at 700 K[5.42] (the resistance at lower temperatures is too high to measure reliably the resistivity and Hall effect). The electron drift mobility in w-ZnS is 165 cm^2/V·sec and 280 cm^2/V·sec along and across the c axis, respectively. The hole mobility in w-ZnS varies from 100 cm^2/V·sec to 800 cm^2/V·sec.[5.7] It has to be mentioned that in spite of the long and successful use of ZnS in the electrolumonophor industry, its electrical properties are still insufficiently studied. The current growing interest in application of the high-energy-gap semiconductors may give a new stimulus to the investigation of electrical properties of the ZnS crystals and films.[5.44,5.45]

The ZnS thermal conductivity was discussed by Slack,[5.6] Ioffe,[5.46] and Eucken and Kuhn.[5.47] The room temperature data in all three reports for c-ZnS are in reasonable agreement (the magnitude in Reference 5.46 is 0.256 W/cm·K), whereas the magnitudes in References 5.6 and 5.47 are sufficiently different at lower temperatures (Eucken and Kuhn used natural c-ZnS crystals for measurements; Slack used high purity crystals with controlled impurity content). The results of References 5.6 and 5.47 are shown in Figure 5.5 (the thermal conductivity of the w-ZnS samples was measured along the c axis). The 300-K magnitude for c-ZnS in Reference 5.6 is of 0.27 W/cm·K.

The molar magnetic susceptibility of ZnS is $-49.6 \cdot 10^{-5}$ SI units/g-mol (Reference 5.4, p. 106). The static dielectric constant of c-ZnS is from 8.0 to 8.4, depending on temperature (8.08 at 77 K and 8.37 at 298 K and 10 kHz [Reference 5.22, p. 12-53]). The high-frequency limit is 5.13[5.48] (in Reference 5.7, p. 26, the range of 5.1 to 5.7 is cited). The static dielectric constant of w-ZnS is 9.6, while the data for high-frequency limits vary from 5.13 to 5.7 (Reference 5.7, p. 27). From these data, it would be reasonable to expect, for the visible part of spectrum, the index of refraction in the range of 2.2 to 2.4 in satisfactory agreement with the available experimental data (2.368 for c-ZnS and 2.356 for w-ZnS [Reference 5.4, p. 106]). For detailed information regarding the dependence of the ZnS index of refraction on photon energy see Reference 5.49.

Up to now, several methods have been utilized for the ZnS crystal growth. The conditions under which the growing crystals are either only c-ZnS or only w-ZnS were analyzed in References 5.51 and 5.52. The first report on the chemical transport growth of ZnS was, apparently, published by Nitsche.[5.52] This method and the metal-organic chemical vapor deposition are the most widely used techniques for the epitaxial and polycrystalline ZnS growth.[5.45,5.53] Polycrystalline, 99.99% purity ZnS in the form of 15-mm to 25-mm lumps, is available commercially for $0.63 to $1.12 per gram (1995).

5.1.3. Zinc Selenide

Zinc selenide, ZnSe, has the molecular weight of 144.34, mean atomic weight of 72.17, average atomic number of 32, and the enthalpy of formation of -163.0 kJ/mol (Reference 5.22, p. 5-22).

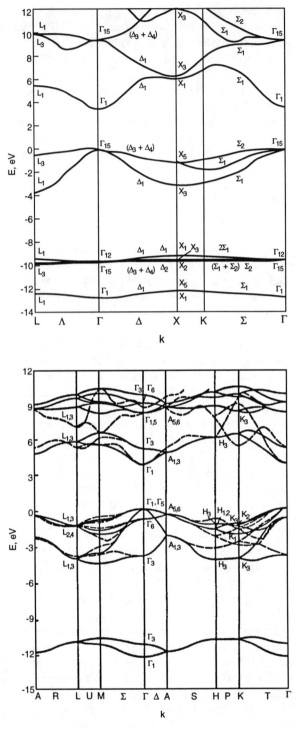

Figure 5.3 Band structure of (a) cubic and (b) hexagonal zinc sulfide. (After (a) D.V. Farberovich, S.I. Kurganskii, and E.P. Domashevskaya, *Phys. Stat. Sol.*, 97, 631, 1980 and (b) A. Kobayashi, O.F. Sankey, S.M. Volz, and J.D. Dow, *Phys. Rev.*, B28, 935, 1983. With permission.)

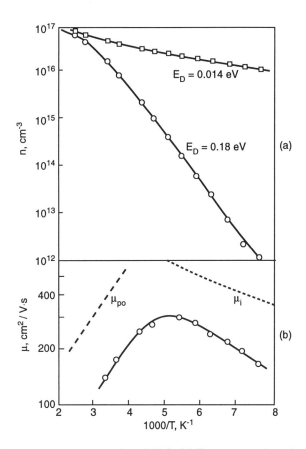

Figure 5.4 The electrical transport properties of ZnS. (a) Temperature dependence of the electron concentration. Circles: Al-doped hexagonal ZnS fired at 1050°C; squares: I-doped cubic ZnS fired at 950°C. (b) Temperature dependence of the Hall mobility of an Al-doped hexagonal ZnS sample fired at 1050°C. The dashed curve represents the calculated mobility for predominant optical phonon scattering; the dotted curve is the mobility calculated for scattering by ionized impurities. (After S.S. Devlin, *Physics and Chemistry of II-VI Compounds,* M. Aven and J.S. Prener, Eds., John Wiley & Sons, 1967. With permission.)

There are two ZnSe allotropes: the sphalerite type with the lattice parameter of 0.56685 ± 0.00005 nm and X-ray density of 5.2630 ± 0.0014 g/cm^3 (Reference 5.22, p. 4-144), and the wurtzite-type allotrope with the lattice parameters a = 0.4003 nm and c = 0.6540 nm[5.54] and the X-ray density of 5.282 g/cm^3. The thermal expansion of c-ZnSe was measured by Novikova[5.55] (Figure 5.6) and Semiletov.[5.56] The coefficient of linear thermal expansion, α, at 300 K is close to $7.2 \cdot 10^{-6}$ 1/K. The magnitudes at 373 and 473 K are $7.8 \cdot 10^{-6}$ and $8.3 \cdot 10^{-6}$ 1/K, respectively.[5.67] With temperature decrease, α decreases, becomes equal to zero at 64 K, and, being negative at lower temperatures, passes through the minimum magnitude of $-3.10 \cdot 10^{-6}$ 1/K at 36 K.

With the exception of optical data, information regarding the properties of hexagonal ZnSe is very limited, so the following data will relate mainly to the cubic ZnSe allotrope.

The ZnSe melting point is 1793 K (M.R. Lorenz in Reference 5.29, p. 85), the 0-K magnitude of Debye temperature is 400 K,[5.34] and the specific heat at 298 K is 0.339 J/g·K.[5.67] The authors of Reference 5.57 studied the Vickers microhardness, H, of ZnSe poly- and single crystals and found that increase of the average grain size, d, from 24 to 33 μm results in H decrease according to the power law H = 1860d$^{-0.07}$ (H in MPa, d in μm, 20-g load). The

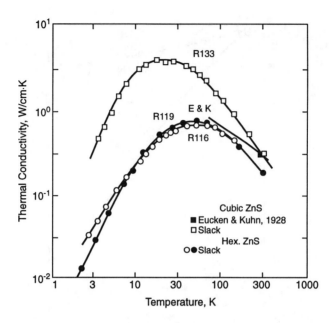

Figure 5.5 Temperature dependence of thermal conductivity of cubic (squares) and hexagonal (circles) ZnS crystals. The data presented by solid squares are from Reference 5.47. (After G.A. Slack, *Phys. Rev.*, B6, 3791, 1972. With permission.)

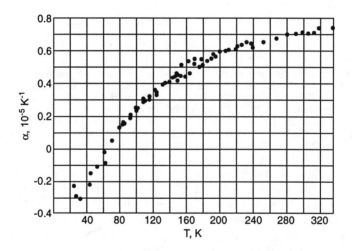

Figure 5.6 The experimental values of the ZnSe coefficient of linear thermal expansion. (After S.I. Novikova, *Sov. Phys. Solid State*, 3, 129, 1961. With permission.)

temperature rise decreases microhardness of ZnSe single crystals from about 1400 MPa at room temperature to 500 MPa at 600 K. Goryunova (Reference 5.24, p. 122) reported a magnitude of 1350 ± 50 MPa at room temperature and 50-g load. The ZnSe second-order elastic constants, $c_{11} = 81.0$ GPa, $c_{12} = 48.8$ GPa, and $c_{44} = 44.1$ GPa, reported by Berlincourt et al.,[5.35] are in reasonable agreement (±10%) with the magnitudes $c_{11} = 90.0$ GPa, $c_{12} = 53.4$ GPa, and $c_{44} = 39.6$ GPa cited in Reference 5.7, p. 27. The Young's modulus and Poisson's ratio magnitudes are 67.2 GPa and 0.28, respectively.[5.67]

Both cubic and hexagonal ZnSe are direct semiconductors. The energy band structures of c-ZnSe calculated by the pseudopotential (see B. Segall in Reference 5.29, p. 39) and first-principle OPW (see Figure 5.7) methods basically are not contradictive. The minimum

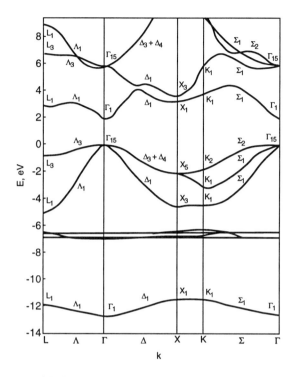

Figure 5.7 Band structure of ZnSe calculated using the first-principles method. Spin-orbit interaction is not included. (After C.S. Wang and M.M. Klein, *Phys. Rev.*, B24, 3293, 1981. With permission.)

direct energy gap (transition $\Gamma_{15v} \rightarrow \Gamma_{1c}$) is equal to 2.82 eV at 10 K and 2.70 eV at 293 K (Reference 5.7, p. 27) with the electron and hole effective mass of 0.21 and 0.6 m_o, respectively (Reference 5.58, p. 632) (in Reference 5.7, the magnitudes are $m_n^*/m_o = 0.13$ to 0.17 and $m_p^*/m_o = 0.57$ to 0.75, while Marple[5.59] found $m_n^*/m_o = 0.17 \pm 0.02$ and Hite et al.[5.60] reported the experimental value of $m_p^*/m_o \simeq 0.75$). Park et al.'s[5.61] results of the photoluminescence analysis of the high-purity ZnSe epitaxial layers are in agreement with the data in Reference 5.7 (Figure 5.8).

The direct energy gap in w-ZnSe at 300 K evaluated from the photoconductivity measurements by Park and Chan[5.62] is 2.795 eV (the light polarized normal to the c axis). With the light polarized parallel to the c axis, a photoconductivity peak appears at the photon energy of 2.834 eV.

From different sources (see, e.g., Hartmann et al. in Reference 5.8, p. 101) the room temperature mobility in c-ZnSe is 530 cm^2/V·sec and 28 cm^2/V·sec for electrons and holes, respectively (Shur [Reference 5.58, p. 632] quotes 500 cm^2/V·sec and 30 cm^2/V·sec, respectively). The temperature dependence of the electron mobility and concentration (Aven and Kennicott,[5.63] quoted by S.S. Devlin in Reference 5.29, p. 597 is shown in Figure 5.9. Fujita et al.[5.64] studied the electrical properties of undoped n-ZnSe epilayers and found that depending on the growth temperature, the electron mobility varies from less than 9.0 cm^2/V·sec to 625 cm^2/V·sec. Raiskin[5.68] reports the large ZnSe single crystals' dark resistivity from 10^0 Ohm·cm to 10^4 Ohm·cm and the electron mobility of about 380 cm^2/V·sec (300 K).

The thermal conductivity of c-ZnSe was measured on single crystals by Ioffe,[5.46] who found the room temperature magnitude of 125 mW/cm·K (or 130 mW/cm·K[5.34]); the 300-K magnitude determined by Slack[5.6] on two ZnSe single crystals is 190 mW/cm·K. The results

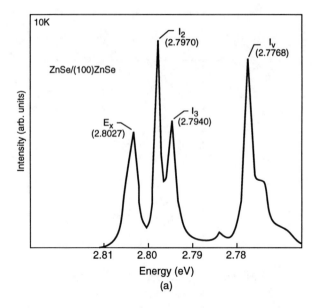

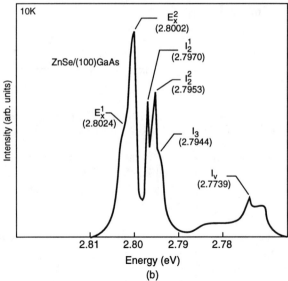

Figure 5.8 Typical 10-K photoluminescence spectra recorded from 2-μm-thick ZnSe layers grown by molecular beam epitaxy on (a) (100) ZnSe and (b) (100) GaAs substrates. (After R.M. Park, C.M. Rouleau, M.B. Troffer, T. Koyama, and T. Yodo, *J. Mater. Res.*, 5, 475, 1990. With permission.)

of ZnSe thermal conductivity vs. temperature measurements[5.6] are presented in Figure 5.10 together with the results of Ladd[5.65] (sample R164 has a higher impurity content).

ZnSe is a diamagnetic with the gram-molar magnetic susceptibility of $-58.4 \cdot 10^{-5}$ SI units.[5.66] The 300-K static and optical dielectric constants of c-ZnSe are 9.1 and 6.3, respectively (Reference 5.58, p. 632).

The typical room temperature values of the index of refraction of the ZnSe layers produced by the chemical vapor deposition method are shown in Table 5.1.[5.67] The ZnSe absorption coefficient at 1.3 μm is $5.0 \cdot 10^{-3}$ cm^{-1}; the thermo-optical coefficient is $7.0 \cdot 10^{-5}$ 1/K.

ZnSe has a transmitting band between 500 nm and 14 μm. This makes it a very suitable material for many applications, first of all for high-power laser windows. The single crystals

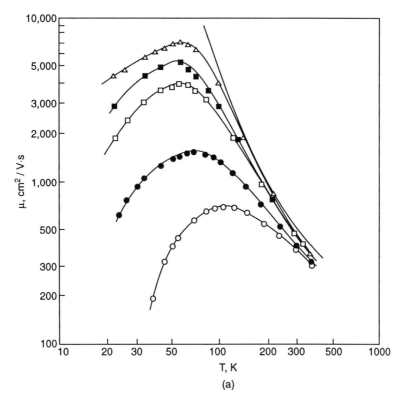

Figure 5.9 The electrical transport properties of ZnSe. (a) Temperature dependence of the Hall mobility in n-type samples. The donor and acceptor concentrations (in units of 10^{16} cm^{-3}) are triangles: $N_D = 0.34$, $N_A = 0.13$; full squares: $N_D = 1.8$, $N_A = 0.5$; open squares: $N_D = 1.05$, $N_A = 0.75$; full circles: $N_D = 3.7$, $N_A = 0.5$; open circles: $N_D = 7.4$, $N_A = 3.4$. The solid line is the drift mobility calculated using the temperature dependence of the static dielectric constant and the magnitude of the electron effective mass of 0.17 m_o. (b) Temperature dependence of the carrier concentration in these samples. The average donor ionization energy calculated from these data is 21 meV. (After S.S. Devlin, *Physics and Chemistry of II-VI Compounds,* M. Aven and J.S. Prener, Eds., John Wiley & Sons, 1967, 597. With permission.)

and polycrystals of ZnSe are available commercially from several manufacturers (see, e.g., References 5.67 and 5.68). The primary material for vacuum deposition in the form of 3-mm to 6-mm pieces of 99.999% purity may be acquired for $5.20 to $8.00 per gram (1995).

5.1.4. Zinc Telluride

Zinc telluride, ZnTe, has the molecular weight of 192.98, average atomic weight of 96.49, average atomic number of 41, and the heat of formation of –125.6 kJ/mol (Reference 5.24, p. 122).

The ZnTe allotrope with sphalerite structure has a lattice parameter at room temperature of 0.61020 ± 0.00006 nm (Reference 5.22, p. 4-144) (W.L. Roth in Reference 5.29, p. 127 quotes a magnitude of 0.61037 nm). Chistyakov and Krucheanu[5.69] reported the wurtzite ZnTe allotrope lattice parameters of a = 0.427 nm and c = 0.699 nm. Because the properties of w-ZnTe are virtually unknown and even its existence at standard conditions is questionable, the following information is relevant to the cubic ZnTe. The X-ray density of ZnTe at 300 K is 5.642 g/cm^3. The coefficient of thermal expansion (Reference 5.58, p. 633) magnitudes are $2.94 \cdot 10^{-6}$ 1/K and $8.19 \cdot 10^{-6}$ 1/K at 75 and 283 K, respectively. The temperature dependence of the thermal expansivity between 20 and 340 K was reported by Novikova and Abrikosov[5.70] (Figure 5.11). Near 46 K, the thermal expansivity (coefficient of thermal expansion) passes

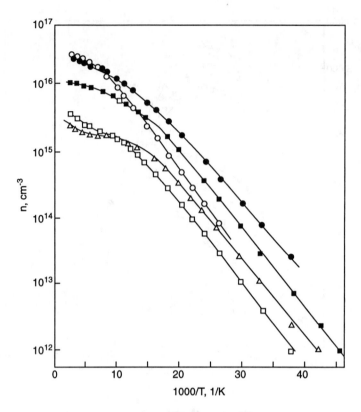

Figure 5.9 (continued)

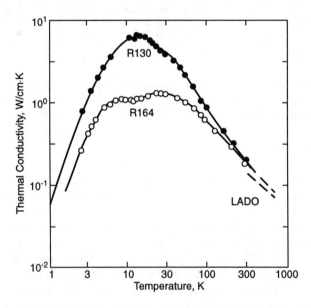

Figure 5.10 Experimental values for thermal conductivity vs. temperature for two ZnSe single crystals. The results of Ladd[5.65] are for polycrystalline ZnSe. (After G.A. Slack, *Phys. Rev.*, B6, 3791, 1972. With permission.)

Table 5.1 Index of Refraction of CVD Zinc Selenide

Wavelength (nm)	Index of refraction	Wavelength (nm)	Index of refraction	Wavelength (nm)	Index of refraction	Wavelength (nm)	Index of refraction
540	2.6754	1,800	2.4496	7,400	2.4201	13,000	2.3850
580	2.6312	2,200	2.4437	7,800	2.4183	13,400	2.3816
620	2.5994	2,600	2.4401	8,200	2.4163	13,800	2.3781
660	2.5755	3,000	2.4376	8,600	2.4143	14,200	2.3744
700	2.5568	3,400	2.4356	9,000	2.4122	14,600	2.3705
740	2.5418	3,800	2.4339	9,400	2.4100	15,000	2.3665
780	2.5295	4,200	2.4324	9,800	2.4077	15,400	2.3623
820	2.5193	4,600	2.4309	10,200	2.4053	15,800	2.3579
860	2.5107	5,000	2.4295	10,600	2.4028	16,200	2.3534
900	2.5034	5,400	2.4281	11,000	2.4001	16,600	2.3487
940	2.4971	5,800	2.4266	11,400	2.3974	17,000	2.3438
980	2.4916	6,200	2.4251	11,800	2.3945	17,400	2.3387
1,000	2.4892	6,600	2.4235	12,200	2.3915	17,800	2.3333
1,400	2.4609	7,000	2.4218	12,600	2.3883	18,200	2.3278

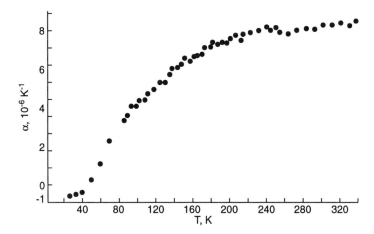

Figure 5.11 The experimental values of the ZnTe coefficient of linear thermal expansion. (After S.I. Novikova and N.Kh. Abrikosov, *Sov. Phys. Solid State*, 5, 1558, 1964. With permission.)

through zero and becomes negative below this temperature. It may be safe to assume that the expansivity passes through its minimum at about 20 K.

The ZnTe melting point is 1568 K.[5.71] This magnitude is about 56 K higher than the one cited in Reference 5.4, p. 106; 5.22, p. 4-113, and 5.82, p. 82. This difference may be a result of the melting point evaluations at uncontrolled pressure (this problem is discussed by M.R. Lorenz in Reference 5.29, p. 73). The specific heat of ZnTe was measured by Kelemen et al.[5.72] in the range of 293 to 813 K. Their magnitude of the specific heat at room temperature (about 0.264 J/g·K) is very close to that cited in Reference 5.73. The Debye temperature of ZnTe at 293 K is 204.5 K (from the thermal conductivity data[5.72]; Leroux-Hugon et al.[5.74] cite a magnitude of $\theta_D = 250$ K; Slack[5.6] prefers $\theta_D = 225$ K.

The ZnTe microhardness (square pyramid, 20-g load) is 900 ± 50 MPa (Reference 5.4, p. 106). The second-order elastic constants at room temperature are $c_{11} = 71.3$ GPa, $c_{12} = 40.7$ GPa, and $c_{44} = 31.2$ GPa.[5.35]

Like the other IIb-VIb compounds, ZnTe is a direct semiconductor. The energy gap from the electrical measurements by Tubota et al.[5.75] on the ZnTe:In samples is 2.12 eV. The optical

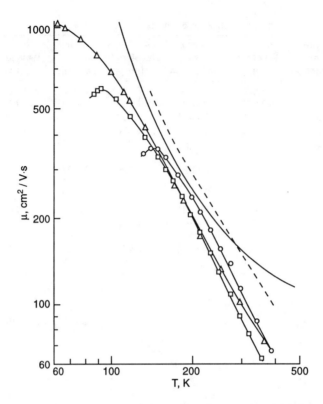

Figure 5.12 Temperature dependence of the Hall mobility of two undoped (circles and squares) and silver-doped (triangles) p-type ZnTe crystals. The solid curve is the mobility due to pure optical mode scattering; the dashed curve represents the effect of an assumed temperature dependence of the static dielectric constant. (After M. Aven and B. Segall, *Phys. Rev.*, 130, 81, 1963. With permission.)

energy gap is 2.26 eV,[5.76] 2.37 eV,[5.77] 2.381 eV,[5.78] and 2.3941 eV (Reference 5.7, p. 28) at 300, 77, 4, and 1.6 K, respectively. The ZnTe band structure from OPW calculations is presented in Reference 5.7, p. 111. The carrier effective mass in ZnTe, calculated by Cardona[5.38] ($m_n^* = 0.17\ m_o$, $m_p^* = 1.0\ m_o$), is in somewhat reasonable agreement with the experimental data ($m_n^* = 0.20\ m_o$ [Reference 5.58, p. 633]; $m_p^* = 0.6\ m_o$[5.79]).

The electrical transport properties of ZnTe strongly depend on the method of synthesis and crystal growth, material chemical purity, and defect content.[5.80] The undoped ZnTe crystals usually have the p-type conductivity which varies at room temperature between 10^{-12} and 10^0 S/cm.[5.68] The authors of Reference 5.72 report on the ZnTe crystals with the conductivity of 10^{-2} S/cm and the hole concentration of $2.5 \cdot 10^{16}$ cm^{-3} at room temperature.

The room temperature electron mobility reported by Fisher et al.[5.81] is 340 cm^2/V·sec (measured on illuminated samples). The temperature dependence of the hole mobility in p-ZnTe is shown in Figure 5.12.[5.79] The room temperature Hall mobility of holes is close to 100 cm^2/V·sec to 120 cm^2/V·sec (Delvin in Reference 5.29, p. 603; Shur in Reference 5.58, p. 633).

The electrical transport in the ZnTe melt is discussed in Reference 5.82. The electrical conductivity of ZnTe in the temperature range that contains the melting point is shown in Figure 5.13. The electrical behavior in both solid and liquid states is typical for a semiconductor. At higher temperatures, conductivity shows a fast increase with temperature, possibly related to some qualitative changes in the melt structure.

A number of papers contains information regarding the ZnTe thermal conductivity. Improvement of the synthesis methods almost always is followed by the increase of the reported magnitude of κ. Thus, Ioffes (1960, cited in Reference 5.74) reported the room temperature magnitude of 110 mW/cm·K, Kelemen et al.[5.72] found κ = 118 mW/cm·K, while

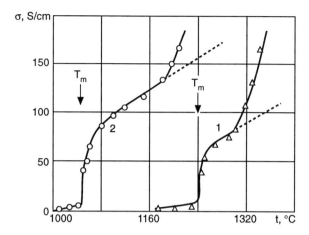

Figure 5.13 Electrical conductivity of ZnTe (curve 1) and CdTe (curve 2) in the solid and liquid states. (After V.M. Glazov, S.N. Chizhevskaya, and N.N. Glagoleva, *Liquid Semiconductors,* Nauka Publishing House, Moscow, 1967. With permission.)

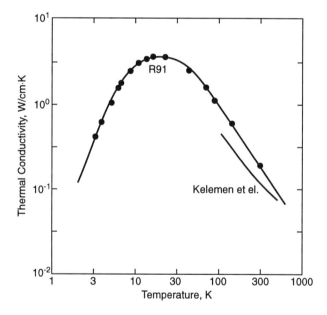

Figure 5.14 Temperature dependence of ZnTe thermal conductivity. (After G.A. Slack, *Phys. Rev.,* B6, 3791, 1972. With permission.)

Reference 5.6 contains the magnitude of 180 mW/cm·K. The temperature dependence of κ is presented in Figure 5.14 (after References 5.6, 5.72, and 5.84). The κ passes through a maximum of about 3.7 W/cm·K near 16 K.

ZnTe is diamagnetic with the gram-molar magnetic susceptibility of about -58.3×10^{-5} SI units at room temperature (extrapolation of the data of Reference 5.82). The static dielectric constant of ZnTe at room temperature is 10.10[5.35] (9.67 in Reference 5.7, p. 28), the optical magnitude is 7.28;[5.59] this gives for the index of refraction a magnitude of 2.70, close to the one (2.72) listed in Reference 5.58, p. 633.

Synthesis of ZnTe can be performed by heating a stoichiometric mixture of the elemental components in the evacuated and sealed silica ampules with the inner walls of the ampules covered with graphite deposits to prevent reaction of the ampule content with SiO_2. The process of ZnTe synthesis in a sealed ampule consists of reaction

$$Te_{liq.} + Zn_{vap.} \rightarrow ZnTe_{solid}$$

at and near the surface of melted tellurium. Being heavier than Te, the formed molecules go down to the bottom of the ampule forming there a sponge-like mass. Glazov et al. (Reference 5.82, p. 94) recommend stepped heating process in which the mixture of Zn and Te is held during 2 to 3 hr at temperatures 873 K, 1073 K, and 1273 K. After that, the products of reaction are crushed, again immersed into an ampule, and exposed to the temperature slightly higher than the ZnTe melting point for 3 hr with periodic intense vibrational stirring, followed by very slow (5 K/hr) decrease of temperature to the ambient. This process gives ZnTe polycrystals with the large (up to 5 mm) grains.

Methods of ZnTe single crystal growth are summarized in Reference 5.85; the single crystals are available commercially,[5.68] and 3-mm to 12-mm ZnTe lumps of 99.999% purity can be acquired for $2.56 to $4.10 per gram (1995).

5.1.5. Cadmium Sulfide

Cadmium sulfide, CdS, has the molecular weight of 144.47, average atomic weight of 72.24, average atomic number of 32, and the standard enthalpy of formation of –161.9 kJ/mol (Reference 5.22, p. 5-9).

CdS has either the sphalerite or wurtzite structure. Both allotropes exist in nature as minerals hawleyite (similar to sphalerite), with lattice parameter at room temperature of 0.5833 ± 0.0002 nm and the X-ray density of 4.835 ± 0.005 g/cm^3, and greenockite (wurtzite crystal structure), with parameters a = 0.41354 ± 0.00010 nm and c = 0.67120 ± 0.0001 nm and density of 4.8261 ± 0.024 g/cm^3 (Reference 5.22, p. 4-143). Artificially grown c-CdS (see, e.g., de Melo et al.[5.86] who reported the growth of c-CdS films with the lattice parameter at room temperature of 0.5794 nm after annealing in S_2, H_2, and In at elevated temperatures) and w-CdS have a relatively wide variety of the lattice parameters depending on defect and impurity concentration. The cubic CdS is an allotrope stable at ambient temperatures; w-CdS is a high-temperature allotrope, metastable at lower temperatures. Yeh et al.[5.87] paid attention to the small difference of the thermodynamic parameters of the c-CdS and w-CdS and, consequently, at a possibility of preparation of either allotrope at ambient pressure (the same can be said about SiC, ZnS, CdSe, CuCl, and CuBr). The crystals of CdS can be grown from the vapor phase, but the growth of one or another CdS allotrope depends basically on the material and orientation of the substrate chosen, temperature, and chemical composition of the gas phase in the growth zone (see, e.g., Reference 5.88).

The coefficient of thermal expansion of w-CdS (300 K) is $4.7 \cdot 10^{-6}$ 1/K and $3.0 \cdot 10^{-6}$ 1/K along and across the c axis, respectively; for c-CdS, the room temperature magnitude is close to $3.6 \cdot 10^{-6}$ 1/K.[5.23,5.90]

The liquidus of the system Cd-S was studied by Woodbury,[5.91] who reported the CdS melting point of 1748 K. Goryunova (Reference 5.24, p. 122) noted that CdS melts congruently at 1750 K under the pressure of 10 MPa. In accordance with Reference 5.68, CdS melts at 1670 K. The attempts to explain the discrepancy in the published data using the Clausius equation describing the dependence of melting point on pressure,

$$dT/dp = (T_m/\lambda)\left(\frac{1}{D_L} - \frac{1}{D_S}\right)$$

where p, T_m, λ, D_L, and D_S are pressure, melting temperature at standard pressure, enthropy of fusion, density of melt, and density of solid (both near the melting point), respectively, have failed; the calculations show a decrease of the melting point by about 3 K under the pressure of 10 MPa. It is possible to assume that the differences in the reported T_m data rather relate to deviation from the stoichiometric composition, presence of impurities, etc.

The specific heat of CdS near room temperature is 0.33 J/g·K;[5.92] the Debye temperature is 219 K.[5.93]

The hardness of natural w-CdS (greenockite) is between 3.0 to 3.5 units of the Mohs scale. The microhardness of w-CdS is 0.90 GPa, 0.50 GPa, and 1.25 GPa being measured at 300 K in the directions normal to the planes {0001}, {10$\bar{1}$0}, and {11$\bar{2}$0}, respectively.[5.94] The w-CdS elastic moduli are c_{11} = 53.34 GPa, c_{13} = 46.15 GPa, c_{33} = 93.7 GPa, c_{44} = 14.87 GPa, c_{55} = 15.33 GPa, and c_{66} = 16.3 GPa[5.106] (the magnitudes cited in Reference 5.7, p. 29 are c_{11} = 83.1 GPa, c_{12} = 50.4 GPa, c_{13} = 46.2 GPa, c_{33} = 94.8 GPa, c_{55} = 15.33 GPa, and c_{66} = 16.3 GPa). The speed of longitudinal and transversal sound waves in CdS is 3.35 km/sec and 1.79 km/sec, respectively (Reference 5.109, p. 84).

Both w-CdS and c-CdS are direct semiconductors. From the optical data,[5.95-5.98] the minimum energy gap of w-CdS at 300 K is 2.41 ± 0.01 eV (Ghosh et al.[5.99] found the lowest 300-K energy gap of w-CdS equal to 2.380 eV). The optical energy gap of c-CdS at 300 K is 2.31 eV.[5.86] The temperature coefficient of energy gap between 77 and 300 K is close to −4.2·10^{-4} eV/K for both allotropes. The band structures of w-CdS and c-CdS are shown in Figure 5.15. The w-CdS electron effective mass (in the units of m_o) is equal to 0.205 ± 0.01, both along and across the c axis[5.96b] (the range of magnitudes cited in Reference 5.7, p. 29 is from 0.20 to 0.25); the hole effective mass is close to 5.0 and 0.7 along and across the A-direction in the w-CdS Brillouin zone, respectively.[5.96b,5.105] The calculated electron and hole effective mass in c-CdS is equal to 0.14 and 0.51, respectively (Reference 5.7, p. 29).

The room temperature electrical conductivity, measured on c-CdS thin films grown by different techniques, varies between 10^{-9} S/cm and 10^1 S/cm (Reference 5.86, 5.100). The electron mobility in bulk w-CdS evaluated from photogalvanomagnetic measurements by Fujita et al.[5.103] is shown in Figure 5.16 (adapted from Reference 5.29, p. 582). The room temperature Hall mobility in undoped n-type w-CdS is 300 cm^2/V·sec to 350 cm^2/V·sec; it is equal to 5000 cm^2/V·sec at 80 K (Delvin in Reference 5.29, p. 581). The maximum observed dark mobility of electrons in w-CdS at any temperature is 1.1·10^4 cm^2/V·sec;[5.104] the hole mobility varies at 300 K between 15 cm^2/V·sec and 40 cm^2/V·sec, depending on the sample quality.

The thermal conductivity of w-CdS along the c axis at 300 K is 0.20 W/cm·K.[5.107] The temperature dependence of κ is presented in Figure 5.17.

CdS is a diamagnetic with the molar magnetic susceptibility of −67.0·10^{-5} SI units per gram-mol (Reference 5.4, p. 106) at room temperature. The static dielectric constants of w-CdS are 9.12 and 8.45 along and across the c axis, respectively; the optical dielectric constant is equal to 5.32 in both directions.[5.108] More detailed information regarding the CdS dielectric constant is compiled by Frederikse in Reference 5.22, p. 12-45.

The index of refraction of w-CdS is 2.32;[5.110] the thermooptical coefficient (dn/dT) is 1.106·10^{-4} 1/K.[5.99] Hartman et al.[5.8] cite n_o(1400 nm) = 2.304, n_o(10600 nm) = 2.226, n_e(1400 nm) = 2.321, and n_e(10600 nm) = 2.239.

The two most common ways to grow CdS crystals are the growth from the melt of the same compound in the sealed fused silica ampule moving through the furnace with a temperature gradient (Bridgman-Stockbarger method[5.68]), and the growth of crystals or films deposited on a substrate from the stream of volatile compounds containing the components of CdS (chemical vapor deposition, CVD). A method of the epitaxial growth from volatile alkils, such as $(CH_3)_2Cd$ reacting with H_2S, was first reported by Manasevit and Simpson[5.112] and is widely used in manufacturing photovoltaic devices (see in References 5.86, 5.113, and 5.114).

CdS of 99.999% purity in the form of a 1-μm to 3-μm powder or 3-mm to 12-mm lumps is available commercially (1995 price: from $0.62 to $2.24 per gram).

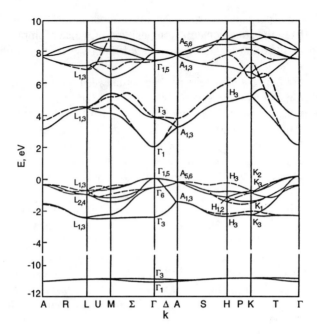

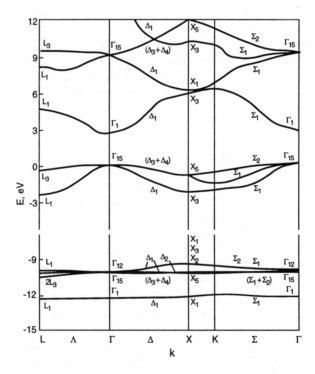

Figure 5.15 (a) Band structure of hexagonal CdS from empirical tight-binding model (solid lines) and pseudopotential method (dashed lines) calculations. (b) Band structure of cubic CdS calculated using the first-principle method. (After A. Kobayashi, O.F. Sankey, S.M. Volz, and J.D. Dow, *Phys. Rev.*, B28, 935, 1983 (a) and D.V. Farberovich, S.I. Kurganskii, and E.P. Domashevskaya, *Phys. Stat. Sol.*, 97, 631, 1980 (b). With permission.)

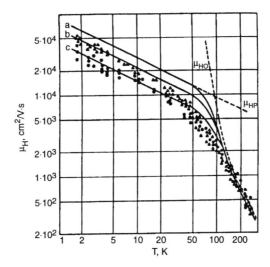

Figure 5.16 Temperature dependence of the Hall mobility of photoexcited electrons in CdS. The triangles and the circles indicate measurements taken with the current across and along the c axis, respectively. Line a represents the combination of optical mode scattering with piezoelectric scattering; curves b and c represent a best fit to the triangles and to the circles, respectively. (After S.S. Devlin, *Physics and Chemistry of II-VI Compounds*, M. Aven and J.S. Prener, Eds., John Wiley & Sons, New York, 1967, 582. With permission.)

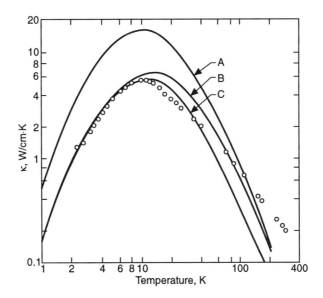

Figure 5.17 Temperature dependence of the CdS thermal conductivity along the c axis. Curve A is the result of theoretical calculation; curves B and C are obtained by fitting to the experimental data. (After S.S. Devlin, *Physics and Chemistry of II-VI Compounds*, M. Aven and J.S. Prener, Eds., John Wiley & Sons, New York, 1967, 578. With permission.)

5.1.6. Cadmium Selenide

Cadmium selenide, CdSe, has the molecular weight of 191.37, mean atomic weight of 95.69, mean atomic number of 41, and the enthalpy of formation equal to -136.4 kJ/mol[5.115] (or -136.0 kJ/mol[5.8,5.116]).

There are three known CdSe allotropes. The hexagonal CdSe has the wurtzite crystal structure with lattice parameters (300 K) a = 0.42977 ± 0.00010 nm and c = 0.7002 ±

0.00010 nm; the cubic allotrope, c-CdSe, possesses the sphalerite structure with the lattice constant at 300 K of 0.64805 ± 0.00006 nm (Reference 5.22, p. 4.143). The cubic allotrope is metastable and converts into w-CdSe upon heating; the transition starts at about 400 K and is complete in about 18 hr as a result of exposure to the temperature close to 1000 K (W.L. Roth in Reference 5.29, p. 132). There is a noticeable variation of the data on the CdSe lattice parameters reported by different authors. Roth (Reference 5.29, p. 128) has analyzed available data and come to the conclusion that the most reliable magnitudes for 300 K are a = 0.42985 nm and c = 0.70150 nm for w-CdSe and a = 0.428 nm for c-CdSe. The w-CdSe density (300 K) is 5.6738 ± 0.0028 g/cm^3 (Reference 5.22, p. 4-143). The third CdSe allotrope is a high-pressure phase with the rock salt structure and the lattice parameter of 0.554 nm (Reference 5.4, p. 99). The coefficient of thermal expansion of w-CdSe at room temperature is 2.45·10^{-6} 1/K along and 4.40·10^{-6} 1/K across the c axis.[5.35]

The maximum melting point of CdSe at the pressure of 41 kPa is 1512 K (M.R. Lorenz in Reference 5.29, p. 86) (Raiskin[5.68] found it equal to 1531 K; the magnitude cited in Reference 5.7, p. 29 is 1514). The CdSe heat of fusion is 43.93 kJ/mol.[5.117] The w-CdSe specific heat at room temperature and constant pressure and its Debye temperature are 0.255 J/g·K and 181 K.[5.92]

The CdSe microhardness is between 0.9 and 1.3 GPa (20-g load), depending on the crystal orientation (Reference 5.24, p. 122). The elastic moduli (in GPa, at room temperature) are

c_{11}	c_{12}	c_{13}	c_{33}	c_{44}	c_{66}	Ref.
74.0	45.2	39.3	83.6	13.17	14.45	5.35
74.90	46.09	39.26	84.51	13.15	14.41	5.7

The hexagonal w-CdSe is a direct semiconductor with the minimum energy gap at 300 K equal to 1.72 eV;[5.68] there is a wide variety of the reported magnitudes: from 1.670[5.118] to 1.85 eV [5.4] (E_g = 2.70 eV in Reference 5.58, p. 632 is, apparently overstated). The optical measurements by Gupta and Doh[5.111] on polycrystalline CdSe films gave E_g = 1.75 eV. The room temperature minimum energy gap of c-CdSe (also a direct semiconductor) is 1.9 eV (Reference 5.7, p. 30). The band structure of w-CdSe is shown in Figure 5.18. The band structure of c-CdSe in most of its features is similar to that of CdS. The electron effective mass in CdSe is 0.13 m_o.[5.119,5.120] (the magnitude listed in Reference 5.7 is 0.12 m_o, Böer (Reference 5.109, p. 216) prefers m_n^* = 0.112 m_o, while Kubo and Onuki[5.121] found it equal to (0.15 ± 0.1)m_o from the reflectivity and absorption measurements). The effective mass of holes in CdSe is 2.5 m_o and 0.45 m_o along and across the k_z axis of the Brillouin zone.[5.8] The relatively low electron effective mass explains the persistent n-type conductivity in the samples of CdSe.

The Hall mobility of electrons in CdSe at room temperature varies between 450 and 900 cm^2/V·sec and the hole mobility between 10 cm^2/V·sec and 50 cm^2/V·sec. From Figure 5.19 (S.S. Delvin in Reference 5.29, p. 587) one can see that at liquid nitrogen temperatures the Hall mobility may be as high as 5000 cm^2/V·sec.

The CdSe thermal conductivity at room temperature is 0.043 W/cm·K;[5.46] Slack[5.6] cites a magnitude close to 0.09 W/cm·K (extrapolated from the low temperature data).

CdSe is a diamagnetic with the molar magnetic susceptibility at room temperature of −80.3·10^{-5} SI units per gram-mol. The 300-K dielectric constants of w-CdSe are

Static		Optical		Ref.
Along c	Across c	Along c	Across c	
10.20	9.33			5.35
		5.96	6.05	Devlin in Ref. 5.29
10.16	9.29	6.2	6.3	5.7
	9.3[a]		6.1[a]	5.109

[a] Direction is not indicated.

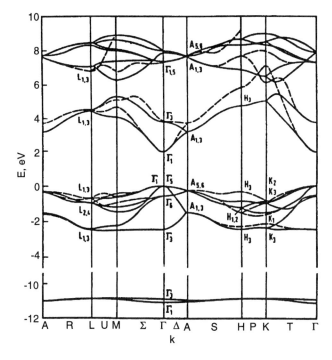

Figure 5.18 Band structure of hexagonal CdSe calculated using a semiempirical tight-binding method. 4d-states of cadmium and spin-orbit coupling are excluded from consideration. (After A. Kobayashi, O.F. Sankey, S.M. Volz, and J.D. Dow, *Phys. Rev.*, B28, 935, 1983. With permission.)

The index of refraction in CdSe[5.8] is 2.50 and 2.430 at 1.4 μm and 10.6 μm, respectively, for the o-ray and 2.38 and 2.448 at 1.4 μm and 10.6 μm, respectively, for the e-ray. The CdSe index of refraction, extinction coefficient, and absorption coefficient as functions of wavelength (in the high absorption regime) is presented in Figure 5.20 (adapted from Reference 5.111).

The first successful attempts of CdSe crystal growth were reported, probably, by Frerichs,[5.122] who used a gas-phase reaction between cadmium and selenium hydride. Until now, the majority of the results germane to the CdSe single crystal fabrication describes the synthesis processes in which at least one of the compound components is transported to the reaction zone by a gaseous entity, while the other component may be in either solid, liquid, or gas state (see, e.g., Reference 5.111, 5.123). From the author's point of view, the best methods for the high-quality CdSe single crystal synthesis remain one or another of the bulk growth methods, first of all, the Bridgman-Stockbarger technique. Using this method, Rais Enterprises of San Diego, CA provides[5.68] CdSe single crystals 6 to 7 mm in diameter and up to 35 to 40 mm in length. CdSe is available also in the form of 99.999% purity powder or 99.995% purity 3-mm to 12-mm lumps for $2.20 to $29.55 per gram (1995).

5.1.7. Cadmium Telluride

Cadmium telluride, CdTe, has the molecular weight of 240.01, mean atomic weight of 120.0, average atomic number of 50, and the enthalpy of formation of –92.47 kJ/mol.[5.115]

CdTe crystallizes in the sphalerite structure with lattice constant at room temperature equal to 0.64805 ± 0.00006 nm (Reference 5.22, p. 4-143) (0.64822 ± 0.00001 nm in Reference 5.4, p. 99). Stuckes and Farrell,[5.124] after having analyzed the lattice parameters of some hexagonal II-VI solid solutions, estimated probable lattice parameters of hexagonal CdTe as a = 0.457 nm and c = 0.747 nm, albeit the pure wurtzite modification in bulk CdTe specimens

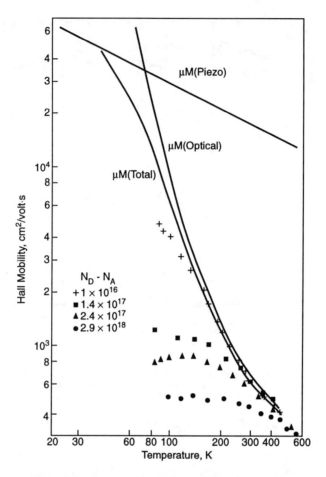

Figure 5.19 Temperature dependence of the Hall mobility in n-CdSe. The samples denoted with squares, triangles, and circles were doped with Ga; the fourth sample was undoped. The solid lines indicate the theoretical Hall mobility due to the effect of piezoelectric and optical mode scattering. (After S.S. Devlin, *Physics and Chemistry of II-VI Compounds*, M. Aven and J.S. Prener, Eds., John Wiley & Sons, New York, 1967, 587. With permission.)

has not been as yet reported. The X-ray density of CdTe at room temperature is 5.8569 ± 0.00006 nm (Reference 5.22, p. 4-143).

The coefficient of thermal expansion, α, of CdTe at 300 K is $4.90 \cdot 10^{-6}$ 1/K;[5.8] the temperature dependence of α was reported by Novikova,[5.125] who found that $\alpha = 0$ at 71.5 K and remains negative below this temperature, passing through a minimum of $-3.34 \cdot 10^{-6}$ 1/K at about 37.5 K. The experimental magnitudes of $\alpha = \alpha(T)$ from Reference 5.125 are shown in Figure 5.21.

In accordance with M.R. Lorenz (Reference 5.29, p. 82), CdTe is the only compound in the Cd-Te binary system with the melting point at 1365 K. Comparison of available data gives reason to assume that the best fit for the melting point is 1365 ± 5 K. The CdTe specific heat at room temperature and constant pressure is 205 mJ/g·K.[5.92] The CdTe Debye temperature is 200 K,[5.126] although Slack[5.6] has found that a magnitude of 158 K better suits the experimental data on thermal properties.

The microhardness of CdTe (20-g load) is 0.60 ± 0.05 GPa (Reference 5.24, p. 122). The CdTe elastic moduli are $c_{11} = 53.3$ GPa, $c_{12} = 36.5$ GPa, and $c_{44} = 20.4$ GPa (Reference 5.7, p. 30) (Delvin in Reference 5.29, p. 603 cites $c_{11} = 53.51$ GPa, $c_{12} = 36.81$ GPa, and $c_{44} = 19.94$ GPa).

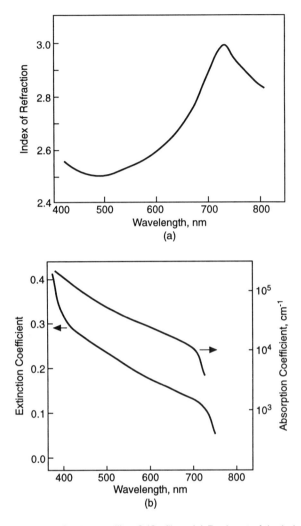

Figure 5.20 Optical properties of polycrystalline CdSe films. (a) Real part of the index of refraction in the strong absorption regime. (b) Extinction and absorption coefficients in the strong absorption regime for electroless deposited films. (After T.K. Gupta and J. Doh, *J. Mater. Res.*, 7, 1245, 1992. With permission.)

CdTe is a direct semiconductor with the minimum energy gap at room temperature of 1.47 eV.[5.4] The CdTe energy gap evaluated by van Doorn and de Nobel[5.127] from the results of the optical absorption spectrum and photo-e.m.f. measurements is $E_g(0\ K) \simeq 1.57$ eV, $E_g(300\ K) = 1.51$ eV, and $E_g(500\ K) = 1.17$ eV. Cohen and Bergstresser's results[5.128] of the energy band calculations by the pseudopotential method are presented in Figure 5.22a. The CdTe band structure reported by Chelikowsky and Cohen is shown for comparison in Figure 5.22b. The electron effective mass (in units of m_o) in CdTe at the bottom of the conduction band is 0.096 ± 0.005 (from the cyclotron resonance measurements by Kanazawa and Brown[5.129]). The hole effective mass is (1) heavy holes: $m_{ph}^*(100) = 0.72\ m_o$, $m_{ph}^*(110) = 0.81\ m_o$, and $m_{ph}^*(111) = 0.84\ m_o$; (2) light holes: $m_{pl}^* = 0.13\ m_o$ (Reference 5.109, p. 219).

The conductivity type in CdTe can be changed without substantial difficulties. The room temperature electron mobility in CdTe bulk samples varies from 500 cm²/V·sec to 1000 cm²/V·sec; the hole mobility varies between 70 cm²/V·sec and 120 cm²/V·sec. The mobility vs. temperature data for n-CdTe and p-CdTe samples of different purity are shown in Figures 5.23 and 5.24, respectively (from Segall et al.[5.130] and Yamada,[5.131] discussed by

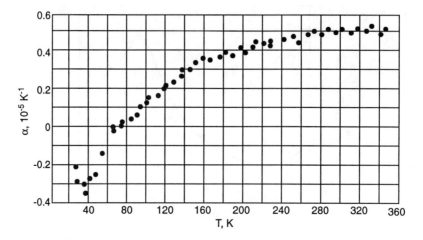

Figure 5.21 Temperature dependence of the CdTe coefficient of linear thermal expansion. (After S.I. Novikova, *Sov. Phys. Solid State*, 2, 2087, 1961. With permission.)

S.S. Delvin in Reference 5.29, p. 590). The room temperature electron mobility in the CdTe epitaxial layers, reported by Sitter et al.,[5.97] is 600 cm^2/V·sec (see also Reference 5.132).

Electrical properties of the CdTe melt were studied by Glazov et al.[5.82] (see Figure 5.13). The electrical conductivity is increasing by the order of magnitude during the melting process; the melt retains the semiconductor character of conductivity.

The room temperature thermal conductivity, κ, of CdTe is 75 mW/cm·K.[5.6] The temperature dependence of κ was reported by Slack and Galginaitis[5.133] and Slack.[5.6] At about 7 K to 10 K, κ passes through a maximum, the magnitude of which decreases with the purity decrease. "Pure" CdTe has an estimated maximum of about 9 W/cm·K, while the sample containing 1.3 at% Fe possesses a maximum only of about 0.13 W/cm·K near 75 K.

CdTe is a diamagnetic with the room temperature molar magnetic susceptibility of $-106 \cdot 10^{-5}$ SI units per gram-mol (Reference 5.4, p. 106). The static dielectric constant (300 K) of CdTe is 10.6; its optical limit is 7.21.[5.134,5.135] It is slightly different from the data listed in Reference 5.7 (10.2 and 7.1) and in Reference 5.8 (10.9 and 7.2). Dependence of the dielectric constant on frequency in the range from UV to IR parts of spectrum is compiled in Reference 5.49.

The index of refraction of CdTe is equal to 2.64 and 2.75 at 10.6 μm and 1.4 μm, respectively,[5.8] in good agreement with the magnitude calculated from its magnetic and dielectric properties (n = 2.68).

Similar to ZnTe, cadmium telluride may be synthesized by melting the stoichiometric mixture of the elemental components in an evacuated and sealed fused silica ampule, the inner surface of which is covered with graphite to prevent reaction of Cd and CdTe with silica. Glazov et al.[5.82] recommend a step-shaped increase of temperature in the process of synthesis with 2- to 3-hr stoppages at 773, 973, and 1173 K. This part of the synthesis is followed by cooling the ampule, breaking it, crushing the received CdTe, loading it into another ampule, sealing the ampule, heating it to the temperature close to 1420 to 1430 K, holding at this temperature for 3 hr with intensive stirring, and then cooling it at the rate of 5 K/hr almost to the ambient temperature.

The most effective technique of CdTe single crystal growth is, probably, the Bridgman method (M.R. Lorenz in Reference 5.29, p. 92) or its variant, the Bridgman-Stockbarger method.[5.68] Among the CdTe thin film growth techniques, the metal-organic chemical vapor deposition (MOCVD) as a result of pyrolysis of compounds such as dymethylcadmium and dymethyltelluride (see a review by Dapkus[5.137]) is, apparently, the most promising. Another process developed for CdTe epitaxial growth is the hot-wall epitaxy.[5.97,5.132,5.138]

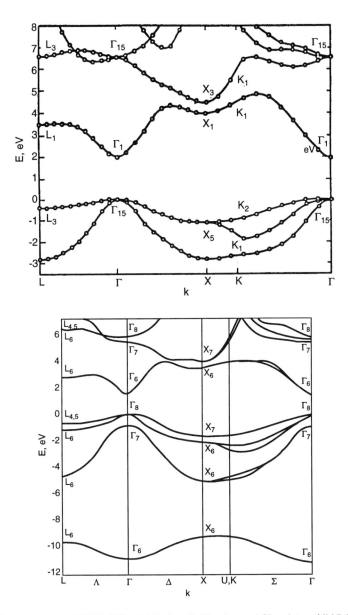

Figure 5.22 Band structure of CdTe. (After (a) B. Segall, *Physics and Chemistry of II-VI Compounds*, M. Aven and J.S. Prener, Eds., John Wiley & Sons, New York, 1967; and (b) J.R. Chelikowsky and M.L. Cohen, *Phys. Rev.*, B14, 556, 1976. With permission.)

Cadmium telluride is available commercially in the form of single crystals as well as the random-size polycrystalline pieces of 99.99999+% purity (1995 price of the latter is from $11.36 to $20.40 per gram).

5.1.8. Mercury Sulfide

Mercury sulfide, HgS, has the molecular weight of 232.65, mean atomic weight of 116.33, mean atomic number of 48, and the enthalpy of formation of −57.7 kJ/mol.[5.116]

The allotrope α-HgS, stable at room temperature, crystallizes in a dihedrally coordinated structure with hexagonal symmetry (the cinnabar structure), space group $P3_121$-D_3^4 or $P3_221$-D_3^6, with lattice parameters a = 0.4149 nm and c = 0.9495 nm under standard

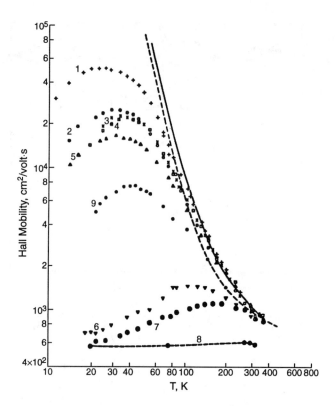

Figure 5.23 Temperature dependence of the Hall mobility in n-CdTe. The sets of data denoted with 1, 2, 3, 4, and 5 relate to multiple zone refined undoped CdTe. The data denoted with 6, 7, and 8 were received on the samples doped with Nd, In, and I, respectively. The solid line is the theoretical drift mobility due to optical mode scattering taking into account the temperature dependence of the static dielectric constant. The dashed line next to it results when this temperature dependence is neglected. The set of data denoted with 9 is received on the multiple zone refined CdTe annealed in Cd. (After S.S. Devlin, *Physics and Chemistry of II-VI Compounds*, M. Aven and J.S. Prener, Eds., John Wiley & Sons, New York, 1967. With permission.)

conditions[5.139] (Choe et al.[5.140] found a = 0.4142 nm and c = 0.9489 nm). Above 617 K, the stable allotrope is one similar to the mineral metacinnabar, with the sphalerite-like structure and with the lattice parameter at standard conditions of 0.58517 ± 0.00010 nm. The density of α-HgS and the cubic β-HgS is 8.187 g/cm^3 and 7.712 g/cm^3, respectively (Reference 5.22, p. 4-144).

Under pressure of about 12 MPa, HgS can be melted at about 1820 K; in an evacuated and sealed ampule, HgS sublimates at 719 K (Reference 5.24, p. 122). The constant pressure specific heat of α-HgS at room temperature is 0.21 J/g·K.[5.142]

The hardness of β-HgS at room temperature is equal to 3 Mohs units (Reference 5.24, p. 122). The bulk modulus of α-HgS at standard conditions is 19.4 ± 0.5 GPa.[5.139] The second-order elastic moduli of the α-phase (in GPa) are c_{11} = 35.0, c_{33} = 48.6, and c_{66} = 13.0; for the β-phase they are c_{11} = 81.3, c_{12} = 62.2, and c_{44} = 26.4 (Reference 5.7, p. 31).

The cinnabar allotrope is a direct semiconductor with the energy gap of 2.10 eV at 300 K and 2.275 eV at 4.2 K[5.7] (Choe et al.[5.140] report a 298-K magnitude of 2.030 eV); β-HgS is a semimetal: the conduction band minimum has the energy level 0.2 to 0.5 eV below the level of the valence band maximum (both extrema are of Γ-symmetry[5.1]).

The electrical transport in HgS is characterized by the electron mobility of 30 to 45 and 10 to 13 cm^2/V·sec along and across the c axis at 300 K in α-HgS (Reference 5.7, p. 31) and 250 cm^2/V·sec in β-HgS (Reference 5.4, p. 106).

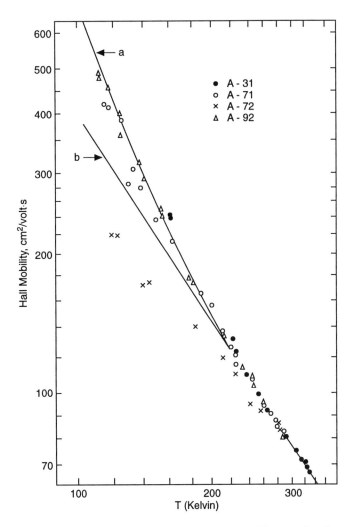

Figure 5.24 Temperature dependence of the Hall mobility in p-CdTe. The samples denoted A-31, A-71, A-72, and A-92 have room temperature volume resistivity of 2.2, 0.15, 0.19, and 0.21 kOhm·cm, respectively. Curves a and b are calculated from $\mu_H = 57[\exp(252/T) - 1]$ and $\mu_H = 4 \cdot 10^5 \, T^{-3/2}$, respectively. (After S.S. Devlin, *Physics and Chemistry in II-VI Compounds*, M. Aven and J.S. Prener, Eds., John Wiley & Sons, New York, 1967. With permission.)

The static dielectric constant in α-HgS is 23.5 and 18.2 along and across the c axis; its optical magnitude is 7.9 and 6.25 in the same directions, respectively.[5.4] The latter magnitudes are in reasonable agreement with the index of refraction of 2.85.[5.4]

Although neither one of the HgS allotropes has as yet been studied sufficiently (see, e.g., References 5.143 to 5.148), the relatively high photoelectric sensitivity of α-HgS in the visible part of the spectrum presents a certain interest for its application in optoelectronics.

Low melting and boiling points of mercury (234.3 K and 630 K) and sulfur (386 K and 718 K) permit synthesis of HgS in evacuated and sealed silica ampules, placed in protective steel vessels, at temperatures not exceeding 900 to 930 K, much lower than the HgS melting point, although the reaction of the HgS formation has a low rate. Usually, the time period of exposure to these temperatures is 360 to 480 hr; it is followed by a slow (2 K/hr) cooling. If cooling takes place to the ambient temperature, the synthesized material has the α-HgS structure. Quenching of the material to room temperature from 650 K gives predominantly β-HgS crystals. Substantial progress has been achieved in growth of HgS thin films (see, e.g., References 5.144 to 5.146).

Mercury sulfide is available commercially: the red powder of α-HgS for $0.25 to $0.40 per gram and the black, 99.999% purity powder of β-HgS for $15.84 to $21.20 per gram (1995 prices).

5.1.9. Mercury Selenide

Mercury selenide, HgSe, has the molecular weight of 279.55, average atomic weight of 139.78, average atomic number of 57, and the enthalpy of formation of −58.6 kJ/mol.[5.116]

HgSe crystallizes in the sphalerite structure with the lattice parameter at room temperature of 0.60853 ± 0.00050 nm and the X-ray density of 8.239 ± 0.020 g/cm^3 (Reference 5.22, p. 4-144). The HgSe coefficient of thermal expansion near room temperature is 1.46·10^{-6} 1/K.[5.149]

HgSe melts at 1072 K.[5.150] Its heat capacity at constant pressure is 49.8 J/mol·K at 300 K and 36.1 J/mol·K at 80 K;[5.151] the Debye temperature is 151 K[5.6] (Kelemen et al.[5.72] cite θ_D = 137.2 K, while the magnitude of 242 K was received by the authors of Reference 5.151).

The HgSe hardness for both mineral tiemannite and synthetic crystals is 2.5 ± 0.5 Mohs units; the second-order elastic moduli (stiffness constants) are c_{11} = 61, c_{12} = 44, and c_{44} = 22 GPa (Reference 5.7, p. 31).

HgSe is a semimetal. The overlapping of the conduction band minimum anf valence band maximum is 0.061 eV at 300 K and 0.205 eV at 80 K.[5.7,5.152,5.153] From electrical measurements, the HgSe energy gap at 300 K is 0.12 eV (Reference 5.4, p. 106). This difference may be related to the specifics of the carrier distribution between the states; a strong dependence of the electron effective mass in HgSe on electron concentration (in accordance with Whitsett et al.,[5.154] the 4.2-K magnitude of the electron effective mass, calculated from the Shubnikov-de Haas effect measurements, varies between 0.035 m_o and 0.070 m_o) may be considered as a justification of this point of view. The hole effective mass in HgSe is 0.78 m_o (Reference 5.7, p. 31).

The electron mobility in HgSe single crystals is relatively high, from about 5000 cm^2/V·sec at 300 K to 6·10^4 cm^2/V·sec at 77 K (T.C. Harman in Reference 5.29, p. 780 indicates the range of mobility at 77 K from 1.7·10^4 cm^2/V·sec to 8.0·10^4 cm^2/V·sec).

The HgSe thermal conductivity, κ, at 300 K is 0.017 W/cm·K.[5.6] The temperature dependence of κ was reported by Kelemen et al.,[5.72] Aliev et al.,[5.155] and Wasim et al.[5.156] The typical κ vs. T data from the three references are shown in Figure 5.25.

In accordance with Reference 5.22, p. 12-47, the dielectric constant of HgSe at room temperature and 10 kHz to 1 MHz is 25.6; the high frequency limit is between 12 and 21 (the most probable magnitude is 15.9) (Reference 5.7, p. 31). The index of refraction of HgSe is close to 4.0.

Two methods are most commonly used for bulk HgSe synthesis, namely, growth from the vapor (see, e.g., References 5.157 and 5.158) and the Bridgman technique (see T.C. Harman in Reference 5.29, p. 777). For some applications, pressed and sintered powder specimens of HgSe may be successfully used (Nikolskaya and Regel[5.159] report electron mobility up to 12,000 cm^2/V·sec in the samples received by this method). Extensive information regarding the growth and structure of HgSe epitaxial films is compiled by Holt.[5.160]

HgSe of the 99.999% purity is available commercially in the form of 100-mesh powder for $29.20 to $39.50 per gram (1995 prices).

5.1.10. Mercury Telluride

Mercury telluride, HgTe, has the molecular weight of 328.19, average atomic weight of 164.10, average atomic number of 66, and the enthalpy of formation of −50.2 kJ/mol[5.116] (the authors of Reference 5.162 report a magnitude of −33.9 kJ/mol).

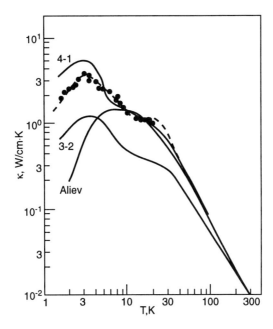

Figure 5.25 Temperature dependence of HgSe thermal conductivity. The solid circles and the fitting dash line represent the data of Reference 5.156. The curves denoted 4-1 and 3-2 are from Reference 5.154; the curve denoted "Aliev" is from Reference 5.155. (After S.M. Wasim, B. Fernandez, and R. Aldana, *Phys. Status Solidi,* A76, 743, 1983. With permission.)

The normal pressure HgTe allotrope has the sphalerite structure with the lattice parameter 0.6453 nm at 300 K[5.139] (0.64600 ± 0.00006 nm in Reference 5.22, p. 4-144; 0.64623 ± 0.00001 nm in Reference 5.4, p. 100) and the X-ray density of 8.0855 ± 0.0023 g/cm³ (the density of the natural HgTe, the mineral coloradoite, is 8.07 g/cm³). The HgTe coefficient of thermal expansion, α, is $4.6 \cdot 10^{-6}$ 1/K at 300 K. The temperature dependence of α between 20 and 340 K was reported in Reference 5.70 (Figure 5.26). At 62 K, α becomes equal to zero and is negative at lower temperatures. At the pressure of 1.4 GPa, HgTe undergoes a transition into the hexagonal, cinnabar phase with the lattice parameters (at 2.6 GPa) a = 0.445 nm and c = 0.989 nm. This phase exists between 1.4 GPa[5.139,5.161] and 8 GPa. At 8.0 ± 0.3 GPa, HgTe transforms into a phase with the rock salt structure and the lattice constant of 0.583 nm (at 8.2 GPa). Above 11.0 GPa and at least to 20 GPa, the HgTe structure is one of white tin with lattice parameters a = 0.568 nm and c = 0.304 nm at 11.7 GPa.[5.139]

The HgTe maximum melting point is 943 K at the equilibrium mercury vapor pressure 1.25 MPa.[5.163] The specific heat at 293 K is close to 0.164 J/g·K.[5.72] There is a substantial variation of the data on the HgTe Debye temperature. Mavroides and Kolesar[5.164] have reported θ_D = 105 K, Kelemen et al.[5.72] have found θ_D = 114.3 K, while in Reference 5.4, θ_D = 242 K. Slack[5.6] has shown that the magnitude of θ_D = 141 K is in fairly good agreement with some theoretical predictions.[5.165-5.167]

The HgTe hardness is 2.5 units of the Mohs scale; the microhardness at 20-g load is close to 280 MPa. The HgTe second-order elastic moduli are (300 K) c_{11} = 53.61 GPa, c_{12} = 36.60 GPa, and c_{44} = 21.23 GPa.[5.7]

Like HgSe, mercury telluride is a semimetal. The band structure is characterized by overlapping the conduction band minimum and the valence band maximum at the Brillouin zone center. The overlapping of the states is clear from Figure 5.27 (after Groves and Paul[5.168] and T.C. Harman in Reference 5.29, p. 804), where subscripts "c" or "v" relate to the conduction or valence band, respectively. The electron effective mass in bulk HgTe is 0.029 m_o (with n = $2 \cdot 10^{15}$ cm⁻³);[5.169] the hole effective mass is substantially higher (0.3 m_o cited by

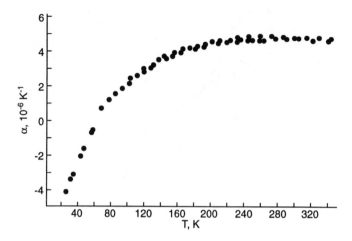

Figure 5.26 Temperature dependence of the HgTe coefficient of linear thermal expansion. (After S.I. Novikova and N.Kh. Abrikosov, *Sov. Phys. Solid State*, 5, 1558, 1964. With permission.)

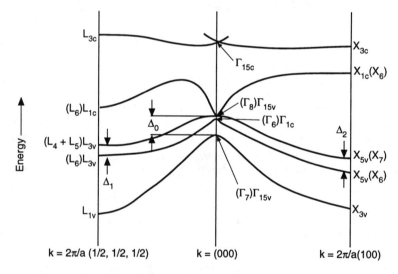

Figure 5.27 Qualitative presentation of the HgSe and HgTe band structures with single and double (in parentheses) group representations. (After T.C. Harman, *Physics and Chemistry of II-VI Compounds*, M. Aven and J.S. Prener, Eds., John Wiley & Sons, New York, 1992, 804. With permission.)

T.C. Harman in Reference 5.29, p. 810, 0.42 m_o is included in the tables in Reference 5.7, p. 31).

The study of the electric transport in HgTe shows that the material may have conductivity of both n- and p-type depending on the dopant(s) used. The intrinsic carrier concentration at 4.2 K is $3 \cdot 10^{16}$ cm^{-3} (Harman in Reference 5.29, p. 797). The electron mobility varies in the wide limits, depending on the samples' purity and synthesis method. Galazka[5.170] reported the mobility of $7.91 \cdot 10^5$ cm^2/V·sec and $2.65 \cdot 10^4$ cm^2/V·sec at 4.2 and 300 K; the results received by Irvine et al.[5.171] on the epitaxial HgTe films are $3.2 \cdot 10^4$ cm^2/V·sec and $2.1 \cdot 10^4$ cm^2/V·sec at the same temperatures. The hole mobility in HgTe at 90 K to 120 K is about 700 cm^2/V·sec[5.172a] (Dziuba and Szlenk[5.172b] found μ_p (4.2 K) $\leq$ 5000 cm^2/V·sec). The volume resistivity of both bulk samples and thin films of HgTe at 300 K is usually in the limits 10^{-2} to 10^{-3} Ohm·cm (see, e.g., Reference 5.169).

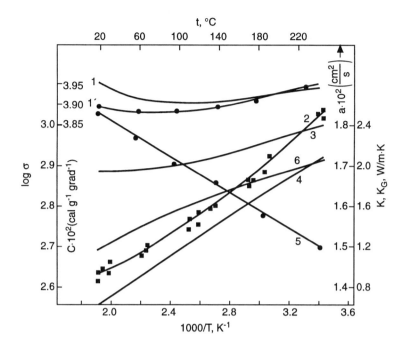

Figure 5.28 Temperature dependence of HgTe thermal properties and electrical conductivity. 1 — calculated specific heat, $C_{calc} = f(t)$; 1' — experimental specific heat, $C_{exp} = f(t)$; 2 — thermal diffusivity, $a = f(1/T)$; 3 — thermal conductivity, $K = f(1/T)$; 4 — lattice thermal conductivity, $K_G = f(1/T)$; 5 — electrical conductivity, $\lg \sigma = f(1/T)$, σ in S/cm; 6 — calculated lattice thermal conductivity, $K_{G,calc} = f(1/T)$. (After F. Kelemen, E. Crucheanu, and D. Niculescu, *Phys. Status Solidi*, 11, 865, 1965. With permission.)

The thermal conductivity, κ, of HgTe at 300 K is 19 mW/cm·K.[5.6] The temperature dependence of κ between 2 and 300 K was reported by Whitsett and Nelson.[5.173] The common feature of several κ vs. T curves is the presence of a smooth maximum of about 1 to 3 W/cm·K near 15 K and near linear dependence of κ on T^{-1} up to 300 K. A detailed analysis of the[5.173] data is presented in Reference 5.174. Kelemen et al.[5.72] measured κ of HgTe between 293 and 513 K. They reported the 293-K magnitude of 21 mW/cm·K. The results of measurements of κ, together with the data on thermal diffusivity, specific heat, and electrical conductivity of the HgTe samples tested, are presented in Figure 5.28.

The static dielectric constant of HgTe is between 20.0 and 21.0; the optical dielectric constant is equal to 14.[5.175]

Single crystals of HgTe may be grown by the horizontal melting zone[5.176] and Bridgman[5.177] techniques. Epitaxial layers of HgTe were grown on CdTe substrates by Alekseenko et al.,[5.178] Irvine et al.,[5.171] and others. The 99.999% purity HgTe 3-mm to 12-mm fused lumps are available commercially for $23.80 to $31.50 per gram (1995).

5.2. BINARY I-VII SEMICONDUCTORS

Among the halides of the group Ib elements, five, namely, CuF, CuCl, CuBr, CuI, and AgI, have allotropes with the coordination number equal to four. The other silver halides — AgF, AgCl, and AgBr — crystallize in the rock salt lattice. The latter materials are indirect semiconductors and have a noticeable ionic component in their electrical conductivity. This section is devoted to the halides with adamantine crystal structure.

5.2.1. Copper Fluoride

Copper fluoride, CuF, has the molecular weight 82.54, mean atomic weight 41.27, and average atomic number of 19. The compound is unstable and readily dissolves in muriatic, sulfuric, and nitric acids.

CuF (γ-phase) crystallizes in the sphalerite structure with the room temperature lattice constant of 0.426 nm.[5.179]

5.2.2. Copper Chloride

Copper chloride, CuCl, has the molecular weight of 99.0, mean atomic weight of 49.5, average atomic number 23, and the enthalpy of formation −137.2 kJ/mol (Reference 5.22, p. 5-9).

CuCl crystals produced from its melt have the wurtzite structure with lattice parameters a = 0.391 nm and c = 0.642 nm (Reference 5.4, p. 100). Upon cooling to 680 K, CuCl transforms into its low temperature modification with the sphalerite structure and lattice parameter at 298 K equal to 0.5416 ± 0.0003 nm and the X-ray density 4.139 ± 0.007 g/cm^3 (Reference 5.22, p. 4-149) (natural CuCl, mineral nantockite, has density 3.93 g/cm^3). At high pressure and standard temperature, CuCl transforms from the sphalerite into a tetragonal phase (at 4.46 GPa) with the lattice parameters a = 0.5191 nm and c = 0.4688 nm; at the pressure of about 8.2 GPa and up to 12 GPa, the tetragonal phase coexists with a rock salt phase; at 12.3 GPa and higher, only the rock salt phase (a = 0.490 nm at 12.3 GPa) is present.[5.180] The results of Reference 5.180 are in reasonable agreement with the data of Skelton et al.[5.181] The T-p section of the CuCl phase diagram has been studied by Rusakov et al.[5.182] The CuCl coefficient of thermal expansion at 300 K is 1.2·10^{-5} 1/K (Hanson et al.[5.184] cite α(300 K) = 1.54·10^{-5} 1/K); with the temperature decrease its magnitude decreases and passes through zero at about 100 K,[5.183a] becoming negative below this temperature. The results of Reference 5.183a, received on the pressed samples of CuCl, are in good agreement with the magnitude of the coefficient of thermal expansion measured on single crystalline samples grown by the traveling heater method (α[300K] = 12.1·10^{-6} 1/K, T[α = 0] = 105 K)[5.185] (see, also, Reference 5.186).

The hardness of CuCl is 2.3 units of the Mohs scale; the elastic moduli (300 K) are c_{11} = 47.0 GPa, c_{12} = 36.2 GPa, and c_{44} = 14.5 GPa (Reference 5.7, p. 15); the compressibility is 2.58·10^{-11} m^2/N (Hanson et al.[5.184] have found c_{11} = 45.4 GPa, c_{12} = 36.3 GPa, and c_{44} = 15.35 GPa, the bulk modulus of 39.3 GPa and compressibility of 2.54·10^{-11} m^2/N).

The CuCl melting point is 695 K,[5.184] the specific heat at room temperature is 0.49 J/g·K, and the Debye temperature is 240 K (at 300 K)[5.183b] (the 300-K magnitude, calculated from the elastic parameters, is 141 K[5.184]).

CuCl is a direct semiconductor with the experimentally evaluated energy gap of 3.4 eV.[5.188] The CuCl band structure parameters were calculated, in particular, by Song,[5.189] Kahn,[5.190] Calabrese and Fowler,[5.191] Zunger and Cohen,[5.192] Kleinman and Mednick,[5.193] and Ren et al.[5.194] The CuCl band structure (from Reference 5.197) is shown in Figure 5.29. The minimum direct energy gap (transition Γ_{7v} to Γ_{6c}) is 3.395 and 3.17 eV at 4.2 and 300 K, respectively (the authors of Reference 5.193 found it equal to 3.09 eV). Whereas there is satisfactory agreement between the data on the electron effective mass reported by different authors (e.g., 0.415[5.194] and 0.44 m$_o$[5.195]), the magnitudes of the hole effective mass reported vary quite widely (e.g., 0.913 m$_o$;[5.193] 20.4 m$_o$;[5.195] and 3.6 m$_o$[5.196]). The authors of Reference 5.193 performed the CuCl energy band calculations and found m$_n^*$ (Γ_6) = 0.417 m$_o$, m$_p^*$(Γ_7) = 1.477 m$_o$, and m$_p^*$(Γ_8) — from 0.969 m$_o$ to 3.101 m$_o$. The wide scattering of the data suggests that the solution of the problem of the CuCl band parameters needs additional effort.

CuCl is partially an ionic conductor; this complicates the measurements of the electrical transport properties. The experimental volume resistivity of single crystalline CuCl at room

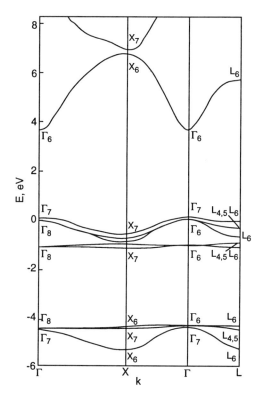

Figure 5.29 Band structure of CuCl. (After H. Overhof, *Phys. Stat. Sol.*, B97, 267, 1980. With permission.)

temperature is close to $7 \cdot 10^6$ Ohm·cm, while pressed samples show a magnitude of about $8 \cdot 10^5$ Ohm·cm.[5.181] The authors of Reference 5.181 found that the volume resistivity increases with the pressure increase, and under the pressure of about 6 GPa, the volume resistivity magnitude (about $4 \cdot 10^6$ Ohm·cm) virtually does not depend on the choice of the electrode material (platinum, copper, or graphite) and on the method of the sample synthesis. Perner[5.187] found the room temperature resistivity of cubic CuCl single crystals, grown by the traveling heater method from different solvents, varying between $7 \cdot 10^5$ Ohm·cm and $1.2 \cdot 10^8$ Ohm·cm. Divakar et al.[5.205] found a noticeable dependence of resistivity on the DC voltage applied to the thin layers of CuCl. The resistance slowly decreases with the voltage increase; at a certain voltage it decrease is avalanche-like, by several orders of magnitude; at higher voltage, resistance does not depend on voltage.

In accordance with Ioffe,[5.34] the thermal conductivity of copper chloride at 290 K is 8 mW/cm·K.

CuCl is a diamagnetic with the molar susceptibility of $-88 \cdot 10^{-5}$ SI units per gram-mol (Reference 5.4, p. 106). The static and optical dielectric constants of CuCl are 7.9 and 3.61, respectively (Reference 5.7, p. 15) (Rusakov[5.198] cites $\varepsilon[0] = 9.5$ and $\varepsilon[\infty] = 3.7$); the CuCl index of refraction at 550 nm is 1.93, in good agreement with its dielectric parameters.

As mentioned before, the cubic CuCl allotrope exists at temperatures below 680 K. Above this temperature and up to the melting point of 695 K, CuCl possesses the wurtzite structure. This makes it difficult to grow the c-CuCl crystals from the melt of pure CuCl. To grow crystals at lower temperature, it is convenient to use as a source the solutions of CuCl and halides of either alkali or alkaline-earth metals.[5.187,5.199-5.201] Kaifu and Kawate[5.199] have grown CuCl single crystals by the Bridgman technique from the solutions CuCl-KCl; Soga and Imaizumi[5.200] have used solutions of CuCl with KCl, RbCl, CsCl, $SrCl_2$, $BaCl_2$, and $PbCl_2$. The traveling heater method (see, e.g., Reference 5.202) has been successfully used by

Perner[5.187] to grow c-CuCl single crystals 7 mm to 8 mm in diameter and up to 40 mm long from CuCl-KCl solutions at 583 K, 603 K, 659 K, and 643 K. He also found that the good optical quality and resistivity up to $1.3 \cdot 10^8$ Ohm·cm single crystals of c-CuCl could be grown from solution of CuCl with $Ba_{0.3}Sr_{0.7}Cl_2$ as a solvent at 679 K. Hanson et al.[5.203] also reported on successful application of the traveling heater method for c-CuCl crystal growth. Wilcox and Corley[5.201] applied the Czochralski technique to grow c-CuCl crystals from solutions with KCl, but the crystals obtained were optically inhomogeneous. Neuhaus and Recker[5.206] produced CuCl crystals by the growth from the vapor phase on a surface heated below 680 K.

The 99.999% purity CuCl in the form of crystalline aggregates is available commercially for $1.32 to $15.60 per gram (1995).

5.2.3. Copper Bromide

Copper bromide, CuBr, has the molecular weight 143.45, mean atomic weight 71.73, mean atomic number 32, and the enthalpy of formation –104.6 kJ/mol.

At the low temperature and up to 658 K, the stable allotrope of CuBr has the sphalerite structure and lattice parameter (at 300 K) equal to 0.5692 nm[5.179] (Goryunova [Reference 5.24, p. 123] cites 0.5692 nm). Between 658 K and 743 K, the stable allotrope of CuBr is one with the wurtzite structure and lattice parameters at 300 K a = 0.406 nm and c = 0.666 nm (Reference 5.4, p. 100). Between 743 K and the melting point lies the domain of the third allotrope, α-CuBr, which, in accordance with the Phillips[5.204] classification, may possess the rock salt crystal structure. The CuBr coefficient of thermal expansion, α, at 300 K is $15.4 \cdot 10^{-6}$ 1/K;[5.183b] it decreases with temperature, reaches the magnitude $\alpha = 0$ at 95 K, and remains negative below this temperature.

The CuBr melting point is between 760 and 777 K from different reports; this variation may be related to the difficulties in controlling the vapor pressure upon the samples at the elevated temperatures. The CuBr constant pressure specific heat at 300 K is 0.381 J/g·K (Reference 5.22, p. 5-7); the Debye temperature is 207 K (at 300 K).[5.183b]

CuBr is a relatively soft material. Its microhardness (at 10-g load) is 212 MPa; the volume compressibility is $2.56 \cdot 10^{-11}$ m²/N.[5.186] The CuBr elastic moduli are c_{11} = 43.5 GPa, c_{12} = 34.9 GPa, and c_{44} = 14.7 GPa (Reference 5.7, p. 15).

Copper bromide is a direct semiconductor with the minimum energy gap at 300 K equal to 2.94 eV (Reference 5.4, p. 106). The CuBr band structure was studied, in particular, by Overhof[5.197] and Ves et al.[5.179] The energy band structure calculated in Reference 5.197 is shown in Figure 5.30 (adapted from Reference 5.197). The CuBr energy gap cited in Reference 5.7, p. 15 is 2.91 and 3.077 eV at 300 and 4.2 K, respectively (the magnitudes in Reference 5.179 are 2.91 and 3.016 eV). The electron effective mass in CuBr is equal to (or slightly greater than) 0.21 m_o; the light and heavy holes have effective mass close to 1.1 m_o and 1.5 m_o, respectively.[5.7]

Similar to CuCl, copper bromide is, most probably, a material with a noticeable ionic conductivity component. Continuous passage of a direct current (at high-enough electric field strength) through a CuBr sample will be followed by deposition of copper on the electrodes and formation of highly conductive copper microfilaments in the sample. The CuBr thermal conductivity at room temperature is 11 mW/cm·K.[5.34]

CuBr is a diamagnetic with the molar magnetic susceptibility $-70.6 \cdot 10^{-5}$ SI units per gram-mol (Reference 5.4, p. 106). The static and optical magnitudes of the CuBr dielectric constant are 9.9 and 4.4, respectively[5.198] ($\varepsilon[0]$ = 7.9 and $\varepsilon[\infty]$ = 4.062 cited in Reference 5.7, p. 15); ε (0.5 MHz, 293 K) = 8.0 is cited in Reference 5.22, p. 12-47). The index of refraction, n = 2.116, in Reference 5.24, p. 123, relates closer to the optical dielectric constant magnitude cited in Reference 5.198 than to the data in Reference 5.7. Detailed absorption and reflectance measurements on the CuBr samples were performed by Cardona.[5.188] Fundamental UV-absorption spectra of CuBr above 3 eV were reported by Ishii et al.[5.207] The results of

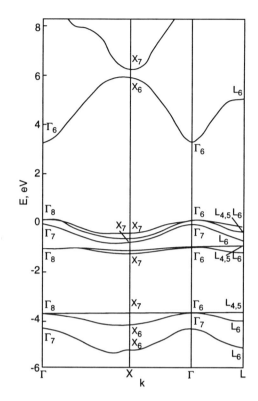

Figure 5.30 Band structure of CuBr. (After H. Overhof, *Phys. Stat. Sol.*, B97, 267, 1980. With permission.)

investigation of the CuBr optical properties were reported by Martzen and Walker.[5.208] The pressure dependence of the optical absorption edge in CuBr was studied by Ves et al.[5.179]

The majority of the experimental data on the CuBr properties was received using pressed powder samples and thin films. A brief description of preparation of the CuBr single crystalline samples about 0.5 mm thick, grown from the vapor, is given in Reference 5.208. The traveling heater method was used in Reference 5.183b to grow single crystals of c-CuBr with CsBr as a solvent.

5.2.4. Copper Iodide

The copper iodide, CuI, has the molecular weight of 190.45, average atomic weight 95.23, average atomic number 41, and the enthalpy of formation –67.8 kJ/mol.

At temperatures below 642 K, CuI has the sphalerite crystal lattice with the 300-K lattice parameter 0.6052 nm (Reference 5.2, p. 100) (0.60507 ± 0.00010 nm in Reference 5.22, p. 4-149, 0.60427 nm in Reference 5.7, p. 15) and the X-ray density 5.710 ± 0.003 g/cm^3. Between 642 K and 680 K, a wurtzite allotrope (β-CuI or w-CuI), with lattice parameters a = 0.431 nm and c = 0.7510 nm, is stable.[5.2] Above 680 K and up to the melting point, a cubic phase, possibly with the rock salt structure (α-CuI) is stable. The room temperature coefficient of thermal expansion, α, of CuI measured on a single crystal grown by the traveling heater method is $19.2 \cdot 10^{-6}$ 1/K.[5.183b] The temperature dependence of the CuI thermal expansivity is shown in Figure 5.31 (adapted from Reference 5.183a). At 80 K, α = 0 and is negative below this temperature.

With increasing pressure, CuI transforms into a rhombohedral modification at 2.0 GPa to 2.5 GPa; at higher pressures, up to 9.0 GPa to 9.5 GPa, CuI has the tetragonal structure; at higher pressures it has the rock salt crystal structure.[5.179]

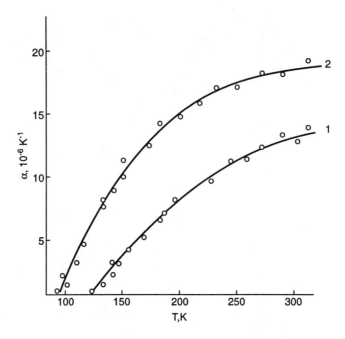

Figure 5.31 Temperature dependence of the CuCl (1) and CuI (2) coefficients of linear thermal expansion. (After L.I. Berger, N.L. Rapoport, and A.N. Novikova, *Trudy IREA (Proc. IREA) (Moscow)*, 33, 251, 1971. With permission.)

The CuI melting point is 875 K (Reference 5.7, p. 15). The specific heat at 298 K and constant pressure is 0.28 J/g·K (Reference 5.22, p. 5-12). The CuI Debye temperature is 181 K.[5.183b]

The Mohs hardness of natural CuI (mineral marshite) is 2.5. The square pyramide microhardness (10-g load) is 192 MPa (Reference 5.24, p. 123). The elastic moduli of c-CuI are c_{11} = 45.1 GPa, c_{12} = 30.7 GPa, and c_{44} = 18.2 GPa (Reference 5.7, p. 15).

Copper iodide is a direct semiconductor with the minimum energy gap (transition Γ_8 to Γ_6) of 2.94 eV at 300 K and 3.115 eV at 4.2 K.[5.4,5.7] The results of calculations of the CuI band structure[5.197] are presented in Figure 5.32. The electron effective mass in CuI is 0.33 m_o; the hole effective mass from the data of different reports varies between 1.4 and 2.4 m_o.[5.7]

Even from the early reports (see, e.g., References 5.209 and 5.210), it is known that CuI has a mixed electronic-ionic conductivity. Ionic conductivity dominates at higher temperatures, whereas the electrical transport at lower temperatures is controlled by electrons.

The thermal conductivity of CuI at room temperature is 17 mW/cm·K.[5.34]

CuI is a diamagnetic with the molar magnetic susceptibility –87.9·10^{-5} SI units per gram-mol.[5.4] The CuI static and optical dielectric constant is 6.5 and 4.84, respectively,[5.7] in reasonable agreement with its index of refraction (2.35 in Reference 5.24). The real and imaginary parts of the CuI dielectric constant vs. photon energy, calculated using the results of the reflectance measurements, are shown in Figure 5.33.[5.208]

There are several ways of c-CuI bulk single-crystalline sample preparation. The simplest one consists of pressing CuI powder at elevated temperatures (pressure between 15 MPa and 1 GPa,[5.179,5.183b] temperature about 550 K). Another growth method suitable for CuI is the traveling heater method[5.187] using some alkali and alkaline-earth metal iodides. Among thin film deposition techniques, the laser-assisted molecular beam deposition may be considered as a promising method (see, e.g., Wijekoon et al.[5.211]). DiCenzo et al.[5.212] have grown epitaxial layers of c-CuI on the (111) surface of single-crystalline copper.

The CuI 99.999% purity powder is available commercially for $2.09 to $2.24 per gram (1995).

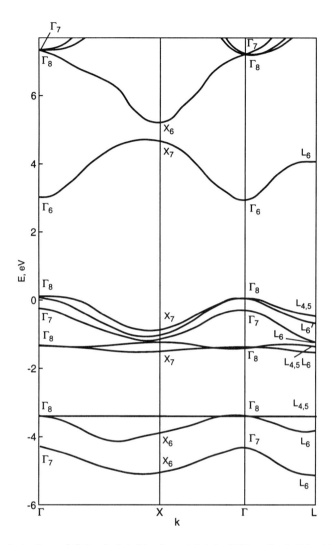

Figure 5.32 Band structure of CuI calculated by the relativistic KKR method. (After H. Overhof, *Phys. Stat. Sol.*, B97, 267, 1980. With permission.)

5.2.5. Silver Iodide

Silver iodide, AgI, has the molecular weight 234.77, average atomic weight 117.39, average atomic number 50, and the enthalpy of formation –61.8 kJ/mol.

There are three AgI crystallographic modifications known: (1) the high-temperature phase, α-AgI, with the b.c.c. structure, which exists at the temperatures above 419 K (or 420 K[5.213]); (2) the wurtzite phase, β-AgI (or w-AgI), with the area of stability between 419 and 410 K (Reference 5.4, p. 100); and (3) the δ-AgI (or c-AgI) phase, stable below 410 K with the sphalerite structure. Bienenstock and Burley[5.214] assume that δ-AgI is metastable over its entire range of existence and the only stable phase below 419 K is the wurtzite allotrope; the two structures are related by a stacking difference and have virtually the same lattice energy (both c-AgI and w-AgI are common minerals, marshite and iodyrite [or iodargyrite], respectively).

The room temperature lattice constant of c-AgI is 0.6473 nm (Reference 5.7, p. 16) (0.64963 ± 0.00010 nm in Reference 5.22, p. 4-149); the w-AgI phase has lattice parameters a = 0.45955 ± 0.00010 nm and c = 0.75005 ± 0.00033 nm. The X-ray density (298 K) is 5.688 ± 0.003 and 5.683 ± 0.004 g/cm^3 for c-AgI and w-AgI, respectively.

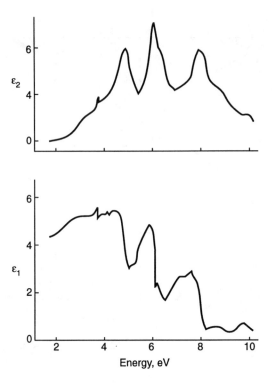

Figure 5.33 Dielectric functions ε_1 and ε_2 for CuI. (After P.D. Martzen and W.C. Walker, *Phys. Rev.*, B21, 2562, 1980. With permission.)

The coefficient of linear thermal expansion, α, of c-AgI, as it was first reported by Fizeau,[5.215] is negative at room temperature ($\alpha = -4.1 \cdot 10^{-6}$ 1/K between 288 and 324 K;[5.215] $\alpha = -1.6 \cdot 10^{-6}$ 1/K between 293 and 333 K;[5.216] $\alpha = -4.0 \cdot 10^{-6}$ 1/K at 273 K;[5.217] $\alpha = -1.6 \cdot 10^{-6}$ 1/K (room temperature — extrapolated);[5.214] $\alpha = -2.5 \cdot 10^{-6}$ 1/K.[5.183b] In accordance with Reference 5.217, $\alpha = -5.7 \cdot 10^{-6}$ 1/K at 333 K, while in Reference 5.183b it is stated that α passes through zero at 335 K. This difference may be explained by presence in the samples of an uncontrolled amount of either β-AgI or α-AgI (or both) at elevated temperatures. The synthesis methods also may influence the results of α measurements.

The melting point of AgI is 831 K[5.7] (the best fit for the various reports is 830 ± 5 K). The specific heat at constant pressure and 298.15 K is 0.24 J/g·K (Reference 5.22, p. 5-4) (Kelley[5.218] has reported 0.23 J/g·K). The Debye temperature at 300 K is 135 K;[5.183b] Burley (cited in Reference 5.213) found it close to 120 K.

The microhardness of c-AgI is 65 MPa (Reference 5.24, p. 124), the Mohs hardness is close to 2.5, and the bulk modulus is 9.890 GPa.[5.214] The elastic moduli (at 298 K) of c-AgI are $c_{11} = 26.2$ GPa, $c_{12} = 22.9$ GPa, and $c_{44} = 8.17$ GPa; the data for w-AgI are $c_{11} = 29.3$ GPa, $c_{12} = 21.3$ GPa, $c_{13} = 19.6$ GPa, $c_{33} = 35.4$ GPa, $c_{44} = 3.7$ GPa, and $c_{66} = 3.99$ GPa.[5.219]

Both c-AgI and w-AgI are direct semiconductors. The room temperature minimum energy gap is 2.82 eV (300 K) and 2.91 eV (4 K) for c-AgI and 3.023 eV (4.5 K) for w-AgI (Reference 5.7, p. 17). The band structure of c-AgI and w-AgI is shown in Figures 5.34 and 5.35.

The electrical transport in AgI is sufficiently ionic at higher temperatures, but the electronic conductivity is also presented, especially at room and lower temperatures, where a noticeable photoelectric effect is observed. The electron mobility in c-AgI at room temperature is 30 cm^2/V·sec.[5.220]

The thermal conductivity, κ, of c-AgI at room temperature has been initially measured on pressed-powder samples[5.34] and is close to 4.2 mW/cm·K. The temperature dependence of κ between 120 and 500 K was reported by Goetz and Cowen,[5.213] who also measured it

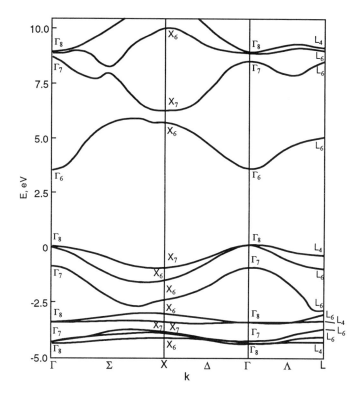

Figure 5.34 Band structure of δ-AgI (sphalerite structure) calculated using the relativistic KKR method. (After H. Overhof, *J. Phys. Chem. Sol.*, 38, 1214, 1977. With permission.)

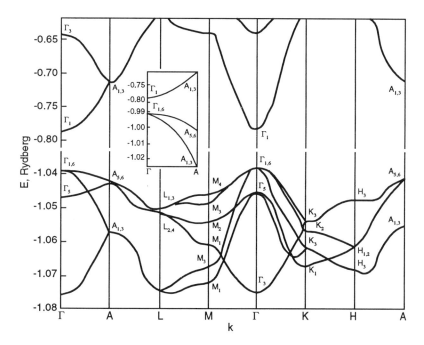

Figure 5.35 Band structure of β-AgI (wurtzite structure) calculated using the tight-binding method; spin-orbit interaction is excluded from consideration. (After P.V. Smith, *J. Phys. Chem. Sol.*, 37, 588, 1976. With permission.)

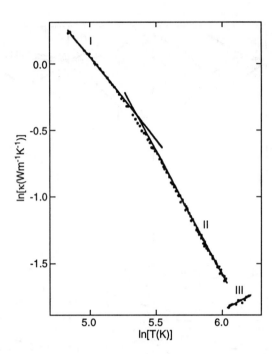

Figure 5.36 Logarithm-logarithm plot of the temperature dependence of AgI thermal conductivity (see description in the text). (After M.C. Goetz and J.A. Cowen, *Solid State Commun.*, 41, 293, 1982. With permission.)

on pressed pellets. Their results are presented in Figure 5.36. The samples were annealed for a week at 410 K to eliminate c-AgI. After annealing, the samples contain less than 1% of c-AgI. There are three distinguishable parts of the κ vs. T plot, namely, part I with $\kappa \sim T^{-1.3}$, part II with $\kappa \sim T^{-1.8}$, and part III where $\kappa \sim T^{0.5}$. The discontinuity of the curve relates to the phase transition ($\beta \rightarrow \alpha$) at 420 K. The authors conclude that the heat transport by the mobile cations does not make a sizable contribution to the thermal conductivity.

AgI is a diamagnetic with the room temperature magnetic susceptibility equal to $-113.9 \cdot 10^{-5}$ SI units per gram-mol (Reference 5.4, p. 106). The w-AgI static and optical dielectric constants are 7.0 and 4.1 to 4.9, respectively (Reference 5.7, p. 17), which is in good agreement with the magnitude of the index of refraction (2.22) presented in Reference 5.4.

The technology of the polycrystalline samples and single crystals of AgI production is similar to that of the copper halides. The 99.999% purity AgI powder is available commercially for $1.33 to $1.78 per gram (1955).

REFERENCES

5.1. Landolt-Börnstein, New Series, Vol. III/17b, O. Madelung, M. Schultz, and H. Weiss, Eds.
5.2. D. R. Lide, Ed., *Handbook of Chemistry and Physics,* 74th ed., CRC Press, Boca Raton, FL, 1993–1994.
5.3. N. D. Kumar, M. N. Kamalasanan, and S. Chanda, *Appl. Phys. Lett.*, 65, 1373, 1994.
5.4. N. A. Goryunova, *Slozhnyye Almazopodobnyye Poluprovodniki (Multinary Diamond-Like Semiconductors),* Soviet Radio, Moscow, 1968.
5.5. R. A. Montalvo and D. W. Langer, *J. Appl. Phys.*, 41, 4101, 1970.
5.6. G. A. Slack, *Phys. Rev.*, B6, 3791, 1972.
5.7. Data in Science and Technology, R. Poerschke, Ed., *Semiconductors Other Than Group IV Elements and III-V Compounds,* O. Madelung, Ed., Springer-Verlag, New York, 1992.

5.8. H. Hartman, R. Mach, and B. Selle, *Current Topics of Materials Science,* Vol. 9, E. Kaldis, Ed., North Holland, Amsterdam, 1982.

5.9. F. Quaranta, A. Valentini, F. R. Rizzi, and G. Gasamassima, *J. Appl. Phys.,* 74, 244, 1993.

5.10. K. Mukae, K. Tuda, and I. Nagasawa, *Jpn. J. Appl. Phys.,* 16, 1361, 1977.

5.11. V. Balek, *Thermochim. Acta,* 192, 1, 1991.

5.12. M. Leskelä, T. Leskelä, and L. Niinistö, *J. Therm. Anal.,* 40, 1077, 1993.

5.13. J.-L. Huang and K.-B. Li, *J. Mater. Res.,* 9, 1526, 1994.

5.14. Y. Yano, Y. Shirakawa, and H. Morooka, *J. Mater. Res.,* 9, 669, 1994.

5.15. D. M. Kolb, *Current Topics in Materials Science,* Vol. 7, E. Kaldis, Ed., North-Holland, Amsterdam, 1981, 277.

5.16. S. Harrison, *Phys. Rev.,* 93, 52, 1954.

5.17. S. N. Qiu, C. X. Qiu, and I. Shih, *Mater. Lett.,* 5, 263, 1987; C. X. Qiu and I. Shih, *Solar Energy Mater.,* 13, 75, 1986.

5.18. L. I. Berger, V. A. Kovalev, L. I. Vesker, and V. B. Kuznetsov, *Elektron Tekh. Mater.,* 5, 73, 1973.

5.19. A. V. Ioffe and S. S. Sinani, *Sov. Phys. Tech. Phys.,* 25, 1659, 1955.

5.20. W. D. Kingery, J. Francl, R. L. Coble, and T. Vasilos, *J. Am. Ceram. Soc.,* 37, 107, 1954.

5.21. J. J. Martin and M. W. Wolf, *Bull. Am. Phys. Soc.,* 7, 282, 1972.

5.22. D. R. Lide, Ed., *Handbook of Chemistry and Physics,* 75th ed., CRC Press, Boca Raton, FL, 1994.

5.23. T. F. Smith and G. K. White, *J. Phys.,* C8, 2031, 1975.

5.24. N. A. Goryunova, *The Chemistry of Diamond-Like Semiconductors,* MIT Press, Lancaster, England, 1965.

5.25. C. Frondel and C. Palache, *Am. Mineral.,* 35, 29, 1950.

5.26. D. M. Seaman and H. Hamilton, *Am. Mineral.,* 35, 43, 1950.

5.27. H. T. Evans and E. T. McKnicht, *Am. Mineral.,* 44, 1210, 1959.

5.28. D. C. Buck and L. W. Strock, *Am. Mineral.,* 39, 318, 1954; 40, 192, 1955.

5.29. M. Aven and J. S. Prener, Eds., *Physics and Chemistry of II-VI Compounds,* John Wiley & Sons, 1967.

5.30. A. Addamiano and P. A. Dell, *J. Phys. Chem.,* 61, 1020, 1957.

5.31. D. B. Averbukh, E. A. Vetrenko, and G. I. Chufarov, *Zh. Prikl. Khim. (J. Appl. Chem.),* 32, 1255, 1959.

5.32. C. F. Cline, H. L. Dunegan, and G. W. Henderson, *J. Appl. Phys.,* 38, 1944, 1967.

5.33. A. Zarembovitch, *J. Phys. Radium,* 24, 1097, 1963; *Bull. Soc. Fr. Mineral. Crist.,* 88, 17, 1965.

5.34. A. V. Ioffe, *Sov. Phys. Solid State,* 5, 2446, 1963.

5.35. D. A. Berlincourt, H. Jaffe, and L. R. Shiozawa, *Phys. Rev.,* 129, 1009, 1963.

5.36. M. Cardona and G. Harbeke, *Phys. Rev.,* 137, A1467, 1965.

5.37. A. Lempicki, J. Birman, H. Samelson, and G. Neumark, Proc. 1960 Int. Conf. on Semicond. Phys., Prague, Chechosl. Acad. Sci., 1960, 768.

5.38. M. Cardona, *J. Phys. Chem. Solids,* 24, 1543, 1963; 26, 1351, 1965.

5.39. R. G. Wheeler and J. C. Miklosz, Proc. 7th Int. Conf. on Phys. of Semicond., M Hulin, Ed., Dunod, Paris, 1964, 873.

5.40. S. Narita and K. Nagasaka, *J. Phys. Soc. Jpn.,* 20, 1728, 1965.

5.41. M. Aven and C. A. Mead, *Appl. Phys. Lett.,* 7, 8, 1965.

5.42. M. Aven, *Extend. Abstr. Meet. Electrochem. Soc.,* 11, 46, 1962.

5.43. L. I. Berger and V. A. Kovalev, *Electronnaya Promyshlennost; Materialy (Electron. Ind. Mater.)* No. 5, 73, 1973; No. 11, 84, 1973.

5.44. K. Okamoto, T. Yoshimi, and S. Miura, *Appl. Phys. Lett.,* 53, 678, 1988.

5.45. A. Kato, M. Katayama, A. Mizutani, Y. Hattori, N. Ito, and T. Hattori, *J. Appl. Phys.,* 76, 3206, 1994.

5.46. A. V. Ioffe and A. F. Ioffe, *Sov. Phys. Solid State,* 2, 719, 1960.

5.47. A. Eucken and G. Kuhn, *Z. Phys. Chem.,* 134, 193, 1928.

5.48. H. Yoshinaga, *Phys. Rev.,* 100, 753, 1955.

5.49. E. D. Palik, Ed., *Handbook of Optical Constants of Solids,* Academic Press, New York, 1985.

5.50. H. Samelson and V. A. Brophy, *J. Electrochem. Soc.,* 108, 150, 1961.

5.51. H. Hartman, *Phys. Status Solidi,* 2, 585, 1962.

5.52. R. Nitsche, *J. Phys. Chem. Solids,* 17, 163, 1960.

5.53. T. Matsumoto, Y. Nagahiro, Y. Nasu, K. Oki, and M. Okabe, *Appl. Phys. Lett.*, 65, 1549, 1994.
5.54. F. L. Chan and Y. S. Park, *Bull. Am. Phys. Soc.*, 9, 296, 1964.
5.55. S. I. Novikova, *Sov. Phys. Solid State*, 3, 129, 1961.
5.56. S. A. Semiletov, *Kristallografiya (Crystallography)*, 1, 306, 1956.
5.57. L. F. Komolova, L. A. Grosheva, L. L. Mukhina, V. V. Sakharov, V. V. Proyanenkov, I. V. Razumovskaya, and N. F. Shevtsova, *Inorg. Mater.*, 19, 1590, 1983.
5.58. M. Shur, *Physics of Semiconductor Devices*, Prentice-Hall, Englewood Cliffs, NJ, 1990.
5.59. D. T. F. Marple, *J. Appl. Phys.*, 35, 1879, 1964.
5.60. G. E. Hite, D. T. F. Marple, and M. Aven, cited by B. Segall in *Physics and Chemistry of II-VI Compounds*, M. Aven and J. S. Prener, Eds., John Wiley & Sons, New York, 1967, 48.
5.61. R. M. Park, C. M. Rouleau, M. B. Troffer, T. Koyama, and T. Yodo, *J. Mater. Res.*, 5, 475, 1990.
5.62. Y. S. Park and F. L. Chan, *J. Appl. Phys.*, 36, 800, 1965.
5.63. M. Aven and P. R. Kennicott, Semiconductor Device Concept. Sci. Rep. No. 8, Contract No. AF-19(628)-329, U.S. Air Force Cambridge Res. Labs., Bedford, MA, 1964, 17.
5.64. S. Fujita, Y. Matsuda, and A. Sasaki, *J. Cryst. Growth*, 68, 231, 1984; *Jpn. J. Appl. Phys.*, 23, L360, 1984.
5.65. L. S. Ladd, quoted by Y. S. Touloukian, R. W. Powell, C. Y. Ho, and P. G. Klemens, *Thermophysical Properties of Matter*, Vol. 1, 1970, 1371.
5.66. F. Bailly and P. Manca, quoted in *Slozhnyye Almazopodobnyye Poluprovodniki (Multinary Diamond-Like Semiconductors)*, Soviet Radio, Moscow, 1968, 106.
5.67. Morton Intern. Inc. Advance Materials Techn. Bull. #110, 1994.
5.68. E. Raiskin, Rais Enterprises, Inc. Catalog, 1994.
5.69. I. D. Chistyakov and E. Cruceanu, *Rev. Roman. Phys.*, 6, 211, 1961.
5.70. S. I. Novikova and N. Kh. Abrikosov, *Sov. Phys. Solid State*, 5, 1558, 1964.
5.71. J. Carides and A. G. Fisher, *Solid State Commun.*, 2, 217, 1964.
5.72. F. Kelemen, E. Cruceanu, and D. Niculescu, *Phys. Status Solidi*, 11, 865, 1965.
5.73. A. F. Demidenko and A. K. Maltsev, *Inorg. Mater.*, 5, 158, 1969.
5.74. P. Leroux-Hugon, N. X. Xinh, and M. Rodot, *J. Phys.*, 28, Cl-73, 1967.
5.75. H. Tubota, H. Suzuki, and K. Hirakawa, *J. Phys. Soc. Jpn.*, 16, 1038, 1961; see, also, H. Tubota, *Jpn. J. Appl. Phys.*, 2, 259, 1963.
5.76. S. Larach, R. E. Shrader, and C. F. Stocker, *Phys. Rev.*, 108, 587, 1957.
5.77. M. Cardona and D. L. Greenaway, *Phys. Rev.*, 131, 98, 1963.
5.78. A. C. Aten, C. Z. Van Doorn, and A. T. Vink, Proc. Int. Conf. Phys. Semicond., Exeter, 1962. Phys. Soc. London, 1962, 696.
5.79. M. Aven and B. Segall, *Phys. Rev.*, 130, 81, 1963.
5.80. R. T. Lynch, D. G. Thomas, and R. E. Dietz, *J. Appl. Phys.*, 34, 706, 1963.
5.81. A. G. Fischer, J. N. Carides, and J. Dresher, *Solid State Commun.*, 2, 157, 1964.
5.82. V. M. Glazov, S. N. Chizhevskaya, and N. N. Glagoleva, *Zhidkiye Poluprovodniki (Liquid Semiconductors)*, Nauka, Moscow, 1967.
5.83. L. I. Berger, R. V. Bakradze, and M. V. Kutsikovich, *Inorg. Mater.*, 7, 1851, 1971.
5.84. F. Kelemen, A. Neda, E. Cruceanu, and D. Niculesku, *Phys. Status Solidi*, 28, 421, 1968.
5.85. R. T. Lynch, D. G. Thomas, and R. E. Dietz, *J. Appl. Phys.*, 34, 706, 1963.
5.86. O. de Melo, L. Hernández, O. Zelaya-Angel, R. Lozada-Morales, M. Becerill, and E. Vasco, *Appl. Phys. Lett.*, 65, 1278, 1994.
5.87. C.-Y. Yeh, Z. W. Lu, S. Froyen, and A. Zunger, *Phys. Rev.*, B45, 12130, 1992.
5.88. J. Chikawa and T. Nakayama, *J. Appl. Phys.*, 35, 2493, 1964.
5.89. O. P. Astakhov, L. I. Berger, K. Dovletov, and K. Tashliev, *Dokl. Akad. Nauk Turkm. SSR Ser. Phys.*, No. 5, 101, 1973.
5.90. R. R. Reeber and B. A. Kulp, *Trans. Met. Soc. AIME*, 233, 698, 1965.
5.91. H. H. Woodbury, *J. Phys. Chem. Solids*, 24, 881, 1963.
5.92. A. F. Demidenko, *Inorg. Mater.*, 5, 252, 1969.
5.93. R. K. Kirby, *AIP Handbook*, D. E. Gray, Ed., McGraw-Hill, New York, 1963, 4-68.
5.94. R. V. Bakradze, A. E. Balanevskaya, L. I. Berger, and A. P. Nesynova, *Trudy IREA (Proc. IREA)*, 29, 238, 1966.
5.95. F. A. Kröger, H. J. Vink, and J. Volger, *Physics*, 20, 1095, 1950.

5.96. (a) D. G. Thomas and J. J. Hopfield, *Phys. Rev.,* 116, 573, 1959; (b) J. J. Hopfield and D. G. Thomas, *Phys. Rev.,* 122, 35, 1961.
5.97. H. Sitter, J. Humenberger, W. Huber, and A. Lopez-Otero, *Solar Energy Mater.,* 9, 199, 1983.
5.98. H. Okamoto, J. Matsuoka, H. Nasu, K. Kamiya, and H. Tanaka, *J. Appl. Phys.,* 75, 2251, 1994.
5.99. D. K. Ghosh, L. K. Samanta, and G. C. Bhar, *Pramana,* 23, 485, 1984.
5.100. I. Kaur, D. K. Pandya, and K. L. Chopra, *J. Electrochem. Soc.,* 127, 943, 1980.
5.101. S. Koele, S. K. Kulkarni, and A. S. Nigavekav, *Solar Energy Mater.,* 10, 47, 1984.
5.102. O. Zelaya-Angel, J. J. Alvarado-Gil, R. Lozada-Morales, H. Vargas, and F. da Silva, *Appl. Phys. Lett.,* 64, 291, 1994.
5.103. H. Fujita, K. Kobayashi, T. Kawai, and K. Shiga, *J. Phys. Soc. Jpn.,* 20, 109, 1965.
5.104. M. Aven and H. H. Woodbury, *Appl. Phys. Lett.,* 1, 53, 1962.
5.105. M Balkanski and J. J. Hopfield, *Phys. Status Solidi,* 2, 623, 1962.
5.106. H. J. McSkimin and T. B. Bateman, *J. Acoust. Soc. Am.,* 33, 34, 1961.
5.107. M. G. Holland, *Phys. Rev.,* 134, A471, 1964.
5.108. Landolt-Börnstein, *New Series,* Vol. III-22a, O. Madelung, Ed., 1987.
5.109. K. W. Böer, *Survey of Semiconductor Physics,* Van Nostrand Reinhold, New York, 1990.
5.110. M. P. Lisitsa, L. F. Gudimenko, V. N. Malinko, and S. F. Terekhova, *Phys. Status Solidi,* 31, 389, 1969.
5.111. T. K. Gupta and J. Doh, *J. Mater. Res.,* 7, 1243, 1992.
5.112. H. M. Manasevit and W. I. Simpson, *J. Electrochem. Soc.,* 118, 644, 1971.
5.113. J. B. Mullin, S. J. C. Irvine, and D. J. Ashen, *J. Cryst. Growth,* 55, 92, 1981.
5.114. L. P. Deshmukh, S. G. Holikatti, and P. P. Hankare, *J. Phys.,* D27, 1786, 1994.
5.115. A. Aresti, L. Garbato, and A. Rucci, *J. Phys. Chem. Solids,* 45, 361, 1984.
5.116. P. Goldfinger and M. Jeunehomme, *Trans. Faraday Soc.,* 59, 2851, 1963.
5.117. M. Ilegems and G. L. Pearson, *Annu. Rev. Mater. Sci.,* 5, 345, 1975.
5.118. M. Grynberg, Proc. 7th Int. Conf. Phys. Semicond., M. Hulin, Ed., Dunod, 1964, 135.
5.119. R. G.Wheeler and J. O. Dimmock, *Phys. Rev.,* 125, 1805, 1962.
5.120. U. Dolega, *Z. Naturforsch.,* 18A, 809, 1963.
5.121. S. Kubo and M. Oniki, *J. Phys. Soc. Jpn.,* 20, 1280, 1965.
5.122. R. Frerichs, *Phys. Rev.,* 72, 549, 1947.
5.123. Z. Deng, M. Cinquino, and M. F. Lawrence, *J. Mater. Res.,* 6, 1293, 1991.
5.124. A. D. Stuckes and G. Farrell, *J. Phys. Chem. Solids,* 25, 477, 1964.
5.125. S. I. Novikova, *Sov. Phys. Solid State,* 2, 2087, 1961.
5.126. P. Aigrain and M. Balkanski, *Constantes Sélectionnées des Semiconducteurs,* Gauthier-Villars, Paris, 1961.
5.127. C. Z. van Doorn and D. de Nobel, *Physica,* 22, 338, 1956.
5.128. M. H. Cohen and T. K. Bergstresser, cited by B. Segall in *Physics and Chemistry of II-VI Compounds,* M. Aven and J. S. Prener, Eds., John Wiley & Sons, New York, 1967, 40.
5.129. K. K. Kanazawa and F. C. Brown, *Phys. Rev.,* 135A, 1757, 1964.
5.130. B. Segall, M. R. Lorenz, and R. E. Halsted, *Phys. Rev.,* 129, 2471, 1963.
5.131. S. Yamada, *J. Phys. Soc. Jpn.,* 15, 1940, 1960.
5.132. J. Humenberger, M. Sadeghi, E. Gruber, G. Elsinger, H. Sitter, and A. Lopez-Otero, *J. Phys.,* 43, C5-405, 1982.
5.133. G. A. Slack and S. Galginaitis, *Phys. Rev.,* 133, A253, 1964.
5.134. P. Fisher and H. Y. Fan, *Bull. Am. Phys. Soc.,* 4, 409, 1959.
5.135. D. T. F. Marple, *J. Appl. Phys.,* 35, 539, 1964.
5.136. L. I. Berger and V. M. Petrov, *Inorg. Mater.,* 7, 1905, 1971.
5.137. P. D. Dapkus, *Annu. Rev. Mater. Sci.,* 12, 243, 1982.
5.138. A. Lopez-Otero, *Thin Solid Films,* 49, 3, 1978.
5.139. A. Werner, H. D. Hochheimer, K. Strössner, and A. Jayaraman, *Phys. Rev.,* B28, 3330, 1983.
5.140. S.-H. Choe, K.-S. Yu, Y.-E. Kim, H. Y. Park, and W.-T. Kim, *J. Mater. Res.,* 6, 2677, 1991.
5.141. F. W. Dickson and G. Tupell, *Am. Mineral.,* 44, 471, 1959.
5.142. E. G. King and W. W. Weller, U.S. Bureau of Mines Rep. RI-6001, 1, 1962.
5.143. G. G. Roberts, E. L. Lind, and E. A. Davis, *J. Phys. Chem. Solids,* 30, 833, 1969.
5.144. T. Nakada and A. Kunioka, *Jpn. J. Appl. Phys.,* 10, 518, 1971.
5.145. T. Nakada, *J. Appl. Phys.,* 48, 3405, 1977.

5.146. M. Perakh and H. Ginsburg, *Thin Solid Films*, 52, 195, 1978.
5.147. E. Doni, L. Resca, S. Rodriguez, and W. M. Becker, *Phys. Rev.*, B20, 1663, 1979.
5.148. T. Huang and A. L. Ruoff, *J. Appl. Phys.*, 54, 5459, 1983.
5.149. V. V. Zhdanova, A. S. Pashinkin, and A. V. Novoselova, *Phys. Status Solidi*, 13, K19, 1966.
5.150. A. J. Strauss and L. B. Farrell, *J. Inorg. Nucl. Chem.*, 24, 1211, 1962.
5.151. P. V. Gul'tyaev and A. V. Petrov, *Sov. Phys. Solid State*, 1, 330, 1959.
5.152. J. G. Broerman, *Phys. Rev.*, 183, 754, 1969.
5.153. S. Bloom and T. K. Bergstresser, *Phys. Status Solidi*, 42, 191, 1970.
5.154. C. R. Whitsett, D. A. Nelson, J. G. Broerman, and E. C. Paxhia, *Phys. Rev.*, B7, 4625, 1973.
5.155. S. A. Aliev, L. L. Korenblit, and S. S. Shalyt, *Sov. Phys. Solid State*, 8, 565, 1966.
5.156. S. M. Wasim, B. Fernández, and R. Aldana, *Phys. Status Solidi*, A76, 743, 1983.
5.157. D. R. Hamilton, *Brit. J. Appl. Phys.*, 9, 103, 1958.
5.158. E. Krucheanu, D. Nikulesku, and A. Vanku, *Kristallografia*, 9, 537, 1964.
5.159. E. Ya. Nikolskaya and A. R. Regel, *Zh. Tekh. Fiz.*, 25, 1347, 1955; 25, 1351, 1955.
5.160. D. B. Holt, *Thin Solid Films*, 24, 1, 1974.
5.161. P. J. Ford, A. J. Miller, G. A. Saunders, Y. K. Hogurthi, J. K. Furdyna, and M. Jaczynski, *J. Phys.*, C15, 657, 1982.
5.162. E. Ratajczak and J. Terpilowski, *Rocz. Chem.*, 43, 1609, 1969.
5.163. R. F. Brebrick and A. J. Strauss, *J. Phys. Chem. Solids*, 26, 989, 1965.
5.164. J. G. Mavroides and D. F. Kolesar, *Solid State Commun.*, 2, 263, 1964.
5.165. Y. K. Vekilov and A. P. Rusakov, *Sov. Phys. Solid State*, 13, 956, 1971.
5.166. E. F. Steigmeier, *Appl. Phys. Lett.*, 3, 6, 1963.
5.167. T. Alper and G. A. Saunders, *J. Phys. Chem. Solids*, 28, 1637, 1967.
5.168. S. Groves and W. Paul, Proc. 7th Int. Conf. on Phys. of Semicond., M. Hulin, Ed., Dunod, Paris, 1964, 41.
5.169. G. A. Antcliffe and H. Kraus, *J. Phys. Chem. Solids*, 30, 243, 1969.
5.170. R. R. Galazka, *Phys. Lett.*, 32A, 101, 1970.
5.171. S. J. C. Irvine, J. B. Mullin, and A. Royle, *J. Cryst. Growth*, 57, 15, 1982.
5.172. (a) J. E. Lewis and D. A. Wright, *Br. J. Appl. Phys.*, 17, 783, 1966; (b) Z. Dziuba and K. Szlenk, *J. Phys. Chem. Solids*, 45, 97, 1984.
5.173. C. R. Whitsett and D. A. Nelson, *Phys. Rev.*, B5, 3125, 1972.
5.174. A. Noguera and S. M. Wasim, *Phys. Rev.*, B32, 8046, 1985.
5.175. D. H. Dickey and J. G. Mavroides, *Solid State Commun.*, 2, 213, 1964.
5.176. T. C. Harman, M. J. Logan, and H. L. Goering, *J. Phys. Chem. Solids*, 7, 228, 1958.
5.177. T. C. Harman, *J. Electrochem. Soc.*, 106, 205C, 1959.
5.178. L. I. Alekseenko, B. F. Bilen'kii, Z. G. Grechukh, and V. G. Savitskii, *Inorg. Mater.*, 10, 1215, 1974.
5.179. S. Ves, D. Gloetzel, M. Cardona, and H. Overhof, *Phys. Rev.*, B24, 3073, 1981.
5.180. G. J. Piermarini, F. A. Mauer, S. Block, A. Jayaraman, T. H. Geballe, and G. W. Hull, *Solid State Commun.*, 32, 275, 1979.
5.181. E. F. Skelton, A. W. Webb, F. J. Rachford, P. C. Taylor, S. C. Yu, and I. L. Spain, *Phys. Rev.*, B21, 5289, 1980.
5.182. A. P. Rusakov, S. G. Grigoryan, A. V. Omel'chenko, and A. E. Kadyshevich, *Sov. Phys. JETF*, 45, 380, 1977.
5.183. (a) L. I. Berger, N. L. Rapoport, and A. N. Novikova, *Trudy IREA (Proc. IREA)*, 33, 251, 1971; (b) L. I. Berger, *Bull. Am. Phys. Soc.*, 27, 882, 1982.
5.184. R. C. Hanson, K. Helliwell, and C. Schwab, *Phys. Rev.*, B9, 2649, 1974.
5.185. J. N. Plendl and L. C. Mansur, *Appl. Optics*, 11, 1194, 1972.
5.186. J. Shanker, P. S. Bakhshi, and L. P. Sharma, *J. Inorg. Nucl. Chem.*, 41, 1285, 1979.
5.187. B. Perner, *J. Cryst. Growth*, 6, 86, 1969.
5.188. M. Cardona, *Phys. Rev.*, 129, 69, 1963.
5.189. K. S. Song, *J. Phys. Chem. Solids*, 28, 2003, 1967.
5.190. M. A. Kahn, *J. Phys. Chem. Solids*, 31, 2309, 1970.
5.191. E. Calabrese and W. B. Fowler, *Phys. Status Solidi*, B57, 135, 1973.
5.192. A. Zunger and M. L. Cohen, *Phys. Rev.*, B20, 1189, 1979.
5.193. L. Kleinman and K. Mednick, *Phys. Rev.*, B20, 2487, 1979.

5.194. S.-Y. Ren, R. E. Allen, J. D. Dow, and I. Lefkowitz, *Phys. Rev.*, B25, 1205, 1982.
5.195. J. Ringeissen and S. Nikitine, *J. Phys.*, C28, 3, 1967.
5.196. Y. Kato, T. Goto, T. Fuji, and M. Ueta, *J. Phys. Soc. Jpn.*, 36, 169, 1974.
5.197. H. Overhof, *Phys. Status Solidi*, B97, 267, 1980.
5.198. A. P. Rusakov, *Phys. Status Solidi*, B72, 503, 1975.
5.199. Y. Kaifu and Y. Kawate, *J. Phys. Soc. Jpn.*, 22, 517, 1967.
5.200. M. Soga and R. Imaizumi, *J. Electrochem. Soc.*, 114, 388, 1967.
5.201. W. R. Wilcox and R. A. Corley, *Mater. Res. Bull.*, 2, 571, 1967.
5.202. G. A. Wolff and H. E. LaBelle, *J. Am. Ceram. Soc.*, 48, 441, 1965.
5.203. R. C. Hanson, J. R. Hallberg, and C. Schwab, *Appl. Phys. Lett.*, 21, 490, 1972.
5.204. J. C. Phillips, *Rev. Mod. Phys.*, 42, 317, 1970; *Phys. Rev. Lett.*, 27, 1197, 1971; *Bonds and Bands in Semiconductors,* Academic Press, New York, 1973.
5.205. C. Divakar, M. Mohan, and A. K. Singh, *Solid State Commun.*, 34, 385, 1980.
5.206. A. Neuhaus and K. Recker, *Crystal Growth,* H. S. Peiser, Ed., Pergamon Press, Elmsford, NY, 1967, 235.
5.207. I. Ishii, S. Sato, T. Matsukawa, and Y. Sakisaka, *J. Phys. Soc. Jpn.*, 32, 1440, 1972.
5.208. P. D. Martzen and W. C. Walker, *Solid State Commun.*, 28, 171, 1978; *Phys. Rev.*, B21, 2562, 1980.
5.209. G. V. Hevesy, *Z. Phys. Chem.*, 101, 337, 1922.
5.210. K. Nagel and C. Wagner, *Z. Phys. Chem.*, 25, 71, 1934.
5.211. W. M. K. P. Wijekoon, M. Y. M. Lyktey, P. N. Prasad, and J. F. Garvey, *J. Appl. Phys.*, 74, 5767, 1993; *J. Phys.*, D27, 1548, 1994.
5.212. S. B. DiCenzo, G. K. Wertheim, and D. N. E. Buchanan, *Appl. Phys. Lett.*, 40, 888, 1982.
5.213. M. C. Goetz and J. A. Cowen, *Solid State Commun.*, 41, 293, 1982.
5.214. A. Bienenstock and G. Burley, *J. Phys. Chem. Solids*, 24, 1271, 1963.
5.215. H. Fizeau, *Ann. Phys. 2te. Folge*, 132, 292, 1867.
5.216. G. Jones and F. C. Jelen, *J. Am. Chem. Soc.*, 57, 2532, 1935.
5.217. K. H. Lieser, *Z. Phys. Chem.*, 5, 125, 1955.
5.218. K. K. Kelley, *U.S. Bur. Mines Bull.*, 434, 1941.
5.219. T. A. Fjeldly and R. C. Hanson, *Phys. Rev.*, B10, 3569, 1974.
5.220. H. Welker, *Ergebn. Exakt. Naturwiss.*, 25, 275, 1956.

CHAPTER 6

Ternary Adamantine Semiconductors

The ternary adamantine compounds are the crystallochemical analogs of either III-V or II-VI compounds; they are formed in accordance with the schemes:

$$2(III + V) \rightarrow (II + IV) + V_2 \tag{6.1}$$

$$3(III + V) \rightarrow (I + IV_2) + V_3 \tag{6.2}$$

$$2(II + VI) \rightarrow (I + III) + VI_2 \tag{6.3}$$

$$3(II + VI) \rightarrow (I_2 + IV) + VI_3 \tag{6.4}$$

$$4(II + VI) \rightarrow (I_3 + V) + VI_4 \tag{6.5}$$

or

$$2(III + V) \rightarrow III_2 + (IV + VI) \tag{6.6}$$

The substances of the 6.1 to 6.5 types are the single-anion compounds, while the compounds of 6.6 type are the single-cation ones. One can see from Table 1.2 (Chapter 1) that there are other types of ternary compounds with the average valence electron concentration equal to four (II_4-III-VII_3, II_3-IV-VII_2, II_2-V-VII, and III_3-IV_2-VII), but not one of the compounds of these types has yet been synthesized.

A part of the compounds of the III_2-VI_3 and IV_3-V_4 types crystallizes in the structure similar to that of sphalerite, as was first noticed by Hahn and Klinger.[6.1] A specific of the structure is that one third and one fourth of the cation positions in the sphalerite-type lattices of III_2-VI_3 and IV_3-V_4, respectively, are not occupied. Essentially, III_2-VI_3- and IV_3-V_4-type compounds have the valence electron concentration per atom equal to 4.8 and 4.57, respectively, but if we consider, after Suchet[6.2] and Palatnik et al.,[6.3] the cation vacancies as atoms of zero valency, the "ternary" compounds (vac_1-III_2-VI_3 and vac_1-IV_3-V_4) formed this way have the electron concentration equal to four and, consequently, the tetrahedral coordination of atoms in the lattice. These quasiternary compounds are called the binary tetrahedral defect compounds, although a more logical term would be the ternary tetrahedral compounds with defect structure, since the binary compounds do not have any "deficiency" — the number of valence electrons in cations is equal to the number of electrons needed for completion of the valence electron octets in anions. This consideration justifies inclusion in this chapter of available information relevant to these compounds.

6.1. TERNARY ANALOGS OF THE III-V ADAMANTINE SEMICONDUCTORS

6.1.1. The II-IV-V_2 Compounds (Ternary Pnictides)

The vast information regarding the properties of II-IV-V_2 compounds is compiled in References 6.4 to 6.8. All known compounds of this type (their molecular and average atomic weight, mean atomic number, and the enthalpy of formation are presented in Table 6.1) have the crystal structure similar to either sphalerite or wurtzite. Two of them, $BeSiN_2$ and, possibly, $CaCN_2$, have the hexagonal wurtzite crystal structure. The others possess the crystal structure of mineral chalcopyrite, $CuFeS_2$, in which the cations are distributed in an orderly way among the positions in the cation part of the crystal lattice shown in Figure 6.1. The space group for the chalcopyrite structure is $I\bar{4}2d$-D_{2d}^{12}; the Structurbericht and Pearson's symbols are El_1 and tI16, respectively (Reference 6.11, p. 12-1). At elevated temperatures, some of the II-IV-V_2 compounds transfer from the tetrahedral chalcopyrite structure to the cubic structure similar to that of sphalerite. In the latter, the Group II and IV atoms in the cation sublattice are distributed randomly. In the tetragonal modification, the c parameter of the unit cell is smaller than 2a — the lattice is compressed in the c-direction in comparison with the sphalerite structure. The measure of the cell contraction may be a dimensionless parameter ("tetragonality" or "tetragonal contraction") (Reference 6.5, p. 121) $B = bV_0^{-n}$. (The tetragonal lattice becomes similar to a cubic when c = 2a or the tetragonality is equal to zero.)

Table 6.1 II-IV-V_2 Compounds. Molecular and Mean Atomic Weight, Mean Atomic Number, and Enthalpy of Formation

Compound	Molecular weight	Mean atomic weight	Mean atomic number	Enthalpy of formation (kJ/mol)	Comments
Nitrides					
$BeSiN_2$	65.11	16.28	8		
$CaCN_2$	80.10	20.03	10		Cyanamide
$ZnGeN_2$	165.98	41.50	23		
Phosphides					
$MgSiP_2$	114.34	28.58	14		
$MgGeP_2$	158.84	39.71	18.5		
$ZnSiP_2$	155.41	38.85	18.5	318[a]	
$ZnGeP_2$	199.92	49.98	23	292.7[a]	
$ZnSnP_2$	246.02	61.50	27.5	274.8[a]	
Arsenides					
$ZnSiAs_2$	243.31	60.83	27.5	289.7[a]	
$ZnGeAs_2$	287.81	71.95	32	387.0[b]	
$ZnSnAs_2$	333.91	83.48	36.5	251.8[a]	
				307.8[b]	
$CdSiAs_2$	290.34	72.58	32		
$CdGeAs_2$	334.84	83.71	36.5	327.1[b]	
$CdSnAs_2$	380.94	95.24	41	235.9[b]	
Antimonides					
$ZnSnSb_2$	427.57	106.89	45.5		

[a] Reference 6.9.
[b] Reference 6.24.

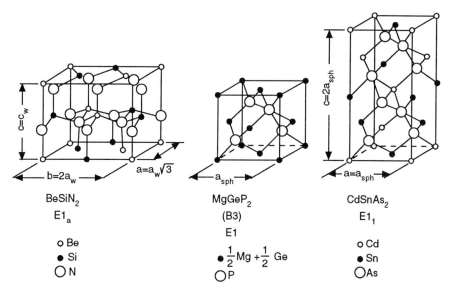

Figure 6.1 Structure types of II-IV-V_2 compounds (w — wurtzite, sph — sphalerite). (After E. Parthe, *Crystal Chemistry of Tetrahedral Structures*, Gordon and Breach, 1964. With permission.)

An interesting specifics of the II-IV-V_2 compounds is that at least some of them ($CdGeP_2$ and $CdGeAs_2$) can be prepared in the amorphous state, which is unusual for the pure adamantine compounds in the bulk form (see Reference 6.5, p. 22). The data on the crystal structure and lattice parameters of the II-IV-V_2 compounds are presented in Table 6.2. There were numerous attempts to analyze the distances between the II and V and IV and V atoms in the II-IV-V_2 crystals using the experimental lattice parameters, covalent radii of the components,[6.18-6.20] and the electronegativity of the participating elements (the data on the covalent radii and electronegativity of the elements are compiled in Table 6.3 with the data on electronegativity adapted from Reference 6.7, p. 10). A reasonable agreement with experimental data was reached by Abrahams and Bernstein[6.15] and Palatnik et al.[6.19]

The temperature dependence of the coefficient of thermal expansion of the II-IV-V_2 compounds is, apparently, similar to that of the elemental and binary adamantine substances; namely, it is negative at low temperatures.[6.17,625] The difference in the energy of II-V and IV-V interatomic bonds causes an anisotropy of the coefficient of thermal expansion, α, in the a and c directions in the chalcopyrite-type structure. These specifics of α of II-IV-V_2 compounds are shown in Figure 6.2 (adapted from Reference 6.17). The transition from positive to negative α of $ZnSiP_2$ takes place at relatively high temperatures (between 70 and 90 K, depending on crystallographic direction). The stronger, more covalent bond between Si and P is a cause of a smaller α_{Si-P} in comparison with α_{Zn-P}. The available data on room temperature magnitudes of density, gram-atomic volume, and coefficient of thermal expansion of the II-IV-V_2 compounds are presented in Table 6.4.

Some of the compounds are formed by peritectic reactions (see, e.g., Reference 6.27). The temperature at which a melt of a compound begins to crystallize strongly depends on pressure; this may be an explanation of a noticeable variation of the melting point magnitudes, as the majority of the reported data has been received at uncontrolled pressure with the material sealed in the fused silica ampules. For a review of the phase relationships in II-IV-V systems, see Reference 6.7, p. 38 to 44. The results of investigation of the composition of the gas phase upon the melts of some II-IV-As_2 compounds,[6.28] the mass-spectrometric data on the gas phase composition upon solid $ZnSnAs_2$,[6.29] some data on dissociation of $CdSiP_2$ and

Table 6.2 Crystal Structure of the II-IV-V$_2$ Compounds at Room Temperature

Compound	Structure prototype[a]	Structurbericht type	Space group	Lattice parameters				Ref.
				a (nm)	b (nm)	c (nm)	$2 - \frac{c}{a}$	
BeSiN$_2$	w	El$_a$	C_{6v}^4-P6$_3$mc or C_{2v}^9-Pna2$_1$	0.287$_2$		0.4674		6.5, 6.16
ZnGeN$_2$	Monocl., w			0.3167	0.5194	0.3167	118°53'	6.13
MgSiP$_2$	c	El$_1$	D_{2d}^{12}-I$\bar{4}$2d	0.5718		1.0114	0.23	6.5
				0.572		1.012		6.13
				0.5721		1.0095		6.7
MgGeP$_2$	s	El	T_d^2-F$\bar{4}$3m	0.5652				6.5, 6.7, 6.14
ZnSiP$_2$	c	El$_1$	D_{2d}^{12}-I$\bar{4}$2d	0.5399		1.0435	0.07	6.5, 6.15
				0.5400		1.0438		6.13
				0.546			0.03	6.6
				0.5400		1.0440		6.14
				0.5401		1.04395		6.17
ZnGeP$_2$	c	El$_1$	D_{2d}^{12}-I$\bar{4}$2d	0.5465		1.0771	0.04	6.5
				0.5463		1.074		6.13
				0.546			0.03	6.6
				0.5491		1.0800		6.14
				0.546		1.071		6.159
ZnSnP$_2$	c	El$_1$	D_{2d}^{12}-I$\bar{4}$2d	0.5651		1.1302	0.0	6.5
				0.5652		1.1305		6.13
CdSiP$_2$	s	El	T_d^2-F$\bar{4}$3m	0.5652				6.14
	c	El$_1$	D_{2d}^{12}-I$\bar{4}$2d	0.5678		1.0431	0.16	6.5, 6.6
CdGeP$_2$	c	El$_1$	D_{2d}^{12}-I$\bar{4}$2d	0.5684		1.0442		6.14
				0.5741		1.0775	0.12	6.5
				0.5740		1.0776		6.13
				0.5768		1.0823		6.14
				0.574				6.6
	s	El	T_d^2-F$\bar{4}$3m	0.556			0.12	6.5

Compound							
CdSnP$_2$	c	EI$_1$	D_{2d}^{12}-I$\bar{4}$2d	0.5900	1.1518	0.05	6.5
				0.5901	1.1514		6.13
					1.1512		6.14
ZnSiAs$_2$	c	EI$_1$	D_{2d}^{12}-I$\bar{4}$2d	0.5606	1.0890	0.06	6.5
				0.5606	1.0866		6.13
				0.5610	1.0884		6.14
ZnGeAs$_2$	c	EI$_1$	D_{2d}^{12}-I$\bar{4}$2d	0.5672	1.1153	0.03	6.5
				0.5671	1.1153		6.13
				0.5670			
ZnSnAs$_2$	c	EI$_1$	D_{2d}^{12}-I$\bar{4}$2d	0.5852	1.1703	0.03	6.6
				0.5815	1.1703	0	6.5
							6.13
CdSiAs$_2$	s	EI	T_d^2-F$\bar{4}$3m	0.581			6.5
	c	EI$_1$	D_{2d}^{12}-I$\bar{4}$2d	0.5884	1.0882	0.15	6.5
				0.5885	1.0881		6.13
				0.5883	1.0880		6.14
CdGeAs$_2$	c	EI$_1$	D_{2d}^{12}-I$\bar{4}$2d	0.5943	1.1217	0.11	6.5
				0.5943	1.1220		6.13
				0.59427	1.12172		6.6
CdSnAs$_2$	c	EI$_1$	D_{2d}^{12}-I$\bar{4}$2d	0.6094	1.1918	0.04	6.5
				0.6089	1.1925		6.13
				0.6092		0.04	6.6
ZnSnSb$_2$	s	EI	T_d^2-F$\bar{4}$3m	0.6051			6.5
	c	EI$_1$	D_{2d}^{12}-I$\bar{4}$2d	0.6273	1.2546	0	6.13
	s	EI	T_d^2-F$\bar{4}$3m	0.6281			6.5

[a] c — chalcopyrite, s — sphalerite, w — wurtzite.

Table 6.3 The Covalent Tetrahedral Radii (in pm) and Electronegativity of Some Elements of the Periodic Table

					Group					
	1A	1B	2A	2B	3B	4B	5B	6B	7b	Ref.
Element	Li		Be		B	C	N	O	F	
Covalent			97.5		85.3	77.4	71.9	67.8	67.2	6.20
Radius							70			6.19
Electronegativity	1.00		1.50		2.00	2.50	3.00	3.50	4.00	6.21
	1.0		1.5		2.0	2.5	3.0	3.0	4.0	6.22
	0.95		1.5		2.0	2.5	3.0	3.0	3.95	6.23
Element	Na		Mg		Al	Si	P	S	Cl	
Covalent			140		131.7	117	110	104		6.19
Radius			130.1		123.0	117.3	112.8	112.7	112.7	6.20
Electronegativity	0.72		0.95		1.18	1.41	1.64	1.87	2.10	6.21
	0.9		1.2		1.5	1.8	2.1	2.5	3.0	6.22
	0.9		1.2		1.5	1.8	2.1	2.5	3.0	6.23
Element		Cu	Ca	Zn	Ga	Ge	As	Se	Br	
Covalent		136		139	131	122	117	114	120	6.19
Radius		122.5	133.3	122.5	122.5	122.5	122.5	122.5	122.5	6.20
Electronegativity		0.79		0.91	1.13	1.35	1.57	1.79	2.01	6.21
		1.9		1.6	1.6	1.8	2.0	2.4	2.8	6.22
		1.8		1.5	1.5	1.8	2.0	2.4	2.8	6.23
Element		Ag	Sr	Cd	In	Sn	Sb	Te	I	
Covalent		155.5		156	148	140	135	128	113	6.19
Radius		140.5	168.9	140.5	140.5	140.5	140.5	140.5	140.5	6.20
Electronegativity										
Element		Au		Hg	Tl	Pb	Bi			
Covalent				154	154					6.19
Radius										
Electronegativity		0.64		0.79	0.94	1.09	1.24			6.21
		2.4		1.9	1.8	1.8	1.9			6.22
		2.3		1.8	1.5	1.9	1.8			6.23

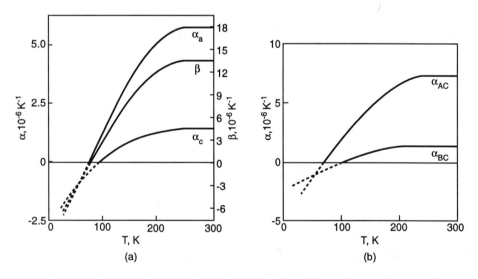

Figure 6.2 Thermal expansion of ZnSiP$_2$. (a) Coefficients of linear thermal expansion along the a (α_a) and c (α_c) axes and of volume thermal expansion (β) vs. temperature; (b) the temperature dependence of the coefficient of thermal expansion along the Zn-P (α_{AC}) and Si-P (α_{BC}) bond directions. (After P. Deus and H.A. Schneider, *Phys. Stat. Solidi*, A79, 411, 1983. With permission.)

Table 6.4 II-IV-V$_2$ Compounds. Room Temperature Density, Mean Atomic Volume, and Coefficient of Linear Thermal Expansion

Compound	Density (g/cm³) X-ray	Density (g/cm³) Pycnometric	Mean atomic volume (cm³/g-atom)	Coefficient of linear thermal expansion in 10⁻⁶ 1/K Along a axis	Along c axis	Polycrystal
BeSiN$_2$	3.24	3.12	5.02			
MgSiP$_2$	2.30		12.43			
MgGeP$_2$	2.92		13.60			
ZnSiP$_2$	3.39	3.35	11.46	5.9	1.5	4.4
ZnGeP$_2$	4.18	4.12	11.96			
		4.04				
ZnSnP$_2$	4.52		13.59			
CdSiP$_2$	4.00	3.97	12.65			
CdGeP$_2$	4.66	4.54	13.25			
CdSnP$_2$	4.85		15.09			
ZnSiAs$_2$	4.72	4.69	12.89			
ZnGeAs$_2$	5.325	5.29	13.51			1.0
		5.26				
ZnSnAs$_2$	5.533	5.53	15.09			2.3
CdSiAs$_2$	5.12		14.18			
CdGeAs$_2$	5.612	5.61	14.92			3.5
CdSnAs$_2$	5.716	5.66	16.66			4.7
ZnSnSb$_2$	5.67	5.72				
	5.73		18.65			

From References 6.5, 6.6, 6.13, 6.17, and 6.26.

ZnSiP$_2$[6.30,6.31] and ZnGeAs$_2$, ZnSnAs$_2$, CdGeAs$_2$, and CdSnAs$_2$[6.24] present the limited information pertinent to the mechanism of formation and decomposition of II-IV-V$_2$ compounds.

The specific heat of several II-IV-V$_2$ compounds was reported in References 6.32 to 6.34; its low-temperature magnitudes for CdGeAs$_2$ and CdSnAs$_2$ are cited in Reference 6.35. The available information regarding the melting point, enthalpy and entropy of formation, heat capacity, and Debye temperature of II-IV-V$_2$ compounds is compiled in Table 6.5. The mechanical properties of the compounds were reported in References 6.5, 6.6, 6.26, and 6.37 to 6.42.

It was established by various authors (see, e.g., References 6.42 to 6.45) that within a given group of compounds with the same crystal structure, chemical bonding properties, molecular formula, and valence electron concentration there is a correlation between the bulk modulus, B, and the unit cell volume, V_o, which can be presented as

$$B = bV_o^{-n} \qquad (6.7)$$

where b and n have particular magnitudes for each compound group. Neuman[6.42] found that for II-IV-V$_2$ compounds, b = 9.80 and n = 2.1035 if B is expressed in GPa and V_o in nm³. The available data on mechanical properties of these compounds are presented in Table 6.6.

The electronic structure of II-IV-V$_2$ differs from that of III-V compounds as a consequence of the presence of two sorts of atoms in the cation sublattice of the ternary compounds, distortion of the "adamantine" tetrahedrons in the lattice, and increase of the unit cell volume. The latter feature is responsible for the much smaller Brillouin zone dimensions of the II-IV-V$_2$ chalcopyrite-type semiconductors in comparison with their III-V sphalerite analogs. To show their relative features, the two zones are shown in the same k-coordinate system in Figure 6.3. A compact but comprehensive analysis of the electronic structure of the chalcopyrite

Table 6.5 II-IV-V$_2$ Compounds. The Melting and Transition Temperatures, Heat of Atomization, Heat Capacity, and Debye Temperature

Compound	Melting point (K)	Phase transition temperature (K)	Heat of atomization (kJ/g-atom)[6.36]	Heat capacity[a] (J/mol·K)[6.34]		Debye temperature (K)		
				C_p	C_v	Elastic[6.3]	At 0 K[6.34]	At 290 K[6.34]
ZnSiP$_2$	1520–1560[b]		312.5					
ZnGeP$_2$	1300[b]	1223[c]	293.7					
ZnSnP$_2$	1203[c]	993[c]	274.9					
CdSiP$_2$	1393[c]							
CdGeP$_2$	1073[c]		289.1					
CdSnP$_2$	843[c]		270.3					
ZnSiAs$_2$	1372[c]		290.0					
ZnGeAs$_2$	1148[c]	812(?)[c]	271.1			263		
ZnSnAs$_2$	1051[c]	923[c]	252.3	24.1	1.97	270	319	232
CdSiAs$_2$	>1123[c]							
CdGeAs$_2$	943[c]	903(?)[c]	266.5	24.0	2.59		302	
CdSnAs$_2$	869[c]	840[c]	247.7	24.4	1.88	223	281	195
ZnSnSb$_2$	870[d]							

[a] From Reference 6.34.
[b] From Reference 6.13.
[c] From Reference 6.7.
[d] From Reference 6.5.

Table 6.6 II-IV-V$_2$ Compounds, Unit Cell Volume, Microhardness, Speed of Ultrasound, and Bulk and Young's Moduli

Compound	Unit cell volume (nm³)	Microhardness (GPa)[6.5]	Speed of USW (km/sec)[6.26,6.37]		Moduli (GPa)	
			Longitudinal	Transversal	Bulk	Young's[6.26]
ZnSiP$_2$	0.304	11.00				
ZnGeP$_2$	0.322	8.10				
ZnSnP$_2$	0.361	6.50				
CdSiP$_2$	0.336	10.50			97[6.40]	
CdGeP$_2$	0.355	5.65				
CdSnP$_2$	0.401	5.00				
ZnSiAs$_2$	0.342	9.20				
ZnGeAs$_2$	0.359	6.80	4.77–4.28	2.73		128.9
ZnSnAs$_2$	0.401	4.55	4.20–4.28	2.49		97.7
CdSiAs$_2$	0.377	6.85				
CdGeAs$_2$	0.396	4.70	3.78		69.7[6.39]	76.2
CdSnAs$_2$	0.443	3.95	3.51–3.66	2.06		75.3
ZnSnSb$_2$	0.494	2.50				
	0.496*					

Notes: (1) The asterisk sign denotes the volume of the sphalerite modification doubled unit cell; (2) as in the case of the elemental and binary adamantine substances, both elastic moduli and hardness change antibatically with the unit cell volume.

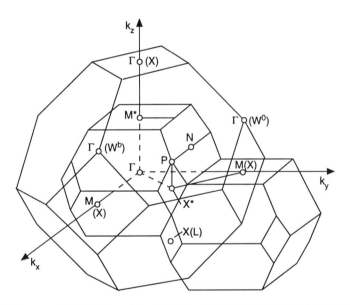

Figure 6.3 Brillouin zone of sphalerite and two Brillouin zones of chalcopyrite. (After A. Miller, A. MacKinnon, and D. Weaire, Springer-Verlag, New York, 1992. With permission.)

semiconductors is contained in Reference 6.7. The models relating (1) the lowest direct energy gaps in the substances such as InP (a direct semiconductor) and in its ternary analog, CdSnP$_2$; and (2) the lowest indirect energy gap in the compounds such as GaP (an indirect semiconductor) and in its ternary analog, ZnGeP$_2$, are shown in Figure 6.4 a and b, respectively (adapted from Reference 6.7). In passing from sphalerite to chalcopyrite (case a), the Γ_{15} valence band splits into a nondegenerate level Γ_4 laying above a doubly degenerate level Γ_5 and separated from it by the crystal-field splitting of magnitude Δ_{cf}. Transitions $\Gamma_{4v} \rightarrow \Gamma_{1c}$ are allowed only for $\vec{E} \| \vec{z}$ and $\Gamma_{5v} \rightarrow \Gamma_{1c}$ — only for $\vec{E} \perp \vec{z}$. When the spin-orbit interaction is

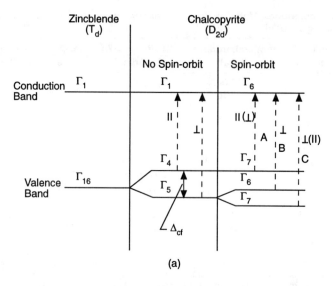

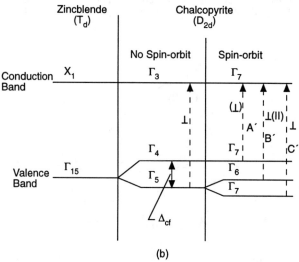

Figure 6.4 Band structure and selection rules for (a) the transitions in a chalcopyrite crystal derived from the $\Gamma_{15} \to \Gamma_1$ energy gap in a sphalerite structure crystal and (b) for pseudodirect transitions in a chalcopyrite structure crystal derived from the $\Gamma_{15} \to X_1$ indirect transitions in a sphalerite structure crystal. (After J.L. Shay and J.H. Wernick, *Ternary Chalcopyrite Semiconductors: Growth, Electronic Properties and Applications*, Pergamon Press, Elmsford, NY, 1975. With permission.)

included in consideration, the Γ_{5v} level splits into Γ_{6v} and Γ_{7v} levels and the polarization selection rules are not as restrictive, i.e., the vector $\vec{E}$ may be either parallel to the z direction or not. This model agrees with the experimental data, in particular with the results of the reflectance and electroreflectance measurements.

In passing from the sphalerite structure to the chalcopyrite one in the case of an indirect binary semiconductory (case b), there is the possibility that the formed ternary compound has a direct energy gap, which is commonly called "pseudodirect" to reflect the fact that the pseudopotentials of two cations in the chalcopyrite-type crystal are, in general, different. The mapping of energy levels in the sphalerite Brillouin zone into the chalcopyrite Brillouin zone results in the change of the band structure which consists of transformation of one of the

Table 6.7 II-IV-V$_2$ Compounds. The Direct, $E_{g,dir}$, Indirect, $E_{g,ind}$, and Pseudodirect, $E_{g,p}$, Energy Gaps, Crystal-Field Splitting, Δ_{cf}, and Spin-Orbit Splitting, Δ_{so}, in eV

Compound	$E_{g,dir}$			E_g^a		$E_{g,p}$			$-\Delta_{cf}$	Δ_{so}	T (K)
	A	B	C	$E_{g,ind}$	Experim.	A'	B'	C'			
ZnGeN$_2$		2.67									
MgSiP$_2$		2.82		2.26			2.03		0.08	0.16	300
ZnSiP$_2$	2.98	3.03	3.10	1.95		2.06	2.11	2.18			
		2.96					2.07		0.12	0.06	300
ZnGeP$_2$	2.365	2.43	2.50		1.99		2.05		0.08	0.09	300
ZnSnP$_2$	1.66	1.66	1.75		1.66				0	0.09	300
CdSiP$_2$		2.75			2.10		2.2		0.20	0.07	300
	2.945	2.71	2.75								90
CdGeP$_2$	1.72	1.90	1.99		1.72				0.20	0.11	300
CdSnP$_2$	1.17	1.25	1.33		1.17				0.10	0.48	300
ZnSiAs$_2$	2.12	2.22	2.47						0.13	0.28	300
	2.13	2.23	2.69		1.74		1.74		0.13	0.286	300
ZnGeAs$_2$	1.15	1.19	1.48		1.15				0.06	0.31	300
ZnSnAs$_2$	0.745	0.745	1.185		0.73				0	0.34	300
CdSiAs$_2$	1.55	1.74	1.99		1.55				0.24	0.297	300
CdGeAs$_2$	0.57	0.73	1.02		0.57				0.21	0.33	300
CdSnAs$_2$	0.26	0.30	0.79		0.26				0.06	0.48	300
ZnSnSb$_2$		0.4									300
		0.7									77

^{a} Reference 6.48.

X conduction band levels into a Γ_{3c} level (Figure 6.4b) and splitting of the Γ_{15} level into the Γ_4 and Γ_5 levels. Inclusion of spin-orbit interaction lifts the double degeneracy of the Γ_{5v} band and makes $\Gamma_{4v} \to \Gamma_{3c}$ (now $\Gamma_{7v} \to \Gamma_{7c}$) transitions weakly allowed. From Figures 6.4 a and b, the similarities between the direct electroreflectance A, B, and C peaks and the peaks due to the pseudodirect A', B', and C' transitions are evident. The available data on the band structure parameters of the II-IV-V$_2$ compounds, mainly from References 6.6, 6.7, 6.13, and 6.46, are gathered in Table 6.7. As an example, the ZnSiP$_2$ band structure is shown in Figure 6.5.[6.46]

The temperature dependence of the A, B, and C (or A', B', and C') transition energies of some II-IV-V$_2$ compounds was discussed in References 6.49 to 6.55. As a typical example, the transition energies vs. temperature for ZnSiP$_2$ are shown in Figure 6.6 (adapted from Reference 6.54).

The data on the carrier effective mass in II-IV-V$_2$ compounds are compiled in Table 6.8.

From the first reports on electrical properties of II-IV-V$_2$ compounds (see, e.g., Reference 6.55), it was shown that the properties strongly depend on the sample composition. Usually, deviation from stoichiometry has a very strong effect on the carrier mobility in the samples of compounds. The published data on the mobility vary in the wide limits. The typical data on the electrical parameters of the compounds are presented in Table 6.9. It has to be noted that the improvement in the methods of synthesis and crystal growth, as a rule, is followed by the increase in carrier mobility.

The known data on electrical conductivity of the II-IV-V$_2$ compounds, which melt without decomposition, in solid and liquid state, show that their melt possesses the semiconductor character of conductivity. The temperature dependence of the electrical conductivity of a sample of CdSnAs$_2$ in the solid and liquid state is shown in Figure 6.7 (adapted from Reference 6.58). The energy gap estimated from the slope of the conductivity polyterm in both solid and liquid state is close to 0.20 eV in reasonable agreement with the available data (see, e.g., Reference 6.59 and Table 6.7), taking into consideration the information on temperature dependence of the CdSnAs$_2$ energy gap reported by Matias and Höschl.[6.60]

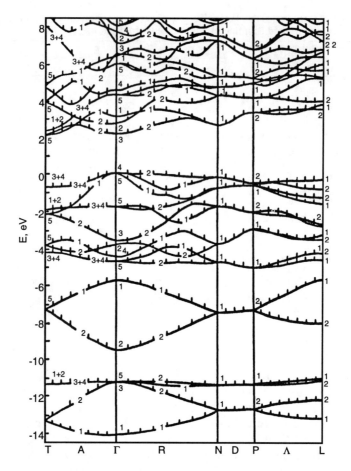

Figure 6.5 Band structure of ZnSiP$_2$. (After N.A. Zakharov and V.A. Chaldyshev, *Sov. Phys. Semicond.*, 19, 842, 1985. With permission.)

The thermal conductivity of II-IV-V$_2$ compounds was reported by Leroux-Hugon[6.61] (ZnGeAs$_2$, CdGeAs$_2$, and CdSnAs$_2$), Ping-Hsi et al.[6.38] (CdGeP$_2$, ZnGeAs$_2$, and CdGeAs$_2$), Masumoto et al.[6.63] (ZnGeP$_2$, CdGeP$_2$, and ZnSiAs$_2$), Annamamedov et al.[6.26] (ZnGeAs$_2$, ZnSnAs$_2$, CdGeAs$_2$, and CdSnAs$_2$), Borshchevskii et al.[6.64] (ZnSnAs$_2$), Amma and Jeffrey,[6.65] Baranov and Goryunova,[6.66] and Spitzer et al.[6.67] Because thermal conductivity is also a parameter very sensitive to the concentration and character of the lattice defects, there is a substantial variation of its magnitude reported by different authors (see Table 6.10). The temperature dependence of thermal conductivity of ZnGeAs$_2$, CdGeAs$_2$, and CdSnAs$_2$ is shown in Figure 6.8 (adapted from Reference 6.61); κ vs. T for the samples of ZnSnAs$_2$ with ordered and disordered structure is presented in Figure 6.9 (adapted from Reference 6.64).

The dielectric constant and magnetic susceptibility data for the II-IV-V$_2$ compounds are compiled in Table 6.11, together with the available data on the refractive index for the ordinary and extraordinary rays. In the reported limits of the wavelengths, the refractive indices are monotonically decreasing with the wavelength increase.

The optical properties of II-IV-V$_2$ compounds are considered interesting for their application in the nonlinear optical devices. The dependence of the optical absorption coefficient of several compounds of this type on the photon energy is shown in Figure 6.10 (adapted from Reference 6.71). The spectral distribution of the ZnSnSb$_2$ absorption coefficient is shown in Figure 6.11 (adapted from Reference 6.47). The results of a comprehensive study of the exiton absorption spectrum in ZnSiP$_2$ were reported by Madelon et al.;[6.54] a detailed

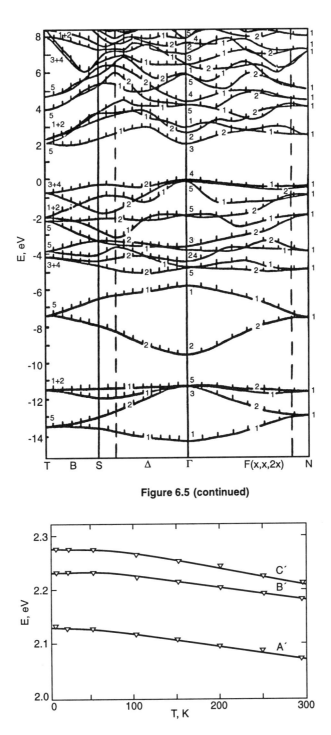

Figure 6.6 Temperature dependence of the transition energies for ZnSiP$_2$. (After R. Madelon, J. Hurel, and E. Paumer, *Phys. Stat. Solidi*, B136, 679, 1986. With permission.)

investigation of the photodielectric effect in CdGeP$_2$ was performed by Baitenev et al.[6.12] The literature analysis pertaining to the luminescence spectra of II-IV-V$_2$ compounds is presented in Reference 6.7, p. 129. The photoluminescence data for ZnGeP$_2$ were reported by Hobgood et al.[6.57] This compound presents a substantial interest as a material for midinfrared optical paramagnetic oscillators.[6.157,6.158]

Table 6.8 Carrier Effective Mass in II-IV-V$_2$ Compounds

Compound	Effective mass (in m_o units)							
	$m_n^{*\,6.13}$	m_{pA}^{*}	$m_p^{*\,6.13}$	m_{pB}^{*}	$m_n^{*\,6.6}$	$m_{nT}^{*\,6.7}$	$m_{nL}^{*\,6.7}$	$m_{pl}^{*\,6.7}$
ZnSiP$_2$	0.110		0.4		0.096			
ZnGeP$_2$						0.108	0.105	0.15
CdSiP$_2$	1.068($\parallel$); 0.124($\perp$)				0.092			
CdGeP$_2$						0.088	0.080	0.099
CdSnP$_2$						0.060	0.056	
ZnSiAs$_2$					0.071	0.103	0.098	0.15
ZnCeAs$_2$						0.060	0.058	0.084
ZnSnAs$_2$			0.35		0.029			
CdSiAs$_2$						0.084	0.74	0.096
CdGeAs$_2$	0.26		0.035		0.020	0.039	0.030	0.67
CdSnAs$_2$	0.05					0.018	0.016	0.10
ZnSnSb$_2$	0.025$^{6.47}$	0.031$^{6.47}$		0.25$^{6.47}$				

Note: The data in the last three columns are the calculated magnitudes.

Table 6.9 Electrical Properties of II-IV-V$_2$ Compounds at Room Temperature

Compound	Conductivity Type	Magnitude (S/cm)	Carrier Concentration (cm^{-3})		Mobility (cm^2/V·sec)		Ref.
			Electrons	Holes	Electrons	Holes	
ZnGeN$_2$			10^{18}–10^{19}		0.5–5		6.13
MgSiP$_2$	n	10^{-4}–10^{-1}					6.13
ZnSiP$_2$			10^{15}		210		6.5
			10^{17}	6·10^{13}	70–100	4–11	6.13
ZnGeP$_2$				10^{11}		20	6.13
	p			10^{16}		7	6.57
ZnSnP$_2$				3·10^{16}–1·10^{17}		10–60	6.13
CdSiP$_2$			10^{14}–10^{15}		80–150		6.13
CdGeP$_2$			10^{12}–10^{15}	1·10^{7}–2·10^{7}	100	30–60	6.13
			10^{12}		300		6.5
CdSnP$_2$			10^{15}–10^{18}	10^{14}	400–2,000	90–150	6.13
ZnSiAs$_2$				10^{14}		45	6.5
			10^{9}	8.2·10^{14}	40	100	6.13
ZnGeAs$_2$				7·10^{18}–1·10^{19}		20–24	6.13
ZnSnAs$_2$				5·10^{17}–1·10^{20}		80–90	6.13
CdSiAs$_2$				6·10^{15}		300–500	6.13
CdGeAs$_2$			4·10^{16}–1·10^{18}	7·10^{15}–1·10^{16}	1,000–4,000	140–400	6.13
CdSnAs$_2$			9·10^{16}		22,000		6.5
					11,000–15,000	36–540	6.13
	n	240					6.56
ZnSnSb$_2$	p			10^{20}		70	6.13

6.1.2. The I-IV$_2$-V$_3$ Compounds

In the I-IV-V concentration triangle, the compound I-IV$_2$-V$_3$ is located at the intersection of the "normal valency" line I$_3$V-IV$_3$V$_4$ and the "four-valence-electron-per-atom" line, IV-(I)(V)$_3$. As an example, the concentration triangle of the system Cu-Ge-P is shown in Figure 6.12.

There are only three compounds of this type known. Copper-germanium phosphide, CuGe$_2$P$_3$, was first reported in References 6.73 and 6.74; later, Folbert and Pfister[6.75] reported the results of synthesis and investigation of CuGe$_2$P$_3$ and CuSi$_2$P$_3$. The third known compound of this type, AgGe$_2$P$_3$, does not crystallize in a tetrahedral structure.[6.76] The energy gap of CuGe$_2$P$_3$ was evaluated from the results of investigation of the coefficient of diffuse reflection from the powder samples of the ternary compound. The curve of the coefficient vs. wavelength

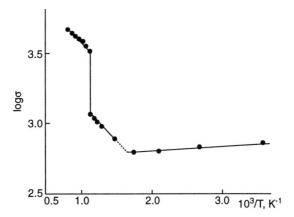

Figure 6.7 Temperature dependence of the CdSnAs$_2$ electrical conductivity in the solid and liquid states. (After G.F. Nikolskaya, L.I. Berger, I.V. Evfimovskii, G.I. Kagirova, I.K. Shchukina, and I.S. Kovaleva, *Inorg. Mater.*, 2, 1876, 1966. With permission.)

Table 6.10 Thermal Conductivity of II-IV-V$_2$ Compounds

Compound	Thermal Conductivity at Room Temperature (mW/cm·K)							
	Ref. 6.63	Ref. 6.62	Ref. 6.66	Ref. 6.65	Ref. 6.61	Ref. 6.67	Ref. 6.26	Ref. 6.64
ZnGeP$_2$	180		42					
CdGeP$_2$	110	84	83					
ZnSiAs$_2$	140							
ZnGeAs$_2$		113		114	105		67	71–142
ZnSnAs$_2$					74		58	
CdGeAs$_2$	42				35		49	
CdSnAs$_2$					92		40	

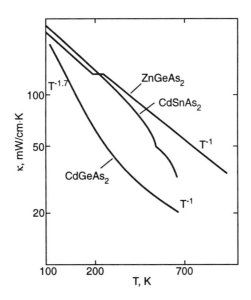

Figure 6.8 Temperature dependence of the thermal conductivity of ZnGeAs$_2$, CdGeAs$_2$, and CdSnAs$_2$. (After P. Leroux-Hugon, *Compt. Rend. Acad. Sci.*, 256, 118, 1963. With permission.)

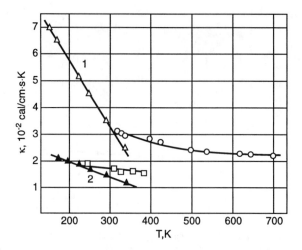

Figure 6.9 Temperature dependence of the p-ZnSnAs$_2$ thermal conductivity for samples with ordered (1) and disordered (2) structure. (After A.S. Borshchevskii, N.A. Goryunova, F.P. Kesamanly, and D.N. Nasledov, *Phys. Stat. Solidi*, 21, 9, 1967. With permission.)

Table 6.11 II-IV-V$_2$ Compounds. Dielectric Constant, Magnetic Susceptibility, and Index of Refraction at 300 K

	Dielectric constant				Atomic magnetic susceptibility (-10^{-5} SI units) (Ref. 6.68)	Index of refraction (Ref. 6.7, 6.69, 6.70)	
	Static		Optical				
Compound	$\parallel$c (Ref. 6.13)	$\perp$c (Ref. 6.13)	$\parallel$c (Ref. 6.13)	$\perp$c (Ref. 6.13)		n_o	n_e
ZnSiP$_2$	11.15	11.7	9.68	9.26		3.11[a]	
ZnGeP$_2$	15	12				3.5052 (640 nm)	3.5802 (640 nm)
						3.0552 (12 µm)	3.0949 (12 µm)
ZnSnP$_2$	10.0	8.08				3.01[a]	
CdGeP$_2$						3.4403 (850 nm)	3.4942 (850 nm)
						3.1165 (12.5 µm)	3.1332 (12.5 µm)
CdSnP$_2$	11.8	10.0				3.16[a]	
ZnSiAs$_2$						3.5579 (700 nm)	3.6201 (700 nm)
						3.1685 (11.5 µm)	3.1953 (11.5 µm)
ZnGeAs$_2$					18.1	3.5973 (2.4 µm)	3.7545 (2.4 µm)
						3.5604 (11.5 µm)	3.5871 (11.5 µm)
ZnSnAs$_2$	15.6				23.1		
CdGeAs$_2$	15.4	15.2			29.4		
CdSnAs$_2$	12.1				34.6	3.5[a]	

[a] Estimated value.

is shown in Figure 6.13 (adapted from Reference 6.77). Measurements were performed on the CuGe$_2$P$_3$ polycrystalline samples with electrical conductivity of about 20 S/cm. The energy gap estimated from the reflection measurements is 0.90 ± 0.05 eV.

The elastic properties of CuGe$_2$P$_3$ were studied by Tu Hailing et al.,[6.78] who also reported the lattice parameters and coefficient of thermal expansion of the compound. The thermal

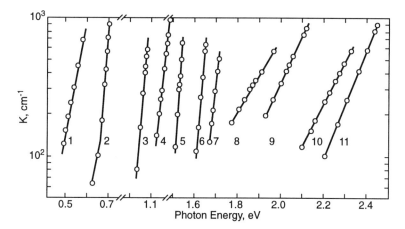

Figure 6.10 II-IV-V$_2$ compounds. The absorption coefficient vs. photon energy. 1 — CdGeAs$_2$, 2 — ZnSnAs$_2$, 3 — ZnGeAs$_2$, 4 — CdSnAs$_2$, 5 — CdSiAs$_2$, 6 — ZnSnAs$_2$, 7 — CdGeP$_2$, 8 — ZnSiAs$_2$, 9 — ZnGeP$_2$, 10 — ZnSiP$_2$, 11 — CdSiP$_2$. (After R. Bendorius, V.D. Prochukhan, and A. Shileika, *Phys. Stat. Solidi*, B53, 745, 1972. With permission.)

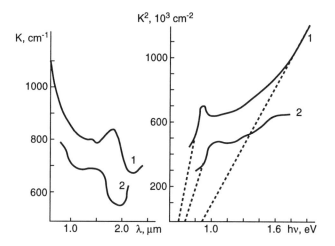

Figure 6.11 Spectral distribution of the ZnSnSb$_2$ absorption coefficient (in different coordinate systems at 300 K (1) and 77 K (2). (After L.I. Berger, L.V. Kradinova, V.M. Petrov, and V.D. Prochukhan, *Inorg. Mater.*, 9, 1258, 1973. With permission.)

conductivity of CuGe$_2$P$_3$ and some alloys of the CuGe$_2$P$_3$-Ge system were reported in Reference 6.79. The infrared reflectivity spectra of CuSi$_2$P$_3$ were studied by Neumann et al.[6.80]

The available data on the properties of the I-IV$_2$-V$_3$ compounds are compiled in Table 6.12. One can see from the table that the information pertinent to the compounds is very limited. Goryunova (Reference 6.5, p. 146) assumes that the number of compounds of this type has to be small because one of the criteria for a ternary compound formation — existence of chemical interaction in the binary systems that form the concentration triangle (I-IV, I-V, and IV-V) — is not satisfied. The authors of Reference 6.76 have reported on the numerous attempts to synthesize ternary nitrides, arsenides, and antimonides with copper, silver, and other first-group elements.

The compounds CuSi$_2$P$_3$ and CuGe$_2$P$_3$ have never been synthesized with the ordered distribution of atoms in the cation sublattice, although this might give materials with optical properties interesting for practical applications similar to the properties of some II-IV-V$_2$ and I-III-VI$_2$ compounds.

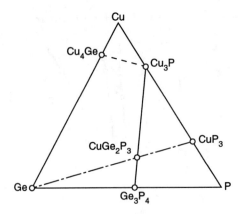

Figure 6.12 The concentration triangle of the system Cu-Ge-P.

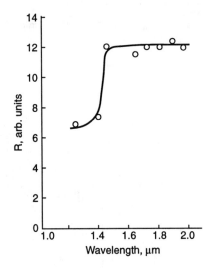

Figure 6.13 Spectral distribution of the diffusive reflection coefficient of a $CuGe_2P_3$ powder sample. (After L.I. Berger, V.I. Sokolova, and V.M. Petrov, *Trudy IREA Proc. IREA*, 30, 406, 1967. With permission.)

6.2. TERNARY ANALOGS OF II-VI ADAMANTINE SEMICONDUCTORS

Three groups of two-cation compounds, namely, I-III-VI_2, I_2-IV-VI_3, and I_3-V-VI_4, belong to this family of semiconductor materials.

6.2.1. I-III-VI$_2$ Compounds

The known compounds of this type crystallize in the tetragonal structure, space group $\bar{I}42d$, with 16 atoms (four formula units) per unit cell. This structure is close to that of mineral chalcopyrite, $CuFeS_2$, in which the iron atoms are assumed to be trivalent. There are 25 I-III-VI_2 compounds known with a general formula (Cu, Ag) (Al, Ga, In, Tl, Fe) (S, Se, Te)$_2$.

The first report on synthesis and lattice parameter data of the compounds (Cu, Ag) (Al, Ga, In, Tl) (S, Se, Te)$_2$ was published by Hahn et al.[6.81] The information regarding the lattice parameters and density of I-III-VI_2 compounds was published later by Goryunova (Reference 6.4, p. 142), Wooley and Williams,[6.83] Berger and Balanevskaya,[6.84] Berger and Prochukhan[6.85]

Table 6.12 Properties of the I-IV$_2$-V$_3$ Compounds

Parameter	CuSi$_2$P$_3$	CuGe$_2$P$_3$	AgGe$_2$P$_3$
Molecular weight	212.64	301.65	345.97
Mean atomic weight	35.44	50.28	57.66
Mean atomic number	17	23	26
Crystal structure	E1-F$\bar{4}$3m	E1-F$\bar{4}$3m	
Lattice parameter, nm	0.525 [6.5]	0.5375 [6.5]	
		0.53678 [6.78]	
Coefficient of linear thermal expansion, 10^{-6} 1/K		8.21 ± 0.02 [6.78]	
Vickers microhardness, GPa		8.50 ± 0.20 [6.78]	6.15 ± 0.30 [6.5]
		6.1 [6.79]	
Elastic constants (291 K), GPa			
c_{11}		136.6 [6.78]	
c_{12}		61.7 [6.78]	
c_{44}		66.6 [6.78]	
Bulk modulus, GPa		86.7 [6.78]	
Melting point, K		1113 [6.76]	1015 [6.76]
Energy gap (300 K), eV		0.90 ± 0.05 [6.77]	
Electrical conductivity (300 K) in S/cm		~20 (p-type) [6.77]	
Carrier mobility (holes), cm^2/V·sec		1–2 [6.77]	
Thermal conductivity at 300 K, mW/cm·K		32 [6.79]	

(see, also, Reference 6.6), Goryunova,[6.5] Berger et al.,[6.86] Belova et al.,[6.87] Brandt et al.,[6.88] Bodnar,[6.89] Shay and Wernick,[6.7] and, lately, Wada et al.[6.90] (thin films). A large volume of information regarding the X-ray diffraction data pertinent to I-III-VI$_2$ compounds is collected by the JCPDS.[6.91]

The information on the lattice parameters, density, and coefficient of thermal expansion of the compounds is compiled in Table 6.13. Apparently, only one of I-III-VI$_2$ compounds, AgInS$_2$, may crystallize in either chalcopyryte or wurtzite-type crystal lattice. The compounds with a general formula (Li, Na, K, Rb, Cs) (B, Al, Ga, In, Tl) (O, S)$_2$ and (Cu, Ag) (Al, Ga, In)O$_2$ do not crystallize in the diamond-like structure and therefore are not included in this chapter. A few I-III-VI$_2$ compounds with the lanthanides as the third group components, such as AgLaSe$_2$,[6.92] will be considered later.

The temperature dependence of the coefficient of thermal expansion, α, of CuInTe$_2$ at temperatures between 20 and 340 K was studied by Novikova[6.95] using a quartz dilatometer. She found that with temperature decrease, α decreases and reaches zero magnitude at ~43 K and is negative at lower temperatures. At 30 K, α passes through a minimum of $-1.15 \cdot 10^{-6}$ 1/K (see Figure 6.14). The thermal expansion of CuInTe$_2$ at elevated temperatures was reported by the authors of Reference 6.86 who found that α_{av} (293 to 473 K) = $8.2 \cdot 10^{-6}$ 1/K, α_{av} (473 to 673 K) = $9.36 \cdot 10^{-6}$ 1/K, and α_{av} (293 to 673 K) = $8.81 \cdot 10^{-6}$ 1/K.

The information regarding the crystal structure of other I-III-VI$_2$ compounds is limited to data on some oxygen- and iron-containing substances (see Table 6.14).

The thermal properties of I-III-VI$_2$ compounds were analyzed in References 6.5 to 6.7. The heat of formation and atomization energy were measured mass-spectrometrically by Berger et al.[6.96] The data on melting point (or the crystallization beginning temperature) and some thermodynamic parameters of the compounds are compiled in Table 6.15.

The information regarding mechanical properties of I-III-VI$_2$ compounds includes the elastic constants of AgGaS$_2$,[6.101] Young's modulus of CuInSe$_2$, CuGaTe$_2$, and CuInTe$_2$,[6.26] and bulk modulus of some of the compounds.[6.101-6.104] Several attempts have been made to calculate the bulk modulus of some compounds using data on the Debye temperature, surface energy, unit cell volume, etc. (see Reference 6.42), but agreement with the experimental data is quite poor. The microhardness of the compounds is reported, in particular, in References 6.5 and 6.56. The data on mechanical properties are presented in Table 6.16.

Table 6.13 I-III-VI$_2$ Compounds. The Molecular and Mean Atomic Weight, Average Atomic Number, Lattice Parameters, Density, and Coefficient of Thermal Expansion

Compound	Molecular weight	Mean atomic weight	Mean atomic number	Lattice Parameters a (nm)	c (nm)	$(2 - \frac{c}{a}) \times 10^3$	Coeff. of thermal expansion 10^{-6} 1/K	Density (g/cm^3)[6.6] X-ray	Pycnom.
CuAlS$_2$	154.65	38.66	18.5	0.5336[6.88]	1.0444[6.88]	43		3.47	3.45
CuAlSe$_2$	248.45	62.11	26.5	0.56035[6.93]	1.0977[6.93]	41		4.70	4.69
CuAlTe$_2$	345.73	86.43	36.5	0.596[6.13]	1.177[6.13]	25		5.50	5.47
CuGaS$_2$	197.39	49.35	23	0.535[6.13]	1.046[6.13]	45		4.35	4.29
CuGaSe$_2$	291.19	72.80	31	0.5607[6.100]	1.1054[6.100]	29	5.4[6.94]	5.56	5.45
CuGaTe$_2$	388.47	97.12	41	0.600[6.13]	1.193[6.13]	12	11.2[6.6]	5.99	5.87
CuInS$_2$	242.49	60.62	27.5	0.5523[a]	1.1141[a]	−17		4.75	4.71
CuInSe$_2$	336.29	84.07	35.5	0.5773[6.5]	1.155[6.5]	0	7.1[6.6]	5.77	5.65
CuInTe$_2$	433.57	108.39	45.5	0.6179[6.86]	1.2358[6.86]	0	8.6[6.6]	6.10	6.00
CuTlS$_2$	332.05	83.01	35.5	0.5580[6.5]	1.117[6.5]	0		6.32	6.07
CuTlSe$_2$	425.85	106.46	43.5	0.5832[6.5]	1.163[6.5]	6		7.11	7.08
AgAlS$_2$	198.97	49.74	23	0.5695[6.5]	1.026[6.5]	198		3.94	3.93
AgAlSe$_2$	292.77	73.19	31	0.5956[6.5]	1.075[6.5]	195		5.07	4.99
AgAlTe$_2$	390.05	97.51	41	0.6296[6.5]	1.183[6.5]	121		6.18	6.15
AgGaS$_2$	241.71	60.43	27.5	0.5754[6.89]	1.030[6.89]	210		4.72	4.58
AgGaSe$_2$	335.51	83.88	35.5	0.59920[6.93]	1.08863[6.93]	183		5.84	5.71
AgGaTe$_2$	432.79	108.20	45.5	0.6283[6.5]	1.194[6.5]	100	4.2[6.6]	6.05	5.96
AgInS$_2$ (chalcopyrite)	286.81	71.70	32	0.5880[6.89]	1.120[6.89]	95		5.00	4.97
AgInS$_2$ (wurtzite)				0.4121[6.5]	0.6674[6.5]			4.83	
AgInSe$_2$	380.61	95.15	40	0.60913[6.93]	1.17122[6.93]	77		5.81	5.80
AgInTe$_2$	477.89	119.47	50	0.643[6.13]	1.259[6.13]	42	5.5[6.6]	6.12	6.08

[a] From Reference 6.91, Card No. 27-01.59.

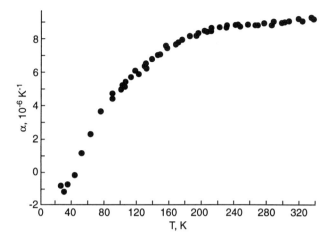

Figure 6.14 Temperature dependence of the $CuInTe_2$ coefficient of linear thermal expansion. (After S.I. Novikova, *Sov. Phys. Solid State*, 7, 2170, 1966. With permission.)

Table 6.14 The Crystal Structure of $I-III-O_2$ and $I-Fe-VI_2$ Compounds

Compound	Hexagonal cell a (nm)	c (nm)	Z	Rhombohedral cell a (nm)	α	Z	Chalcopyrite cell a (nm)	c (nm)	Tetragonal cell a (nm)	c (nm)	Density (g/cm³) X-ray	Pycnom.
$CuAlO_2$	0.2864	1.695	3	0.5884	28.1°	1					5.104	4.897
$CuGaO_2$	0.3027	1.709	3	0.5954	29.4°	1					6.034	5.753
$CuFeS_2$							5.29	10.41				
$AgFeS_2$									6.58	8.96		

Table 6.15 $I-III-VI_2$ Compounds. The Melting Point and Thermodynamic Properties

Compound	Melting point (K) (Ref. 6.13)	Heat of formation (kJ/mol) (Ref. 6.9)	Heat of atomization (kJ/mol) (Ref. 6.96)	Specific heat (calc.) (J/g·K) (Ref. 6.34)	Debye temperature (K) (Ref. 6.98)
$CuAlS_2$	1570				
$CuAlSe_2$	1470				
$CuAlTe_2$	1163 [6.97]				
$CuGaS_2$	1550				
$CuGaSe_2$	1340	316.6	333.5	0.343	280
$CuGaTe_2$	1140	168.2	312.5	0.257	183
$CuInS_2$	1270–1320				
$CuInSe_2$	1260	267.4	312.5	0.297	218
$CuInTe_2$	1050	107.1	268.2	0.230	167
$CuTlSe_2$	678 [6.7]				
$AgAlSe_2$	1223 [6.5]			0.339	
$AgAlTe_2$	1002 [6.5]				
$AgGaS_2$	1253 [6.5]				
$AgGaSe_2$	1130	446.0	352.3	0.293	
$AgGaTe_2$	978 [6.5]	140.2	271.1	0.230	212 [6.99]
$AgInS_2$	1150				
$AgInSe_2$	1055	241.8	301.2	0.259	
$AgInTe_2$	960	123.0	258.6	0.209	
$CuFeS_2$	1120–1150				
$CuFeSe_2$	850				
$CuFeTe_2$	1015				
$AgFeSe_2$	1009				
$AgFeTe_2$	953				

Table 6.16 Mechanical Properties of I-III-VI$_2$ Compounds

Compound	Microhardness (GPa)	Bulk modulus (GPa)	Young's modulus (GPa)
CuAlS$_2$		94[6.102]	
CuGaS$_2$		94;[6.102] 96[6.103]	
CuGaSe$_2$	4.3[6.94]	71[6.104]	
CuGaTe$_2$	3.5[6.94]	44[6.104]	60.4[6.26]
CuInS$_2$		71.1[a]	
CuInSe$_2$	2.5[6.94]	53.6[a]	68.8[6.26]
CuInTe$_2$	1.9[6.94]	36.0[a]	64.0[6.26]
CuTlSe$_2$	0.9[6.6]		
CuTlTe$_2$	1.0[6.6]		
AgGaS$_2$		66.8[6.101]	
AgGaSe$_2$	3.1[6.94]	54.8[a]	
AgGaTe$_2$	2.6[6.94]	35.7[a]	
AgInS$_2$		54.1[a]	
AgInSe$_2$	2.3[6.6]	42.0[a]	
AgInTe$_2$	1.9[6.5]	29.6[a]	

[a] Estimated by Neuman.[6.42]

The general features of the band structure of the I-III-VI$_2$ chalcopyrite-type semiconductors are similar to those of the II-IV-V$_3$ compounds. A diagram of the features is shown in Figure 6.15. The interest in the electronic properties of I-III-VI$_2$-type materials is stimulated by very strong absorption known for the solar spectrum[6.105,6.106] in some of the compounds, in particular, Cu(In, Ga) (S, Se)$_2$ and their solid solutions, in comparison with many other adamantine semiconductors. The electronic band structure of six I-III-VI$_2$ compounds is shown in Figure 6.16. Some numerical data regarding the band structure parameters are presented in Table 6.17. One can see that there is a substantial variation in the data reported by different authors; our knowledge relevant to the band structure of the I-III-VI$_2$ compounds may be considered not as yet sufficient.

Electrical properties of I-III-VI$_2$ compounds strongly depend on the synthesis and crystal growth methods. These factors are especially important for compounds that are formed by peritectic reactions or have phase transformations in the solid state. Tell et al.[6.116] found that annealing of samples may be used to change the concentration and mobility of the current carriers in the samples of compounds in the wide limits.

The earlier reports on electrical properties of the compounds (see a review in Reference 6.82) indicated that some of them have mobility as high as 1150 cm^2/V·sec (CuInSe$_2$). Detailed reviews of the temperature dependence of electrical and thermoelectrical properties of I-III-VI$_2$ materials are presented in References 6.6 and 6.7. Some data from there are compiled in Table 6.18.

Investigation of electrical conductivity of some Ag-III-VI$_2$ compounds in the liquid state[6.119] showed that AgGaSe$_2$ and AgInTe$_2$ have the semiconductor character of conductivity in the liquid state; while the temperature dependence of conductivity of the liquid AgInSe$_2$ is typical for a metal: the conductivity decreases with the temperature growth. These results contradict the data reported by Abdelghany,[6.160] who has found that AgInSe$_2$ in the solid state behaves as a semiconductor with E_g = 0.06 eV, while the energy gap for the liquid AgInSe$_2$ is 0.37 eV. The temperature dependence of electrical conductivity of several samples of AgGaSe$_2$ in solid and liquid state is shown in Figure 6.17 (adapted from Reference 6.119 [a]). Electrical and thermoelectrical properties of CuGaSe$_2$ and CuInSe$_2$ in the temperature range that includes the melting point have been reported in Reference 6.118. Both materials have the semiconductor character of conductivity ($d\sigma/dT > 0$) in the liquid state.

The thermal conductivity magnitude of several I-III-VI$_2$ compounds near room temperature was reported in Reference 6.98 (see Table 6.19). The measurements were performed on

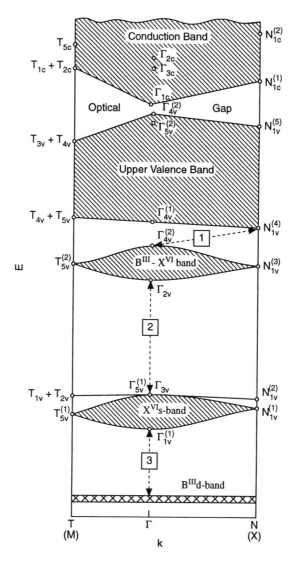

Figure 6.15 General features of the band structure of I-III-VI$_2$ compounds. The shaded and cross-hatched areas denote major subbands; the boxed numbers indicate the internal gaps. (After J.E. Jaffe and A. Zunger, *Phys. Rev.*, B28, 5822, 1983. With permission.)

a large-grain polycrystalline sample. It is reasonable to expect that the thermal conductivity magnitudes of single-crystalline samples are noticeably higher.

Balanevskaya et al.[6.120] have reported the data on magnetic susceptibility of a group of the I-III-VI$_2$ compounds. All investigated substances are diamagnetics. The mean atomic magnetic susceptibility of the investigated compounds is presented in Table 6.19, together with the available data on the dielectric constant of the compounds.

The refraction indices of some I-III-VI$_2$ compounds for ordinary and extraordinary beams were reported in detail by Boyd et al.[6.121,6.122] In the entire wavelength range considered, the investigated compounds have shown normal dispersion. Selected data from References 6.121 and 6.122 are gathered in Table 6.20.

The I-III-VI$_2$ compounds, in particular, CuGaSe$_2$, CuInSe$_2$, and their combinations, present substantial interest for use in photovoltaic devices. Tuttle et al.[6.115] have reported on the method of production of thin-film CuInSe$_2$ which may be used for industrial scale production of the photovoltaic devices with total-area efficiencies of 14.4 to 16.4%, although,

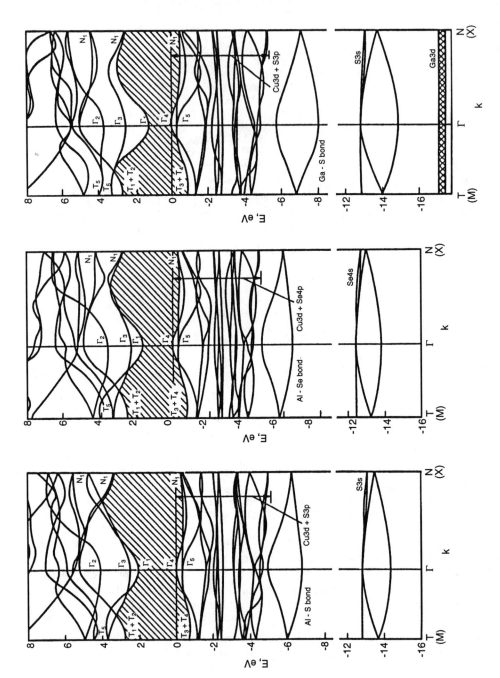

Figure 6.16 The band structure of (a) CuAlS$_2$, (b) CuAlSe$_2$, (c) CuGaS$_2$, (d) CuGaSe$_2$, (e) CuInS$_2$, and (f) CuInSe$_2$. (After J.E. Jaffe and A. Zunger, *Phys. Rev.*, B28, 5822, 1983. With permission.)

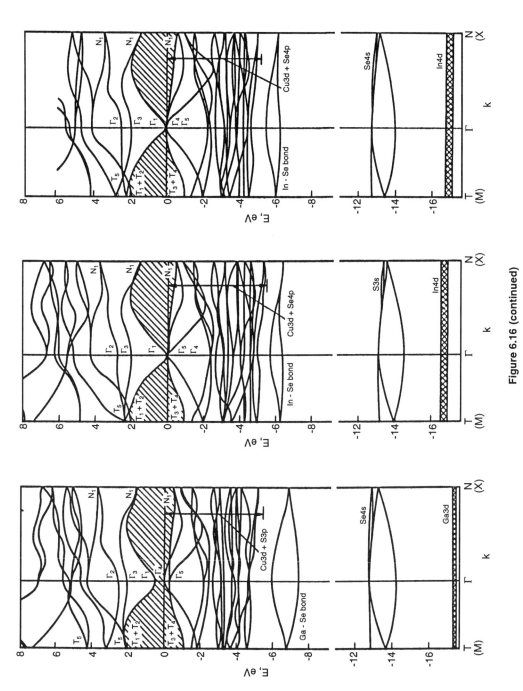

Figure 6.16 (continued)

Table 6.17 Band Parameters of I-III-VI$_2$ Compounds (at 300 K — Except Where Indicated Otherwise)

Compound	Thermal	Energy Gap (eV) Optical A	B	C	Splitting energy $-\Delta_s$(meV) Crystal field	Spin-orbit	Effective mass (m_o) Electrons	Holes	Ref.
CuAlS$_2$	3.5[6.7]	3.49	3.62	3.62	130	0			6.107
CuAlSe$_2$	2.7[6.7]	2.72[a]	2.86[a]	3.01[a]	170[a]	180			6.13
		2.67							6.114
CuAlTe$_2$		2.06							6.13
CuGaS$_2$	2.4	2.43	2.55	2.55	120	0			6.7
		2.502[a]	2.627[a]	2.638[a]	130[a]	−17	0.13	0.69	6.13
CuGaSe$_2$	1.7								6.115
	1.7	1.68	1.75	1.96	90	230			6.7
		1.60							6.108
		1.721	1.806	2.012	110	232			6.109
								1.2	6.111
CuGaTe$_2$		1.227	1.280	1.97	80	710			6.13
		0.95[a]							6.108
CuInS$_2$	1.5	1.53	1.53	1.53	<5	−20			6.7
								1.3	6.13
		1.53							6.112
		1.5							6.115
CuInSe$_2$	1.0	1.04[a]	1.04[a]	1.27[a]	−6[a]	230[a]			6.7
		1.0							6.115
							0.09	0.09	6.13
		0.81							6.108
		1.010 ± 0.001							6.117
CuInTe$_2$		1.064	1.064	1.67	0	610		0.66 (∥)	6.13
								0.85 (⊥)	6.13
CuTlS$_2$	1.39 (est.)								
AgAlSe$_2$	0.70								6.5
AgAlTe$_2$		2.27							6.13
AgGaS$_2$	2.7	2.73[a]	3.01[a]	3.01[a]	280	0			6.7
AgGaSe$_2$	1.8	1.83[a]	2.02[a]	2.29[a]	250[a]	310[a]			6.7
		1.80							6.13
	1.66								6.5
AgGaTe$_2$	0.99								6.5
		1.32							6.13
		1.08–1.16							6.113
AgInS$_2$	1.9	1.87	2.02	2.02	150	70			6.7
AgInSe$_2$	1.04								6.5
	1.2	1.24	1.33	1.60	120	300			6.7
		1.10[a]							6.108
AgInTe$_2$	0.95	1.12							6.13
	0.89								6.5
		1.06[a]							6.108
AgLaSe$_2$	0.6								6.92

[a] At 77 K.

in the laboratory devices, the efficiency may exceed 17% (see Reference 6.115). One of the important features of the compounds is their high optical absorption coefficient. The absorption coefficient vs. photon energy for CuGaSe$_2$ films of different thickness and substrate temperature is shown in Figure 6.18 (adapted from Reference 6.124). Annapurna and Reddy[6.111] studied the optical absorption of CuGaSe$_2$ thin films prepared by the same technique (flash evaporation) and found that the absorption coefficient on the films two to three times thicker than in Reference 6.124 is about one or two orders of magnitude greater (in both reports, the samples were polycrystalline films deposited on glass substrates).

Table 6.18 I-III-VI$_2$ Compounds. Electrical Conductivity and Thermoelectric Power before and after Annealing

Compound	Electrical conductivity (S/cm)		Thermoelectric power (µV/K)	
	Before annealing	After annealing	Before annealing	After annealing
CuGaSe$_2$	0.02	0.06	+75	+40
CuGaTe$_2$	11	6	+270	+340
CuInSe$_2$	0.14	0.04	+480	+640
CuInTe$_2$	100	65	+140	+260
CuTlSe$_2$	6000	1000	+10	−5
CuTlTe$_2$	2800		+80	
AgGaSe$_2$	8.8	0.12	−70	+90
AgGaTe$_2$	$1.7 \cdot 10^{-4}$	$1.7 \cdot 10^{-5}$	+700	+950
AgInSe$_2$	0.48	$5 \cdot 10^{-5}$	−140	−370
AgInTe$_2$	0.78	$1 \cdot 10^{-5}$	−70	−100
AgTlSe$_2$	$1 \cdot 10^{-5}$		+800	
AgTlTe$_2$	410	1800	+60	+10

From L.I. Berger and V.D. Prochukhan, *Ternary Diamond-Like Semiconductors,* Consultants Bureau, Plenum Press, New York, 1969.

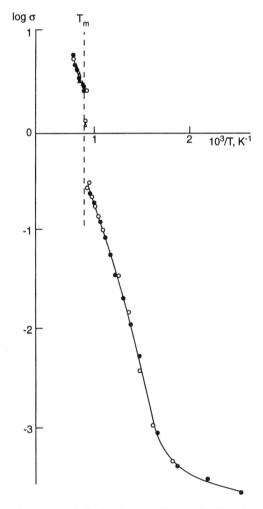

Figure 6.17 Electrical conductivity of AgGaSe$_2$ in the solid and liquid states (σ in S/cm). (After A.E. Balanevskaya, L.I. Berger, and S.A. Isaev, *Inorg. Mater.,* 7, 2084, 1971. With permission.)

Table 6.19 I-III-VI$_2$ Compounds. Thermal Conductivity, Magnetic Susceptibility, and Dielectric Constant

Compound	Thermal conductivity (W/cm·K)[6.98]	Magnetic susceptibility (-10^4 SI units/g-mol)[6.120]	Dielectric constant[6.13]			
			$\varepsilon(0)\|$	$\varepsilon(0)\perp$	$\varepsilon(\infty)\|$	$\varepsilon(\infty)\perp$
CuGaS$_2$			7.6	8.9	6.1	6.2
CuGaSe$_2$	4.18	11.1				
AgGaSe$_2$	2.65	13.3	10.48	9.05	6.95	6.80
CuInSe$_2$	2.9	14.1	15.2	16.0	8.5	9.5
AgInSe$_2$	1.93	16.1	10.73	11.94	7.16	7.20
CuGaTe$_2$	2.7	16.6				
AgGaTe$_2$	1.85	17.6	14.5	15.0	11.0	11.86
CuInTe$_2$	2.5	18.6	12.9	10.5	8.7	11.0
AgInTe$_2$	1.55	21.1	7.68	8.10	6.38	6.48
AgGaS$_2$			8.21	8.51	5.50	5.90

From the I-III-VI$_2$ compounds, only CuInSe$_2$ is available commercially in the form of lumps of purity 99.9% and priced from $16.40 to $22.40 per gram (1995).

6.2.2. I$_2$-IV-VI$_3$ Semiconductors

In the I-IV-VI concentration triangle, the I$_2$-IV-VI$_3$ compound lies at the intersection of the sections (I$_2$-VI)-(IV-VI$_2$) (e.g., Cu$_2$Se-GeSe$_2$) and IV-(I$_2$-VI$_3$) (e.g., Ge-Cu$_2$Se$_3$) — see Figure 6.19. It makes sense to notice that there is a possibility of existence, in the I-IV-VI concentration triangle, compounds of the type I$_{14}$-IV-VI$_9$ and I$_2$-IV$_7$-VI$_{15}$ with the average valence electron concentration per atom equal to three and five, respectively. A similar prediction regarding the I-III-VI system has led Mikkelsen[6.126] to synthesis of the compound Ag$_9$GaSe$_6$ with the average valence electron concentration equal to three.

The ternary I$_2$-IV-VI$_3$ compounds were first mentioned by Goodman[6.127] and later by Hahn.[6.128] Averkieva and Vaipolin[6.129] reported the results of synthesis and investigation of Cu$_2$SnS$_3$, Cu$_2$SnSe$_3$, and Cu$_2$SnTe$_3$; they found that the compounds have the sphalerite structure. The available data on crystal structure, density, and coefficient of thermal expansion of I$_2$-IV-VI$_3$ compounds are presented in Table 6.21.

The first reports regarding synthesis of Ag$_2$-IV-VI$_3$ compounds[6.129,6.133] indicated that Ag$_2$SnSe$_3$ and Ag$_2$SnSe$_3$ have a cubic structure different from that of sphalerite and that Ag$_2$SnTe$_3$ contained two phases, one of which (up to 60% of the crystal) has the sphalerite structure. It has to be mentioned that the problem of synthesis of I$_2$-IV-VI$_3$ compounds, and particularly silver-containing compounds, in the form of large single crystals, is not solved as yet. The majority of the published data on the properties of the compounds were received on polycrystalline samples. The difficulties in synthesis of some of the I$_2$-IV-VI$_3$ compounds are related to the fact that many of the compounds are formed by peritectic reactions and have a tendency toward decomposition at elevated temperatures. The information regarding thermodynamic properties of the compounds is compiled in Table 6.22.

The data on mechanical properties of I$_2$-IV-VI$_3$ compounds are limited by reports on speed of ultrasonic waves in a few compounds and calculated magnitudes of elastic moduli,[6.135] and on the microhardness.[6.139] The available data are presented in Table 6.23.

The band structure of the I$_2$-IV-VI$_3$ compounds is virtually unknown. The general tendency among ternary analogs of II-VI semiconductors is a decrease of the energy gap with increase of number of atoms in the formula unit of compounds with the same mean atomic weight (or mean atomic number). The data on the energy gap of the compounds were received from the optical, photoelectric and electrical measurements. As an example, the dependence of photoconductivity and photo-emf at 300 K of a Cu$_2$SnS$_3$ sample on the wavelength is shown in Figure 6.20 (adapted from Reference 6.132). The data on the energy gap and effective mass

Table 6.20 I-III-VI$_2$ Compounds. Ordinary (n_o) and Extraordinary (n_e) Refractive Indices

Wavelength (nm)	CuAlSe$_2$ [6.122]		CuGaS$_2$ [6.121]		CuGaSe$_2$ [6.122]		CuInS$_2$ [6.121]		AgGaS$_2$ [6.121]		AgGaSe$_2$ [6.122]		AgInSe$_2$ [6.122]	
	n_o	n_e	n_o	n_e	n_o	n_e	n_o	n_e	n_o	n_e	n_o	n_e	n_o	n_e
500	2.7794	2.7886												
600	2.6560	2.6489	2.6983	2.7019										
700	2.5973	2.5886	2.6293	2.6266										
800	2.5650	2.5556	2.5925	2.5886	2.9365	2.9759								
900	2.5437	2.5327	2.5681	2.5630	2.8716	2.8925	2.7907	2.7713	2.6916	2.6867				
1,000	2.5293	2.5179	2.5517	2.5464	2.8358	2.8513	2.7225	2.7067	2.5748	2.5303				
2,000	2.4851	2.4734	2.5051	2.4991	2.7430	2.7510	2.6020	2.5918	2.5205	2.4706	2.7849	2.7866		
3,000	2.4759	2.4638	2.4945	2.4880	2.7273	2.7344	2.5838	2.5741	2.4989	2.4395	2.7406	2.7275		
4,000	2.4712	2.4586	2.4884	2.4816	2.7211	2.7276	2.5760	2.5663	2.4716	2.4192	2.7132	2.6934		
5,000	2.4659	2.4533	2.4843	2.4772	2.7170	2.7232	2.5699	2.5598	2.4582	2.4053	2.6376	2.6071	2.6761	2.6838
6,000			2.4774	2.4694	2.7133	2.7192	2.5645	2.5539	2.4164	2.3637	2.6245	2.5925	2.6542	2.6592
7,000			2.4714	2.4621	2.7101	2.7158	2.5587	2.5474	2.4080	2.3545	2.6189	2.5863	2.6463	2.6504
8,000			2.4639	2.4539	2.7060	2.7111	2.5522	2.5401	2.4024	2.3488	2.6144	2.5819	2.6416	2.6451
9,000			2.4539	2.4435	2.7021	2.7065	2.5448	2.5311	2.3953	2.3419	2.6113	2.5784	2.6381	2.6414
10,000			2.4429	2.4311	2.6974	2.7014	2.5366	2.5225	2.3908	2.3369	2.6070	2.5743	2.6358	2.6379
11,000			2.4311	2.4179	2.6926	2.6981	2.5274	2.5112	2.3827	2.3291	2.6032	2.5704	2.6318	2.6343
12,000			2.4094		2.6872	2.6898	2.5166	2.4987	2.3757	2.3219	2.5988	2.5659	2.6286	2.6310
13,000			2.3999						2.3663	2.3121	2.5939	2.5608	2.6251	2.6274
									2.3548	2.3012	2.5890	2.5555	2.6210	2.6229
									2.3417	2.2880	2.5837	2.5505	2.6167	2.6183
									2.3266	2.2716	2.5771	2.5439		
									2.3076					

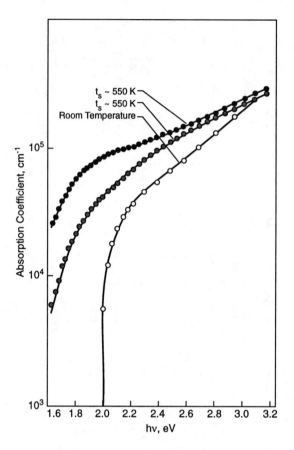

Figure 6.18 Spectral distribution of the CuGaSe$_2$ absorption coefficient for the films of different thicknesses, d, deposited at different substrate temperatures, t$_s$. Open circles — d = 73.5 nm; dotted circles — d = 74.0 nm; solid circles — d = 105.2 nm. (After C.D. George, M. Kapur, and A. Mitra, *Thin Solid Films*, 135, 35, 1986. With permission.)

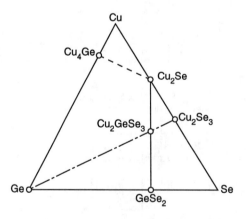

Figure 6.19 Concentration triangle of the Cu-Ge-Se system.

of some compounds are compiled in Table 6.24. As can be seen from the table, there is a substantial difference between the magnitudes of E$_g$ presented in different reports.

The electrical properties of Cu$_2$-IV-VI$_3$ compounds were reported first by Palatnik et al.;[6.131] the authors of Reference 6.135 reported the results of measurements of electrical

Table 6.21 I_2-IV-VI_3 Compounds. The Molecular and Mean Atomic Weight, Mean Atomic Number, Crystal Structure Parameters, Density, and Coefficient of Thermal Expansion

Compound	Molecular weight	Mean atomic weight	Mean atomic number	Crystal structure	Lattice parameters a (nm)	Lattice parameters c (nm)	Lattice parameters c/a	Density (300 K) (g/cm^3)	Coefficient of thermal expansion (10^{-6} 1/K) (300 K)
Cu_2SiS_3 [a]	251.36	41.89	20	Wurtzite	0.3684 [6.5]	6.004 [6.5]	1.63	3.81 [6.5] 3.89 [6.130]	
Cu_2SiS_3 [b]				Tetragonal	0.5290 [6.5]	1.0156 [6.5]	1.92	3.63 [6.5]	
Cu_2SiTe_3	537.98	89.67	38	Cubic or tetragonal	0.593 [6.5]			5.47 [6.5] 5.69 [6.130]	
Cu_2GeS_3 [a]	295.86	49.31	23	Cubic	0.5317 [6.5] 0.530 [6.131]			4.45 [6.5] 4.36 [6.130]	7.2 [6.26]
Cu_2GeS_3 [b]				Tetragonal [c]	0.5327 [6.5]	0.5215 [6.5]	0.98	4.46 [6.5] 4.43 [6.130]	
Cu_2GeSe_3	436.56	72.76	32	Tetragonal	0.5589 [6.5]	0.5485 [6.5]	0.98	5.57 [6.5] 5.63 [6.130]	8.4 [6.26]
Cu_2GeTe_3	582.48	97.08	41	Cubic Tetragonal	0.555 [6.131] 0.5958 [6.5] 0.5956 [6.130]	0.5935 [6.5] 1.1852 [6.130]	0.995 1.990	5.95 [6.5] 6.13 [6.130]	
Cu_2SnS_3	341.96	56.99	26	Cubic	0.5436 [6.5] 0.543 [6.131]			5.02 [6.5]	7.8 [6.26]
Cu_2SnSe_3	482.66	80.44	35	Cubic	0.5687 [6.5,6.86] 0.568 [6.131]			5.02 [6.5]	9.12 [6.86] 8.9 [6.26]
Cu_2SnTe_3	628.58	104.76	44	Cubic	0.6048 [6.5] 0.604 [6.131]			6.51 [6.5]	
Ag_2GeSe_3	525.21	87.54	38					4.66 [6.132]	
Ag_2SnSe_3	571.31	95.22	41					4.86 [6.132]	
Ag_2GeTe_3	671.13	111.86	47					5.12 [6.132]	
Ag_2SnTe_3	717.23	119.54	50					5.76 [6.132]	

Note: The magnitudes of density from Reference 6.131 are calculated from the X-ray measurements.

[a] High-temperature allotrope.
[b] Low-temperature allotrope.
[c] Without analysis of distribution of the copper and germanium atoms in the unit cell.

Table 6.22 I_2-IV-VI$_3$ Compounds. The Melting Point, Heat of Atomization (Sublimation), Specific Heat, and Debye Temperature

Compound	Melting point[a] (K)	Heat of sublimation (kJ/mol)	Heat of atomization (kJ/mol)	Specific heat at RT (J/g·K)	Debye temperature (K)
Cu$_2$Si$_3$ (wurtzite)	1198[6.130]				
Cu$_2$SiS$_3$ (tetragonal)	1113[6.125] (transition)				
Cu$_2$GeS$_3$	1229[6.132] 1221[6.134] 1228[6.125] 1206[6.130] 1220[6.13] 943[6.140] (transition)				252[6.135]
Cu$_2$GeSe$_3$	1056[6.132] 1043[6.134] 1061[6.125] 1033[6.130]	194.1[6.138]	211.7[6.137]	0.34[b]	170[6.135]
Cu$_2$GeTe$_3$	765[6.125] 777[6.134] 868[6.130]	172.8[6.138]		0.26[b]	
Cu$_2$SnS$_3$	1147[6.132] 1118[6.134] 1128[6.125] 1111[6.130]				218[6.135]
Cu$_2$SnSe$_3$	971[6.132] 969[6.134] 970[6.125] 958[6.130]	126.4[c][6.136] 220.7[6.138]	190.4[6.137]		157[6.135]
Cu$_2$SnTe$_3$	680[6.131] 684[6.125] 683[6.130]				
Ag$_2$GeSe$_3$	813[6.132]				
Ag$_2$GeTe$_3$	603[6.132]				
Ag$_2$SnSe$_3$	763[6.132]				
Ag$_2$SnTe$_3$	588[6.132]				

[a] Or the temperature of peritectic decomposition.
[b] Calculated.
[c] Enthalpy of melting.

Table 6.23 Room Temperature Mechanical Properties of Some I_2-IV-VI$_3$ Compounds

Compound	Speed of ultrasonic waves (km/sec)		Moduli (GPa)		Microhardness (GPa)	
	Longitudinal[6.135]	Transversal[6.137]	Young's[6.135]	Shear[6.137]	Ref. 6.139	Ref. 6.137
Cu$_2$GeS$_3$	5.66		14.0		4.55	4.6 ± 0.7
Cu$_2$GeSe$_3$	4.02	2.41	9.1	3.28	3.84	3.9 ± 0.3
Cu$_2$GeTe$_3$					2.89	3.0 ± 0.2
Cu$_2$SnS$_3$	4.95		11.6		2.77	2.8 ± 0.3
Cu$_2$SnSe$_3$	3.61	2.17	7.5	2.70	2.51	2.5 ± 0.2
Cu$_2$SnTe$_3$					1.97	2.0 ± 0.2

conductivity, Hall effect, and Seebeck effect on Cu$_2$(Ge, Sn)(S, Se)$_3$. Rivet et al.[6.140] reported that the compounds Cu$_2$(Si, Ge, Sn)Te$_3$ have metallic conductivity. The electrical conductivity and Seebeck effect measurements reported in Reference 6.142 confirmed the metallic character of conductivity of Cu$_2$GeTe$_3$ and Cu$_2$SnTe$_3$, but it was found that being melted the compounds acquire the semiconductor conductivity with the energy gap of 0.15 eV and

TERNARY ADAMANTINE SEMICONDUCTORS

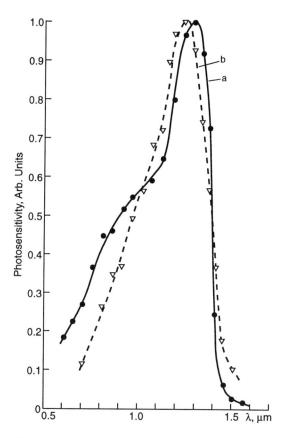

Figure 6.20 Spectral distribution of the photoconductivity (a) and photo-emf (b) for a Cu_2SnS_3 sample at 300 K. (After A.E. Balanevskaya, L.I. Berger, and V.M. Petrov, *Inorg. Mater.*, 2, 810, 1966. With permission.)

Table 6.24 Band parameters of the I_2-IV-VI_3 Compounds

Compound	Thermal	Optical[6.132] At 300 K	At 77 K	Hole effective mass (in the units of m_o)[6.141]
Cu_2SiTe_3	Metal[6.140]			
Cu_2GeS_3	1.1 [6.132]			
	0.3 [6.131]			
Cu_2GeSe_3	0.8 [6.132]	0.94	0.77	0.40
Cu_2GeTe_3	Metal[6.140]			
Cu_2SnS_3	0.8 [6.132]	0.91	0.93	1.8
	0.59 [6.13]			
Cu_2SnSe_3	0.6 [6.132]	0.66	0.78	0.30
	0.7 [6.131]	0.96 [6.13]		
	0.22 [6.140]			
	0.6–0.83 [6.13]			
Cu_2SnTe_3	Metal[6.140]			
Ag_2GeSe_3	0.9 [6.132]			
Ag_2GeTe_3	0.25 [6.132]		0.91	
Ag_2SnS_3	0.5 [6.132]			
Ag_2SnSe_3	0.7 [6.132]	0.81	0.84	
Ag_2SnTe_3	0.08 [6.132]			

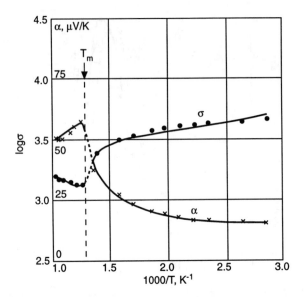

Figure 6.21 Dependence of the electrical conductivity (in S/cm) and thermal emf of Cu_2GeTe_3 on reciprocal temperature. (After O.P. Astakhov, L.I. Berger, K. Dovletov, and E.I. Kovaleva, *Izv. Akad. Nauk Turkmen. SSR Phys. Tech. Series*, No. 6, 111, 1967. With permission.)

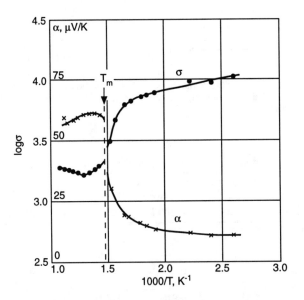

Figure 6.22 Dependence of electrical conductivity (in S/cm) and thermal emf of Cu_2SnTe_3 on reciprocal temperature. (After O.P. Astakhov, L.I. Berger, K. Dovletov, and E.I. Kovaleva, *Izv. Akad. Nauk Turkmen. SSR Phys. Tech. Series*, No. 6, 111, 1967. With permission.)

0.10 eV, respectively. The temperature dependence of the electrical conductivity and thermal-emf of the compounds is shown in Figures 6.21 and 6.22. The p-type conductivity has been observed in the entire temperature range.

Measurements of electrical conductivity and Seebeck effect on Cu_2GeSe_3 and Cu_2SnSe_3 in the solid and liquid state[6.143] have shown that these materials melt in accordance with the scheme "semiconductor-semiconductor" (see Figures 6.23 and 6.24).

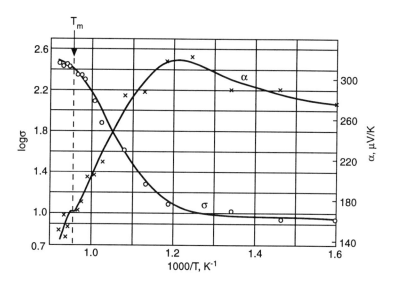

Figure 6.23 Dependence of electrical conductivity (in S/cm) and thermal emf of Cu_2GeSe_3 on reciprocal temperature. (After K. Dovletov, O.P. Astakhov, L.I. Berger, and E.I. Kovaleva, *Izv. Akad. Nauk Turkmen. SSR Phys. Tech. Series*, No. 3, 106, 1968. With permission.)

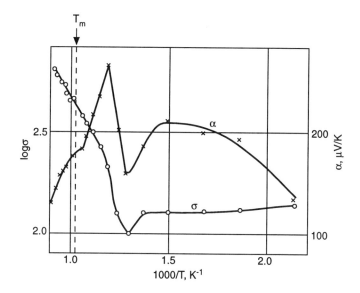

Figure 6.24 Dependence of electrical conductivity (in S/cm) and thermal emf of Cu_2SnSe_3 on reciprocal temperature. (After K. Dovletov, O.P. Astakhov, L.I. Berger, and E.I. Kovaleva, *Izv. Akad. Nauk Turkmen. SSR Phys. Tech. Series*, No. 3, 106, 1968. With permission.)

The alloys of composition Ag_2GeSe_3, Ag_2GeTe_3, Ag_2SnSe_3, and Ag_2SnTe_3 are most probably of eutectic type; the X-ray analysis of the alloy samples shows complicated systems of maxima which have not been explained up to now. Nevertheless, mainly in connection with the search of materials for high-temperature thermoelectric generators, the electrical and thermoelectric properties of the alloys have been investigated by Astakhov et al.[6.144,6.145] at elevated temperatures in solid and liquid states. The four alloys retain the semiconductor conductivity being melted. Of particular interest is the alloy Ag_2GeSe_3, which is the only

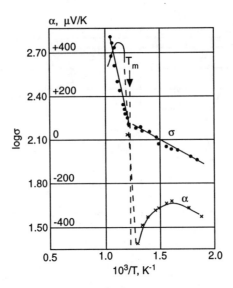

Figure 6.25 Dependence of electrical conductivity (in S/cm) and thermal emf of the alloy of composition of Ag_2GeSe_3 on reciprocal temperature. (After O.P. Astakhov, L.I. Berger, and K. Dovletov, *Izv. Akad. Nauk Turkmen. SSR Phys. Tech. Series*, No. 3, 104, 1968. With permission.)

Table 6.25 I_2-IV-VI_3 Compounds. The Room Temperature Electrical Transport Parameters and Seebeck Coefficient

Compound	Electrical conductivity (S/cm)	Current carrier concentration (10^{17} cm^{-3})	Hole mobility (cm^2/V·sec)	Seebeck coefficient (μV/K)	Ref.
Cu_2SiS_3	123[a]			36[a]	6.140
Cu_2SiTe_3	400				6.140
Cu_2GeS_3	17.3	3.0	360	439	6.132
	0.53[a]			168	6.140
	1.9			300	6.131
Cu_2GeSe_3	5.71	1.5	238	143	6.132
	10			130	6.140
	50	1.7	1,870	100	6.131
Cu_2GeTe_3	3,900			13	6.140
	1,400			10	6.131
Cu_2SnS_3	39.6	6.1	405	467	6.132
	5	6	5.5	400	6.140
	0.49			600	6.131
Cu_2SnSe_3	71.0	5.1	870	137	6.132
	192	90	30	83	6.140
	91	14	400	250	6.131
Cu_2SnTe_3	13,000	>30,000	<20	6	6.140
	14,000			30	6.131
Ag_2GeSe_3	27	2	850 (electrons)		6.132
Ag_2GeTe_3	92	8	720		6.132
Ag_2SnS_3	3				6.132
Ag_2SnSe_3	146	10	910		6.132
Ag_2SnTe_3	48	5	600		6.132

[a] High-temperature allotrope.

material among I_2-IV-VI_3 compounds that has the n-type conductivity in solid state and p-type conductivity in the liquid state. At the melting point (or at the liquidus temperature of the alloy), the Seebeck coefficient drop between the liquid and solid parts of the alloy is close

Table 6.26 I_2-IV-VI_3 Compounds. Room Temperature Thermal Conductivity

Compound	Cu_2SiS_3	Cu_2GeS_3	Cu_2GeSe_3	Cu_2GeTe_3	Cu_2SnS_3	Cu_2SnSe_3	Cu_2SnTe_3
Thermal conductivity (mW/cm·K)	23a [6.140] 12b [6.140]	7.7 [6.135] 12 [6.140]	6.9 [6.135] 24 [6.140]	130 [6.140]	7.33 [6.135] 28 [6.140]	6.6 [6.135] 35 [6.140]	144 [6.140]

a High-temperaure allotrope.
b Low-temperature allotrope.

Table 6.27 I_2-IV-VI_3 Compounds. Magnetic Susceptibility at Room Temperature

	Magnetic susceptibility		
Compound	Specific (-10^{-5} SI units/g)	Molar (-10^{-3} SI units/g-mol)	Atomic (-10^{-4} SI units/g-atom)
Cu_2GeS_3	4.78	1.41	2.35
Cu_2SnS_3	4.02	1.37	2.28
Cu_2GeSe_3	2.64	1.15	1.92
Cu_2SnSe_3	3.27	1.58	2.63
Cu_2GeTe_3	3.39	1.98	3.30
Cu_2SnTe_3	3.39	2.13	3.55
Ag_2GeSe_3	4.27	2.24	3.73
Ag_2SnSe_3	3.90	2.23	3.72
Ag_2GeTe_3	3.52	2.36	3.93
Ag_2SnTe_3	3.27	2.34	3.90

From L.I. Berger, V.I. Chechernikov, A.E. Balanevskaya, A.V. Pechennikov, and I.K. Shchukina, *Trudy IREA (Proc. IREA)*, 31, 435, 1969.

to 0.9 mV/K.[6.146] The temperature dependence of the electrical conductivity and Seebeck coefficient of Ag_2GeSe_3 is shown in Figure 6.25 (adapted from Reference 6.144). The available data on electrical conductivity and the concentration and mobility of the current carriers are presented in Table 6.25.

The thermal conductivity of the I_2-IV-VI_3 compounds was reported in References 6.135 and 6.140. In both reports, the samples were polycrystalline, but the authors of Reference 6.140 added a certain excessive amount of the Group VI component in the mixture of components prepared for synthesis to compensate the losses related to the high volatility of chalcogenes. The results of the measurements are compiled in Table 6.26.

The data on magnetic properties of a group of I_2-IV-VI_3 compounds were reported in Reference 6.68. All investigated materials are diamagnetics. The room temperature magnitudes of magnetic susceptibility are compiled in Table 6.27. The dielectric permeability and optical properties of the I_2-IV-VI_3 compounds have not been studied systematically, first of all, because of the lack of availability of relatively large single crystals. Photoelectric properties of Cu_2SnSe_3 single crystals were reported in Reference 6.147; the photoconductivity and photo-emf of the Cu_2SnS_3 polycrystals were studied by the authors of Reference 6.132 (see Figure 6.20).

In the process of investigation of the I_2-IV-VI_3 compounds,[6.110] two phases, Cu_2CeSe_3 and Cu_2CeTe_3, were synthesized and investigated. Both materials are semiconductors with the energy gap, evaluated from the electrical conductivity and spectral distribution of the diffuse reflection, equal to 0.22 eV for Cu_2CeSe_3 and 0.40 eV for Cu_2CeTe_3. The semiconductor materials with lanthanides present a certain theoretical and practical interest; participation of transition elements in compounds may provide interesting combinations of electrophysical and, especially, optical properties (see, e.g., Reference 6.13, p. 89 to 91).

Table 6.28 I_3-V-VI$_4$ Compounds. Molecular Weight, Average Atomic Weight and Number, Crystal Structure, and Lattice Parameters

Compound	Molecular weight	Mean atomic weight	Mean atomic number	Symmetry group	Lattice parameters a (nm)	b (nm)	c (nm)	V_o (nm³)	Ref.
Cu_3PS_4	349.85	43.73	20.75	C_{2v}^7-Pnm2$_1$	0.7296	0.6319	0.6072		6.13
Cu_3AsS_4	393.80	49.23	23	C_{2v}^7-Pnm2$_1$	0.646	0.743	0.618	0.2965	6.154, 6.155
					0.6433	0.7404	0.6154		6.13
					0.6406	0.7399	0.6140		6.161
				D_{2d}^{11}-I$\bar{4}$2m	0.5289		1.0440		6.161
					0.5290		1.0465		6.5
Cu_3AsSe_4	581.40	72.68	32	D_{2d}^{12}-I$\bar{4}$2d	0.550		1.0966		6.161
					0.5570		1.0957		6.5, 6.13
				T_d^2-F$\bar{4}$3m	0.5535^a				6.161
Cu_3SbS_4	440.63	55.08	25.25	O_h^5-Fm3m	0.528				6.162
				T_d^2-F$\bar{4}$3m	1.075				6.162
				D_{2d}^{11}-I$\bar{4}$2d	0.5382		1.0957		6.161
Cu_3SbSe_4	628.23	78.53	34.25	D_{2d}^{11}-I$\bar{4}$2d	0.5654		1.1256		6.163
					0.556		1.095		6.161
				O_h^5-Fm3m	0.5637^a				6.161
Cu_3VS_4	369.82	46.23	21.75	O_h^5-Fm3m	1.075^b				6.149
				T_d^1-P$\bar{4}$3m	0.537				6.148
Cu_3VSe_4	557.42	69.68	30.75	O_h^5-Fm3m	0.5569				6.163

^{a} High-temperature phase. The transition temperature to the chalcopyrite phase is 715 K and 698 K for Cu_3AsSe_4 and Cu_3SbSe_4, respectively.[6.161]
^{b} Z = 8.

6.2.3. I_3-V-VI$_4$ Compounds

The I_3-V-VI$_4$ compounds are located in the concentration triangle I-V-VI on the intersection of the "normal valency" line, I_2-VI to V_6-VI$_5$, and the "four-valence electrons-per-atom" line, I-V$_3$ to I_2VI$_3$. In principle, the I-V-VI system may contain the ternary analogs of the elements of the third group, I_{19}-V-VI$_{12}$, and the fifth group, I_5-V$_7$-VI$_{20}$. Unfortunately, the author of this book does not have information about the phase relations in the sections I_2-VI to V_6-VI$_5$ of the systems (Cu, Ag)-(P, As, Sb)-(S, Se, Te).

Nature is generous regarding the I_3-V-VI$_4$ substances: the minerals Cu_3PS_3,[6.148] Cu_3VS_4 (sulvanite), Cu_3AsS_4 (lusonite, enargite), and Cu_3SbS_4 (famatinite) are quite abundant. The crystal structure of enargite and famatinite was first described by de Jong[6.149] and Herman et al.[6.150] (see, also, References 6.151 to 6.156). Crystals of Cu_3AsS_4, Cu_3SbS_4, and some other I_3-V-VI$_4$ compounds were synthesized first by Wernick and Benson[6.156] (see also U.S. Patent No. 2,882,192). The data on crystal structure and lattice parameters of the compounds are gathered in Table 6.28.

The mineral sulvanite is not the only known I_3-V-VI$_4$ compound containing a transition element as a component. Hulliger[6.164] synthesized the compounds of this type with vanadium, niobium, and tantalum (sulfides, selenides, and tellurides). All synthesized materials are semiconductors.

According to the results reported in Reference 6.156, the melting temperatures of Cu_3AsS_4 and Cu_3SbS_4 are 928 K and 828 K, respectively. The phase diagrams of the I_3-V-VI$_4$ have not been studied in detail, so the data in Table 6.29 on melting points may refer as well to the temperatures of the beginning of peritectic decomposition. Some available thermodynamic properties of the compounds are also included in the table.

The information regarding mechanical properties of I_3-V-VI$_4$ compounds is presented in Table 6.30.

Table 6.29 I$_3$-V-VI$_4$ Compounds. Density, Melting Temperature, Lattice Energy, Specific Heat, Coefficient of Thermal Expansion and Debye Temperature

Compound	Density (g/cm³) R.T. (pycnometric)	Melting point (K)	Lattice energy (kJ/mol)	Specific heat, R.T. (J/g·K)	Coefficient of thermal exp. (10⁻⁶ K⁻¹)	Debye temperature (K)
Cu$_3$AsS$_4$	4.478[a] (lusonite) 4.37	931[6.13] 928[6.156] 923[6.137]	269.9[6.137]	0.51[6.34]	3.2[6.137]	
Cu$_3$AsSe$_4$	5.74	7.33[6.137]	161.5[6.137]		9.5[6.137]	169[6.6]
Cu$_3$SbS$_4$	4.687[a] (famatinite) 4.90 (X-ray)	828[6.156]				
Cu$_3$SbSe$_4$	6.0	721[6.6] 723[6.137] 700[6.13]	127.2[6.137]	0.32[6.34]	12.4[6.137]	131[6.6]

[a] Minerals.

Table 6.30 I$_3$-V-VI$_4$ Compounds. Hardness and Elastic Parameters at Room Temperature

Compound	Hardness (Mohs)	Vickers microhardness (GPa)	Speed of ultrasound waves (km/sec)		Modulus (GPa)	
			Longitudinal	Transversal	Young's	Shear
Cu$_3$AsS$_4$	3	2.00 ± 0.05	5.5	3.3	132	47.5
Cu$_3$AsSe$_4$		3.00 ± 0.10	3.5	2.1	70	25.2
Cu$_3$SbS$_4$	3–4					
Cu$_3$SbSe$_4$		2.00 ± 0.10	3.0	1.8	53	19

From L.S. Palatnik, V.M. Koshkin, L.P. Galchinetskii, V.I. Kolesnikov, and Yu.F. Komnik, *Sov. Phys. Solid State*, 4, 1430, 1962.

Table 6.31 Energy Gap of I$_3$-V-VI$_4$ Compounds

Compound	Energy gap (eV)		
	From electrical measurements	From optical measurements at 77 K[6.14]	at 300 K
Cu$_3$PS$_4$	2[6.13]		
Cu$_3$AsS$_4$	0.80 ± 0.08[6.137]	1.48[a]	1.24[a 6.141]
Cu$_3$AsSe$_4$	0.76[6.6]	0.65[a]	0.88[b 6.6] 0.86[6.166] 0.74[a 6.141]
Cu$_3$SbSe$_4$	0.37 ± 0.04[6.137] 0.13[6.13] 0.42[6.6]		0.31[a 6.6, 6.141]

[a] From photoconductivity measurements.
[b] From diffuse optical reflection measurements on powder samples.

The band structure of the I$_3$-V-VI$_4$ compounds is virtually unknown. The available data relate only to the energy gap. The electrical and thermal transport properties have been studied mainly on polycrystals, so the magnitudes of the carrier mobility and thermal conductivity are apparently lower than the ones expected for single-crystalline samples. The available data on the energy gap are compiled in Table 6.31. From the table, one can see that Cu$_3$AsS$_4$ presents a certain interest for some of the currently designed optical and electro-optical devices.

The temperature dependence of electrical conductivity of the samples Cu$_3$AsSe$_4$ and Cu$_3$SbSe$_4$ + 1.0% Zn is shown in Figures 6.26 and 6.27. The data on the Hall constant vs.

Table 6.31 Energy Gap of I_3-V-VI_4 Compounds

Compound	Energy gap (eV)		
	From electrical measurements	From optical measurements	
		at 77 K[6.14]	at 300 K
Cu_3PS_4	2[6.13]		
Cu_3AsS_4	0.80 ± 0.08[6.137]	1.48[a]	1.24[a 6.141]
Cu_3AsSe_4	0.76[6.6]	0.65[a]	0.88[b 6.6]
			0.86[6.166]
Cu_3Sb_4			0.74[a 6.141]
Cu_3SbSe_4	0.37 ± 0.04[6.137]		0.31[a 6.6,6.141]
	0.13[6.13]		
	0.42[6.6]		

[a] From photoconductivity measurements.
[b] From diffuse optical reflection measurements on powder samples.

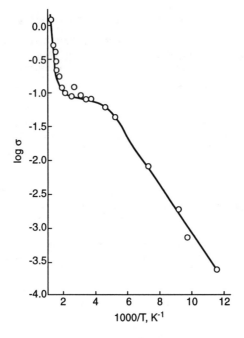

Figure 6.26 Dependence of electrical conductivity (in S/cm) of Cu_3AsS_4 on reciprocal temperature. (After R. Annamamedov, L.I. Berger, V.M. Petrov, and V.S. Slobodchikov, *Inorg. Mater.*, 3, 1370, 1967. With permission.)

1000/T are displayed in Figures 6.28 and 6.29 (adapted from Reference 6.161). The scarce information regarding the electrical and transport parameters is compiled in Table 6.32.

The temperature dependence of the thermal conductivity of a sample of Cu_3AsSe_4 and a sample of 99%Cu_3AsSe_4.1%Zn is exhibited in Figure 6.30 (adapted from Reference 6.161).

The Cu_3-V-VI_4 compounds are diamagnetics; their magnetic susceptibility is presented in Table 6.33.

The optical absorption coefficient, spectral distribution of the diffuse reflection, and spectral distribution of photoconductivity for I_3-V-VI_4 compounds were reported in References 6.161, 6.165, and 6.166. The photoconductivity of Cu_3AsS_4 at 77 K and 293 K is shown in Figure 6.31 (adapted from Reference 6.161); the photoconductivity of Cu_3AsSe_4

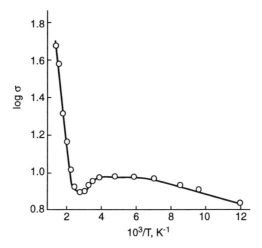

Figure 6.27 Dependence of electrical conductivity (in S/cm) of a Cu_3SbSe_4 sample doped with 1% Zn on temperature. (After R. Annamamedov, L.I. Berger, V.M. Petrov, and V.S. Slobodchikov, *Inorg. Mater.*, 3, 1370, 1967. With permission.)

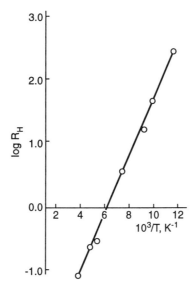

Figure 6.28 Logarithm of the Hall constant (R_H in cm^3/C) vs. reciprocal temperature for a Cu_3AsS_4 sample. (After R. Annamamedov, L.I. Berger, V.M. Petrov, and V.S. Slobodchikov, *Inorg. Mater.*, 3, 1370, 1967. With permission.)

at 77 K vs. wavelength is displayed in Figure 6.32 (adapted from Reference 6.165). The spectral distribution of the coefficient of diffuse reflection and the absorption coefficient of Cu_3AsSe_4 at room temperature is exhibited in Figure 6.33 (adapted from Reference 6.166). The diffuse reflection vs. wavelength for the Cu_3AsSe_4 and Cu_3SbSe_4 samples at room temperature is shown in Figure 6.34.

There are other known I_3-V-VI$_4$ compounds, such as lithium orthophosphate, Li_3PO_4, and lithium orthoaresenate, Li_3AsO_4, which are not considered in this monograph, first of all, because the information relevant to them is very limited (see, e.g., Reference 6.11, p. 4.69 and 4.70; and Reference 6.5, p. 158), and second, they apparently do not belong to the materials of the adamantine group.

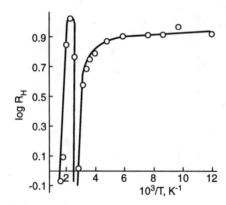

Figure 6.29 Logarithm of the Hall constant (R_H in cm³/C) vs. reciprocal temperature for a Cu_3SbSe_4 sample doped with 1% Zn. (After R. Annamamedov, L.I. Berger, V.M. Petrov, and V.S. Slobodchikov, *Inorg. Mater.*, 3, 1370, 1967. With permission.)

Table 6.32 I_3-V-VI_4 Compounds. Room Temperature Electrical and Thermal Transport Parameters

Compound	Conductivity type	Mobility (cm²/V·sec)	Effective mass (m_o units)	Carrier concentration (cm⁻³)	Thermal conductivity (mW/cm·K)
Cu_3AsS_4	n	0.008[6.13]		8·10¹⁹	30.1[6.26]
Cu_3AsSe_4	n	505[6.13]	$m_e^* = 0.22$[6.141]	2.7·10¹⁸	19.2[6.26]
					34.0[6.161]
Cu_3SbSe_4	p	50–60[6.13]	$m_p^* = 0.73$[6.13]	1·10¹⁹	14.6[6.26]

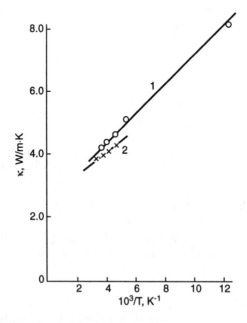

Figure 6.30 Temperature dependence of thermal conductivity in Cu_3AsSe_4. (1) An undoped sample; (2) a sample doped with 1% Zn. (After R. Annamamedov, L.I. Berger, V.M. Petrov, and V.S. Slobodchikov, *Inorg. Mater.*, 3, 1370, 1967. With permission.)

Table 6.33 I_3-V-VI_4 Compounds. Room Temperature Magnetic Susceptibility

Compound	Magnetic susceptibility		
	Specific (10^{-6} SI units/g)	Molar (10^{-4} SI units/g-mol)	Atomic (10^{-4} SI units/g-atom)
Cu_3AsS_4	−4.02	−15.84	−1.98
Cu_3AsSe_4	−2.26	−13.14	−1.64
Cu_3SbS_4	−1.88	−8.31	−1.04
Cu_3SbSe_4	−3.02	−20.57	−2.57

From L.I. Berger, V.I. Chechernikov, A.E. Balanevskaya, A.V. Pechennikov, and I.K. Shchunkina, *Trudy IREA (Proc. IREA)*, 31, 435, 1969.

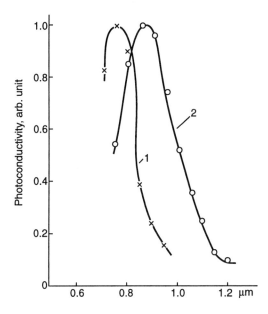

Figure 6.31 Spectral distribution of photoconductivity (photocurrent) for a Cu_3AsS_4 sample at 77 K (1) and 293 K (2). (After R. Annamamedov, L.I. Berger, V.M. Petrov, and V.S. Slobodchikov, *Inorg. Mater.*, 3, 1370, 1967. With permission.)

6.3. DEFECT ADAMANTINE SEMICONDUCTORS

There are two groups of binary compounds which have a diamond-like crystal structure: (1) the IV_3-V_4 compounds and (2) III_2-VI_3 compounds. The first group is presented by two nitrides, Si_3N_4 and Ge_3N_4; the second group includes a relatively large group of compounds (Al, Ga, In)$_2$ (S, Se, Te)$_3$ with the crystal lattice similar to wurtzite or sphalerite. The compounds of these two types may be considered as ternary compounds (vac)-IV_3-V_4 and (vac)-III_2-VI_3 (symbol "(vac)" denotes the vacant positions in the cation part of their crystal lattice); the former are the electron analogs of III-V-type compounds, while the latter are the analogs of the II-VI-type compounds.

6.3.1. IV_3-V_4 Compounds

In the cation sublattice of the IV_3-V_4 compounds, one quarter of the positions is vacant. The available data on the properties of Ge_3N_4 and Si_3N_4 are gathered in Table 6.34.

Stoichiometric Si_3N_4 is an insulator with the resistivity of 10^{12} Ohm·cm to 10^{14} Ohm·cm.[6.167] Si_3N_4 exists in two crystalline forms (α-Si_3N_4 and β-Si_3N_4) and in amorphous state.[6.168-6.170]

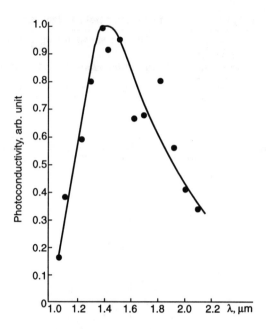

Figure 6.32 Spectral distribution of photoconductivity for a Cu_3AsSe_4 sample at 77 K. (After R. Annamamedov, L.I. Berger, V.M. Petrov, and F.F. Kharakhorin, *Inorg. Mater.*, 2, 772, 1966. With permission.)

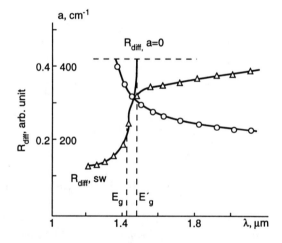

Figure 6.33 Diffusive reflection coefficient, R_{diff}, (triangles) and absorption coefficient, a, (circles) vs. wavelength for powder and thin films of Cu_3AsSe_4 at room temperature. $R_{diff,\,sw}$ — "shortwave" diffusive reflection coefficient a = 0; E_g — estimated magnitude of the energy gap; E'_g — a corrected magnitude of the energy gap. (After L.I. Berger and V.M. Petrov, *Inorg. Mater.*, 7, 1905, 1971. With permission.)

While Ge_3N_4 has not found a wide application, silicon nitride occupies an important position among ceramic materials (in particular, in electronic industry) because of its low density, resistance to reaction with molten metals, high refractory properties, and extremely low coefficient of thermal expansion that gives to Si_3N_4 a high resistance to thermal shock, high hardness, and elastic constants.

The temperature dependence of the Vickers microhardness of polycrystalline β-Si_3N_4 (1-kg load) is shown in Figure 6.35 (adapted from Reference 6.173). The dependence of the

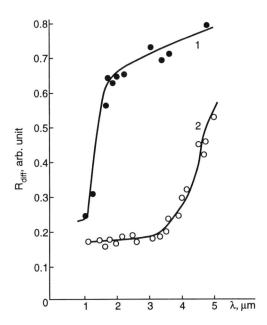

Figure 6.34 Spectral distribution of the diffusive reflection coefficient for Cu_3AsSe_4 (1) and Cu_3SbSe_4 (2) at room temperature. (After R. Annamamedov, L.I. Berger, V.M. Petrov, and F.F. Kharakhorin, *Inorg. Mater.*, 2, 772, 1966. With permission.)

microhardness on load is presented in Figure 6.36 (adapted from Reference 6.175). The transition from α- to β-Si_3N_4 takes place at about 1970 K (see, e.g., Reference 6.173). There is a substantial variation of the data on Young's modulus reported in different publications. Tanaka et al.[6.174] cite a magnitude between 308 and 312 GPa at room temperature; Unal et al.[6.173] indicate a high-temperature (1400°C) Young's modulus of 312 GPa.

Applications of Si_3N_4 are as an insulating diffusion mask in microelectronic technology; the nonelectronic applications are as a matrix material for ceramic composites and a material for machine tools, bearings, and some engine parts. The utilization of many potential applications is still somewhat limited by the lack of toughness of the silicon nitride films and reliability of the technology of their preparation.[6.177]

6.3.2. III$_2$-VI$_3$ Compounds

As was mentioned before, some of the III$_2$-VI$_3$ compounds may be considered as the crystallochemical analogs of II-VI semiconductors. Their crystal lattice is similar to that of sphalerite or wurtzite, but, as has been first discovered by Hahn and Klinger,[6.180] one third of the positions in the cation sublattice of these compounds is vacant. The valence electron concentration in these compounds is 4.80 electrons per atom; however, if we consider the vacancies in the lattice as the particles with zero valency (this idea was discussed in some detail by Suchet[6.2]), the formula of the compound type is (vac)$_1$-III$_2$-VI$_3$ and the valence electron concentration becomes equal to four, like in adamantine family materials. If the vacancies are distributed in the cation sublattice statistically, the broken bonds in a part of the individual VI-III$_4$ tetrahedrons in the unit cell do not distort the symmetry of the cell and the lattice has the structure of either wurtzite type with space group C_{6v}^4 – P6$_3$mc (β-Ga_2S_3) or sphalerite type with space group T_d^2-F$\bar{4}$3m (Ga_2Te_3 and β-In_2Te_3). Vacancy ordering changes crystal structure type to tetragonal, orthorhombic, monoclinic, etc. The data on crystal structure of III$_2$-VI$_3$ compounds are compiled in Table 6.35, which also contains compounds that do not belong to the class of adamantine-type substances.

Table 6.34 Properties of IV$_3$-V$_4$ Compounds

Compound	Molecular weight	Mean atomic weight	Mean atomic number	Density (g/cm^3, 300 K)	Crystal structure	Lattice parameters (nm)[6.172]	Melting point (K)	Dielectric constant at 6 GHz	Coef. of linear thermal expansion (10^{-6} 1/K)	Young's modulus (GPa)
α-Si$_3$N$_4$	140.28	17.54	8.75	3.44	Cubic	0.39	2170 (under pressure)	8.2[6.171]	3.01 (∥a)	544[a][6.179]
β-Si$_3$N$_4$					Hexagonal	a = 0.7607 ± 0.0005 c = 0.2909 ± 0.0005			3.37 (∥c)[6.178]	
Ge$_3$N$_4$	273.80	34.23	15.5	5.25	Cubic	0.42	720 (decompos.)			

[a] Poisson's ratio is equal to 0.24.[6.179]

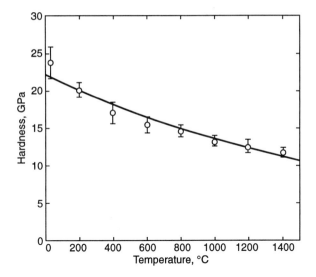

Figure 6.35 Temperature dependence of the Si_3N_4 hardness measured on monolithic samples. (After O. Unal, J.J. Petrovic, and T.E. Mitchell, *J. Mater. Res.*, 8, 626, 1993. With permission.)

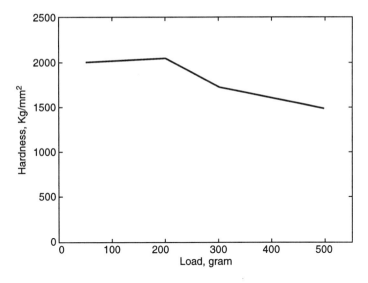

Figure 6.36 Correlation between the microhardness of a Si_3N_4 crystalline film and the load magnitude at room temperature. (From K.J. Grannen, F. Xiong, and R.P. Chang, *J. Mater. Res.*, 9, 2341, 1994. With permission.)

A typical feature of the III_2-VI_3 compounds is existence of many allotropes. It can be a consequence of the "structural" defects which may create a specific stress in the lattice of the compounds at different conditions. In part of the compounds, the allotropes differ only with ordering (in the cation sublattice). The crystal structure of some III_2-VI_3 compounds and, first of all, location of the structural defects can be seen in Figures 6.37 to 6.39.

The information regarding the phase diagrams of some III-VI systems is gathered in Reference 6.183. The data on melting point, density, microhardness, and enthalpy of formation of III_2-VI_3 compounds are shown in Table 6.36.

Table 6.35 Crystal Structure of III$_2$-VI$_3$ Compounds

Compound	Molecular weight	Mean atomic weight	Mean atomic number	Crystal structure	Lattice parameters a (nm)	b (nm)	c (nm)	β (deg.)	Ref.
Al$_2$O$_3$	101.96	20.39	10.0						
Corundum				Rhombohedral, R$\bar{3}$c	0.47591		1.29894		6.11, p. 4.146
α-Al$_2$O$_3$				Rhombohedral	0.513			55.1	6.5
γ-Al$_2$O$_3$				Spinel	0.790				6.5
δ'-Al$_2$O$_3$				Same, with statistic distribution of cations among tetrahedral and octahedral positions	0.395				6.5
Al$_2$S$_3$	150.14	30.03	14.8						
α-Al$_2$S$_3$				Wurtzite (ordered)	0.6423		1.783		6.5
β-Al$_2$S$_3$ (α to β at 1173 K)				Wurtzite (disordered)	0.358		0.583		6.5
γ-Al$_2$S$_3$				Rhombohedral, α-Al$_2$O$_3$ type, Or	0.686			56.27	6.5
				γ-Al$_2$O$_3$ type	0.993				6.5
α-Al$_2$Se$_3$	290.84	58.17	25.6	Hexagonal, wurtzite type	0.389		0.630		6.5
α-Al$_2$Te$_3$	436.76	87.35	36.4	Hexagonal, wurtzite type	0.4078		0.6944		6.5
Ga$_2$S$_3$	235.62	47.12	22.0						
α-Ga$_2$S$_3$				Monoclinic, C$_s^4$-Bb, ordered	1.14	0.641	0.7038	121.22	6.5
					1.1094	0.9578	0.6395	141.25	6.13
β-Ga$_2$S$_3$				Hexagonal, wurtzite type, C$_{6v}^4$-P6$_3$mc					
				Ordered	0.638		1.809		6.5
				Disordered	0.3685		0.6028		6.5
					0.3678		0.6016		6.13

Compound				Structure				
Ga$_2$Se$_3$	376.32	75.26	32.8	Cubic	0.542			6.182
α-Ga$_2$Se$_3$					0.543			6.5
β-Ga$_2$Se$_3$ (α-β trans. 873 K)				Tetragonal, D_{4h}^{20}-I4$_1$/acd	0.5487		0.5411	6.5, 6.188
					2.3235		1.0828	6.13
γ-Ga$_2$Se$_3$ (β-γ trans. 1073 K)				Orthorhombic	0.7760	1.1640	1.082	6.5
δ-Ga$_2$Se$_3$				Sphalerite type T_d^2-F$\bar{4}$3m	0.5463			6.5
Ga$_2$Te$_3$	522.24	104.45	43.6	Sphalerite type	0.5887			6.5
					0.5899			6.13
In$_2$S$_3$	325.82	65.16	29.2					
α-In$_2$S$_3$				Cubic, similar to γ-Al$_2$O$_3$	0.536			6.5
β-In$_2$S$_3$ (α-β trans. 573 K)				Tetragonal, D_{4h}^{19}-I4$_1$/amd	0.762		3.232	6.5
					0.7618		3.233	6.13
				D_{3d}^3-P$\bar{3}$m1	0.38		0.904	6.13
In$_2$Se$_3$	466.52	93.30	40	Layered structure, many phases	0.4025		9.6–28.5	6.13
α-In$_2$Se$_3$				Hexagonal	1.600		1.924	6.5
β-In$_2$Se$_3$ (α-β trans. 473 K)				Wurtzite type	0.711		1.93	6.5
γ-In$_2$Se$_3$ (β-γ trans. 923 K)				Sphalerite type	1.010			6.5
In$_2$Te$_3$	612.44	122.49	50.8					
α-In$_2$Te$_3$				Sphalerite type, ordered	1.840			6.5
β-In$_2$Te$_3$ (α-β trans. 873 K)				Cubic, disordered	0.616			6.5
α-In$_2$Te$_3$				T_d^2-F$\bar{4}$3m	0.6163			6.13
Phase I				Tetragonal, D_{2d}^{11}-I$\bar{4}$2m	0.6173	1.307	1.2438	6.13
Phase II				C_{2v}^{20}-Imm2	0.4358		0.6163	6.13

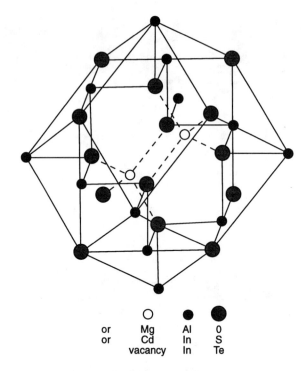

	○	●	●
or	Mg	Al	O
or	Cd	In	S
	vacancy	In	Te

Figure 6.37 Spinel-type crystal structure.

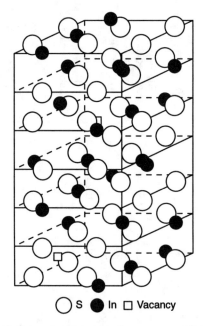

○ S ● In □ Vacancy

Figure 6.38 Unit cell of beta-In_2S_3 (space group D_{4h}^{19}-$I4_1/amd$).

The assumption that many of the III_2-VI_3 compounds are semiconductors was first discussed by Goryunova.[6.186] This assumption was validated by the results of the investigation of the photoelectric properties of these materials reported in Reference 6.187.

An interesting specifics of the electronic properties of these compounds consists of a relatively low sensitivity of the properties to deviation from the stoichiometric composition

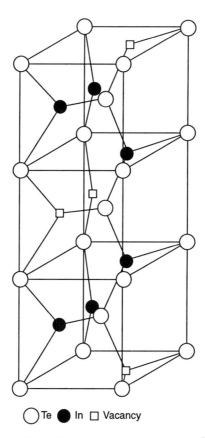

Figure 6.39 Unit cell of alpha-In$_2$Te$_3$ lattice, phase 2 (space group C_{2v}^{20}-Imm2). (After G.L. Bleris, T. Karakostas, N.A. Economou, and R. deRidder, *Phys. Stat. Sol.*, A50, 579, 1978. With permission.)

Table 6.36 III$_2$-VI$_3$ Compounds. Melting Point, Density, Microhardness, and Enthalpy of Formation

Compound	Melting point (K)[6.5]	Density at 300 K (g/cm^3)[6.5]	Vickers microhardness (GPa) (50-g load)[6.5]	Enthalpy of formation (kJ/mol)
α-Al$_2$S$_3$	1403	2.5		
α-Al$_2$Se$_3$	1253	3.9		
Al$_2$Te$_3$	1168	4.5		
β-Ga$_2$S$_3$	1523	3.63	5.00	
	1360[6.13]			
α-Ga$_2$Se$_3$	1293	4.92	3.16	
(or δ-Ga$_2$Se$_3$?)	1278[6.13]			
	1295[6.182]		3.57 ± 0.09[6.182]	
α-Ga$_2$Te$_3$	1063	5.57	2.37	239 ± 30[6.184]
				272[6.183]
β-In$_2$S$_3$	1373	4.63	2.80	
	1363[6.13]			
α-In$_2$Se$_3$	1163	5.48	0.3–0.5	
α-In$_2$Te$_3$		5.79		172[6.184]
β-In$_2$Te$_3$	940	5.73	1.66	199[6.185]

and presence of impurities (Reference 6.5, p. 168) and their anomalously high radiation stability (see Reference 6.182). Some data on electrical and thermal transport parameters of III$_2$-VI$_3$ compounds are presented in Table 6.37.

Table 6.37 III$_2$-VI$_3$ Compounds. Energy Gap and Electrical and Thermal Transport Properties

Compound	Energy gap (eV) Thermal	Energy gap (eV) Optical at 300 K	Cond. type	Resistivity (Ohm·cm) (RT)	Mobility (cm^2/V·sec)	Thermal conductivity (mW/cm·K) (300 K)	Comments
α-Al$_2$S$_3$		4.1 [6.5]	n				
α-Al$_2$Se$_3$		3.1 [6.5]	n				
Al$_2$Te$_3$		2.2 [6.5]	n				
α-Ga$_2$S$_3$		3.438 [6.13]					At 1.6 K
		3.424 [6.13]					At 77 K
β-Ga$_2$S$_3$		2.6–2.9 [6.13]	n		28		At 77 K
		2.5–2.7 [6.5]					
α-Ga$_2$Se$_3$		1.75 [6.5]	p		10 [6.5]	5.1 [6.5]	
			n	4.5·10^{11} [6.182]			
β-Ga$_2$Se$_3$		1.90 [6.5]					
		2.1–2.2 [6.13]					Direct, 77 K
		1.91–2.1 [6.13]					Direct, 290 K
		2.04 [6.13]					Indirect, 77 K
		1.96 [6.13]					Indirect, 300 K
α-Ga$_2$Te$_3$	1.55 [6.13]	1.22 [6.13]		10^1–10^7 [6.13]		4.7 [6.5]	
		1.35 [6.5]	p		50 [6.5]		
β-In$_2$S$_3$		2.03 [6.5]	n				Direct
	2.30 [6.13]	1.1 [6.13]					Indirect
γ-In$_2$S$_3$:As	1.38	1.1 [6.13]					Indirect
γ-In$_2$S$_3$:Sb	1.42	2.03 [6.13]					Direct
α-In$_2$Se$_3$	1.41 [6.13]	1.2–1.5(⊥)	n		10 [6.13]	10.5 [6.5]	
		1.2 [6.5]	n		125 [6.5]		
γ-In$_2$Se$_3$		1.0 [6.5]					
α-In$_2$Te$_3$	1.02–1.55	0.92–1.15	n		5–70	11.2 [6.5]	Ref. 6.13 $m_n^* = 0.7\, m_o$ [6.13] $m_p^* = 1.23\, m_o$ [6.13]
β-In$_2$Te$_3$		1.04	n		50	6.95	Ref. 6.5

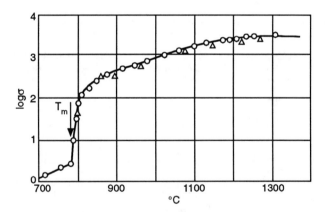

Figure 6.40 Temperature dependence of electrical conductivity (in S/cm) of Ga$_2$Te$_3$ in the solid and liquid states. (After V.M. Glazov, S.N. Chizhevskaya, and N.N. Glagoleva, *Liquid Semiconductors*, Nauka Publishing House, Moscow, 1967. With permission.)

The temperature dependence of the electrical conductivity, magnetic susceptibility, and Seebeck coefficient of Ga$_2$Te$_3$ and In$_2$Te$_3$ in solid and liquid states was reported by Glazov et al.[6.185] Their results are shown in Figures 6.40 to 6.43. As can be seen from the diagrams, both substances retain the semiconductor-type conductivity in the liquid state.

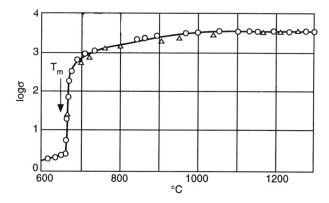

Figure 6.41 Temperature dependence of electrical conductivity (in S/cm) of In_2Te_3 in the solid and liquid states. (After V.M. Glazov, S.N. Chizhevskaya, and N.N. Glagoleva, *Liquid Semiconductors*, Nauka Publishing House, Moscow, 1967. With permission.)

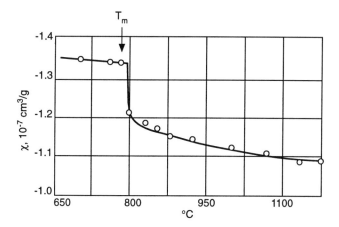

Figure 6.42 Specific magnetic susceptibility (Gauss units) of Ga_2Te_3 in solid and liquid states. (After V.M. Glazov, S.N. Chizhevskaya, and N.N. Glagoleva, *Liquid Semiconductors*, Nauka Publishing House, Moscow, 1967. With permission.)

The available data on the dielectric constant and index of refraction of III_2-VI_3 compounds are collected in Table 6.38.

6.3.3. Defect Ternary Compounds

As can be seen from Equation 1.1 (Chapter 1), the maximum number of components, which belong to different groups of the Periodic table and form a compound with a certain average valence electron concentration, Z_{av}, is equal to three. The binary defect phases III_2-VI_3 described in the previous paragraph may, as was mentioned there, be considered as the adamantine-type substances of composition $(vac)_1$-III_2-VI_3 and $Z_{av} = 4$. Consequently, any combination of elemental, binary, or ternary substances that belong to the adamantine family with a binary defect phase will be a phase (a compound, an ordered or disordered solid solution, etc.) that also belongs to the same family, but with the number of components that belong to the different groups (including the vacancies) greater than three. Of course, formation of a single-phase substance from a particular set of chemical elements has to be determined also by the crystallochemical and thermodynamic principles. It is not difficult to

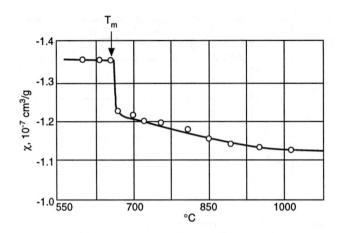

Figure 6.43 Specific magnetic susceptibility (Gauss units) of In_2Te_3 in the solid and liquid states. (After V.M. Glazov, S.N. Chizhevskaya, and N.N. Glagoleva, *Liquid Semiconductors,* Nauka Publishing House, Moscow, 1967. With permission.)

Table 6.38 III_2-VI_3 Compounds. Dielectric Constant and Index of Refraction

Compound	Dielectric constant		Index of refraction $(n = \epsilon_{opt}^{1/2})$	Ref.
	Static limit	Optical limit		
In_2S_3	13.5	6.5	2.55	6.13
In_2Se_3	16.68	9.53	3.09	6.13
In_2Te_3	16			6.13

Table 6.39 Types of Some Binary Compounds and Their Structural Vacancy Containing Ternary Analogs

No.	Binary compound type	Z_{av}	Ternary analog	Z_{av}
1	II-VII_2	5.33	$(vac)_1$-II-VII_2	4.0
2	III-VII_3	6.0	$(vac)_2$-III-VII_3	4.0
3	IV-VII_4	6.4	$(vac)_3$-IV-VII_4	4.0
4	V-VII_5	6.67	$(vac)_4$-V-VII_5	4.0
5	III_2-VI_3	4.8	(vac)-III_2-VI_3	4.0
6	IV-VI_2	5.3	(vac)-IV-VI_2	4.0
7	V_2-VI_5	5.71	$(vac)_3$-V_2-VI_5	4.0
8	IV_3-V_4	4.57	(vac)-IV_3-V_4	4.0

see that any of the binary compound types listed in Table 6.39 may be considered as the ternary compounds with $Z_{av} = 4$ and the structure vacancies as a part of the compound content.

Combination of the binary compounds of the types listed in Table 6.39 with some binary adamantine compounds having common anion (or cation) with them may lead to formation of ternary defect compounds, i.e., compounds with the stoichiometric vacancies in their crystal lattice. Goryunova (Reference 6.4, p. 152 and further) was the first one who predicted existence of the III_3-V-VI_3-type phases; Hahn and Klinger[6.189] pioneered the investigation of the II-III_2-VI_4-type compounds. The type of ternary defect compounds that are known to have synthesized representatives are listed in Table 6.40 (Reference 6.6, p. 79).

The crystal structures of II-III_2-VI_4 compounds are shown in Figure 6.44 (adapted from Reference 6.6).

The limits of this book do not allow a detailed analysis of the available data on the properties of the ternary defect compounds. The number of single-phase substances of the

Table 6.40 Some Ternary Defect Phases and Compounds Forming These Phases[6.5,6.6,6.13]

No.	Valence electron concentration	Ternary phase		Components of the phase	
1	4.57	I_2-II-VII_4	Ag_2HgI_4	2(I-VII) + (II-VII_2)	$2AgI + HgI_2$
2	4.57	II-III_2-VI_4	$ZnIn_2S_4$[a]	(II-VI) + (III_2-VI_3)	$ZnS + In_2S_3$
3	4.57	II_2-IV-VI_4	Zn_2GeS_4	2(II-VI) + (IV-VI_2)	$2ZnS + GeS_2$
4	4.80	II-IV-VI_3	$ZnGeS_3$	(II-VI) + (IV-VI_2)	$ZnS + GeS_2$
5	4.57	III_3-V-VI_3	Ga_3PSe_3	(III-V) + (III_2-VI_3)	$GaP + Ga_2Se_3$
6	5.33	III-V-VI_4	$AlAsSe_4$[b]	(III_2-VI_3) + (V_2-VI_5)	$½(Al_2Se_3 + As_2Se_5)$

[a] There is a large number of single-phase materials of the general formula (Zn, Cd, Hg)(Ga, In)$_2$(S, Se, Te)$_4$, (Zn, Hg)$_2$(Ga, In)$_2$(S, Se, Te)$_5$, and (Zn, Hg)$_3$In$_2$(S, Te)$_6$ (Reference 6.13, p. 145).
[b] The known compounds of this type, (B, Al)(P, As)(S, Se, Te)$_4$, do not crystallize in diamond-like structure.

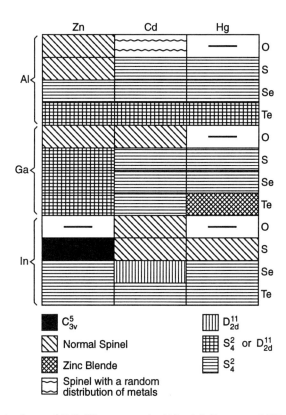

Figure 6.44 Crystal structures of II-III$_2$-VI$_4$ compounds. (After L.I. Berger and V.D. Prochukhan, *Ternary Diamond-Like Semiconductors*, Cons. Bureau/Plenum Press, New York, 1969. With permission.)

composition (II-VI)$_x$(III$_2$-VI$_3$)$_{1-x}$ in a pseudobinary system is different for different combinations of II-VI and III$_2$-VI$_3$ compounds. As an example, from Figure 6.45 (adapted from References 6.6 and 6.201) one can see that in the system CdTe-In$_2$Te$_3$ there exist (1) a continuous series of solid solutions with $0.85 < x < 1$, (2) a phase of composition close to CdIn$_2$Te$_4$, and (3) a phase with formula close to CdIn$_6$Te$_{10}$. The other systems of the kind have different numbers of single-phase entities in them, so the total bulk of information about these substances is quite large (see, e.g., reviews in References 6.5, 6.13, and 6.183). The basic available information regarding II-III$_2$-VI$_4$ materials is compiled in Table 6.41. A majority of these compounds crystallizes either in an ordered structure in the pseudocubic lattice (space group P$\bar{4}$2m), a defect famatinite lattice (space group I$\bar{4}$2m), a defect chalcopyrite

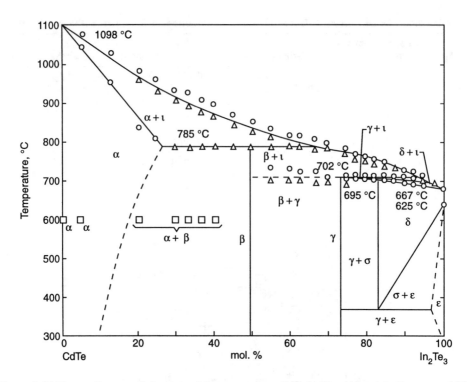

Figure 6.45 Phase diagram of the pseudobinary system CdTe-In$_2$Te$_3$. (After L.I. Berger and V.D. Prochukhan, *Ternary Diamond-Like Semiconductors*, Cons. Bureau/Plenum Press, New York, 1969. With permission.)

lattice (space group $I\bar{4}$), or in partially or totally cation-disordered lattices. These crystal lattices and their relation to the other tetrahedrally coordinated structures are shown in Figure 6.46 (adapted from Reference 6.194). Among compounds of the composition (II-VI)$_x$(III$_2$-VI$_3$)$_{1-x}$, Zn$_2$In$_2$S$_5$, Zn$_3$In$_2$S$_6$, Hg$_5$Ga$_2$Te$_8$, Hg$_3$In$_2$Te$_6$, and Hg$_5$In$_2$Te$_8$ present certain practical interest (see References 6.5 and 6.13).

All tetrahedral II-III$_2$-VI$_4$ compounds are photosensitive, similar to II-VI compounds. Even in the early stages of investigation of these compounds, Beun et al.[6.196] found that the ratio between the resistivity of the samples in the dark and illuminated was very high, reaching 10^6 on the CdGa$_2$Se$_4$ samples. All compounds are also luminophors (see, e.g., References 6.197 and 6.198). The basic data on the electrical, optical, and elastic properties of the compounds are presented in Table 6.42 (compiled mainly from References 6.5, 6.6, and 6.13).

The phase diagrams of some (II-VI)-(III$_2$-VI$_3$) systems are reported by Radautsan and Gavrilitsa[6.199] (HgSe-In$_2$Se$_3$), O'Kane and Mason[6.200] (ZnTe-In$_2$Te$_3$), Thomassen et al.[6.201] (CdTe-In$_2$Te$_3$ — see Figure 6.45), and Ray and Spencer[6.202] (HgTe-In$_2$Te$_3$).

As can be seen from Table 6.42, the II-III$_2$-VI$_4$ compounds possess properties interesting for some optical and electro-optical applications. The chemical vapor deposition methods may be considered as the most appropriate for synthesis of these compounds.

6.4. III$_2$-IV-VI COMPOUNDS

III$_2$-IV-VI compounds are the single-cation ternary analogs of III-V compounds. The first compound of this type, Al$_2$CO, was synthesized by heating a mixture of Al$_2$O$_3$ and Al$_4$C$_3$ to about 2300 K.[6.203] This compound may be considered as an analog of AlN; it has a

Table 6.41 Properties of Tetrahedral Ternary Defect Compounds

Compound	Space group[6.190]	Lattice parameters (nm) a	Lattice parameters (nm) c	Density (g/cm³)	Melting point[a] (K)	Mean heat of atomization (kJ/mol)[6.5,6,36]	Comments
$ZnAl_2S_4$	C_{6v}^4-Pm_3mc	0.376[6.5]	0.614[6.5]	2.6[6.5]			
$ZnAl_2Se_4$	S_4^2-$I\bar{4}$	0.550[6.5]	1.09[6.5]	4.3[6.5]			
$ZnAl_2Te_4$	S_4^2-$I\bar{4}$	0.510[6.5]	1.20[6.5]	4.9[6.5]			
$ZnGa_2S_4$	S_4^2-$I\bar{4}$	0.527[6.5]	1.04[6.5]	3.7[6.5]			
$ZnGa_2Se_4$	S_4^2-$I\bar{4}$	0.549[6.5]	1.09[6.5]	5.1[6.5]			
$ZnGa_2Te_4$	S_4^2-$I\bar{4}$	0.593[6.5]	1.18[6.5]	5.5[6.5]			
$ZnIn_2S_4$	$R3m$[6.191]	0.385[6.13]	n·1.234[6.13]		1393[6.5]	1713	Many polytypes with n = 1,2,..,24
$ZnIn_2Se_4$	S_4^2-$I\bar{4}$	0.571[6.5] 0.569[6.13]	1.142[6.5] 1.14[6.13]	5.36[6.5]	1250[6.5]	1444	
$ZnIn_2Te_4$	S_4^2-$I\bar{4}$	0.612[6.5]	1.22[6.5]	5.8[6.5]	1070[6.5]	1385	
$CdAl_2S_4$	S_4^2-$I\bar{4}$	0.556[6.5]	1.03[6.5]	3.0[6.5]			
$CdAl_2Se_4$	S_4^2-$I\bar{4}$	0.574[6.5]	1.06[6.5]	4.5[6.5]			
$CdAl_2Te_4$	S_4^2-$I\bar{4}$	0.601[6.5]	1.22[6.5]	5.1[6.5]			
$CdGa_2S_4$	S_4^2-$I\bar{4}$	0.553[6.193] 0.557[6.5]	1.0172[6.193] 1.00[6.5]			1795	
$CdGa_2Se_4$	S_4^2-$I\bar{4}$	0.5700[6.193] 0.574[6.5] 0.573[6.13]	1.0805[6.193] 1.073[6.5] 1.076[6.13]	6.2[6.5] 6.28[6.13]	1250[6.13]	1517	
$CdGa_2Te_4$	S_4^2-$I\bar{4}$	0.593[6.5] 0.608[6.13]	1.18[6.5] 1.17[6.13]	5.5[6.5] 5.63[6.13]			
$CdIn_2S_4$	O_h^7-$Fd3m$ or T_d^2-$F\bar{4}3m$	1.0818[6.13]		5.0[6.13]	1378[6.13]	1702	
α-$CdIn_2Se_4$	D_{2d}^1-$P\bar{4}2m$	0.581[6.5] 0.581[6.13]	1.142[6.5]	5.3[6.5] 5.54[6.13]	1188[6.5]	1423	
β-$CdIn_2Se_4$	S_4^2-$I\bar{4}$	0.581[6.13]	1.162[6.13]	5.54[6.13]			
$CdIn_2Te_4$	S_4^2-$I\bar{4}$	0.620[6.5] 0.619[6.13]	1.24[6.5] 1.23[6.13]	5.8[6.5] 5.88[6.13]			
$CdTl_2Te_4$	Hexagonal	0.428[6.13]	0.667[6.13]				
$HgAl_2S_4$	S_4^2-$I\bar{4}$	0.548[6.5]	1.02[6.5]	4.0[6.5]			
$HgAl_2Se_4$	S_4^2-$I\bar{4}$	0.570[6.5]	1.07[6.5]	5.0[6.5]			
$HgAl_2Te_4$	S_4^2-$I\bar{4}$	0.600[6.5]	1.21[6.5]	5.7[6.5]			
	or D_{2d}^1-$P\bar{4}2m$	0.600[6.5]	0.605[6.5]				

Table 6.41 (continued) Properties of Tetrahedral Ternary Defect Compounds

Compound	Space group[6.190]	Lattice parameters (nm) a	Lattice parameters (nm) c	Density (g/cm³)	Melting point[a] (K)	Mean heat of atomization (kJ/mol)[6.5,6.36]	Comments
HgGa$_2$S$_4$	S$_4^2$-I$\bar{4}$	0.549[6.13]	1.02[6.13]	4.95[6.13]		1746	
		0.550[6.5]	1.023[6.5]	4.9[6.5]			
HgGa$_2$Se$_4$	S$_4^2$-I$\bar{4}$	0.5693[6.193]	1.0826[6.193]			1467	
		0.571[6.5]	1.07[6.5]	6.1[6.5]			
		0.570[6.13]	1.07[6.13]	6.10[6.13]			
α-HgGa$_2$Te$_4$	T$_d^2$-F$\bar{4}$3m	0.601[6.5]					
		0.6002[6.193]					
β-HgGa$_2$Te$_4$	D$_{2d}^{11}$-I$\bar{4}$2m	0.6025[6.193]	1.2037[6.193]				
HgIn$_2$Se$_4$	S$_4^2$-I$\bar{4}$	0.576[6.5]	1.18[6.5]	6.2[6.5]	1100[6.5]	1377	
HgIn$_2$Te$_4$	S$_4^2$-I$\bar{4}$	0.618[6.5]	1.23[6.5]	6.3[6.5]	980[6.5]	1318	
MgIn$_2$Se$_4$	R3m[6.192]	0.403[6.192]	3.970[6.192]				
α-CoGa$_2$S$_4$	S$_4^2$-I$\bar{4}$	0.52538[6.193]	1.0393[6.193]				
γ-CoGa$_2$S$_4$	T$_d^2$-F$\bar{4}$3m	0.5221[6.193]					

[a] Or solidus temperature.

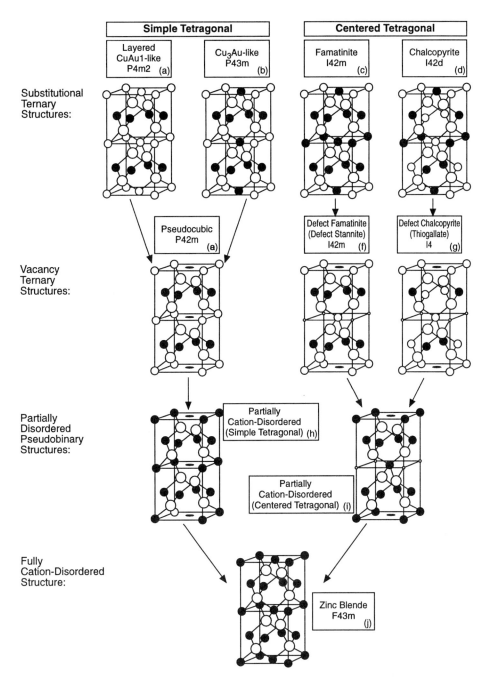

Figure 6.46 Crystal structures (a to d) of ternary $A_nB_{4-n}C_4$ adamantine compounds; (e to g) vacancy structures derived from them; and two stages of cation of the vacancy structures; (h and i) stage I; (j) stage II. (After J.E. Bernard and A. Zunger, *Phys. Rev.*, B37, 6835, 1988. With permission.)

wurtzite-type crystal lattice with parameters a = 0.317 nm, c = 0.506 nm, and c/a = 1.60,[6.204,6.205] slightly greater (about 2%) than the lattice dimensions of AlN. The Al_2CO X-ray density is 3.09 g/cm^3. Amma and Jeffrey[6.65] performed a detail investigation of the Al_2CO crystal structure. As can be expected, Al_2CO forms solid solutions with AlN in the wide range of concentrations (Reference 6.5, p. 162).

Table 6.42 Electrical and Dielectric Properties of II-III$_2$-VI$_4$ Compounds

Compound	Energy Gap (eV)	Electrical Resistivity Ohm·cm Illuminated	Electrical Resistivity Ohm·cm Dark	Wavelength (nm) of the photocurrent curve's Maximum	Wavelength (nm) of the photocurrent curve's Edge	Dielectric Constant $\varepsilon(0)$	Dielectric Constant $\varepsilon(\infty)$	Effective Mass (in m$_o$) m_n^*	Effective Mass (in m$_o$) m_p^*	Mobility cm^2/V·s Electron	Mobility cm^2/V·s Holes
ZnGa$_2$S$_4$	3.4 (optical)	1.5·10^{10}	4·10^{11}	390	360						
GdGa$_2$S$_4$	2.92 (thermal)	2.5·10^8	8·10^{13}	350	358	9.6 (∥) 12.3 (⊥)	6.2 (∥)			60	
	3.44 (direct)										
	3.05 (indirect)										
HgGa$_2$S$_4$	2.84 (direct	7·10^4	1·10^{10}	490	433						
	2.79 (indirect)										
ZnIn$_2$S$_4$	2.6–2.86	2·10^8	1·10^{14}	480	470	6.0 (⊥)				100	
	2.87 (optical)										
CdIn$_2$S$_4$	2.11–2.88 (indirect)	3·10^4	1·10^8	540	540	17	10	0.18	0.3	200	
	2.47–2.62 (direct)										
HgIn$_2$S$_4$	2.0	1.8·10^6	2.7·10^6	620	620					2	
ZnGa$_2$Se$_4$	~2.7	1·10^8	2.5·10^{12}	570							
CdGa$_2$Se$_4$	2.63 (∥); 2.57 (⊥)	1.1·10^5	4·10^{11}	480	515	8.2 (∥) 9.7 (⊥)	6.2 (∥) 6.7 (⊥)			17 33	
	2.43 (thermal)										
HgGa$_2$Se$_4$	1.95 (optical)	2.7·10^4	1.4·10^7	620	677					400	
ZnIn$_2$Se$_4$	1.82–2.6	1.8·10^4	2.4·10^7	580	683					35	
	1.95										
	2.0 (direct)										
CdIn$_2$Se$_4$	1.45–1.72	7·10^5	8·10^5	770	730			0.15		22	
	1.51 (indirect)										
	1.65–1.73 (direct)										
HgIn$_2$Se$_4$	0.6									290	
ZnIn$_2$Te$_4$	1.35 (indirect)										
	1.4–1.8										
	1.87 (direct)										
CdGa$_2$Te$_4$	1.5 (direct)									4000	
CdIn$_2$Te$_4$	1.15 (indirect)										
	0.88–0.9										
	1.25 (direct)										
HgIn$_2$Te$_4$	0.94 (direct)									200	140
	0.9 (indirect)										
	0.8–1.25										

Numerous attempts to synthesize compounds consisting of heavier elements (see, e.g., Reference 6.6, p. 98) have shown that many of the prepared alloys are unstable and often contain noticeable amounts of free elemental components in the products of synthesis. Nevertheless, Goodman[6.127] has synthesized unstable Al_2SnTe, Al_2GeTe, and Al_2GeSe and more stable In_2GeSe and In_2GeTe. These compounds have very complex systems of maxima in the Debye diffraction patterns. Wooley and Williams[6.206] have demonstrated that In_2GeTe, In_2SnTe, and In_2SnSe can crystallize in the sphalerite structure when they form solid solutions with InSb. Some information on the compounds In_2PbTe and Tl_2PbTe, having a distorted sphalerite structure, has been published by Pamplin.[6.207] Unfortunately, the existence of these compounds has not yet been confirmed by other researchers. Goryunova (Reference 6.5, p. 163) assumes that it is possible to synthesize III_2-IV-VI compounds containing boron, scandium, ittrium, lanthanum, and, possibly, other rare-earth elements.

The physical properties of the III_2-IV-VI compounds are virtually unknown; therefore, it is difficult to evaluate their potential applications.

REFERENCES

6.1. H. Hahn and W. Klinger, *Z. Anorg. Allg. Chem.*, 259, 259, 1949.
6.2. J. P. J. Suchet, *Phys. Chem. Solids*, 23, 19, 1962.
6.3. L. S. Palatnik, V. M. Koshkin, and Yu. F. Komnik, *Khimicheskaya Svyaz' v Poluprovodnikakh i Tverdykh Tyelakh (Chemical Bonds in Semiconductors and Solids)*, Nauka i Tekhnika, Minsk, 1965, 301.
6.4. N. A. Goryunova, *The Chemistry of Diamond-Like Semiconductors*, MIT Press, Cambridge, 1965.
6.5. N. A. Goryunova, *Slozhnyye Almazopodobnyye Poluprovodniki (Multinary Diamond-Like Semiconductors)*, Soviet Radio, Moscow, 1968.
6.6. L. I. Berger and V. D. Prochukhan, *Ternary-Diamond-Like Semiconductors*, Consultants Bureau/Plenum Press, New York, 1969.
6.7. J. L. Shay and J. H. Wernick, *Ternary Chalcopyrite Semiconductors: Growth, Electronic Properties, and Applications*, Pergamon Press, Elmsford, NY, 1975.
6.8. M. Rodot, *Les Materiaux Semiconducteur*, Dunod, Paris, 1965.
6.9. A. Aresti, L. Garbato, and A. Rucci, *J. Phys. Chem. Solids*, 45, 361, 1984.
6.10. E. Parthe, *Cristal Chemistry of Tetrahedral Structures*, Gordon and Breach, 1964.
6.11. D. R. Lide, Ed., *Handbook of Chemistry and Physics*, CRC Press, Boca Raton, FL, 1994.
6.12. N. A. Baitenev, A. S. Borshchevskii, S. G. Khusainov, Yu. B. Rud', and Yu. K. Undalov, *Sov. Phys. Semicond.*, 9, 401, 1975.
6.13. O. Madelung, Ed., Semiconductors Other Than Group IV Elements and III-V Compounds, in *Data in Science and Technology*, R. Poerschke, Ed., Springer-Verlag, New York, 1992.
6.14. A. J. Spring-Thrope and B. R. Pamplin, *J. Cryst. Growth*, 3, 313, 1968.
6.15. S. C. Abrahams and J. L. Bernstein, *J. Chem. Phys.*, 52, 5607, 1970.
6.16. A. Rabenau and P. Eckerlin, *Naturwissenschaften*, 46, 106, 1959.
6.17. P. Deus and H. A. Schneider, *Phys. Stat. Solidi*, A79, 411, 1983.
6.18. L. Pauling and M. L. Huggins, *Z. Krist.*, 87, 205, 1934.
6.19. L. S. Palatnik, Yu. F. Komnik, and V. M. Koshkin, *Kristallografia*, 7, 563, 1962.
6.20. J. A. Van Vechten and J. C. Phillips, *Phys. Rev.*, B2, 2160, 1970.
6.21. J. C. Phillips, *Bonds and Bands in Semiconductors*, Academic Press, New York, 1973.
6.22. L. Pauling, *The Nature of the Chemical Bond*, Cornell Press, 1960.
6.23. W. Gordy and W. J. O. Thomas, *J. Chem. Phys.*, 24, 439, 1956.
6.24. S. A. Bondar, V. V. Lebedev, and L. I. Berger, *Elektron. Tekh. Ser. 6 Mater. (Electron. Ind. Ser. 6 Mater.)*, 8, 138, 1972.
6.25. M. Bettini and W. B. Holzapfel, *Solid State Commun.*, 16, 27, 1975.
6.26. R. A. Annamamedov, A. E. Balanevskaya, L. I. Berger, I. A. Dobrushina, and I. K. Shchukina, *Chemical Bonds in Semiconductors and Thermodynamics*, N. N. Sirota, Ed., Consultants Bureau/Plenum Press, New York, 1968, 240.

6.27. H. Borchers and R. G. Maiers, *Metallurgia*, 8, 775, 1963.
6.28. A. S. Borshchevskii and N. D. Roenkov, *Elektron. Tekh. Ser. 14 Mater. (Electron. Ind. Ser. 14 Mater.)*, 1, 55, 1971.
6.29. H. Gutbier, *Z. Naturforsch.*, 16, 268, 1961.
6.30. Yu. A. Valov and R. L. Plechko, *Inorg. Mater.*, 4, 701, 1968.
6.31. Yu. A. Valov and T. N. Ushakova, *Inorg. Mater.*, 4, 1054, 1968; 7, 7, 1971.
6.32. L. I. Berger, V. V. Tarasov, and I. K. Shchukina, *Trudy IREA (Proc. IREA)*, 30, 412, 1967.
6.33. L. I. Berger, V. V. Tarasov, and I. K. Shchukina, *Trudy IREA (Proc. IREA)*, 31, 440, 1969.
6.34. L. I. Berger and I. K. Shchukina, *Trudy IREA (Proc. IREA)*, 32, 309, 1970.
6.35. P. Leroux-Hugon and J. J. Veyssie, *Phys. Stat. Solidi*, 8, 60, 1965.
6.36. V. Sadagopan and H. C. Gatos, Symp. on the Chemical Bond in Semiconductors, Minsk, May 28, 1967; cited in N. A. Goryunova, *Slozhnyye Almazopodobnyye Poluprovodniki (Multinary Diamond-Like Semiconductors)*, Soviet Radio, Moscow, 1968, 144.
6.37. V. S. Arutyunov, L. I. Berger, and I. K. Shchukina, *Trudy IREA (Proc. IREA)*, 31, 427, 1969.
6.38. Chiang Ping-Hsi, I. I. Tychina, E. O. Osmanov, and N. A. Goryunova, Physics — Proc. 21st Sci. Conf., Leningrad Inst. of Construction Engineers, 1963, 8.
6.39. T. Hailing, G. A. Saunders, W. A. Lambson, and R. S. Feigelson, *J. Phys.*, C15, 1399, 1982.
6.40. M. Rubenstein and R. Ure, *J. Phys. Chem. Solids*, 29, 551, 1968.
6.41. P. Deus and H. A. Schneider, *Cryst. Res. Technol.*, 20, 867, 1985.
6.42. H. Neuman, *Phys. Stat. Solidi*, A96, K121, 1986.
6.43. D. B. Sirdeshmukh and K. G. Subhandra, *J. Appl. Phys.*, 59, 276, 1986.
6.44. A. Jayraman, B. Batlogg, R. G. Maines, and H. Bach, *Phys. Rev.*, B26, 3347, 1982.
6.45. R. W. Keyes, *J. Appl. Phys.*, 33, 3371, 1962.
6.46. N. A. Zakharov and V. A. Chaldyshev, *Sov. Phys. Semicond.*, 19, 842, 1985.
6.47. L. I. Berger, L. V. Kradinova, V. B. Petrov, and V. D. Prochukhan, *Inorg. Mater.*, 9, 1258, 1973.
6.48. A. Shaukat, *J. Phys. Chem. Solids*, 43, 407, 1982.
6.49. G. Babonas, G. Ambrazevicius, V. S. Grigoreva, V. Neviera, and A. Shileika, *Phys. Stat. Solidi*, B62, 327, 1974.
6.50. A. V. Raudonis and A. Yu. Shileika, *Sov. Phys. Semicond.*, 9, 1014, 1975.
6.51. H. Nishida, T. Shirakawa, M. Konishi, and J. Nakai, *Jpn. J. Appl. Phys.*, 22, 272, 1983.
6.52. R. G. Humphreys and B. R. Pamplin, *J. Phys.*, 36, C3-159, 1975.
6.53. J. J. Hopfield, *J. Phys. Chem. Solids*, 15, 97, 1960.
6.54. R. Madelon, J. Hurel, and E. Paumier, *Phys. Stat. Solidi*, B136, 679, 1986.
6.55. C. H. L. Goodman, *Nature*, 179, 828, 1957.
6.56. L. I. Berger, *Trudy IREA (Proc. IREA)*, 32, 315, 1970.
6.57. H. M. Hobgood, T. Henningsen, R. N. Thomas, and R. H. Hopkins, *J. Appl. Phys.*, 73, 4030, 1993.
6.58. G. F. Nikolskaya, L. I. Berger, I. V. Evfimovskii, G. N. Kagirova, I. K. Shchukina, and I. S. Kovaleva, *Inorg. Mater.*, 2, 1876, 1966.
6.59. A. J. Straus and A. J. Rosenberg, *J. Phys. Chem. Solids*, 17, 278, 1961.
6.60. M. Matyas and P. Höschl, *Chech. J. Phys.*, 12, 788, 1962.
6.61. P. Leroux-Hugon, *Compt. Rend. Acad. Sci.*, 256, 118, 1963.
6.62. N. A. Goryunova, F. P. Kesamanly, D. N. Nasledov, V. V. Negreskul, Yu. V. Rud', and S. V. Slobodchikov, *Sov. Phys. Tech. Phys.*, 7, 1312, 1965.
6.63. K. Masumoto, S. Isomura, and W. Goto, *J. Phys. Ghem. Solids*, 27, 1939, 1963.
6.64. A. S. Borshchevskii, N. A. Goryunova, F. P. Kesamanly, and D. N. Nasledov, *Phys. Stat. Solidi*, 21, 9, 1967.
6.65. E. L. Amma and G. A. Jeffrey, *J. Chem. Phys.*, 34, 252, 1961.
6.66. B. V. Baranov and N. A. Goryunova, *Dokl. Akad. Nauk SSSR*, 129, 839, 1959.
6.67. W. G. Spitzer, J. H. Wernick, and R. Wolfe, *Solid State Electr.*, 2, 96, 1961.
6.68. L. I. Berger, V. I. Chechernikov, A. E. Balanevskaya, A. V. Pechennikov, and I. K. Shchukina, *Trudy IREA (Proc. IREA)*, 31, 435, 1969.
6.69. G. D. Boyd, E. Buehler, and F. G. Storz, *Appl. Phys. Lett.*, 18, 307, 1971.
6.70. G. D. Boyd, E. Buehler, F. G. Storz, and J. H. Wernick, *IEEE*, QE8, 419, 1972.
6.71. R. Bendorius, V. D. Prochukhan, and A. Shileika, *Phys. Stat. Solidi*, B53, 745, 1972.
6.72. M. Harsy and I. Bertoti, *Phys. Stat. Solidi*, 11, K135, 1965.

6.73. N. A. Goryunova and V. I. Sokolova, *Izv. Moldav. Filiala Akad. Nauk SSSR*, 3, 31, 1960.
6.74. V. I. Sokolova, Abstracts, All-Union Symposium on Semiconductor Compounds. U.S.S.R. Academy of Science, 1961, 21.
6.75. O. G. Folberth and H. Pfister, *Acta Crystallogr.*, 14, 325, 1961.
6.76. V. I. Sokolova and E. V. Tsvyetkova, *Issledovaniya po Poluprovodnikam. Novyye Poluprovodnikovyye Materialy (Investigations in Semiconductors. New Semiconductor Materials)*, Karte Moldoveniasku Publishing House, Kishineu, 1964, 168.
6.77. L. I. Berger, V. I. Sokolova, and V. M. Petrov, *Trudy IREA (Proc. IREA)*, 30, 406, 1967.
6.78. Tu Hailing, G. A. Saunders, M. S. Omar, and B. R. Pamplin, *J. Phys. Chem. Solids*, 45, 163, 1984.
6.79. L. I. Berger, V. I. Sokolova, and N. M. Grinberg, *Trudy IREA (Proc. IREA)*, 30, 408, 1967.
6.80. H. Neumann, W. Kissinger, F. S. Hasoon, B. R. Pamplin, H. Sobotta, and V. Riede, *Phys. Stat. Solidi*, B127, K9, 1985.
6.81. H. Han, G. Frank, W. Klinger, A.-D. Meyer, and G. Störger, *Z. Anorg. Allg. Chem.*, 271, 153, 1953.
6.82. L. S. Palatnik and E. I. Rogachyova, *Inorg. Mater.*, 2, 659, 1966.
6.83. J. C. Wooley and E. W. Williams, *J. Electrochem. Soc.*, 113, 899, 1966.
6.84. L. I. Berger and A. E. Balanevskaya, *Inorg. Mater.*, 2, 1514, 1966.
6.85. L. I. Berger and V. D. Prochukhan, *Troinyye Almazopodobnyye Poluprovodniki (Ternary Diamond-Like Semiconductors)*, Metallurgiya, Moscow, 1968.
6.86. L. I. Berger, M. A. Petrova, Yu. V. Oboznenko, and A. E. Balanevskaya, *Trudy IREA (Proc. IREA)*, 31, 421, 1969.
6.87. I. K. Belova, V. M. Koshkin, and L. S. Palatnik, *Inorg. Mater.*, 3, 617, 1967.
6.88. G. Brandt, A. Rauben, and J. Schneider, *J. Phys. Chem. Solids*, 34, 443, 1973.
6.89. I. V. Bodnar, *Inorg. Mater.*, 15, 1658, 1979.
6.90. T. Wada, T. Negami, and M. Nishitani, *J. Mater. Res.*, 8, 545, 1993; 9, 658, 1994.
6.91. Joint Committee on Powder Diffraction Standards, Powder Diffraction File, Swarthmore, PA, 1984.
6.92. L. I. Berger, V. Ya. Chernykh, and V. M. Petrov, *Trudy IREA (Proc. IREA)*, 31, 431, 1969.
6.93. G. D. Boyd, H. Kasper, J. H. McFee, and F. G. Storz, *J. IEEE*, QE8, 900, 1972.
6.94. L. I. Berger, *Trudy IREA (Proc. IREA)*, 32, 315, 1970.
6.95. S. I. Novikova, *Sov. Phys. Solid State*, 7, 2170, 1966.
6.96. L. I. Berger, S. A. Bondar, V. V. Lebedev, A. D. Molodyk, and S. S. Strelchenko, *Electron. Tekh. Series 6 Mater.*, 1, DE-457, 1972; *Chemical Bonds in Crystals of Semiconductors and Semimetals*, Nauka i Technika, Minsk, 1973, 248; see, also, S. S. Strelchenko, S. A. Bondar, A. D. Molodyk, L. I. Berger, and A. E. Balanevskaya, *Inorg. Mater.*, 5, 593, 1966.
6.97. V. M. Glazov, M. S. Mirgalovskaya, and L. A. Petrova, *Isvestiya Akad. Sci SSSR OTN*, 10, 68, 1957.
6.98. L. I. Berger and A. E. Balanevskaya, *Inorg. Mater.*, 2, 1514, 1966.
6.99. A. V. Petrov and E. L. Shtrum, *Sov. Phys. Solid State*, 4, 1442, 1962.
6.100. S. Shirakata, S. Chichibu, R. Sudo, A. Ogawa, S. Matsumoto, and S. Isomura, *Jpn. J. Appl. Phys.*, 32, L1304, 1993.
6.101. M. H. Grimsditch and G. D. Nolan, *Phys. Rev.*, B12, 4377, 1975.
6.102. M. Bettini and W. B. Holzapfel, *Solid State Commun.*, 16, 27, 1975.
6.103. A. Werner, H. D. Hochheimer, and A. Jayaraman, *Phys. Rev.*, B23, 3836, 1981.
6.104. A. Kraft, G. Kühn, and W. Möller, *Z. Anorg. Allg. Chem.*, 504, 155, 1983.
6.105. J. Jaffe, P. Bendt, and A. Zunger, *Bull. Am. Phys. Soc.*, 27, 248, 1982.
6.106. H. W. Schock, M. Burgelman, M. Carter, L. Stolt, and J. Vedel, 11th Eur. Photovolt. Solar Energy Conf., Mohtreux, 1992, 116.
6.107. J. L. Shay, B. Tell, H. M. Kasper, and L. M. Shiavone, *Phys. Rev.*, B5, 5003, 1972.
6.108. V. M. Petrov, A. E. Balanevskaya, F. F. Kharakhorin, and L. I. Berger, *Inorg. Mater.*, 2, 1874, 1966.
6.109. S. Chichibu, Y. Hagara, M. Uchida, T. Wakiyama, S. Masumoto, S. Shirakata, S. Isomura, and H. Higuchi, *J. Appl. Phys.*, 76, 3009, 1994.
6.110. J. L. Regolini, S. Lewonczak, S. Ringeissen, S. Nikitine, and C. Schwab, *Phys. Stat. Solidi*, B55, 193, 1973.

6.111. J. L. Annapurna and K. V. Reddy, *Indian J. Pure Appl. Phys.*, 24, 283, 1986.
6.112. T. Wada, T. Negami, and M. Nishitani, *J. Mater. Res.*, 8, 545, 1993.
6.113. I. A. Kleinman, A. N. Fedorovskii, and L. I. Berger, *Sov. Phys. Semicond.*, 2, 756, 1968.
6.114. S. Chichibu, S. Shirikata, S. Isomura, Y. Harada, M. Uchida, S. Masumoto, and H. Higuchi, *J. Appl. Phys.*, 77, 1225, 1995.
6.115. J. R. Tuttle, M. Contreras, M. H. Bode, D. Niles, D. S. Albin, R. Matson, A. M. Gabor, A. Tennant, A. Duda, and R. Noufi, *J. Appl. Phys.*, 77, 153, 1995.
6.116. B. Tell, J. L. Shay, and H. M. Kasper, *J. Appl. Phys.*, 43, 2469, 1972.
6.117. H. Neumann and R. D. Tomlinson, *Solid State Commun.*, 57, 591, 1986.
6.118. A. U. Malsagov, *Sov. Phys. Semicond.*, 4, 1213, 1971.
6.119. A. E. Balanevskaya, L. I. Berger, and Z. A. Isaev, (a) *Inorg. Mater.*, 7, 2084, 1971; (b) *Electron. Tekh. Mater.*, 10, DE-480, 1971.
6.120. A. E. Balanevskaya, L. I. Berger, A. V. Pechennikov, and V. I. Checherikov, *Inorg. Mater.*, 1, 2165, 1965.
6.121. G. D. Boyd, H. M. Kasper, and J. H. McFee, *J. IEEE*, QE7, 563, 1971.
6.122. G. D. Boyd, H. M. Kasper, J. H. McFee, and F. G. Storz, *J. IEEE*, QE8, 900, 1972.
6.123. S. W. Wagner, *J. Appl. Phys.*, 45, 246, 1974.
6.124. C. D. George, M. Kapur, and A. Mitra, *Thin Solid Films*, 135, 35, 1986.
6.125. G. K. Averkieva, A. A. Vaipolin, and N. A. Goryunova, *Issledovaniya po Poluprovodnikam. Novyye Poluprovodnikovyya Materialy (Investigations in Semiconductors. New Semiconductor Materials)*, Karte Moldov Publishing House, Kishineu, 1964, 44.
6.126. J. C. Mikkelsen, *J. Cryst. Growth*, 47, 659, 1979.
6.127. C. H. L. Googman, *J. Phys. Chem. Solids*, 6, 305, 1958.
6.128. H. Han, 17th IUPAC Congress, Munich, 1959, 157.
6.129. G. K. Averkieva and A. A. Vaipolin, *1961 All-Union Conf. on Semicond. Compounds*, Abstracts, U.S.S.R. Academy of Science Press, 1961, 4.
6.130. J. Rivet, J. Flahaut, and P. Laruelle, *Compt. Rend. Acad. Sci.*, 251, 161, 1963.
6.131. L. S. Palatnik, V. M. Koshkin, L. P. Galchinestskii, V. I. Kolesnikov, and Yu. F. Komnik, *Sov. Phys. Solid State*, 4, 1430, 1962.
6.132. A. E. Balanevskaya, L. I. Berger, and V. M. Petrov, *Inorg. Mater.*, 2, 810, 1966.
6.133. N. A. Goryunova, A. V. Voitsekhovskii, and V. D. Prochukhan, *Vestnik Leningrad. Gos. Univ. (Proc. Leningr. State Univ.)*, 10, 156, 1962.
6.134. L. S. Palatnik, Yu. F. Komnik, E. K. Belova, and L. V. Atroshchenko, *Kristallografia*, 6, 960, 1961.
6.135. L. I. Berger and N. A. Bulyonkov, *Izv. Akad. Nauk U.S.S.R. Phys.*, 28, 1100, 1964.
6.136. L. I. Berger and E. G. Kotina, *Inorg. Mater.*, 7, 2083, 1971.
6.137. L. I. Berger, R. Annamamedov, and K. Dovletov, *Izv. Akad. Nauk Turkmen. SSR Phys. Tech. Series*, 5, 50, 1969.
6.138. S. A. Bondar, V. V. Lebedev, L. I. Berger, S. S. Strelchenko, V. A. Bondar, and A. D. Molodyk, *Electron. Tekh. Series 14 Mater.*, 7, 74, 1971.
6.139. G. K. Averkieva, A. A. Vaipolin, and N. A. Goryunova, *Sov. Res. on New Semicond. Mater.*, D. N. Nasledov and N. A. Goryunova, Eds., Consultants Bureau/Plenum Press, New York, 1965, 1.
6.140. J. Rivet, O. Gorochov, and J. Flahaut, *Compt. Rend. Acad. Sci.*, 260, 178, 1965.
6.141. L. I. Berger and V. M. Petrov, *Trudy IREA (Proc. IREA)*, 30, 396, 1967.
6.142. O. P. Astakhov, L. I. Berger, K. Dovletov, and E. I. Kovalyova, *Izv. Akad. Nauk Turkmen. SSR Phys. Tech. Series*, No. 6, 111, 1967.
6.143. K. Dovletov, O. P. Astakhov, L. I. Berger, and E. I. Kovalyova, *Izv. Akad. Nauk Turkmen. SSR Phys. Tech. Series*, No. 3, 106, 1968.
6.144. O. P. Astakhov, L. I. Berger, and K. Dovletov, *Izv. Akad. Nauk Turkmen. SSR Phys. Tech. Series*, No. 3, 104, 1968.
6.145. O. P. Astakhov, L. I. Berger, and K. Dovletov, *Izv. Akad. Nauk Turkmen. SSR Phys. Tech. Series*, No. 4, 121, 1968.
6.146. K. Dovletov, O. P. Astakhov, and L. I. Berger, *Izv. Akad. Nauk Turkmen. SSR Phys. Tech. Series*, No. 2, 94, 1971.
6.147. F. F. Kharakhorin and V. M. Petrov, *Sov. Phys. Solid State*, 6, 2867, 1964.
6.148. L. Pauling and R. Hultgren, *Z. Krist.*, 84, 204, 1933.

6.149. W. F. de Jong, *Z. Krist.*, 68, 522, 1928.
6.150. C. Hermann, O. Lohrmann, and H. Philipp, Structurberichte, Vol. 2, Supplement, 1928–1932, 347.
6.151. A. W. Waldo, *Am. Mineral.*, 20, 575, 1935.
6.152. R. W. G. Wyckoff, *Crystal Structures*, Vol. 1, Interscience, New York, 1948.
6.153. A. F. Wells, *Structural Inorganic Chemistry*, 2nd ed., Clarendon Press, Oxford, 1950.
6.154. L. Pauling and S. Weinbaum, *Z. Krist.*, 88, 48, 1934.
6.155. D. Lindquist and A. Westgren, *Swensk. Kem. Tidskr.*, 48, 41, 1936.
6.156. J. H. Wernick and K. E. Benson, *J. Phys. Chem. Solids*, 3, 157, 1957.
6.157. N. P. Barnes, *Int. J. Nonlinear Opt. Phys.*, 1, 639, 1991.
6.158. P. A. Budni, P. G. Schunemann, M. G. Knights, T. M. Pollak, and E. P. Chicklis, *Opt. Soc. Am. Proc. on Adv. Solid State Lasers*, Vol. 13, L. Chase and A. A. Pinto, Eds., OSA, 1992, 380.
6.159. N. C. Giles, L. E. Halliburton, P. G. Schunemann, and T. M. Pollak, *Appl. Phys. Lett.*, 66, 1758, 1995.
6.160. A. Abdelghany, *Appl. Phys.*, A60, 77, 1995.
6.161. R. Annamamedov, L. I. Berger, V. M. Petrov, and S. V. Slobodchikov, *Inorg. Mater.*, 3, 1370, 1967.
6.162. A. G. Alieva and Z. G. Pinsker, *Kristallografia*, 6, 204, 1961.
6.163. G. Busch and F. Hulliger, *Helv. Phys. Acta*, 33, 657, 1960.
6.164. F. Hulliger, *Helv. Phys. Acta*, 34, 5, 1961.
6.165. R. Annamamedov, L. I. Berger, V. M. Petrov, and F. F. Kharakhorin, *Inorg. Mater.*, 2, 772, 1966.
6.166. L. I. Berger and V. M. Petrov, *Inorg. Mater.*, 7, 1905, 1971.
6.167. G. Zimmer and H. Vogt, *IEEE*, ED30, 1515, 1983.
6.168. D. Hardie and K. H. Jack, *Nature*, 180 (4581), 332, 1957.
6.169. H. Kato, Z. Inoue, K. Kijima, T. Yamane, and J. Kawada, *J. Am. Ceram. Soc.*, 58, 90, 1979.
6.170. R. Gruen, *Acta Crystallogr.*, B35, 800, 1979.
6.171. H. Fukushima, T. Yamanaka, and M. Matsui, *J. Mater. Res.*, 5, 397, 1990.
6.172. Y. Hirata, S. Nakagama, and Y. Ishihara, *J. Mater. Res.*, 5, 640, 1990.
6.173. O. Unal, J. J. Petrovic, and T. E. Mitchell, *J. Mater. Res.*, 8, 626, 1993.
6.174. I. Tanaka, G. Pezzotti, T. Okamoto, Y. Miyamoto, and M. Koizumi, *J. Am. Ceram. Soc.*, 72, 1956, 1989; see, also, G. Pezzotti, I. Tanaka, T. Okamoto, M. Koizumi, and Y. Miyamoto, *J. Am. Ceram. Soc.*, 72, 1461, 1989.
6.175. K. J. Grannen, F. Xiong, and R. P. Chang, *J. Mater. Res.*, 9, 2341, 1994.
6.176. H. Lange, G. Wotting, and G. Winter, *Angew. Chemie Int. Ed. Engl.*, 30, 1579, 1991.
6.177. B.-Y. Shewand and J.-L. Huang, *J. Mater. Sci. Eng.*, A159, 127, 1992.
6.178. Y. S. Touloukian, R. K. Kirby, R. E. Taylor, and T. Y. R. Lee, *Thermophysical Properties of Matter. Thermal Expansion — Nonmetallic Solids*, Vol. 13, Plenum Press, New York, 1977.
6.179. M. G. Ellenburg, J. A. Hanigofsky, and W. J. Lackey, *J. Mater. Res.*, 9, 789, 1994.
6.180. H. Hahn and W. Klinger, *Z. Anorg. Allg. Chem.*, 259, 121, 1949.
6.181. D. N. Nasledov and I. A. Feltynsh, *Sov. Phys. Solid State*, 1, 565, 1959; 2, 823, 1960.
6.182. K. V. Savchenko and V. V. Shchennikov, *Can. J. Phys.*, 72, 681, 1994.
6.183. Z. S. Medvedeva, *Chalcogenides of the III-B Group Elements*, Nauka, Moscow, 1968.
6.184. L. I. Berger, S. S. Strelchenko, S. A. Bondar, A. D. Molodyk, A. E. Balanevskaya, and V. V. Lebedev, *Inorg. Mater.*, 5, 872, 1969.
6.185. V. M. Glazov, S. N. Chizhevskaya, and N. N. Glagoleva, *Liquid Semiconductors*, Nauka, Moscow, 1967 (in Russian).
6.186. N. A. Goryunova, *Problems of Theory and Experiments on Semiconductors and Processes of the Semiconductor Metallurgy*, U.S.S.R. Acad. Sci. Press, 1955, 13.
6.187. N. A. Goryunova, V. S. Grigoryeva, B. M. Konovalenko, and S. M. Ryvkin, *Sov. Phys. Tech. Phys.*, 25, 1675, 1955.
6.188. D. Lubbers and V. Leute, *J. Solid State Chem.*, 43, 339, 1982.
6.189. H. Hahn and W. Klinger, *Z. Anorg. Allg. Chem.*, 263, 177, 1950.
6.190. H. Hahn, G. Frank, W. Klinger, A. Störger, and J. Störger, *Z. Anorg. Allg. Chem.*, 279, 241, 1955.
6.191. V. P. Dotzel, H. Schäfer, and G. Schön, *Z. Anorg. Allg. Chem.*, 426, 260, 1976.
6.192. L. Gastaldi, A. Maltese, and S. Viticoli, *J. Cryst. Growth*, 66, 673, 1984.
6.193. L. Gastaldi, M. G. Simeone, and S. Viticoli, *Solid State Commun.*, 55, 605, 1985.
6.194. J. E. Bernard and A. Zunger, *Phys. Rev.*, B37, 6835, 1988.

6.195. H. J. Welker, *Annu. Rev. Mater. Sci.,* 9, 1, 1979.
6.196. I. A. Beun, R. Nitsche, and M. Lightensteiger, *Physica,* 26, 647, 1960.
6.197. M. Springford, *Proc. Phys. Soc.,* 82, 1029, 1963.
6.198. S. Shinoya and A. Ecbina, *J. Phys. Soc. Jpn.,* 19, 1142, 1964.
6.199. S. I. Radautsan and E. I. Gavrilitsa, *Izv. Mold. Akad. Sci.,* 10, 95, 1961.
6.200. D. F. O'Kane and D. R. Mason, *Trans. Met. Soc. AIME,* 223, 1189, 1965.
6.201. L. Thomassen, D. R. Mason, C. D. Rose, J. C. Sarace, and G. A. Schmitt, *J. Electrochem. Soc.,* 110, 1127, 1963.
6.202. B. Ray and A. M. Spencer, *Solid State Commun.,* 3, 389, 1965; see, also, A. M. Spencer, *Br. J. Appl. Phys.,* 15, 625, 1964.
6.203. L. M. Foster, G. Long, and M. S. Hunter, *J. Am. Ceram. Soc.,* 39, 1, 1956.
6.204. G. A. Jeffrey and H. Linton, *Bull. Am. Phys. Soc.,* 3, 231, 1958.
6.205. J. C. Wooley and B. Ray, *J. Phys. Chem. Solids,* 16, 102, 1960.
6.206. J. C. Wooley and E. W. Williams, *J. Electrochem. Soc.,* 111, 210, 1964.
6.207. B. R. Pamplin, *J. Phys. Chem. Solids,* 25, 675, 1964.

CHAPTER 7

Nonadamantine Semiconductors and Variable-Composition Semiconductor Phases

Any attempt to embrace the world of semiconductors more or less comprehensively is, apparently, doomed to failure, primarily because of a great number and variety of semiconductors. To avoid the failure, the list of semiconductors included in this chapter is limited to those groups of materials the representatives of which have found any practical applications or are considered as having a potential for practical applications.

This chapter includes the description of the 1A-1B compounds, the II_2-IV antifluoride-type compounds, the IV-VI compounds with the galenite structure, V_2-VI_3 compounds, and a few compounds of the platinum group metals. Description of organic semiconductors is limited by the properties of a few phthalocyanines, cyclic hydrocarbons, and dyes. Some of the known compounds, such as KBi_2, Li_4Ba, Na_4Ba, and AgN_3, are not included here because the available information regarding their properties is too scarce.

7.1. IA-IB SEMICONDUCTORS

The intermetallic compounds LiAg, CsAu, and RbAu crystallize into the cesium chloride structure which may be considered as one formed from two equal primitive cubic lattices shifted from each other by one half of the space diagonal of the cubic unit cell (Figure 7.1). This lattice also can be considered as a superstructure with a body-centered unit cell. The lattice parameters of these compounds, together with their molecular weight and atomic number, are presented in Table 7.1.

These compounds are of particular interest because they possess semiconductor properties (i.e., the atoms are connected to each other by covalent bonds) having been formed by the sets of highly conductive metals. The available information regarding electronegativity (see Table 6.3) indicates that the elements forming these compounds have the electronegativity difference (Pauling) from 0.9 (LiAg) to 1.6 (RbAu and CsAu); therefore, they are expected to belong to the group of compounds with polar covalent bonds (with bond ionicity from 9 to 47%). Experiments, however, suggest that both RbAu and CsAu consist of Rb^+ or Cs^+ and Au^- ions formed by the transfer of 5s (Rb) or 6s (Cs) electrons to gold atoms.[7.11-7.13]

The phase diagrams Li-Ag, Rb-Au, and Cs-Au are not known. The samples may be prepared by direct reaction of the elemental components — liquid alkali and solid silver or gold — in the evacuated and sealed vessels or by deposition of the alkali metal from the

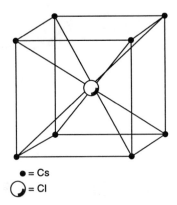

Figure 7.1 Crystal structure of cesium cloride.

Table 7.1 IA-IB Compounds. Molecular Weight and Number, Lattice Parameters, and Density

Compound	Molecular weight	Mean atomic weight	Mean atomic number	Space group	Lattice parameter (nm)	Density (g/cm³)	Ref.
LiAg	114.81	57.40	25	O_h^1-Pm3m	0.3168	5.99	7.2
					0.317		7.21
					0.323		7.35
					0.3169		7.36
RbAu	282.43	141.22	58	O_h^1-Pm3m	0.4105	6.780	7.1, 7.8
					0.4093		7.9
CsAu	329.87	164.94	67	O_h^1-Pm3m	0.4263	7.065	7.8

vapor phase on a film of gold or silver (see, e.g., References 7.8, 7.14, and 7.15). The melting points of RbAu and CsAu are 771 K and 863 K, respectively.[7.9]

The Brillouin zone for the body-centered cubic lattice is shown in Figure 7.2. The band structure for RbAu and CsAu were calculated by Overhof et al.[7.10] using the relativistic Kohn-Korringa-Rostoker (KKR) method.[7.16] Hasegawa and Watanabe[7.17] calculated the CsAu band structure using the augmented plane wave (APW) method (for information, see, e.g., Reference 7.18). The band structures of both aurides reported in Reference 7.10 are presented in Figure 7.3 (adapted from Reference 7.10). Depending on the choice of the crystal potential, the calculated direct energy gap for both RbAu and CsAu is close to 3 eV (ionic potential) or is equal in the case of CsAu to 2.12 eV (at the R point) with an atomic potential.[7.10] In the latter case, there is also an indirect energy gap (transition R → X) of 1.82 eV. The band structures shown in Figure 7.3 are results of calculations with an ionic potential.

The CsAu samples available for electrical transport measurements in the range from 77 K to 300 K[7.14] were polycrystalline films and probably contained excessive cesium atoms. The room temperature magnitudes of electrical resistivity of the n-type samples varied between 0.02 Ohm·cm and 0.23 Ohm·cm; the Hall mobility of electrons was between 30 cm²/V·sec and 40 cm²/V·sec. Hall and Wright[7.15] studied the effect of oxidation on the CsAu films and found that with increase of the oxygen content, samples changed the conductivity type and that the hole mobility initially decreased with the oxygen content increase, reached a minimum of about 2 cm²/V·sec, and then increased up to the magnitude of 9 cm²/V·sec.

Calculations of electric properties of rubidium auride[7.19] predicted that they would be similar to those of cesium auride; however, the reported data are limited (see, e.g., Reference 7.8).

Optical absorption measurements on CsAu thin films at different temperatures (including the liquid state) were performed by Münster and Freyland. It may be concluded from their results (Figure 7.4) that CsAu retains semiconductor properties in the liquid state.

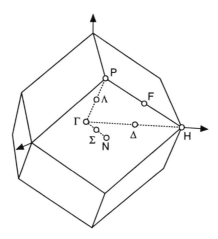

Figure 7.2 Brillouin zone for the body-centered cubic lattice.

7.2. SEMICONDUCTORS IN THE I-V BINARY SYSTEMS

Among the binary compounds in the systems IA-VB, the compounds with the general formula IA-VB and (IA)$_3$-VB attracted attention of physicists and engineers after Görlich's discovery[7.20] of their promising photoelectric properties, first of all their high quantum efficiency in the visible part of spectrum. The early information regarding these compounds (see, e.g., reviews in References 7.5 to 7.7) showed that their energy gap lies within the limits 0.5 eV to 0.8 eV and the carrier mobility magnitudes are of the order 10^2 cm^2/V·sec to 10^3 cm^2/V·sec. Development of the thin film growth technology (see Reference 7.1, p. 135) has led to extension of the number of the compounds and discloser of the wide variety of their electronic properties.

7.2.1. IA-VB Compounds

The known IA-VB compounds are LiAs, LiBi, NaP, NaSb, NaBi, KSb, RbSb, CsSb, and CsBi. The basic physicochemical data regarding these materials are gathered in Table 7.2. The details of the chemical bond in the compounds are discussed in Reference 7.26. The temperature dependence of electrical conductivity of the compounds is reported in References 7.27 to 7.29. The energy gap in the compounds is in the limits 0.5 to 0.8 eV (see Table 7.3). The majority of the results on the properties of IA-VB compounds were attained using thin films grown mainly from the vapor phase. The melting point data are available only for a few of the compounds; some of them melt peritectically.

The optical absorption and photoconductivity of CsSb were studied by Kunze[7.29] in the range of photon energy from 1 eV to 3 eV (Figures 7.5 and 7.6). The insensitivity of the photoconductivity to the temperature is, possibly, a sign that electrical conductivity of the samples at and below room temperature is extrinsic. The IA-VB compounds have been utilized mainly as materials for photocathodes.

7.2.2. (IA)$_3$-VB Compounds

The known (IA)$_3$-VB semiconductors present a certain interest for application as materials for photocathodes and photoresistors, and as such they have been synthesized and investigated mainly in the form of thin film samples. The compounds exist either in hexagonal Na$_3$As-type structure, space group D_{6h}^4-P6$_3$/mmc, with eight atoms (two formula units) in the unit

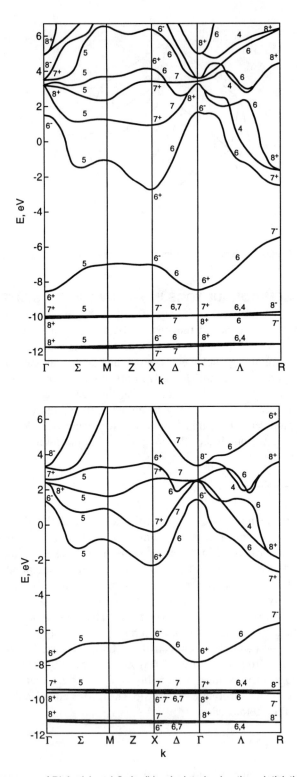

Figure 7.3 Band structure of RbAu (a) and CsAu (b) calculated using the relativistic KKR method. (After H. Overhof, J. Knecht, R. Fischer, and F. Hensel, *J. Phys.*, F8, 1607, 1978. With permission.)

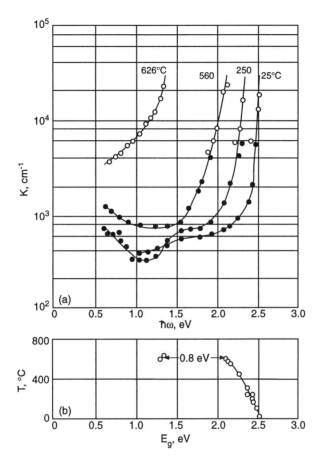

Figure 7.4 Electro-optical properties of CsAu thin films of the thickness of 3 μm (open circles) and 7 μm (solid circles). (a) Absorption coefficient vs. photon energy; (b) correlation between the temperature and the optical energy gap. (After P. Münster and W. Freyland, *Phil. Mag.*, B39, 93, 1979. With permission.)

cell, or in one of two cubic lattices: in BiF$_3$-type structure, space group O_h^5-Fm3m, with 16 atoms in the unit cell, or in the structure similar to that of NaTl, space group O_h^7-Fd3m, also with 16 atoms in the unit cell. The information regarding the crystal structure and lattice parameters of (IA)$_3$-VB compounds is gathered in Table 7.4.

Bates et al.[7.46] assume that the chemical bond in the (IA)$_3$-VB compounds is predominantly ionic, formed by interaction between Cs$^+$ and Sb^{3-} ions, whereas Krebs[7.47] suggests the sp^2-hybrid bonds for the hexagonal structure compounds and Suchet[7.26] considers formation of the resonance sp^3 hybrids between the alkali and pnictogen atoms surrounded by the interstitial compensating cations as a proper bond model.

The Brillouin zone of the I$_3$-V compounds with BiF$_3$-type structure is shown in Figure 7.7. The band structure of hexagonal Na$_3$Sb and K$_3$Sb and cubic Na$_3$Sb, K$_3$Sb, Rb$_3$Sb, and Cs$_3$Sb was calculated by Mostovskii et al.[7.48,7.49] (Figures 7.8 to 7.12). The available data on energy band parameters are compiled in Table 7.5 together with information on density and melting temperature.

The electrical resistance measurements below room temperature of Li$_3$Sb and Li$_3$Bi were performed by Gobrecht,[7.50,7.51] who used film samples, pure or doped with sodium. The doping substantially increased the sample resistance, especially for Li$_3$Sb, although, as was shown

Table 7.2 IA-VB Compounds. Molecular Weight and Number, Structure, and Lattice Parameters

Compound	Molecular weight	Mean atomic weight	Mean atomic number	Structure and space group	Lattice parameters					Z	Ref.
					a (nm)	b (nm)	c (nm)	β°			
LiAs	81.86	40.93	18	Monoclinic C_{2h}^5-$P2_1/c$	0.579	0.524	1.070	117.4	16	7.21
LiBi	215.92	107.96	43	Tetragonal D_{4h}^1-$P4/mmm$	0.3368		0.4256		16	7.21
NaP	53.96	26.98	13	Orthorhombic D_2^4-$P2_12_12_1$						7.21
NaSb	144.74	72.37	31	C_{2h}^5-$P2_1/c$	0.680	0.634	1.248	117.6	16	7.22
NaBi	231.97	115.99	47	D_{4h}^1-$P4/mmm$	0.347		0.481		16	7.23
KSb	160.85	80.42	35	C_{2h}^5-$P2_1/c$	0.718	0.697	1.340	115.1	16	7.24
RbSb	207.22	103.61	44	D_2^4-$P2_12_12_1$	0.7315	0.7197	1.2815		16	7.25
CsSb	254.66	127.33	53	D_2^4-$P2_12_12_1$	0.7575	0.7345	1.3273		16	7.25

Table 7.3 IA-VB Compounds. Density, Melting Temperature, Energy Gap, and Seebeck and Secondary Electron Emission Coefficients

Compound	X-ray density (g/cm³)	Melting point (K)	Energy gap (eV)	Seebeck coefficient (μV/K)	Maximum secondary electron emission coefficient
LiAs	3.77				
LiBi	7.43	6.88[a] 7.21			
NaSb	4.03	738 7.28			
NaBi	6.65	719[a] 7.23			
KSb	3.52	883 7.30	0.82 7.28		
RbSb	4.08	883 7.30	0.88 7.27		
CsSb	4.58	859 7.30,7.33,7.34	0.63 7.29	300[b] 7.32	
			0.61 7.31		
CsBi			Metal		1.97 7.2

[a] Temperature of the peritectic decomposition.
[b] At 313 K.

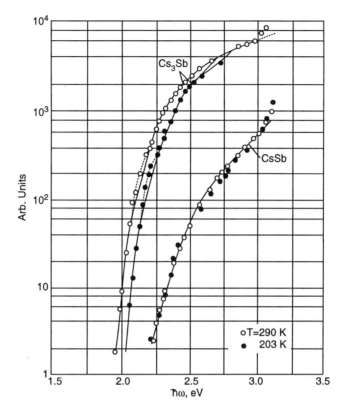

Figure 7.5 Spectral distribution of the photoconductivity for CsSb and Cs_3Sb. (After C. Kunze, *Ann. Phys.*, 6, 89, 1960. With permission.)

by Suhrmann and Kangro,[7.27] deviation from stoichiometry for K_3Sb and (KSb) decreases the resistivity (see Figure 7.13). Imamura[7.52] reports that Na_3Sb has the n-type conductivity. In Sommers' review,[7.53] it is noted that, as a rule, hexagonal allotropes of I_3-V compounds have n-type conductivity, while the undoped samples of the cubic allotropes possess p-type conductivity. The temperature dependence of the resistance for Li_3Sb is presented in Figure 7.14 (adapted from Reference 7.50), the conductance of the Li_3Bi samples is shown in Figure 7.15 (adapted from Reference 7.51), and the (extrinsic) conductivity vs. temperature for K_3Sb and Cs_3Sb is plotted in Figure 7.16 (adapted from Reference 7.27).

The photoelectric and optical properties of I_3-V compounds was reported by Gobrecht[7.50,7.51] for Li_3Sb and Li_3Bi (the spectral response of photoemissive sensitivity), and by Spicer[7.56] and Fisher[7.62] for Na_3Sb (the optical absorption, photoconductivity, and photoemissive yield vs. photon energy). The index of refraction and extinction coefficient of Na_3Sb are shown in Figure 7.17 (adapted from Reference 7.57). Photoconductivity, optical density, and photoemissive yield data for K_3Sb are presented in References 7.53 and 7.57. Optical and photoelectric properties of Rb_3Sb and Cs_3Sb (photoemissive yield, photoconductivity, optical absorption, and index of refraction) have been reported in References 7.53, 7.56, and 7.63 to 7.66. The Cs_3Sb extinction coefficient and index of refraction as the function of the wavelength are presented in Figures 7.18 [7.64,7.65] and 7.19.[7.66] The optical and photoelectric properties of Cs_3Bi were reported in Reference 7.61 (absorption coefficient, Figure 7.20, photoemissive yield, Figure 7.21.

In the conclusion of Section 7.2, it can be noted that some of the considered binary compounds in IA-VB systems present a substantial theoretical and practical interest. Their current application is limited by the photomultiplier and photoresistor industry. But their high optical absorption and its strong dependence on the photon energy in combination with low

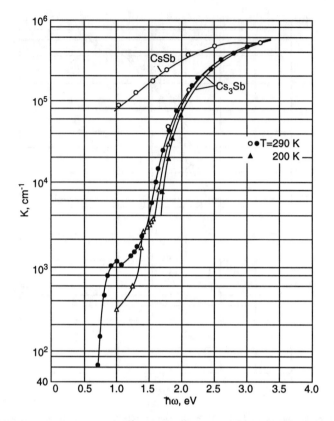

Figure 7.6 Spectral distribution of the absorption coefficient for CsSb and Cs₃Sb. Open triangles — data from other sources. (After C. Kunze, *Ann. Phys.*, 6, 89, 1960. With permission.)

electrical conductivity and relatively high optical energy gap suggest that they can be effectively used in optical filters and shutters. The fact that some of the compounds change the conductivity sign in the process of allotropic transformations is also of substantial interest. In spite of the relatively long history of studying and utilizing these compounds, many of their physical and physicochemical properties have not yet been investigated.

7.3. I-VI COMPOUNDS

The oxides and chalcogenides of the elements of IB group are materials whose semiconducting properties had been discovered and reported even before introduction of the term "semiconductor" by Königsberg and Weiss.[7.67] In 1833, Faraday[7.68] discovered that Ag_2S has a negative temperature coefficient of resistivity and relatively high magnitude of resistivity in comparison with metals. In 1874, Braun[7.69] reported the rectifying effect in natural copper pyrites, and Shuster[7.70] discovered the same effect on a contact of two copper wires, one of which had an oxidized surface. In 1926, Grondahl[7.71] reported upon, patented, and built for technical applications the first industrial scale rectifier, which was based on Cu_2O. In 1933, Zhuze and Kurchatov,[7.72] using the results of their measurements of electrical conductivity of Cu_2O, discovered the phenomenon of intrinsic conductivity in semiconductors.

The oxides and chalcogenides of copper and silver are widely presented in nature by, e.g., cuprite (chalcotrichite), Cu_2O; tenorite (paramelaconite), CuO; chalcocite, Cu_2S; covellite, CuS; berzelianite, Cu_2Se (or $Cu_{1.9}Se$); klockmannite, $CuSe$; umangite, Cu_2Se_3; athabaskite, Cu_5Se_4; weissite (rickardite), Cu_5Te_3 (or $Cu_{2-x}Te$ or Cu_3Te_2); argentite (acanthite), Ag_2S;

Table 7.4 (IA)$_3$-VB Compounds. Molecular Weight and Number and Lattice Parameters

Compound	Molecular weight	Mean atomic weight	Mean atomic number	Structure and space group	Lattice parameters a (nm)	c (nm)	References and comments
Li$_3$P	51.80	12.95	6	Na$_3$As type D_{6h}^4-P6$_3$/mmc	0.4273	0.7549	7.37, p. 717
Li$_3$As	95.74	23.94	10.5	Same	0.4377	0.7801	7.38
					0.4387	0.7810	7.39
α-Li$_3$Sb (HT)	142.57	35.64	15	Same	0.4710	0.8326	Above 923 K 7.40
β-Li$_3$Sb (LT)				BiF$_3$ type O_h^5-Fm3m	0.6572		Below 923 K 7.40
Li$_3$Bi	229.80	57.45	23	Same	0.6722		7.21
Na$_3$P	99.94	24.99	12	Na$_3$As type	0.4980	0.8797	7.39
Na$_3$As	143.89	35.97	16.5	Same	0.5088	0.8982	7.39
					0.5098	0.900	7.40
Na$_3$Sb	190.72	47.68	21	Same	0.5355	0.9496	7.39
Na$_3$Bi	277.95	69.49	29	Same	0.5459	0.9674	7.40
					0.5448	0.9655	7.39
K$_3$P	148.27	37.07	18	Same	0.569	1.005	7.41
K$_3$As	192.22	48.05	22.5	Same	0.5794	1.0242	7.40
h-K$_3$Sb	239.04	59.76	27	Same	0.6025	1.0693	7.42
c-K$_3$Sb				BiF$_3$ type	0.8493		7.43
K$_3$Bi (HT)	326.28	81.57	35	Same	0.8805		7.1 Above 673 K
K$_3$Bi (LT)				Na$_3$As type	0.6191	1.0955	7.40
					0.6178	1.0933	7.39 Below 673 K
Rb$_3$As	331.33	82.83	36	Same	0.605	1.173	7.41
Rb$_3$Sb	378.15	94.54	40.5	Same	0.629	1.117	7.44
				BiF$_3$ type	0.884		7.42
				Cs$_3$Sb type O_h^7-Fd3m	0.884		7.39
Rb$_3$Bi	465.38	116.35	48.5	Na$_3$As type	0.642	1.146	7.44 Below 503 K
				BiF$_3$ type	0.898		7.41 Above 503 K
Cs$_3$Sb	520.47	130.12	54	Cs$_3$Sb type (or BiF$_3$ type)	0.9128 0.9147		7.39 7.45
Cs$_3$Bi	607.70	151.92	62	BiF$_3$ type Cs$_3$Sb type	0.931 0.9305		7.41 7.39

naumannite, Ag$_2$Se; hessite, Ag$_2$Te; stützite, Ag$_4$Te; aguilarite, Ag$_2$(S,Se); and eucairite (stromeyrite), (Cu,Ag)$_2$Se.

7.3.1. Copper and Silver Oxides

There are several compounds in the Cu-O and Ag-O systems: Cu$_4$O,[7.2] Cu$_2$O, Cu$_3$O$_2$, CuO, Ag$_2$O, Ag$_4$O$_3$, AgO, Ag$_2$O$_3$ (or Ag$_7$O$_{11}$). CuO is stable at standard conditions, while Cu$_2$O is metastable below 648 K.[7.73,7.74] The data on the molecular weight and crystal structure of some of the oxides are gathered in Table 7.6.

The unit cell of Cu$_2$O is shown in Figure 7.22. Each copper atom has two oxygen neighbors, whereas each oxygen atom is surrounded by four copper neighbors. The thermal and mechanical properties of the Group IB oxides were studied and reviewed in References 7.1 to 7.3, and 7.75 to 7.79. Some of the available data are compiled in Table 7.7.

The CuO band structure is so far not known. The thermal energy gap evaluated from the electrical measurements on thin films[7.80] is 0.14 eV. The band structure of Cu$_2$O calculated

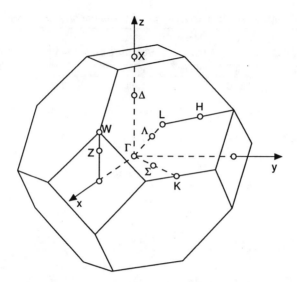

Figure 7.7 Brillouin zone of cubic potassium antimonide.

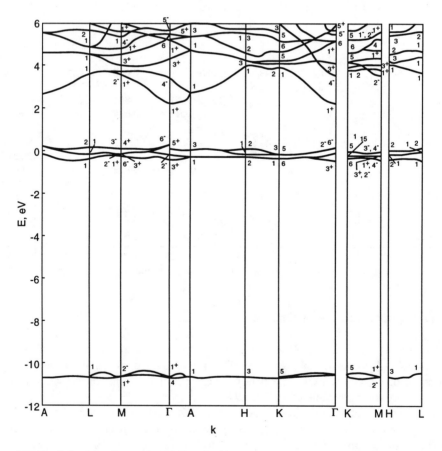

Figure 7.8 Band structure of hexagonal K_3Sb; spin-orbit interaction is not included. (After A.A. Mostovskii, V.A. Chaldyshev, V.P. Kiselev, and A.I. Klimin, *Izv. Akad. Nauk SSSR, Ser. Fiz.*, 40, 2490, 1976. With permission.)

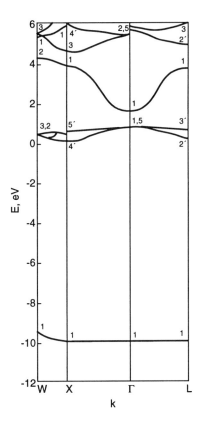

Figure 7.9 Band structure of cubic K$_3$Sb. (After A.A. Mostovskii, V.A. Chaldyshev, V.P. Kiselev, and A.I. Klimin, *Izv. Akad. Nauk SSSR, Ser. Fiz.*, 40, 2490, 1976. With permission.)

by Kleinman and Mednick[7.81] is shown in Figure 7.23. The minimum (direct) energy gap (transition $\Gamma_{7v}^+ \rightarrow \Gamma_{6c}^+$) is 2.17 eV at 4.2 K;[7.82] the early optical measurements[7.83] give the energy gap in good agreement with this magnitude (1.9 eV) and with the data received from electrical measurements (from 1.8 eV to 2.1 eV[7.84]); although in some works, magnitudes such as 1.56 eV[7.85] and 2.3 eV[7.86] are listed, the latter is, possibly, the $\Gamma_{8v}^+ \rightarrow \Gamma_{6c}^+$ transition discussed by Grun et al.[7.87] The effective mass of electrons is 0.93 m_o; that of holes is 0.56 m_o.[7.88]

The information regarding electric transport properties of CuO is limited by the data on electrical resistivity of thin films (from 10^{-1} to 10^{-3} Ohm·cm), thermal energy gap (0.14 eV),[7.80] and the magnitude of the Seebeck coefficient (from 180 µV/K to 250 µV/K).[7.89]

Cuprous oxide, Cu$_2$O, always has only the p-type conductivity, although the effective mass of holes in it is only twice smaller than that of electrons. Dünwald and Wagner[7.90] were, apparently, the first to suggest that excess of oxygen in Cu$_2$O determines p-type conductivity. The assumption that the Cu$_2$O crystal is formed by the Cu$^+$ and O^{2-} ions requires two Cu$^+$ per each O^{2-}. An excess of oxygen is caused by formation of the cation vacancies in the amount of four vacancies per each excessive O$_2$ molecule. It is followed from the requirement of electrical neutrality that any vacancy has to form with the simultaneous formation of a free hole. However, this explanation was not essentially confirmed by the results of experiments by Bloem,[7.74] who studied dependence of the Cu$_2$O conductivity on the temperature and partial pressure of oxygen. Under any of the conditions, the n-type material was not produced, although the conductivity depended synbatically on the oxygen pressure. In accordance with Henisch,[7.84] the average magnitudes of the hole mobility are about 200 cm^2/V·sec

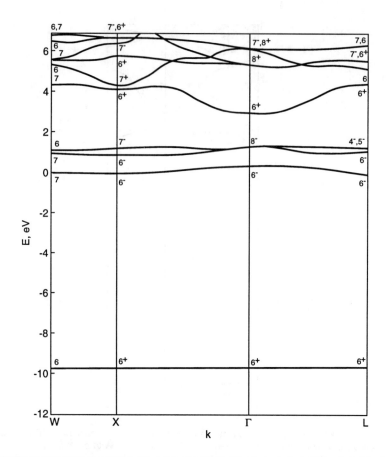

Figure 7.10 Band structure of cubic K_3Sb; spin-orbit interaction is included. (After A.A. Mostovskii, V.A. Chaldyshev, V.P. Kiselev, and A.I. Klimin, *Izv. Akad. Nauk SSSR, Ser. Fiz.*, 40, 2490, 1976. With permission.)

at 170 K and 50 cm^2/V·sec at 300 K. These magnitudes do not contradict the results of Pollack and Trivich[7.91] presented in Figure 7.24. The electrical properties of Cu_2O were reviewed in References 7.84, 7.91, and 7.92.

CuO is a paramagnetic with the magnetic susceptibility (in SI units) at room temperature of $2.8 \cdot 10^{-4}$. O'Keeffe and Stone[7.93] report a magnitude of $1.65 \cdot 10^{-4}$ for the temperature range of 14 to 130 K. The authors of Reference 7.93 found that, depending on the quenching atmosphere, the magnetic susceptibility of Cu_2O varies between $+9.95 \cdot 10^{-6}$ and $-11.5 \cdot 10^{-6}$ when the temperature changes from 14 to 293 K. They note that pure Cu_2O ought to be diamagnetic; the paramagnetic component of susceptibility is determined by excessive oxygen atoms (or impurities). Ag_2O is a diamagnetic with the magnetic susceptibility varying from $-41.6 \cdot 10^{-6}$ to $-50.3 \cdot 10^{-6}$ SI units at room temperature.[7.75]

The dielectric constant of CuO (at room temperature and 2 MHz) is 18.1.[7.94] The static dielectric constant of Cu_2O is 7.5 at 293 K,[7.95] and the high frequency limit is 6.46;[7.96] the authors of Reference 7.94 cite a magnitude of 7.60 ± 0.6 at room temperature and 100 kHz. The dependence of the components of the Cu_2O dielectric constant on wave number at 300 K[7.97] is shown in Figure 7.25. The dielectric constant of Ag_2O at room temperature is equal to 8.8.[7.94]

The optical properties of CuO (index of refraction and extinction coefficient) were reported in References 7.80, 7.98, and 7.99. Investigation of the optical parameters of Cu_2O was done by O'Keeffe,[7.96] Noguet et al.,[7.95] Wieder and Czanderna,[7.98] Dawson et al.,[7.97] Prévot et al.,[7.100] Drobny and Pulfrey,[7.80] Bairamov et al.,[7.101] and Wautelet et al.[7.102] The absorption

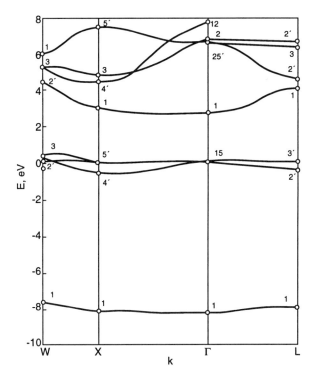

Figure 7.11 Band structure of cubic Rb$_3$Sb calculated using the empirical pseudopotential method. (After A.A. Mostovskii, V.A. Chaldyshev, G.F. Karachaev, A.I. Klimin, and I.N. Ponomarenko, *Izv. Akad. Nauk SSSR, Ser. Fiz.*, 38, 195, 1974. With permission.)

and reflection spectra for Cu$_2$O samples are shown in Figure 7.26 (adapted from Nikitine's review[7.103]). The optical absorption and transmission in Ag$_2$O was reported by Kreingold and Kulinkin[7.104] and Kreingold et al.[7.105]

The copper and silver oxides are available commercially, although the purity of the materials is relatively low. Cu$_2$O in the form of the 325 mesh powder, 99.5% purity, has the price from $1.90 to $2.50 per gram; the Ag$_2$O powder of 99% purity is available for $0.83 to $1.50 per gram (1995–1996 prices).

7.3.2. Copper and Silver Chalcogenides

Consideration of the copper and silver chalcogenides is limited here by the materials with composition I$_2$-VI, although there is a vast number of semiconducting phases known in the Cu-VI and Ag-VI systems. The data pertinent to the basic chemical properties and crystal structure are presented in Table 7.8. Some thermal, thermodynamic, and mechanical properties of the compounds are gathered in Table 7.9.

The minimum (indirect) optical energy gap of β-Cu$_2$S from the different reports[7.115,7.116] varies between 1.0 eV and 1.4 eV at room temperature. The hole effective mass estimation from the Seebeck coefficient measurements gives magnitudes between 1.6 m$_o$ and 1.82 m$_o$ below room temperature.[7.1] The Cu$_2$Se (direct) energy gap, in accordance with Voskanian et al.,[7.117] lies between 1.0 eV and 1.1 eV at room temperature. The hole effective mass at 300 K is close to 0.5m$_o$[7.117] (see, also, Reference 7.107). The energy gap of CdTe from the temperature dependence of electrical conductivity measurements by Glazov and Burkhanov[7.118] is close to 0.5 eV (see, also, Reference 7.119). The optical energy gap of Cu$_2$Te is 1.08 eV.[7.120] Goswami and Rao[7.121] suggest that Cu$_2$Te possesses an indirect energy gap. The reported magnitude of the hole effective mass varies between 0.5 m$_o$ and 0.85 m$_o$

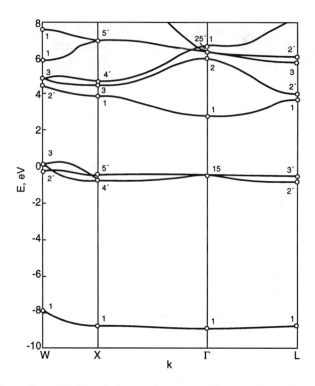

Figure 7.12 Band structure of Cs_3Sb calculated using the empirical pseudopotential method. (After A.A. Mostovskii, V.A. Chaldyshev, G.F. Karachaev, A.I. Klimin, and I.N. Ponomarenko, *Izv. Akad. Nauk SSSR, Ser. Fiz.*, 38, 195, 1974. With permission.)

Table 7.5 $(IA)_3$-VB Compounds. Density, Melting Temperature, Energy Gap, Photoemissive Threshold, and Electrical Conductivity

Compound	X-ray density (g/cm³)	Melting point (K)	Energy gap (eV) Optical at 300 K	Thermal	Photoemissive threshold energy (eV)	Conductivity S/cm (at T)
Li_3As	2.45	1273[7.38]				
α-Li_3Sb	2.98	1420–1570[7.3]		~1.0[7.50]	3.9[7.51]	10^{-3} (663 K)[7.54]
β-Li_3Sb	3.34	923[a][7.1]				
Li_3Bi	5.03	1418[7.54]		0.7[7.51]	3.6[7.51]	10^{-1} (653 K)[7.54]
Na_3P	1.75					
Na_3As	2.37					
Na_3Sb	2.69	1132[7.55]	1.15 (calc.)[7.49] 1.0–1.1[7.56,7.58]			
Na_3Bi	3.70					
h-K_3Sb	2.36	1085[7.58]	1.1–1.3[7.56,7.59] 2.06 (calc.)[7.49]	0.75–0.79[7.59]		~10^{-5} (300 K)[7.27]
c-K_3Sb	2.59		(Ind.) 1.39[7.4] 1.4[7.1] 1.44 (calc.)[7.49] (Dir.) 0.56[7.4]			
c-Rb_3Sb	3.64	1006[7.1]	1.0[7.53]	0.4[7.60]		
Cs_3Sb	4.54	998[7.3]	1.6[7.53] (Ind.) 17.5[7.4] (Dir.) 1.02[7.4]		1.2 (at 300 K)	
c-Rb_3Bi	4.27	915[7.30]		0.04–0.4[7.60]		
Cs_3Bi	5.00	908[7.30]	~0.7[7.61]	0.04–0.4[7.60]		

[a] The temperature of peritectic decomposition.

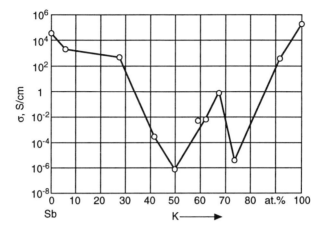

Figure 7.13 The 273-K electrical conductivity vs. composition diagram for the K-Sb system. (After R. Suhrmann and C. Kangro, *Naturwiss*, 40, 137, 1953. With permission.)

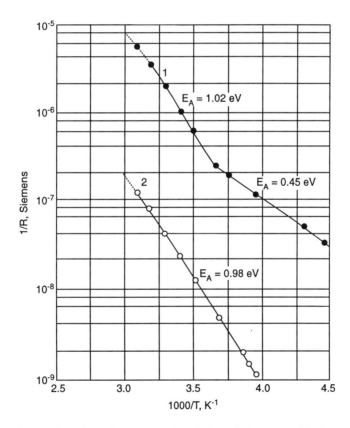

Figure 7.14 The temperature dependence of the electrical conductance of a Li$_3$Sb sample doped with sodium (1) and an undoped Li$_3$Sb sample (2). (After R. Gobrecht, *Phys. Stat. Sol.*, 13, 429, 1966. With permission.)

in the temperature range of 100 K to 640 K;[7.107] the authors of Reference 7.122 found from the Seebeck coefficient measurements that between 300 K and 570 K, the hole effective mass is virtually independent of temperature and is 1.5 m_o to 1.6 m_o.

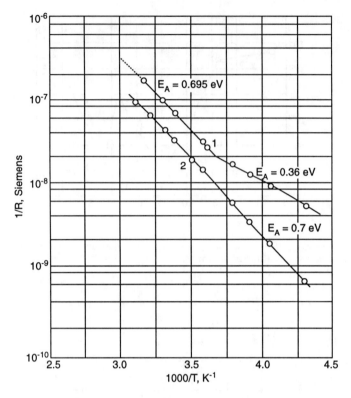

Figure 7.15 The temperature dependence of the electrical conductance of sodium-doped (1) and undoped (2) samples of Li$_3$Sb. (After R. Gobrecht, *Ann. Phys.*, 20, 262, 1967. With permission.)

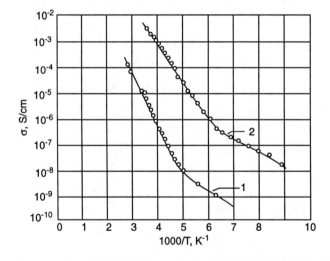

Figure 7.16 Temperature dependence of K$_3$Sb (1) and Cs$_3$Sb (2) electrical conductivity. (After R. Suhrmann and C. Kangro, *Naturwiss*, 40, 137, 1953. With permission.)

The energy gap of α-Ag$_2$S at different temperatures, evaluated from the optical absorption measurements by Junod et al.[7.114] (see Figure 7.27), varies from 0.78 eV to 1.00 eV. The authors of Reference 7.114 suggest the existence of an indirect energy gap of 0.85 eV at 300 K. The effective mass of electrons and holes in α-Ag$_2$S is equal to 4.65 m$_o$ and 7.59 m$_o$, respectively, at 365 K.[7.108] The energy gap of β-Ag$_2$S evaluated from the optical absorption

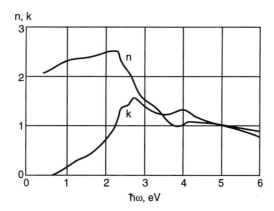

Figure 7.17 Index of refraction, n, and extinction coefficient, K, of Na_3Sb (at 300 K) vs. photon energy. (After A. Ebina and T. Takahashi, *Phys. Rev.*, B7, 4712, 1973. With permission.)

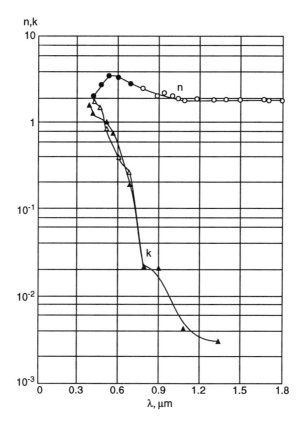

Figure 7.18 Index of refraction, n, and extinction coefficient, k, of Cs_3Sb vs. wavelength. (After K. Hirschberg and K. Deutscher, *Phys. Stat. Sol.*, 26, 527, 1968. With permission.)

measurements by Junod et al.[7.114] at room temperature is 0.4 eV; the carrier effective mass is 0.23 m_o for both electrons and holes.[7.123-7.125]

Silver selenide, Ag_2Se, belongs to the narrow band semiconductors. Its optical energy gap at 293 K is 0.15 eV;[7.114] the effective mass of electrons and holes in accordance with Junod[7.108] is 0.32 m_o and 0.54 m_o, respectively. The known electronic properties of Ag_2Te are limited by the optical energy gap of about 0.67 eV[7.126] and the electron effective mass of 0.026 m_o to 0.034 m_o (63 to 83.4 K) evaluated from the Hall and Seebeck effects measurements.[7.127]

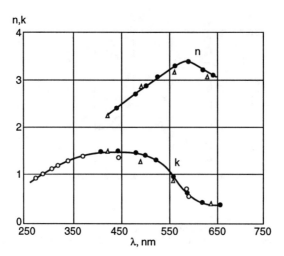

Figure 7.19 Index of refraction, n, and extinction coefficient, k, of Cs_3Sb in the UV and visible parts of the spectrum. (After K. Hirschberg and K. Deutscher, *Phys. Stat. Sol.*, 26, 527 1968. With permission.)

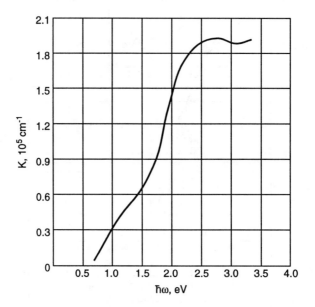

Figure 7.20 Optical absorption coefficient in Cs_3Bi at 300 K vs. photon energy. (After A.H. Sommer and W.E. Spicer, *J. Appl. Phys.*, 32, 1036, 1961. With permission.)

The room temperature electric transport properties of the copper and silver chalcogenides are compiled in Table 7.10. One can see from the table that some of these materials (e.g., Ag_2Se and Ag_2Te) possess the mobility and Seebeck coefficient magnitudes which, together with the materials' ability to change the conductivity type, present a certain interest for various applications. The temperature hysteresis of electrical conductivity of Cu_2S, reported by Okamoto and Kawai and related to the β-γ transition (see Figure 7.28) may be used in current-controlled contactless switches. The room temperature thermoelectric figure of merit of Cu_2Se ($6 \cdot 10^{-3}$ 1/K[7.128]) and the rapid increase of its Seebeck coefficient with temperature (Figure 7.29) make this material prospective for thermoelectric applications.

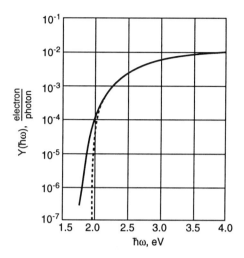

Figure 7.21 Photoemissive yield in electrons per photon for Cs_3Bi at 300 K vs. photon energy. Solid line is based on experimental data, dashed line — a theoretical fit. (After A.S. Sommer, *Photoemissive Materials*, John Wiley & Sons, 1968. With permission.)

The temperature dependence of the magnetic susceptibility of Ag_2S and Ag_2Se was reported in Reference 7.114; its room temperature magnitude is about $-29 \cdot 10^{-6}$ and $-34 \cdot 10^{-6}$ SI units, respectively. The high frequency limit of the Cu_2Se dielectric constant at 300 K is between 11.0 and 11.6.[7.107] The spectral distribution of the Cu_2S index of refraction and imaginary part of the complex dielectric constant are shown in Figure 7.30 (adapted from Reference 7.116). The Cu_2S reflection coefficient vs. photon energy is presented in Figure 7.31 (adapted from Reference 7.129). The spectral distribution of reflectivity in Cu_2Se at different temperatures is shown in Figure 7.32 (adapted from Reference 7.107). The Cu_2Te basic optical properties measured on thin films are presented in Figure 7.33 (adapted from Reference 7.121). The spectral distribution of the absorption coefficient at different temperatures for Ag_2S and Ag_2Se is shown in Figure 7.34 (adapted from Reference 7.114).

Copper and silver chalcogenides are available commercially. The relevant information is gathered in Table 7.11.

7.4. II$_2$-IV COMPOUNDS WITH ANTIFLUORITE STRUCTURE

The history of four compounds of magnesium with the Group IVB elements extends to the beginning of this century when Kurnakov and Stepanov[7.130] first predicted and synthesized Mg_2Sn. The first information regarding the crystal structure of Mg_2-IV compounds was, apparently, reported by Owen and Preston,[7.131] Sacklowski,[7.132] and Friauf.[7.133] Semiconductor properties of the compounds were initially described by Robertson and Uhlig.[7.135]

The Mg_2-IV compounds crystallize in the lattice where the Group IV atoms form a face-centered cube and the magnesium atoms occupy all (eight) tetrahedral voids (Figure 7.35). The lattice is similar to that of fluorite, CaF_2, but because the atoms of magnesium ("cations") occupy the positions of the fluorine atoms (anions), the structure is called the antifluorite. In this lattice, each atom of magnesium is surrounded by four Group IV atoms, while each Group IV atom has eight magnesium atoms, forming a cube, as its nearest neighbors. The space group of the structure is O_h^5-Fm3m; the unit cell contains four formula units. Some basic chemical and crystal structure parameters are compiled in Table 7.12.

Table 7.6 I-O and I$_2$-O Compounds. Molecular Weight and Number, Lattice Parameters, and Density[7.1,7.3]

Compound	Molecular weight	Mean atomic weight	Mean atomic number	Crystal structure and space group	Lattice parameters (RT)				Density (g/cm³)
					a (nm)	b (nm)	c (nm)	β°	
CuO	79.55	39.77	18.5	Monoclinic C_{2h}^6-C2/c	0.4652	0.3410	0.5108	99.48	6.569
Cu$_2$O	143.09	47.70	22	Cubic T_h^2-Pn3 or O_h^4-Pn3m	0.4296				5.75–6.14 5.99 (X-ray)
AgO	123.87	61.93	27.5	Monoclinic C_{2h}^5-P2$_1$/c Or pseudocubic	0.5852 0.458	0.3478 0.483	0.5495 0.483	107.5 88 93 87	7.44
Ag$_2$O	231.74	77.25		Cubic T_h^2-Pn3	0.472				6.9

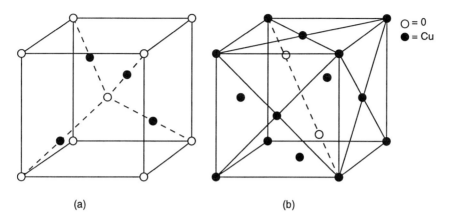

Figure 7.22 Cuprite, Cu_2O, crystal structure. (a) Origin at an O site; (b) origin at a Cu site.

Table 7.7 Copper and Silver Oxides. Thermal, Thermodynamic, and Mechanical Properties (References 7.1 to 7.3, If Not Noted Otherwise)

Parameter	CuO	Cu_2O	AgO	Ag_2O
Melting point, K	1512	1508 (at 80 Pa)		1088
Standard Gibbs energy of formation, kJ/mol	−129.7	−146.0	−27.6	−11.2
Standard enthalpy of formation, kJ/mol	−157.3	−168.6	−24.3	−31.1
Heat capacity (C_p) at 298.15 K (J/mol·K)	42.3	63.6	88.0	65.9
Coefficient of thermal expansion at RT, 10^{-6} 1/K		0.23		
Debye temperature, K (at 298 K)		188 [7.76]		
Elastic moduli (in GPa, at 293 K)				
c_{11}		126.1 [7.77]		
c_{12}		108.6 [7.77]		
c_{44}		13.6 [7.77]		
Young's modulus (in GPa, at 293 K)		30.12 [7.76]		
Adiabatic bulk modulus at 300 K, GPa		110 [7.91]		
Compressibility at 298 K, 10^{-12} m²/N		89.3 [7.76]		
Hardness, Mohs	3–4	3.5–4		

The Mg_2-IV compounds melt congruently; the solubility of the components in the compounds in the solid state is virtually nonexistent.[7.55] The thermochemical properties of the compounds are listed in Reference 7.136. The temperature dependence of their heat capacity in the range from 5 K to 300 K was reported in References 7.139 to 7.142. The detailed data on dependence of volume (or density) on temperature for Mg_2Si, Mg_2Sn, and Mg_2Pb in the range from 415 K to up to 1440 K are gathered in Reference 7.136. The thermal expansion of the Mg_2-IV compounds was also reported in References 7.139 to 7.141. The elastic properties of the compounds were reported in References 7.139 to 7.142, and 7.144 to 7.147. The data on thermochemical, thermal, and elastic properties of the compounds are compiled in Table 7.13.

The results of calculations of the band structure of four Mg_2-IV compounds are presented in References 7.148 and 7.149 (Figure 7.36). The lowest minima of the conduction bands are located at or near to the point X of the Brillouin zone (see Figure 3.15b, Chapter 3). The maxima of the valence bands are located at the Brillouin zone center. Some information relevant to the band structure of the compounds is compiled in Table 7.14.

The electrical and thermal transport properties of the Mg_2-IV compounds were studied by Robertson and Uhlig,[7.135] Busch and Winkler,[7.143,7.152] Morris et al.,[7.153,7.156] Busch and Schneider,[7.154] Blunt et al.,[7.150] Frederikse et al.,[7.155] Heller and Danielson,[7.157] Berger,[7.158]

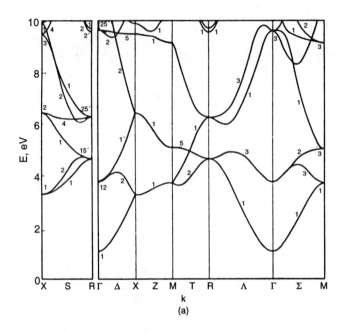

Figure 7.23 Band structure of Cu_2O. (a) Conduction bands; (b) valence bands. (After L. Kleinman and K. Mudnick, *Phys. Rev.*, B21, 1549, 1980. With permission.)

Glazov et al.,[7.136] Shanks,[7.161] Geick et al.,[7.162] and Stringer and Higgins.[7.163] Some information relevant to the thermal and electrical transport properties of Mg_2-IV compounds is presented in Table 7.15.

The temperature dependence of the electrical resistivity and Seebeck coefficient of Mg_2Si is shown in Figure 7.37 (adapted from Reference 7.157). The relative stability and high magnitude of the Seebeck coefficient at temperatures up to 400 to 500 K single out Mg_2Si from other Mg_2-IV compounds (see, e.g., References 7.143 and 7.158). Mg_2Ge has a high (up to 500 to 600 μV/K) Seebeck coefficient, but only in a narrow temperature range between about 300 and 450 K, while its magnitude for Mg_2Sn does not exceed 300 to 350 μV/K and rapidly decreases with temperature growth above 300 K. The temperature dependence of electrical conductivity of four Mg_2-IV compounds in the solid and liquid states was reported in Reference 7.136. The data are presented in Figure 7.38. The authors have come to the conclusion that the melting process is accompanied by dissociation of the compounds and domination of the metallic-type conductivity in the liquid state. The thermal conductivity of Mg_2Si, Mg_2Ge, and Mg_2Sn is shown in Figure 7.39 (adapted from References 7.158 and 7.160).

Glazov et al.[7.136] found that at high temperatures, the Mg_2-IV compounds are diamagnetics in both the solid and liquid state. Korenblit and Kolesnikov,[7.164] who measured the magnetic susceptibility of mg_2Sn samples with different carrier concentrations in the temperature range from 300 to 500 K, have found that most samples are paramagnetics, but their magnetic susceptibility rapidly decreases with temperature increase, and in some samples, the diamagnetic component of susceptibility dominates at temperatures as low as 500 K. The dielectric constant magnitudes and some optical properties of the compounds are gathered in Table 7.16.

Synthesis of the Mg_2-IV compounds can be performed in either alumina or any other crucible that does not react with magnesium (fused silica intensively reacts with mg at elevated temperatures) by melting of the mixture of elemental components in vacuum or in a hydrogen atmosphere.[7.150] Frederikse et al.[7.155] have grown single crystals of Mg_2Sn by the Kyropoulos method. The low vapor pressure of the components (the boiling points are 1363, 3106, 3106, 2022, 3538, and 2875 K for Mg, Ge, Pb, Si, and Sn, respectively) indicates that the single

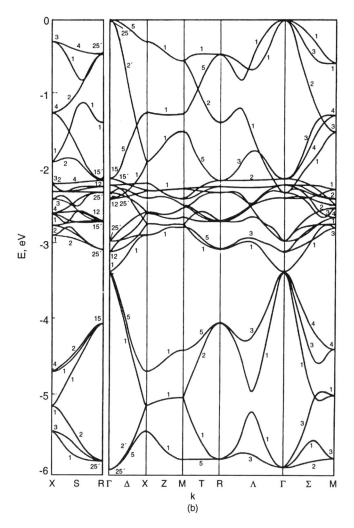

Figure 7.23 (continued)

crystals of the compounds can be grown using Czochralski, Bridgman, and other techniques on the industrial scale.

Mg$_2$Si 20-mesh powder of 99.5 % purity is available commercially for $1.20 to $2.14 per gram (1995).

7.5. IV-VI GALENITE-TYPE COMPOUNDS

The IV-VI compounds are the electron analogs of the Group V elements. One of the materials of this group — mineral galenite, PbS — was widely used at the very dawn of modern radio for rectification of the high-frequency currents induced by electromagnetic waves in the electrical circuits.[7.168]

IV-VI compounds are presented in the world of minerals also by clausthalite, PbSe, massicot (litharge), PbO, and teallite, which may be considered as the material of composition PbS·SnS, herzenbergite, SnS, and altaite, PbTe. Two minerals — cassiterite, SnO$_2$, and plattnerite, PbO$_2$ — that do not belong to IV-VI compounds are sometimes considered as salts of acids (SnO)SnO$_3$ and (PbO)PbO$_3$.

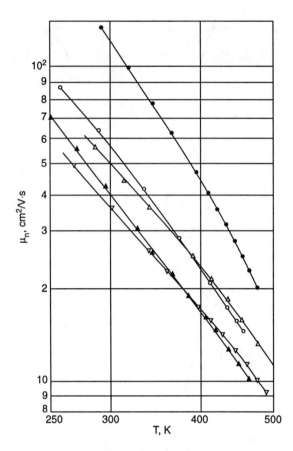

Figure 7.24 Temperature dependence of the carrier mobility in five samples of Cu_2O. (After G.P. Pollack and D. Trivich, *J. Appl. Phys.*, 46, 163, 1975. With permission.)

Semiconductor properties of IV-VI compounds were described first by Eisenmann[7.169] and Hintenberg,[7.170] who reported the parameters of PbS and PbTe such as carrier mobility, Seebeck coefficient, and the thermal energy gap. Properties of PbS, PbSe, and PbTe are also studied by Smith.[7.171]

The IV-VI compounds crystallize in the cubic galenite-type structure or in a distorted NaCl structure, with the coordination number of six and with four formula units in unit cell. The basic chemical and lattice parameters of the compounds are compiled in Table 7.17.

The specifics of many of IV-VI compounds consist of presence of small amounts (0.3 to 0.6 at%) of second phase in the samples of the stoichiometric composition. This problem was discussed, in particular, in References 7.172 to 7.174 (see, also, Reference 7.136, p. 123).

The thermochemical and thermal properties of Group IVB chalcogenides are presented in Table 7.18.

The mechanical properties of IV-VI compounds have not been sufficiently studied as yet. The experiments show that they are relatively soft materials. The available data on their hardness and elastic moduli are gathered in Table 7.19.

The investigation of the electronic properties of IV-VI compounds, especially PbS, has been going on for quite an extended period of time. It is well established that the conductivity type of samples of lead chalcogenides can be changed by introduction either of an excessive amount of lead or chalcogen (to produce n-type or p-type conductivity, respectively). Oxygen, being one of the most common impurities, behaves as an acceptor in the lead chalcogenides.

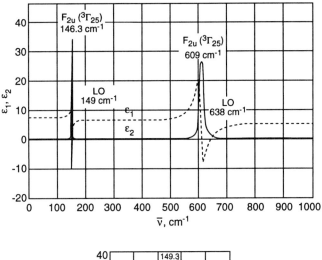
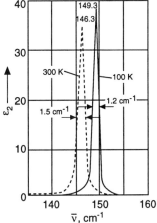

Figure 7.25 Dielectric properties of Cu_2O. (a) Real, ε_1, and imaginary, ε_2, parts of the dielectric constant at 300 K vs. wavenumber; (b) a detail showing the effect of temperature on ε_2 in the region of low-frequency resonance. (After P. Dawson, M.M. Hargreave, and G.R. Wilkinson, *J. Phys. Chem. Solids*, 34, 2201, 1973. With permission.)

The Brillouin zone for the orthorhombic lattice with D_{2h}^{16}-Pnma space group (GeS, GeSe, SnS, and SnSe) is shown in Figure 7.40. The GeS band structure calculated by Grandke and Ley[7.189] is shown in Figure 7.41. The minimum (direct) energy gap of 1.65 eV (at 300 K and $\vec{E} \| a$)[7.1] is related to Λ_{1v}-Λ_{1c} transition. Germanium selenide is an indirect semiconductor with minimum energy gap of 1.075 eV (at 300 K and $\vec{E} \| a$).[7.1] The tin sulfide valence band structure was calculated by Parke and Srivastava.[7.190] Its minimum (indirect) energy gap at 300 K and $\vec{E} \| b$ is 1.095 eV,[7.191] in good agreement with the magnitude (1.087 eV) reported by Nicolic and Vujatovic.[7.192] The band structure of SnSe was calculated by Car et al.[7.193] and is similar to that of GeS. The minimum indirect energy gap at 300 K is 0.9 eV.[7.1]

The electronic structure of the IV-VI compounds with cubic symmetry (space group O_h^5-Fm3m) was studied by Tsu et al.,[7.194] Cohen et al.,[7.195] Herman et al.,[7.196] Tung and Cohen,[7.197] Bahl and Chorpa,[7.198] and Baleva[7.199] (GeTe); Burke et al.,[7.200,7.201] Herman et al.,[7.196] Tung and Cohen,[7.197] Rabii,[7.202] Tsang and Cohen,[7.203] Savage et al.,[7.204] Kemeny and Cardona,[7.205] Melvin and Hendry,[7.206] and Katayama[7.207] (SnTe). The band parameters of PbS, PbSe, and PbTe are reviewed, in particular, in References 7.208 to 7.210. A fully relativistic

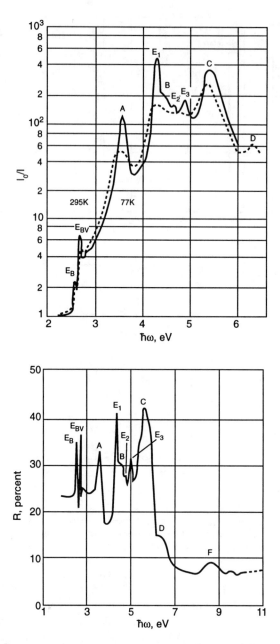

Figure 7.26 Optical properties of Cu$_2$O. (a) The UV absorption spectra of a thin film at 77 and 295 K; (b) reflection from a thick single crystal at 77 K. (After S.N. Kitine, *Excitons: Optical Properties of Solids,* Plenum Press, 1969. With permission.)

band structure of PbS, PbSe, and PbTe was calculated by Valdivia and Barberis.[7.437] Lead chalcogenides constitute a group of semiconductors with unique properties in comparison with a majority of other semiconductors. Their distinguishing features are a narrow fundamental band gap that widens with the temperature increase, high charge carrier mobility, and high dielectric constant. The diluted magnetic materials based on these semiconductors are of substantial interest for practical applications (see, e.g., Reference 7.211). The materials have found wide application in light sources and infrared detectors.

The band structure of GeTe is shown in Figure 7.42 (adapted from Reference 7.197) together with the Brillouin zone for the NaCl structure. The minimum energy gap (transition

Table 7.8 I_2-VI Compounds. Molecular Weight, Mean Atomic Number, Crystal Structure, Lattice Parameters, and Density[7.1–7.3,7.106–7.110]

Compound	Molecular weight	Mean atomic weight	Mean atomic number	Crystal structure and space group	a (nm)	b (nm)	c (nm)	β°	Density (g/cm³) X-ray	Density (g/cm³) Pycnometric
Cu_2S	159.158	53.05	24.7	Hexagonal D_{6h}^5-$P6_3/mmc$ Z = 6 (chalcocite)	0.3961		0.6722		5.79	5.8[7.1] 5.6[7.2]
Cu_2Se	206.05	68.68	30.7	Tetragonal D_{4h}^{17}-$I4/mm$ (crookesite)	1.040		0.393			7.1[7.106]
Cu_2Te	254.69	84.90	36.7	Hexagonal superstructure	1.254		2.171		7.48	7.4[7.107]
α-Ag_2S	247.80	82.60	36.7	Cubic (Z = 4) Monoclinic C_{2h}^5-$P2_1/n$, Z = 4	0.611 0.423 0.4226	0.691 0.6925	0.787 0.7858	99.58 99.69	7.42 7.26	7.1 7.234[7.1] 7.110
β-Ag_2S				Cubic, Z = 2	0.488				7.08	7.1
α-Ag_2Se	294.70	98.23	42.7	Orthorhombic D_2^4-$P2_12_12_1$, Z = 4	0.4333 0.4332	0.7062 0.7065	0.7764 0.7762		8.24	8.25[7.1] 7.110
β-Ag_2Se				Cubic O_h^9-$Im3m$, Z = 2	0.499				7.88	7.108
α-Ag_2Te	343.34	114.45	48.7	Monoclinic (or rhombic) C_{2h}^5-$P2_1/c$	0.6593	0.62	0.6455	88.50	8.20	8.08 to 8.41[7.1]
β-Ag_2Te				Cubic (Z = 4)	0.658				8.00	7.1
γ-Ag_2Te				Cubic (Z = 2)	0.529				7.70	7.3

Table 7.9 I$_2$-VI Compounds. Thermal, Thermodynamic, and Mechanical Properties

Compound	Melting point (K)	Enthalpy of formation (kJ/mol)	Enthalpy of fusion (kJ/mol)	Heat capacity (C_p) (J/mol·K)	Debye temperature (K)	Compressibility (10^{-12} m^2/N)	Hardness Mohs
Cu$_2$S	≈1370[7.2] 1373[7.1]	−79.5[7.2]		76.3[7.2] (298.15 K)			2.5–3.0 (mineral)
Cu$_2$Se	1386[7.2]			73.4 at 193 K 99.5 at 393 K[7.112]			
Cu$_2$Te	1400[7.2]			77.0 at 333 K[7.112]			
Ag$_2$S	1098[7.2]	−32.6[7.2]	14.10[7.2]	76.15 (298.16 K)[7.113] 76.5 (298.15 K)[7.2]		3.207[7.75]	2.0–2.5 (argentite)
Ag$_2$Se[a]	1153[7.2] 1153–1170[7.114]				141 (RT)[7.75]		2.5 (naumannite)
Ag$_2$Te	1228[7.2]						2.5–3.0 (hessite)

[a] The RT coefficient of linear thermal expansion of α-Ag$_2$Se and β-Ag$_2$Se is 35.4·10^{-6} and 18.1·10^{-6} 1/K, respectively.[7.75]

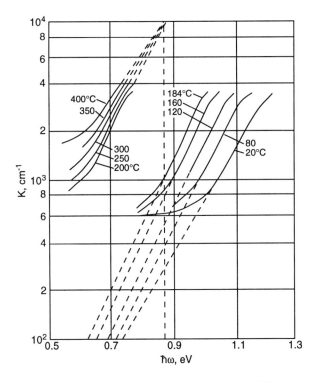

Figure 7.27 Spectral distribution of the absorption coefficient for Ag_2S at different temperatures. (After P. Junod, H. Hediger, B. Kilchër, and J. Wullschleger, *Phil. Mag.*, 36, 941, 1977. With permission.)

$L_{6v}^- \to L_{6c}^+$) from different sources is between 0.1 eV and 0.2 eV. The minimum energy gap of cubic β-SnTe also is related to the transition at point L of the Brillouin zone, is close to 0.19 eV,[7.1,7.206] and strongly depends on the carrier concentration (Burstein shift). The information pertinent to the band structure of the IV-VI compounds is gathered in Table 7.20.

The electrical conductivity measurements on p-type needle-shaped single crystals of GeS by the authors of Reference 7.213 showed that its magnitude at room temperature is about $4 \cdot 10^{-7}$ S/cm; the hole mobility measurements by Stanchev and Vodenicharov[7.214] performed on GeS thin films showed magnitudes between 80 $cm^2/V \cdot sec$ and 90 $cm^2/V \cdot sec$. The mobility increases with temperature in the range of 298 K to 350 K. Yabumoto[7.215] found the GeS Seebeck coefficient up to −2.0 mV/K on polycrystalline samples at temperatures close to 350 K. The thermal conductivity of GeS and GeSe was measured on small single crystals by Okhotin et al.[7.216] Their results are shown in Figure 7.43. The room temperature thermal conductivity of GeS is close to 20 $mW/cm \cdot K$.[7.216]

Kiriakos et al.[7.218] found that electrical conductivity of GeSe along principal crystallographic axes increases with temperature, passes through a maximum near 370 K, and decreases with the following temperature growth (Figure 7.44), while the Hall coefficient is virtually independent of temperature above 370 K. It is possible to assume the existence of a phase transition at this temperature, although it has not been confirmed by other techniques. The electrical conductivity (p-type) at room temperature, reported by different authors, is of the order of 10^0 S/cm; the hole mobility is between 50 $cm^2/V \cdot sec$ and 80 $cm^2/V \cdot sec$. Okhotin et al.[7.216] reported the data on the Seebeck coefficient (about 1 mV/K at room temperature). The thermal conductivity of GeSe (see Figure 7.43) is slightly higher than that of CdS (about 22 $mW/cm \cdot K$ at room temperature).[7.216]

A distinguishing feature of GeTe electrical resistivity is its low magnitude (10^{-4} Ohm·cm) and positive temperature coefficient. Near 670 K (transition point from rhombohedral α-GeTe to cubic β-GeTe), the electrical resistivity vs. temperature curve passes through a sharp bend;

Table 7.10 Copper and Silver Chalcogenides. The Room Temperature Transport Properties

Compound	Conductivity type	Resistivity (Ohm·cm)	Mobility (cm²/V·sec) Electrons	Mobility (cm²/V·sec) Holes	Concentration (10^{19} cm^{-3}) Electrons	Concentration (10^{19} cm^{-3}) Holes	Seebeck coefficient (μV/K)	Thermal conductivity (mW/cm·K)	Ref.
Cu_2S	p	0.06–40		3.02–4.75		0.33–3.5	267–327		7.1
Cu_2Se	p	$2 \cdot 10^{-3}$		21–24		14–17	135	$15.7^{7.26}$	7.128
Cu_2Te ($Cu_2Te_{1.025}$)	p	$5.3 \cdot 10^{-4}$		15–19		65–80	40		7.122
α-Ag_2S	n	$(6-8) \cdot 10^{-4}$	63.5 (365 K)		$n_i = 3.7 \cdot 10^{-4}$				7.122
β-Ag_2S	n		160		$n_i = 4.0$				7.108
Ag_2Se	n	$5 \cdot 10^{-4}$ to $1.2 \cdot 10^{-3\,7.114}$	800–2000	400–600	$n_i \simeq 0.3$				7.108
α-Ag_2Te	n	10^{-3}	$(1-2) \cdot 10^{3\,7.3}$				120		7.127
β-Ag_2Te and γ-Ag_2Te	n		$\simeq 2000$ (extrap.)	$\simeq 25$ (extrap.)					7.1

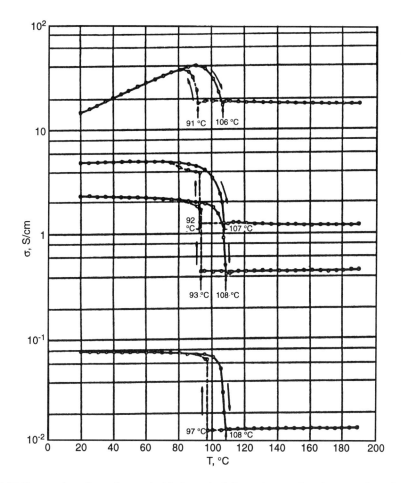

Figure 7.28 Temperature dependence electrical conductivity vs. temperature for several nearly stoichiometric samples of Cu_2S samples. (After K. Okamoto and S. Kwai, *Jpn. J. Appl. Phys.*, 12, 1130, 1973. With permission.)

β-GeTe also has the positive temperature coefficient of resistivity.[7.219] At room temperature, the hole mobility is up to 200 cm^2/V·sec[7.198b,7.220] and strongly depends on the hole concentration, which usually lies in the limits 10^{20} cm^{-3} to 10^{22} cm^{-3}. The majority of the available data on electrical properties of GeTe relates to the rhombohedral alpha-modification. Hein et al.[7.221] observed the transition to the superconducting state on the GeTe samples with the hole concentration greater than $8 \cdot 10^{20}$ cm^{-3}. The thermal conductivity of GeTe has been measured by Kolomoets et al.[7.222] on polycrystalline samples; its magnitude at room temperature is close to 30 mW/cm·K.

Albers et al.[7.223] found room temperature electrical conductivity of SnS single crystals close to 7 S/cm with the hole concentration of about $5 \cdot 10^{18}$ cm^{-3}, the measurements on single crystals by Yabumoto[7.224] gave a magnitude close to $8.5 \cdot 10^{-6}$ S/cm. The authors of Reference 7.223 report that the hole mobility in SnS at 300 K along the a axis is only slightly (about 15%) greater than that along the b axis and is close to 90 cm^2/V·sec, while the mobility magnitude along the c axis is only about 15 cm^2/V·sec.

The results of measurements of the SnSe electrical conductivity and Hall coefficient between 100 and 650 K reported in Reference 7.225 show that the room temperature conductivity of the Sb-doped n-type single crystals varies in the limits from 5 cm^2/V·sec to 60 cm^2/V·sec depending on the heat treatment history of the samples. Both annealing and

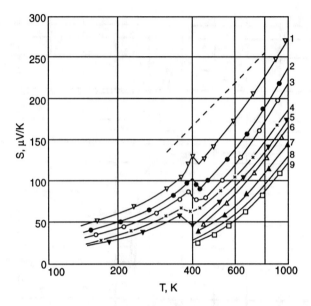

Figure 7.29 Temperature dependence of the Seebeck coefficient in Cu_2Se samples containing excessive Se from $1.9 \cdot 10^{20}$ cm^{-3} (sample 1) to $2.45 \cdot 10^{21}$ cm^{-3} (sample 9). Dashed line — theoretical dependence of S on T. (After A.A. Voskanyan, P.N. Inglizyan, S.P. Lalykin, I.A. Plyutto, and Ya.M. Shevchenko, *Fiz. Techn. Poluprov.*, 12, 2096, 1978. With permission.)

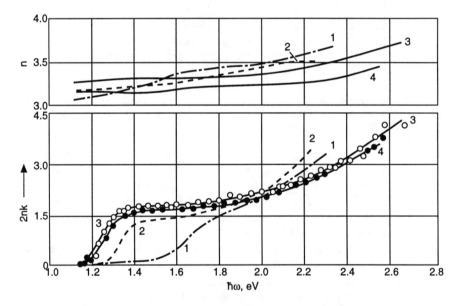

Figure 7.30 Index of refraction, n, and imaginary part of the complex dielectric constant, $\varepsilon_2 = 2nk$, vs. photon energy for platelets of chalcocite, Cu_2S, with different orientation of the crystal lattice relative to the sample surface. (After B.J. Mulder, *Phys. Stat. Sol.*, A18, 633, 1973. With permission.)

quenching increase conductivity. Undoped crystals have p-type conductivity and room temperature mobility between 80 cm^2/V·sec and 120 cm^2/V·sec.[7.226] Asanabe[7.227] found that the Seebeck coefficient of both single-crystalline samples and polycrystals passes through a maximum near 550 K.

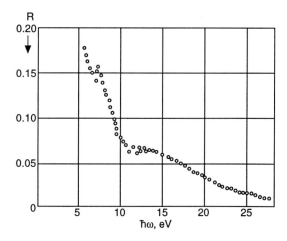

Figure 7.31 The Cu$_2$S reflection coefficient vs. photon energy. (After S. Dushemin and F. Guastavino, *Solid State Comm.*, 26, 187, 1978. With permission.)

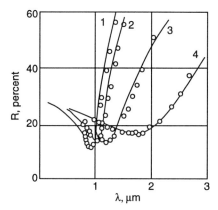

Figure 7.32 Spectral distribution of the Cu$_2$Se reflectivity at 100 K (1), 300 K (2), 370 K (3), and 450 K (4) measured on a sample with p = 2.04·10^{21} cm^{-3} in the region of the plasma minimum. (After V.V. Gorbacher and I.M. Putilin, *Phys. Stat. Sol.*, A16, 553, 1973. With permission.)

Tin telluride, like GeTe, can be transferred into the superconducting state.[7.228] SnTe samples usually have p-type conductivity with hole mobility at 300 K up to 800 cm^2/V·sec, depending on the charge concentration (see Figure 7.45).

Both tetragonal (litharge) and orthorhombic (massicot) allotropes of lead oxide, PbO, are materials with the electrical resistivity at room temperature between 1 Ohm·cm and 100 Ohm·cm and between 10^3 Ohm·cm and 10^5 Ohm·cm, respectively. Keezer et al.[7.230] measured temperature dependence of resistivity and Hall coefficient on PbO single crystals grown by a variant of hydrothermal technique at about 670 K and the oxygen pressure between 10 MPa and 150 MPa (litharge) and between 200 MPa and 310 MPa (massicot); their results are shown in Figure 7.46. The electron mobility varies in the limits of 20 cm^2/V·sec to 80 cm^2/V·sec depending on carrier concentration.

The calculated intrinsic current carrier concentration vs. temperature between 77 K and 350 K in PbS is shown in Figure 7.47 (adapted from Reference 7.208). The electron mobility at room temperature is 700 cm^2/V·sec;[7.1] the conductivity of synthetic and natural PbS crystals is of the order of 10^2 S/cm and 10 S/cm, respectively, in reasonable agreement with the data

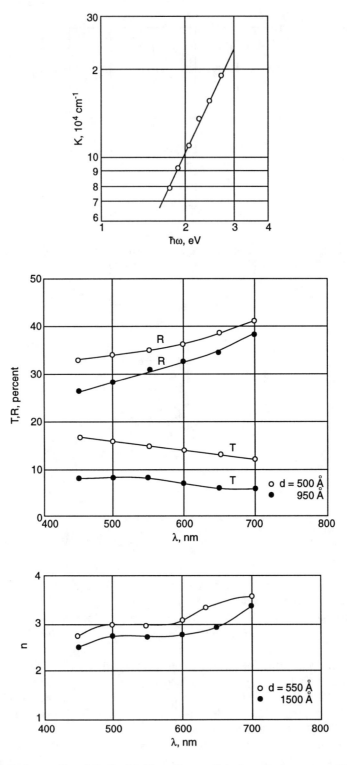

Figure 7.33 Optical properties of Cu_2Te. (a) Absorption coefficient vs. photon energy; (b) transmission, T, and reflectance, R, vs. wavelength (films); (c) index of refraction vs. wavelength. (After A. Goswami and B.V. Rao, *Indian J. Phys.*, 50, 50, 1976. With permission.)

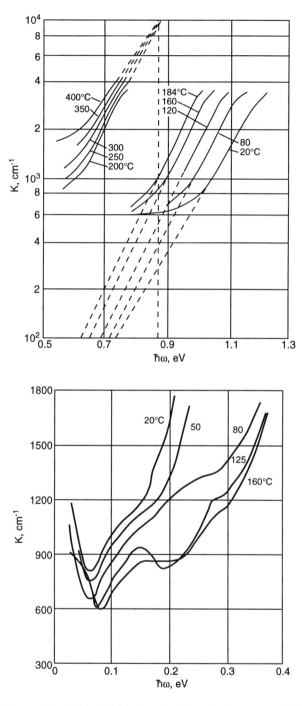

Figure 7.34 Absorption coefficient of Ag_2S (a) and Ag_2Se (b) vs. photon energy at different temperatures. (After P. Junod, H. Hediger, B. Kilchör, and J. Wullschleger, *Phil. Mag.*, 36, 941, 1977. With permission.)

of References 7.136 and 7.186. The natural galenite crystals investigated in Reference 7.186 (source: Pribram, Bohemia) have conductivity at room temperature of 0.1 S/cm to 0.2 S/cm; the samples containing 0.3% Fe have the p-type conductivity and the Seebeck coefficient up

Table 7.11 Availability of Copper and Silver Chalcogenides[7.134]

Material	Form	Purity (%)	1995–1996 price, US $/gram	
			From	To
Cu_2S	Powder	99.5	0.23	0.42
Cu_2Se	Lumps	99	3.34	4.54
Cu_2Te	Pieces up to 3 mm	99.5	5.68	7.74
Ag_2S	Lumps	99.9+	3.00	4.06
Ag_2Te	Powder	99.99	22.20	31.00

From Alfa-Aesar, Johnson Matthey Catalog, 1995–1996.

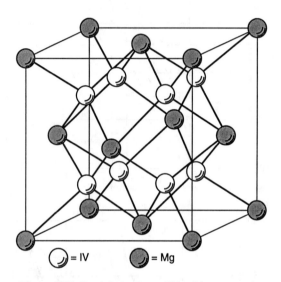

Figure 7.35 Crystal structure of Mg_2-IV compounds.

Table 7.12 Mg_2-IV Compounds. Molecular Weight, Lattice Parameters, and Density (300 K)

Compound	Molecular weight	Mean atomic weight	Mean atomic number	Lattice constant (nm)	Density (g/cm³)
Mg_2Si	76.696	25.57	12.7	0.6338 [7.136]	1.99 [7.2]
					1.88 [7.1]
					2.00 (X-ray)
Mg_2Ge	121.22	40.41	18.7	0.6380 [7.136]	3.09 [7.2]
				0.6393 [7.137]	3.08–3.10 (X-ray)
Mg_2Sn	167.320	55.77	24.7	0.6765 [7.136]	3.60 [7.2]
				0.6762 [7.137]	3.59 (X-ray)
Mg_2Pb	255.81	85.27	35.3	0.6836 [7.136]	5.1 [7.136]
				0.6860 [7.138]	5.54 [7.1]
					5.26–5.32 (X-ray)

to 0.6 mV/K, the dominating impurities in the n-type samples are Cu (0.2%) and Fe (0.1%), and the Seebeck coefficient in these samples reaches 1.1 mV/K at 300 K. The results of the high-temperature electrical conductivity measurements of PbS, PbSe, and PbTe performed by Glazov et al.[7.136] and Ioffe and Regel[7.231] are shown in Figure 7.48 (adapted from

Table 7.13 Mg$_2$-IV Compounds. Thermal, Thermochemical, and Elastic Properties

Parameter	Compound			
	Mg$_2$Si	Mg$_2$Ge	Mg$_2$Sn	Mg$_2$Pb
Melting point, K	1375 [7.1]	1388 [7.3]	1051 [7.3]	828 [7.3]
Standard enthalpy of formation, kJ/mol [7.136]	−79.08		−76.57	−52.72
Enthalpy of fusion, kJ/mol [7.136]	85.35		47.70	38.91
Heat capacity (300 K), J/mol·K	67.87 [7.139]	69.62 [7.140]	72.50 [7.141]	72.43 [7.142]
Coefficient of thermal expansion (300 K), 10^{-6} 1/K	11.5 [7.145]	15.0 [7.146]	9.9 [7.141]	10.0 [7.142]
Debye temperature (300 K), K	450 [7.144]	363 [7.140]	240 [7.141]	244 [7.142]
Elastic moduli, GPa at 300 K				
c_{11}	121 [7.144]	117.9 [7.146]	82.4 [7.147]	
c_{12}	22 [7.144]	23 [7.146]	20.8 [7.147]	
c_{44}	46.4 [7.144]	46.5 [7.146]	36.6 [7.147]	
Compressibility (300 K), 10^{-11} m^2/N	1.81 [7.139]	1.83 [7.140]	2.83 [7.141]	2.59 [7.142]

Reference 7.136). All three materials remain semiconductors in the liquid state. The thermal conductivity of PbS near room temperature is 29 mW/cm·sec.[7.232]

Similar to PbS, lead selenide has very high mobility of both electrons and holes at low temperatures (1.4·10^5 cm^2/V·sec and 1.4·10^4 cm^2/V·sec, respectively, at 4 K and 77.4 K — see e.g., References 7.233 and 7.234). Schlichting and Gobrecht[7.235] found that the mobility is virtually independent of temperature below 10 K. Above 40 K and up to room temperature, the electron mobility is proportional to T$^{-5/2}$, while for holes the temperature dependence of mobility is close to T$^{-3/2}$. This difference may be related to the processes typical for the multivalley semiconductors (see, e.g., Reference 7.236). The authors of Reference 7.235 found that the room temperature mobility of both electrons and holes in PbSe is close to 300 cm^2/V·sec. The PbSe thermal conductivity at room temperature is 17 mW/cm·K.[7.185]

The intrinsic carrier concentration in lead telluride is presented in Reference 7.208 with its magnitude at room temperature close to 1·10^{16} cm^{-3}; the measured electron concentration is about six to seven times greater.[7.237] The room temperature electron mobility in PbTe single crystals is close to 1700 cm^2/V·sec; the hole mobility is 780 cm^2/V·sec.[7.3] Schlicht et al.[7.237] have reported that the room temperature electron mobility in PbTe single-crystalline films is between 600 cm^2/V·sec and 900 cm^2/V·sec; its magnitude at liquid helium temperatures is close to 8·10^5 cm^2/V·sec, in quite good agreement with the data cited in Reference 7.136, p. 128. The room temperature thermal conductivity of PbTe is 20 mW/cm·K.[7.185]

Majority of IV-VI compounds are diamagnetics. The data on their magnetic susceptibility and dielectric constant along with some information on index of refraction are compiled in Table 7.21.

The optical properties of IV-VI compounds, especially lead chalcogenides (see, e.g., References 7.217, 7.240, and 7.248) still attract constant and close attention of researchers, because the optical applications of these materials are among the most promising ones. The spectral distribution of the germanium sulfide index of refraction for two directions of polarization is shown in Figure 7.49 (adapted from Reference 7.249). The GeSe reflectivity as a function of the wave number for three polarization directions is presented in Figure 7.50 (adapted from Reference 7.239); the GeTe reflectivity vs. wavelength is shown in Figure 7.51 (adapted from Reference 7.250).

The optical properties of tin chalcogenides are presented here by the spectral distribution of the SnS refractive index (Figure 7.52, adapted from Reference 7.177), the SnTe reflectivity (Figure 7.53, adapted from Reference 7.251), and the SnTe index of refraction (Figure 7.54, adapted from Reference 7.252).

The spectral distribution of the absorption coefficient of both tetragonal and orthorhombic PbO allotropes has been reported by Keezer et al.[7.230] The dependence of index of refraction

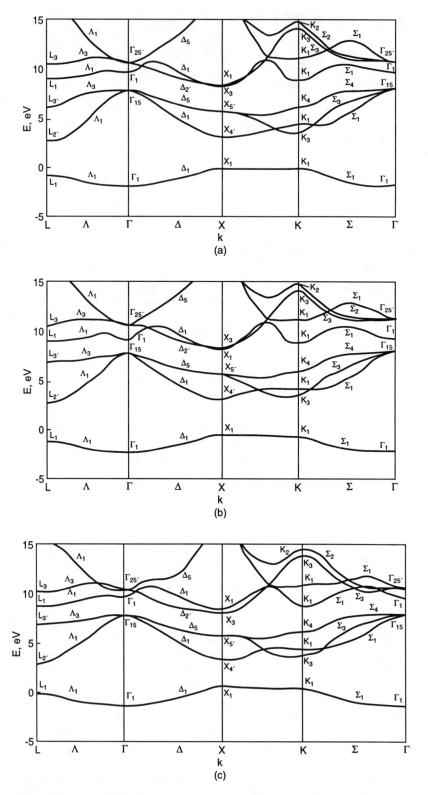

Figure 7.36 Band structure of Mg_2Si (a), Mg_2Ge (b), Mg_2Sn (c), and Mg_2Pb (d). (After F. Aymerich and G. Mula, *Phys. Stat. Sol.*, B42, 697, 1970. With permission.)

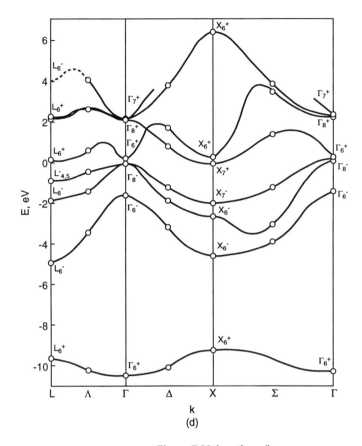

Figure 7.36 (continued)

Table 7.14 Magnesium Thatogenides. Band Structure Parameters (from Reference 7.1 If Not Shown Otherwise)

| Compound | Energy gap (eV) | | Effective mass (in m_o) | | Spin-orbit splitting (eV) |
	Indirect $\Gamma_{15v} \to X_{1c}$	Direct $\Gamma_{15v} \to \Gamma_{1c}$	Electrons	Holes	
Mg_2Si	0.77–0.78 (0 K) 0.60 (90 K)	2.17 (77 K) 2.27 (300 K)	0.46–0.53	0.87–2.2	0.03
Mg_2Ge	0.570 (0 K) 0.532 (300 K)	1.67 (77 K) 1.64 (300 K)	0.18	0.31	0.20
Mg_2Sn	0.36 (0 K) 0.23 (300 K) 0.33 (0 K)[7.150] 0.22 (300 K)[7.150]	1.2 (calculated)[7.148]	0.15–0.2	0.10–1.3	0.48
Mg_2Pb	–0.15	0.1 [7.136]		0.35–0.45 (heavy holes) 0.04 (light holes)	

on wavelength measured on the films of orthorhombic PbO is shown in Figure 7.55 (adapted from Reference 7.243).

Index of refraction and absorption coefficient of PbS, PbSe, and PbTe in the wavelength range from 500 nm to 3.5 μm was measured first by Avery.[7.253] Detailed measurements of the absorption coefficient of these materials were performed by Gibson[7.254] (see, also, Reference 7.240). Because the primary applications of lead chalcogenides (including their

Table 7.15 Magnesium Thatogenides. Electrical and Thermal Transport Properties

Compound	Mobility (cm²/V·sec) at 300 K		Intrinsic carrier concentration (cm⁻³) at 300 K[7.143]	Electrical conductivity near melting point (mS/cm)[7.136]		Thermal conductivity (300 K) (mW/cm·K)[7.158]
	Electrons	Holes		Solid	Liquid	
Mg_2Si	550[7.157] 405[7.136]	70[7.143]	$1 \cdot 10^{14}$	1.12	9.8	
Mg_2Ge	530[7.143]	106[7.143]	$2 \cdot 10^{14}$	1.14	8.4	
Mg_2Sn	320[7.150]	260[7.150]	$3 \cdot 10^{17}$	2.04	10.6	50
Mg_2Pb	12,000[7.142] (at 4.2 K)	14,000 (heavy) 83,000 (light) (both at 4.2 K)[7.142]		3.53	8.6	

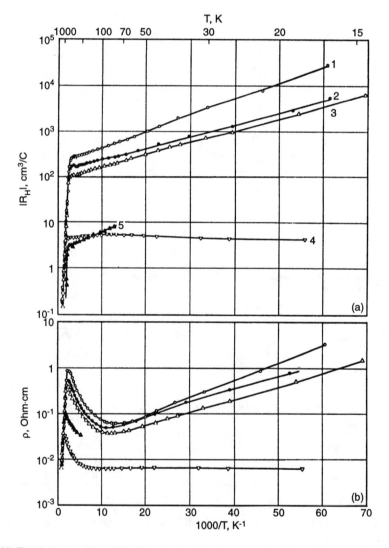

Figure 7.37 Electrical properties of Mg_2Si vs. temperature measured on four n-type samples (1 to 4) and a p-type sample (5) with the room temperature carrier concentration from $n = 2.8 \cdot 10^{16}$ cm⁻³ to $n = 1.7 \cdot 10^{18}$ cm⁻³ and $p = 2.2 \cdot 10^{18}$ cm⁻³. (a) Hall constant, (b) electrical resistivity, and (c) Seebeck coefficient. (After M.W. Heller and G.C. Danielson, *J. Phys. Chem. Solids*, 23, 601, 1962. With permission.)

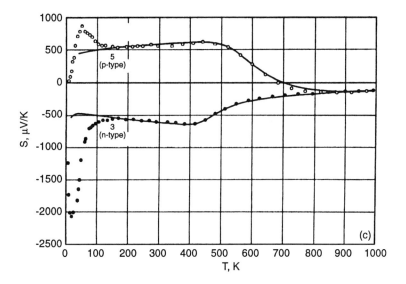

Figure 7.37 (continued)

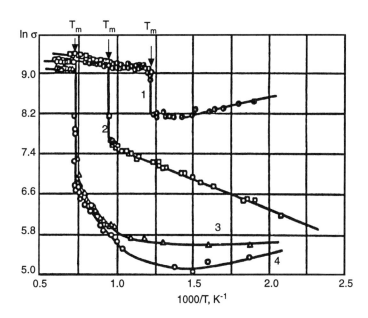

Figure 7.38 Logarithm of electrical conductivity of Mg$_2$-IV compounds vs. reciprocal temperature in the solid and liquid states. 1 and semishaded circles — Mg$_2$Pb; 2 and squares — Mg$_2$Sn; 3 and triangles — Mg$_2$Ge; and 4 and open circles — Mg$_2$Si (σ in S/cm). (After V.M. Glazov, S.N. Chizhevskaya, and N.N. Glagoleva, *Liquid Semiconductors*, Nauka Publishing House, Moscow, 1967. With permission.)

solid solutions) are in the infrared detectors and diode lasers for the range from 3 µm to 30 µm,[7.209,7.254] their optical properties were investigated mainly in this range. The PbS index of refraction at different temperatures was reported by Zemel et al.[7.245] (see Figure 7.56). The PbSe index of refraction as a function of temperature is shown in Figure 7.57 together with the temperature dependence of its lattice parameter (adapted from Reference 7.209). A detailed spectroscopic ellipsometry investigation of the PbSe dielectric function by Suzuki et al.[2.248] enabled them to calculate both the index of refraction and absorption coefficient for

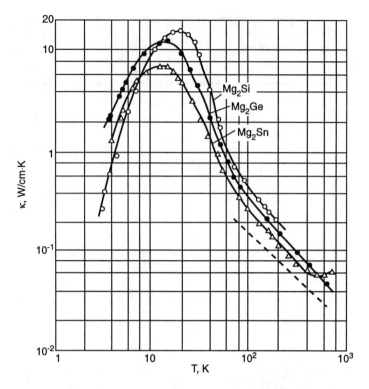

Figure 7.39 Temperature dependence of thermal conductivity for Mg_2-IV compounds. Solid curves from Reference 7.160; the dash line — the data for Mg_2Sn polycrystals.[7.158] (After J.J. Martin, *J. Phys. Chem. Solids*, 33, 1139, 1972. With permission.)

Table 7.16 Mg_2-IV Compounds. Dielectric and Optical Properties

| Compound | Dielectric constant | | Index of refraction | | Extinction coefficient |
	Static	Optical	Calculated	Experim. (550 nm)	(550 nm)[7.167]
Mg_2Si	20[7.165]	13.3[7.165]	3.7	~4.0[7.167]	~5.0
Mg_2Ge	21.7[7.166]	13.9[7.166]	3.7	~3.9[7.167]	~3.0
Mg_2Sn	23.75[7.1]	15.5[7.1]	3.9	~4.2[7.167]	~4.5

300 K and the photon energy from 1.15 eV to 5.4 eV. The PbTe index of refraction was reported in References 7.245 and 7.256 to 7.258. The spectral distribution of the lead telluride index of refraction at different temperatures is shown in Figure 7.58.

Displacement of the fundamental absorption edge in PbS, PbSe, and PbTe in the direction of greater wavelengths with temperature decrease and, consequently, decrease of the direct optical energy gap was measured and discussed in detail by Gibson et al.[7.240] Their data are presented in Figure 7.59. The PbSe absorption coefficient vs. photon energy for the samples with different hole concentration has been reported by Veis et al.[7.259] The information regarding PbSe absorption coefficient for the photon energy between 0.23 eV and 0.8 eV is reported also by Scanlon;[7.261] the data for the range between 1.15 eV and 5.4 eV are published by Suzuki et al.[7.248] The absorption coefficient of PbTe vs. photon energy is shown in Figure 7.60 (compilation of the data of References 7.256, 7.258, and 7.261 to 7.266, adapted from Reference 7.258).

As summarized by Preier,[7.260] the lead chalcogenides have found a wide application in the diode laser spectrometers for high-resolution spectroscopy of molecules, ions, and radicals,

Table 7.17 IV-VI Compounds. Molecular Weight, Crystal Structure, and Density

Compound	Molecular weight	Mean atomic weight	Mean atomic number	Crystal structure and space group	Lattice parameters a (nm)	b (nm)	c (nm)	β°	Density (g/cm³)	Ref.
GeS	104.65	52.33	24	Orthorhombic D_{2h}^{16}-Pnma	1.044	0.365	0.43		4.01	7.1
GeSe	151.55	75.78	33	D_{2h}^{16}-Pnma	1.082	0.385	0.44		5.52	7.1, 7.3
α-GeTe	200.19 (stable below 670 K)	100.10	42	Rhombohedral C_{3v}^{5}-R3m	0.597			88.17	6.2	7.1, 7.3
β-GeTe				Cubic, O_{h}^{5}-Fm3m	0.60121					7.1
					0.598				6.19	7.136
γ-GeTe[a]				D_{2h}^{16}-Pnma	1.215[b]	0.418[b]	0.432[b]			7.1
SnS	150.75	75.38	33	D_{2h}^{16}-Pnma	1.120	0.399	0.434		5.08	7.1
SnSe	197.65	98.83	42	D_{2h}^{16}-Pnma	1.157	0.419	0.446		6.18	7.1, 7.2
α-SnTe	246.29 (stable below 100 K, transition at 97.5 K[7.225])	123.15	51	C_{3v}^{5}-R3m	0.6325			89.895		7.3
β-SnTe				O_{h}^{5}-Fm3m	0.63268				6.445	7.3
					0.629				6.45	7.136
PbO	223.20	111.60	45	Tetragonal D_{4h}^{7}-P4/nmm	0.39759		0.5023		9.334	7.2
				Orthorhombic D_{2h}^{11}-Pbma	0.5489	0.4755	0.5891		9.641	7.2
PbS	239.26	119.63	49	O_{h}^{5}-Fm3m	0.59360				7.5973	7.2
					0.594				7.5	7.136
PbSe	286.16	143.08	58	O_{h}^{5}-Fm3m	0.61255				8.2690	7.2
					0.613				8.1	7.136
PbTe	334.80	167.40	67	O_{h}^{5}-Fm3m	0.6406				8.2459	7.2
					0.645				8.16	7.136

[a] See, also, References 7.182 and 7.183.
[b] At 669 K.

Table 7.18 IV-VI Compounds. Thermochemical and Thermal Properties

Compound	Melting point (K)	Enthalpy of formation (kJ/mol)	Enthalpy of fusion (kJ/mol)	Heat of sublimation (kJ/mol)	Heat capacity (RT) (J/mol·K)	Coefficient of thermal expansion (RT) (10^{-6} K^{-1})	Debye temperature (K)
GeS	938 [7.1]	−76 [7.1] −89.5 [7.187]			47.4 [7.175]		
GeSe	948 [7.175]	−69 [7.175] −82.4 [7.187]			49.9 [7.175]		
GeTe	998 [7.136] 997 for Ge$_{1.0064}$Te [7.1]	−33.5 [7.136]	47.30 [7.136]	195.39 [7.1]	52.51 [7.176] 51.88 [7.175]	23 [7.136]	
SnS	1153 [7.175]	−108 [7.175] −105 [7.187]		220 [7.175]	45 [7.175]	28 [7.177]	270 (at 80 K) [7.1]
SnSe	1134 [7.2]	−88.6 [7.175]			42.25 [7.175]		210 (at 80 K) [7.1]
SnTe	1063 [7.136]	−61.1 [7.136] −60.67 [7.1] −59.4 [7.187]	33.5 [7.136]	222.07 [7.1]		23 [7.136] 21.3 [7.178]	210 (at 80 K) [7.1]
PbO	1163 [7.1] 1161 [7.2]						
PbS	1392 [7.136]	−94.1 [7.136]	36.4 [7.136]		47.70 [7.1] at 200 K	20.4 [7.184] 20 [7.179]	145 [7.1] 227 [7.185]
PbSe	1361 [7.136] 1355 [7.1]	−87.9 [7.136] −75.3 [7.187]	35.56 [7.136]		47.70 [7.1] at 200 K	19.4 [7.181] 19.0 [7.136] 19.5 [7.184]	156 at 0 K [7.1] 138 [7.185]
PbTe	1190 [7.136]	−70.3 [7.136] −69.5 [7.187]	31.38 [7.136]		49.79 [7.1] at 200 K	19.8 [7.179,7.184]	136 at 100 K [7.1] 125 [7.185]

Table 7.19 IV-VI Compounds. Room Temperature Hardness and Elastic Moduli

Compound	Hardness Mohs	Hardness Vickers (MPa)	Elastic constants (GPa) c_{11}	c_{12}	c_{44}	Bulk modulus (GPa)
SnTe			112.50 109.3	7.50 2.1	11.72 [7.2] 9.69 [7.1]	
PbS (galena)	2.5–3.0 [7.2]	328–369 [7.186]	124 113.9	33 28.9	23 [7.1] 27.2 [7.4]	52.88 [7.1]
PbSe			123.7	19.3	15.91 [7.1]	
PbTe (altaite)	3 [7.2]		105.3 107.95 107.2	7.0 7.64 7.7	13.22 [7.188] 13.43 [7.2] 13.0 [7.4]	38.39 [7.188]

sub-Doppler spectroscopy in molecular beams, and in heterodyne and acousto-optic spectroscopy. The tunable diode laser spectrometers are used in automobile engine and power plant exhaust gas analysis, in medicine and environmental control, in chemical, pharmaceutical, and semiconductor industries, and in scientific research.

Some of the IV-VI compounds are available commercially (see Table 7.22).

7.6. V$_2$-VI$_3$ COMPOUNDS

The general formula for the V$_2$-VI$_3$ compounds under consideration in this section may be written in the form (As, Sb, Bi)$_2$(O, S, Se, Te)$_3$. These materials exist in nature as minerals arsenolite (claudetite), As$_2$O$_3$, orpiment, As$_2$S$_3$, senarmontite (valentinite), Sb$_2$O$_3$, stibnite,

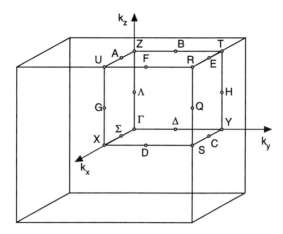

Figure 7.40 Brillouin zone for a simple orthorhombic Bravais lattice of a IV-VI compound. (After T. Grandke and L. Ley, *Phys. Rev.*, B16, 832, 1977. With permission.)

Sb_2S_3, bismite (sillenite), Bi_2O_3, bismuthinite, Bi_2S_3, guanajuatite, Bi_2Se_3, and tetradymite (joseite), from Bi_2Te_2 to Bi_2Te_3.

The crystal structure of V_2-VI_3 compounds has been the subject of numerous reports (see, e.g., References 7.267 to 7.282). The basic chemical information and the data on crystal structure of the compounds are gathered in Table 7.23.

The majority of the V_2-VI_3 compounds, the phase diagrams of which have been reported (see, e.g., References 7.30, 7.55, and 7.283), melt congruently. The solubility of elemental components in the compounds is either nonexistent or does not exceed 1%. The data on melting temperatures and some other thermal and thermochemical properties of the compounds are assembled in Table 7.24.

The V_2-O_3 compounds are, apparently, insulators, while the chalcogenides possess semiconductor properties. The band structure of As_2S_3 is identical to that of As_2Se_3 and, possibly, of As_2Te_3. The band structure of As_2Se_3 was calculated by Althaus et al.[7.285] and Weiser.[7.286] The limited data on the Sb_2S_3 band structure were reported by Khasabov and Nikiforov.[7.287] The Brillouin zone for Sb_2Te_3, Bi_2Se_3, and Bi_2Te_3 is displayed in Figure 7.61 (adapted from Reference 7.288); the Bi_2Te_3 band structure reported by Borghese and Donato[7.289] is shown there also. Bi_2Te_3 is, apparently, an indirect semiconductor (or semimetal). The numerical information pertinent to band structure of V_2-VI_3 compounds appears in Table 7.25.

The electrical transport properties of V_2-VI_3 compounds attracted researchers' attention after high Seebeck coefficient magnitudes were reported in combination with low electrical resistivity and thermal conductivity in the wide range of temperatures including the liquid state (see, e.g., References 7.136 and 7.292 to 7.294).

Shein[7.295] found that the electron mobility in As_2S_3 at room temperature is 1 cm²/V·sec and is about ten times greater than that for holes. The dark electrical resistivity in As_2S_3 samples in the direction normal to the monoclinic lattice layers is $3 \cdot 10^{15}$ Ohm·cm.[7.3] Abkowitz et al.[7.296] found that the dark AC-conductivity of As_2S_3 single crystals is an exponential function of frequency ($\sigma \propto \omega^x$, $-1 < x < 2$). The electron mobility along b axis in As_2Se_3 is between 20 and 80 cm²/V·sec at 300 K.[7.1] The room temperature resistivity of "as-prepared" polycrystals of As_2Te_3 is between 10 Ohm·cm and 100 Ohm·cm;[7.290] the electron and hole mobility in As_2Te_3 is 170 cm²/V·sec and 80 cm²/V·sec, respectively.[7.3] The electrical conductivity of Sb_2S_3 vs. reciprocal temperature is shown in Figure 7.62 (adapted from Reference 7.297), the 300-K conductivity magnitudes cited in Reference 7.1 are $3.3 \cdot 10^{-8}$ S/cm ($\parallel$ c), $2 \cdot 10^{-9}$ S/cm ($\perp$ c), and $7 \cdot 10^{-5}$ S/cm. As is typical for the high-resistivity V_2-VI_3 compounds,

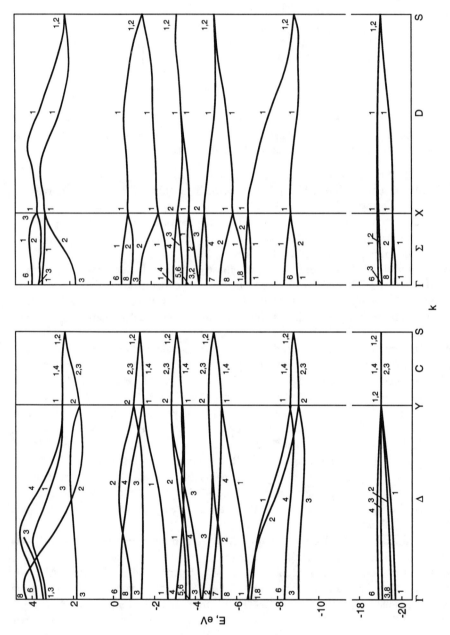

Figure 7.41 The GeS band structure calculated using the empirical pseudopotential method along different directions within the Brillouin zone shown in Figure 7.40. (After T. Grandke and L. Ley, *Phys. Rev.*, B16, 832, 1977. With permission.)

NONADAMANTINE & VARIABLE-COMPOSITION SEMICONDUCTOR PHASES

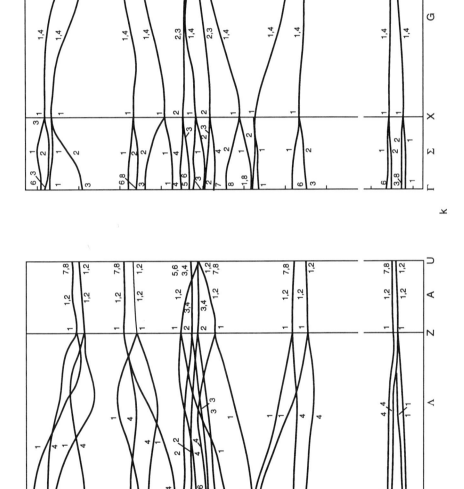

Figure 7.41 (continued)

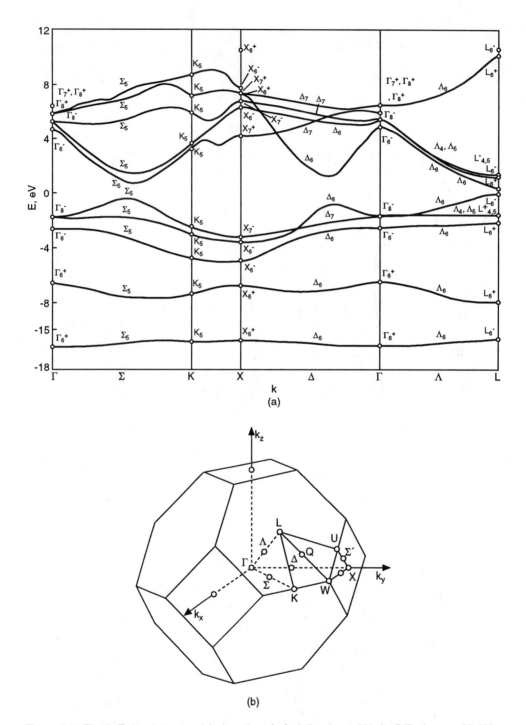

Figure 7.42 The GeTe band structure (a) along the principal directions within its Brillouin zone (b). (After Y. Tung and M.L. Cohen, *Phys. Rev.*, 180, 823, 1969. With permission.)

the Seebeck coefficient of Sb_2S_3 is relatively high, reaching a maximum of 1.1 mV/K near room temperature[7.297] and rapidly decreasing with the following temperature increase (S = 150 μV/K to 200 μV/K at 400 K).

Table 7.20 IV-VI Compounds. Basic Features of the Electronic Structure

Compound	Minimum energy gap (eV) at 300 K		Thermal energy gap (eV)	dE_g/dT, 10^{-4} eV/K		Effective mass in m_o units	
	Direct	Indirect		Direct	Indirect	Electrons	Holes
GeS	1.65[7.3] (E∥a)			−8.1[7.1] (E∥a) −5.8[7.1]			0.3–0.7[7.3]
GeSe		1.075[7.1] (E∥a) 1.080[7.1] (E∥b)			−8.6[7.1] (E∥a)		
α-GeTe	0.73–0.95[7.1]			−4.2[7.1]		0.48–0.78[7.1]	1.15 (light) 5.0 (heavy) $m_{ds} > 1.5$[7.1]
β-GeTe	0.1–0.2[7.1]						
SnS		1.42 (E∥a) 1.095 (E∥b)[7.1]	0.54[7.224]		−4.05[7.1] −4.37[7.1]	≈0.45 (m_{ds})	0.2 (∥a) 0.2 (∥b) ≈1 (∥c)[7.1] 0.15 (⊥c) 0.5[7.1]
SnSe		0.9[7.1]	≈0.45[7.225]				
α-SnTe	2.07 (100 K)[7.1]			≈−2.5[7.212]			
β-SnTe	0.190[7.1]			≈1.3[7.212]			
PbO (tetr.)	2.75–3.18	1.9[7.1]			1.0[7.1]		
PbO (orth.)	2.8–3.36	2.67[7.1]			11.5[7.1]		
PbS	0.41–0.42[7.1] 0.42[7.217]				4.7–5.2[7.208,7.209]		0.075 (⊥) 0.105 (∥) (4 K)[7.1]
PbSe	0.278[7.1] 0.285[7.217]				5.1[7.208]	0.04 (⊥)[7.1] 0.07 (∥)(4 K)[7.1]	0.034 (⊥) 0.068 (∥) (4 K)[7.1]
PbTe	0.310[7.1] 0.306[7.217]	0.360 (above 420 K)[7.1]			4.5[7.208]	0.024 (⊥) 0.024 (∥) (4 K)[7.1] 0.019[7.217]	0.066[7.1]

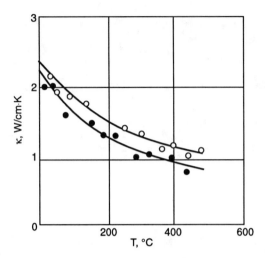

Figure 7.43 Thermal conductivity of GeS (solid circles) and GeSe (open circles) single crystals vs. temperature. (After A.S. Okhotin, A.N. Krestovnikov, A.A. Aivazov, and A.S. Pushkarskii, *Phys. Stat. Sol.*, 31, 485, 1969. With permission.)

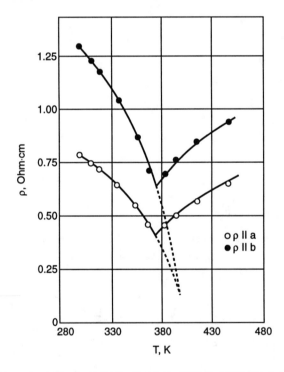

Figure 7.44 Temperature dependence of GeSe electrical resistivity along the a and b axes. (After D.S. Kirakos, O. Valassiades, and N.A. Economov, *AIP Conf. Ser.*, 43, Ch. 8, 1979. With permission.)

The 300-K mobility of electrons and holes in Sb_2Se_3 is 15 cm^2/V·sec and 42 cm^2/V·sec, respectively; the electrical conductivity (300 K) along the c axis is 460 S/cm and it is 2.2 times lower across the c axis.[7.1] The temperature dependence of the Sb_2Se_3 Seebeck coefficient is displayed in Figure 7.63 (adapted from Reference 7.298). The electrical conductivity of Sb_2Te_3, Bi_2Se_3, and Bi_2Te_3 in the solid and liquid state is presented in Figure 7.64 (adapted from Reference 7.136). The increase of resistivity of solid Sb_2Te_3 with the temperature

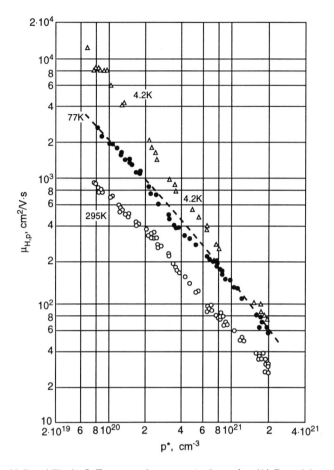

Figure 7.45 The Hall mobility in SnTe vs. carrier concentration, $p^* = (1/eR_{H,77\,K})^{-1}$ at 4.2 K (triangles), 77 K (solid circles), and 295 K (open circles). (After R.S. Allgaier and B. Houston, *Phys. Rev.*, B5, 2186, 1972. With permission.)

increase has been confirmed lately by Eichler and Simon.[7.299] The hole mobility in Sb_2Te_3 at 300 K is 31 cm²/V·sec and 270 cm²/V·sec along and across the c axis, respectively;[7.1] at low temperatures, it is as high as 10^4 cm²/V·sec.[7.300]

The temperature dependence of the Sb_2Te_3 Seebeck coefficient is shown in Figure 7.65; its increase with temperature agrees with the data of Reference 7.299 on temperature dependence of the Sb_2Te_3 Hall coefficient.

The thermoelectric parameters of bismuth sesquioxide, Bi_2O_3 — Seebeck coefficient, electrical conductivity, and thermal conductivity — in the range of temperatures 500 K to about 1400 K (the Bi_2O_3 melting point is 1090 K) were reported in Reference 7.294. The melt of Bi_2O_3 presents a certain practical interest: at 1300 K, the thermoelectric figure of merit is close to $4.1 \cdot 10^{-3}$ K^{-1}.

The electron mobility in the Bi_2S_3 sample with the electron concentration at 300 K of $3 \cdot 10^{18}$ cm^{-3} is equal to 200 cm²/V·sec.[7.187] This gives the magnitudes of electrical conductivity close to 100 S/cm, but the authors of Reference 7.187 cite magnitudes between 10^{-6} S/cm and 10^{-7} S/cm and mention $\sigma = 10$ S/cm only for the samples containing overstoichiometric bismuth. The Seebeck coefficient at 300 K is equal to -0.7 mV/K for stoichiometric samples and -0.140 mV/K for the samples with excess of bismuth.[7.187] The temperature dependence of the Bi_2Se_3 electrical conductivity is exhibited in Figure 7.64; as in Sb_2Te_3, it decreases with the temperature increase up to the melting point, but it increases exponentially with

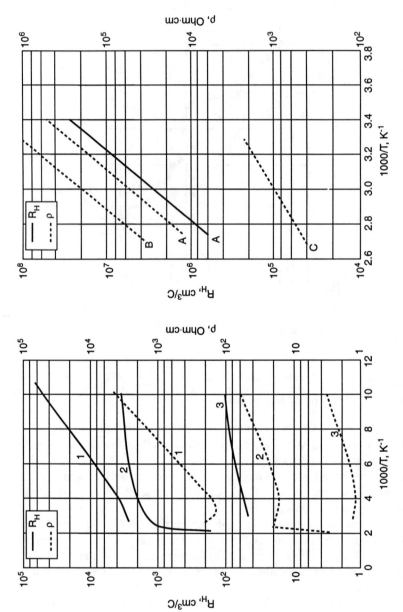

Figure 7.46 Electrical properties of tetragonal (a) and orthorhombic (b) PbO single crystals vs. reciprocal temperature. (After R.C. Keezer, D.L. Bowman, and J.H. Becker, *J. Appl. Phys.*, 39, 2062, 1968. With permission.)

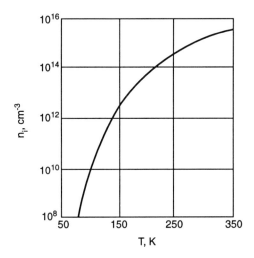

Figure 7.47 Intrinsic carrier concentration in PbS calculated using multiband model. (After G. Nimtz and B. Schlicht in *Springer Tracts in Modern Physics*, Vol. 98, Springer-Verlag, 1983. With permission.)

temperature in liquid Bi_2Se_3. Köhler and Fabricius[7.301] found similar behavior of conductivity in solid Bi_2Se_3 above approximately 200 K. The electron mobility of Bi_2Se_3 at 300 K in a sample with n = $1.8 \cdot 10^{19}$ cm^{-3} is 690 cm^2/V·sec;[7.302] a sample with p = $1.5 \cdot 10^{19}$ cm^{-3} showed the hole mobility of 42 cm^2/V·sec.[7.303] The peak magnitudes of the carrier mobility in Bi_2Se_3 reported in Reference 7.301 are close to 9000 cm^2/V·sec (electrons) and to 7000 cm^2/V·sec (holes), both at about 30 K. The Seebeck coefficient in Bi_2Se_3 at room temperature is between 23 µV/K and 206 µV/K depending on the carrier concentration;[7.1] it does not depend on the direction of the temperature gradient.

The Bi_2Te_3 electrical conductivity along and across the c axis at 293 K is about 700 S/cm and 250 S/cm, respectively,[7.304] in agreement with the magnitudes between 400 S/cm and 900 S/cm cited by Glazov et al.[7.136] (Figure 7.64) from the results of measurements on polycrystalline samples. The maximum room temperature magnitudes of the electron and hole mobility of 1140 cm^2/V·sec and 680 cm^2/V·sec, respectively, cited in Reference 7.136, are in agreement with the data reported by Chamness and Kipling.[7.305] The results of measurements of the Bi_2Te_3 Seebeck coefficient are shown in Figure 7.66 (adapted from Reference 7.305); the data agreed with the Seebeck coefficient vs. temperature results obtained by the authors of References 7.304 and 7.305.

The room temperature thermal conductivity of As_2Se_3 is close to 14 mW/cm·K; that of As_2Te_3 is about 10 mW/cm·K,[7.289] while Abrikosov et al.[7.187] cite a magnitude of 25 mW/cm·K (both results have been obtained using polycrystalline samples). The thermal conductivity of Sb_2Te_3 is between 28 mW/cm·K and 48 mW/cm·K (300 K), depending on carrier concentration;[7.306] its magnitude at 100 K is between 40 mW/cm·K and 73 mW/cm·K. The Bi_2O_3 thermal conductivity in the range from 500 K to 1500 K has been reported in Reference 7.294 (see Figure 7.67). The 300-K thermal conductivity of Bi_2Se_3 across the c axis is 25 mW/cm·K;[7.187] the lattice component of the Bi_2Se_3 thermal conductivity in the same direction, reported earlier[7.308] (13.4 mW/cm·K), seems to be too low. The Bi_2Te_3 thermal conductivity, in accordance with Reference 7.2, is 64 mW/cm·K, 28 mW/cm·K, 36 mW/cm·K, and 46 mW/cm·K at 80 K, 204 K, 303 K, and 370 K, respectively. These magnitudes are in reasonable agreement with the data of Smirnov et al.,[7.309] Sälzer and Nieke,[7.310] and Drabble.[7.311]

The information regarding magnetic susceptibility and dielectric constant of V_2-VI_3 compounds is collected in Table 7.26.

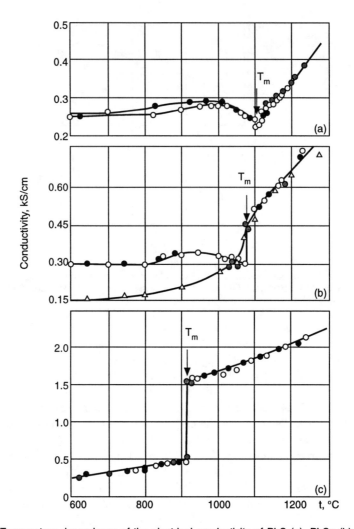

Figure 7.48 Temperature dependence of the electrical conductivity of PbS (a), PbSe (b), and PbTe (c) in the solid and liquid states. (After V.M. Glazov, S.N. Chizhevskaya, and N.N. Glagoleva, *Liquid Semiconductors*, Nauka Publishing House, Moscow, 1967. With permission.)

The As_2O_3 (arsenolite) index of refraction at 589.3 nm is 1.755.[7.1] Claudetite has the index of refraction of 2.01, 1.87, and 1.92 along the a, b, and c axes, respectively.[7.2] The infrared index of refraction of As_2S_3 is 3.02, 2.40, and 2.81 along the same axes;[7.2,7.313] the index of refraction and extinction coefficient of As_2Se_3 is displayed in Figure 7.68 (adapted from Reference 7.314). The index of refraction of Sb_2O_3 is 2.087 (senarmonite) and 2.35, 2.35, and 2.18 along the a, b, and c axes, respectively, in valentinite.[7.2]

The index of refraction of Sb_2S_3 in the far-infrared part of the spectrum (2.5 cm^{-1} to 2.7 cm^{-1}) and across the c axis is close to 2.7.[7.1] Intrinsic Sb_2Se_3 exhibits a sharp maximum of photoconductivity at 1.18 eV.[7.315] The Sb_2Te_3 reflectivity and real and imaginary components of the index of refraction vs. the photon energy are shown in Figure 7.69 (adapted from Reference 7.316).

Natural Bi_2O_3 crystals (bismite) possess the refractive indices for ordinary and extraordinary rays of 2.00 and 1.82, respectively. The optical density of Bi_2O_3 synthetic crystals has been reported in References 7.317 and 7.318. The indices of refraction of Bi_2S_3 in principal crystallographic directions are 1.315, 1.900, and 1.670 for 2.10-eV photons.[7.2] The optical density of Bi_2S_3 single crystals has been reported by Gildart et al.[7.319] The spectral distribution

Table 7.21 IV-VI Compounds. Room Temperature Dielectric Constant, Magnetic Susceptibility, and Index of Refraction

Compound	Dielectric constant		Molar magnetic susceptibility (10^{-5} [SI]cm^3/mol)	Index of refraction	
	Static $\varepsilon(0)$	Optical $\varepsilon(\infty)$		Experimental	Calculated
GeS	25.1	14.8 (E∥a)[7.238]			3.8 [7.240]
	29.5	12.0 (E∥b)[7.238]			
	30.0	10.0 (E∥c)[7.238]			
GeSe	21.9	18.7 (E∥a)[7.239]			4.3 [7.240]
	30.4	21.9 (E∥b)[7.239]			
	25.8	14.4 (E∥c)[7.239]			
GeTe	≈1.6 [7.199]	36 (295 K)[7.1]	−57.9 [7.136]	5.5 (1.5 μm)[7.1]	6.1 [7.240]
		37.5 [7.240]			
SnS	32	14 (E∥a)[7.241]		4.20 (E∥a)	3.9
	48	16 (E∥b)[7.241]		4.55 (E∥b)	
	32	16 (E∥c)[7.241]		(at 1 μm)[7.177]	
SnSe	45	13 (E∥a)[7.193]			
	62	17 (E∥b)[7.193]			
	42	16 (E∥c)[7.193]			
SnTe	1200 ± 200 [7.1]	1770 ± 300 (IR)[7.2]	−61.9 [7.136]	6.5 (1 μm)[7.242]	
PbO (tetr.)	25.9 [7.2]	2.81 (521 nm)[7.1]	−59.0 [7.312]		5.1
		2.74 (600 nm)[7.1]			
PbO (orth.)		2.6 (550 nm)[7.243]	−55.3 [7.312]		
PbS	169 [7.244]	17.2 [7.244]		4.2 (1.9 μm)[7.245]	4.1
	200 ± 35 [7.2]	18.5 [7.4]		4.10 (>3 μm)[7.247]	
		17 [7.247]			
PbSe	210 [7.244]	22.9 [7.245]		4.59 (>3 μm)[7.247]	4.6
	280 [7.4]	25.2 [7.4]			
		21 [7.247]			
PbTe	414 [7.246]	28.5 [7.247]	−71.5 [7.136]	6.1 (2 μm)[7.247]	6.1
	450 [7.4]	36.9 [7.4]		5.35 (>3 μm)[7.247]	

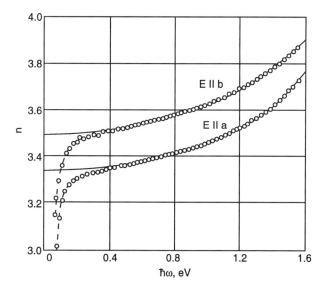

Figure 7.49 The GeS index of refraction vs. photon energy measured by the prism (open circles) and interference (solid circles) methods. The solid curves are a three-parameter fit to the electronic part of dispersion. (After I. Gregora, B. Velicky, and M. Zavetova, J. Phys. Chem. Solids, 37, 785, 1976. With permission.)

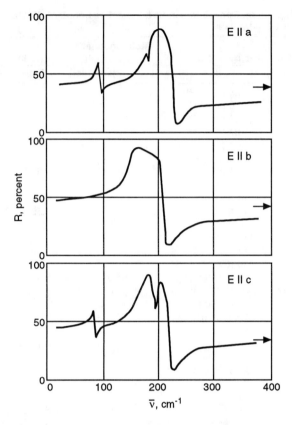

Figure 7.50 The GeSe room temperature reflectivity vs. wavenumber for three principal polarizations. The arrows at the right-hand side indicate the final value of reflectivity measured at 4000 cm^{-1}. (After H.R. Chandrasekhar and U. Zwick, *Solid State Commun.*, 18, 1509, 1976. With permission.)

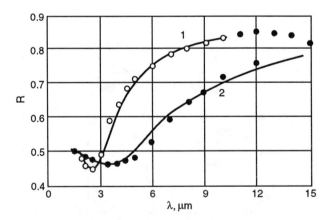

Figure 7.51 Spectral distribution of reflectivity for GeTe samples with $p = 1.48 \cdot 10^{21}$ cm^{-3} (curve 1) and $p = 1.6 \cdot 10^{20}$ cm^{-3} (curve 2). (After I.A. Drabkin, T.B. Zhukova, I.V. Nelson, and L.M. Sysoeva, *Inorg. Mater.*, 15, 936, 1979. With permission.)

of the Bi_2Se_3 index of refraction is presented in Figure 7.70; the information regarding the Bi_2Te_3 index of refraction, extinction coefficient, and the energy loss parameter for E⊥c is presented in Figure 7.71 (adapted from Reference 7.320).

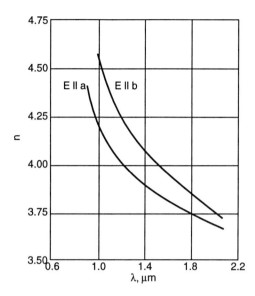

Figure 7.52 Spectral distribution of the SnS index of refraction at 300 K for two polarization directions. (After A.P. Lambros, D. Gerleas, and N.A. Economou, *J. Phys. Chem. Solids*, 35, 537, 1974. With permission.)

Currently, the V_2-VI_3 compounds with tetradymite structure, Sb_2Te_3, Bi_2Se_3, and Bi_2Te_3, and their solid solutions are the most important materials for low-temperature thermoelectric applications on the industrial scale, in the form of both single crystals and ceramics.[7.321-7.324] The commercial availability of the V_2-VI_3 compounds is described in Table 7.27.

7.7. BINARY COMPOUNDS OF THE GROUP VIIIA ELEMENTS

Nine elements of the VIIIA group of the Periodic table form a wide range of binary compounds with semiconductor properties. Some of them (VIIIA-IIIB$_2$-, VIIIA$_2$-IVB$_3$-, VIIIA-VB$_2$, VIIIA-VB$_3$, VIIIA-VIB, VIIIA-VIB$_2$, VIIIA-VIB$_3$, and VIIIA$_3$-VIB$_4$-type compounds) attract interest as novel semiconductor materials for photovoltaic and thermoelectric energy conversion (see, e.g., References 7.325 to 7.327).

The binary compounds of the VIIIA-group elements exist in nature as minerals arsenoferrite (löllingite), $FeAs_2$; hematite, Fe_2O_3; magnetite, Fe_3O_4; pirrhotite (troilite), FeS; pyrite (marcasite), FeS_2; greigite, Fe_3S_4; ferroselite, $FeSe_2$; frohbergite, $FeTe_2$; safflorite (Co-safflorite), $CoAs_2$; Fe-, Co-, and Ni-skutterudites, $FeAs_3$, $CoAs_3$, and $NiAs_3$; linneaite, Co_3S_4; cattierite, CoS_3; trogtalite, $CoSe_2$; niccolite, NiAs; rammelsbergite (pararammelsbergite), $NiAs_2$; millerite (beyrichite), NiS; violarite (vaesite), NiS_2; polydymite, Ni_3S_4; heazelwoodite, Ni_3S_2; melonite, $NiTe_2$; laurite, RuS_2; sperrylite, $PtAs_2$; cooperite, $Pt(As,S)_2$; and some other pnictides, oxides, and chalcogenides. The basic chemical and crystallographic parameters of the compounds are gathered in Table 7.28; their thermal and thermochemical properties are compiled in Table 7.29.

The electronic properties of VIIIA-IIIB compounds have been discussed in References 7.326 and 7.350. Jeitschko[7.350] suggested that VIIIA-IIIB and VIIIA-IVB phases with general formula A_nB_{2n-m} (A — Ru or Os; B — Al, Ga, Si, Ge, or Sn; n and m — integers) are semiconductors, using a band model based on the assumption that A-atoms and B-atoms act as acceptors and donors, respectively. In accordance with this model, the d_{10}-band and four valence bands of the A-atoms are completely filled by 14 valence electrons; therefore, the compounds are expected to be semiconductors (or insulators). The authors of Reference

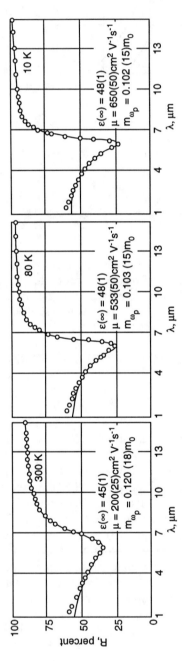

Figure 7.53 Spectral distribution of reflectivity for a SnTe sample with $p = 1.38 \cdot 10^{20}$ cm^{-3} at 300 K, 80 K, and 10 K. Solid lines represent the results of calculations based on classical free carrier dispersion theory and the magnitudes of optical dielectric constant, optical mobility, and effective plasma frequency mass given in the figures. (After R.F. Bis and J.R. Dixon, *Phys. Rev.*, B2, 1004, 1970. With permission.)

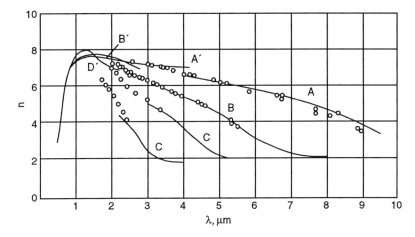

Figure 7.54 Spectral distribution of the index of refraction for SnTe samples with the carrier concentrations of (in 10^{19} cm^{-3}) 3.6 (A), 14 (B), 29 (C), and 68 (D). The solid curves are calculated using classical free-carrier model; curves A′, B′, and D′ represent bound carrier dispersion. (After R.B. Schoolar and J.R. Dixon, *J. Opt. Soc. Amer.*, 58, 119, 1968. With permission.)

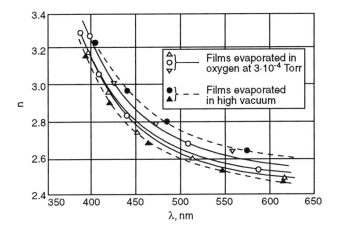

Figure 7.55 Room temperature spectral distribution of the index of refraction for orthorhombic PbO films 200 to 300 μm thick. (After A.E. Ennos, *J. Opt. Soc. Amer.*, 52, 261, 1962. With permission.)

7.326 found that the room temperature electrical resistivity of RuAl$_2$ was about 3 mOhm·cm, but no linear dependence between logarithm of electrical conductivity and inverse temperature was found in the range 293 to 673 K. The authors explained this by the high concentration of lattice defects.

The energy band structure of the group VIIIA tathogenides is discussed in References 7.326, 7.351, and 7.352. Evers et al.[7.326] found that RuGa$_2$ has the band gap of 0.42 ± 0.04 eV; Susz et al.[7.325] reported that Ru$_2$Si$_3$ and Ru$_2$Ge$_3$ are semiconductors with the band gap of 0.44 and 0.34 eV, respectively. The FeAs$_2$ and CoAs$_2$ energy level schemes are shown in Figure 7.72. Although interest in electrical and thermoelectrical properties of the VIIIA-VB$_3$ compounds with skutterudite structure has a long history (see, e.g., References 7.355 to 7.357), the band structure of these compounds has not yet been calculated, possibly because of their intricate crystal structure. The band structure of Group VIIIA chalcogenides has been discussed in Reference 7.358 (NiS) and Reference 7.359 (NiS$_2$ or NiS$_{2-x}$). The

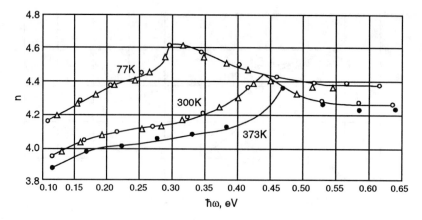

Figure 7.56 Index of refraction vs. photon energy for PbS epitaxial films at 77 K, 300 K, and 373 K. (After J.N. Zemel, J.D. Jensen, and R.B. Schoolar, *Phys. Rev.,* 140, A330, 1965. With permission.)

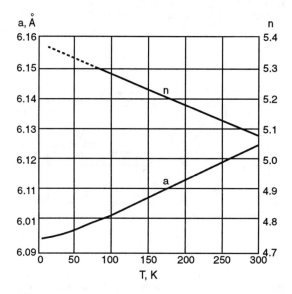

Figure 7.57 Lattice parameter, a, and index of refraction, n, for PbSe single crystal films vs. temperature. (After H. Preier, *Appl. Phys.,* 20, 189, 1979. With permission.)

available data on the band structure of the binary compounds containing Group VIIIA atoms are collected in Table 7.30.

The thermal and electrical transport properties of the Group VIIIA thatogenides, pnictogenides, and chalcogenides were studied primarily in the search for materials for thermoelectric energy conversion. The electrical transport in $RuAl_2$ and $RuGa_2$ was investigated in the course of studying the catalytic properties of ruthenium and osmium. The available information on room temperature electrical and thermoelectric properties is presented in Table 7.31 (the thermal conductivity is shown in Table 7.29).

$RuGa_2$ behaves as a typical semiconductor (see Figure 7.73, adapted from Reference 7.326). An interesting feature of the temperature dependence of the Seebeck coefficient of some thatogenides (Ru_2Ge_3 [7.349]) and pnictogenides ($FeAs_2$, PtP_2, $PtAs_2$, $PtSb_2$;[7.367,7.368] $CoSb_3$ [7.340]) is the existence of relatively strong maxima of the coefficient at certain temperatures. The maxima usually do not correlate with temperature dependence of electrical and

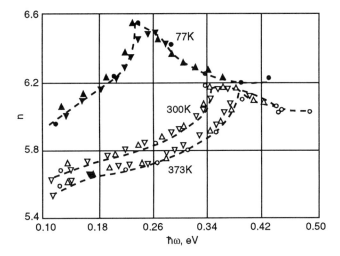

Figure 7.58 Index of refraction vs. photon energy for PbTe single crystal films at 77 K, 300 K, and 373 K. (After J.N. Zemel, J.D. Jensen, and R.B. Schoolar. *Phys. Rev.*, 140, A330, 1965. With permission.)

thermal conductivity and can be used as a criterion for choosing a thermoelectric energy conversion material for particular temperature regions.

At room temperature, the majority of the compounds considered in this section are diamagnetics. Their magnetic susceptibility, together with the limited information on dielectric permittivity and index of refraction, is displayed in Table 7.32. Diamagnetism of the compounds may suggest that the electrons in them form filled valence band, characteristic for semiconductors. Investigation of electrical and magnetic properties of alloys in the system $FeSb_2$-$FeTe_2$ by Yamaguchi et al.[7.370] shows a clear correlation between the conductivity character (metal or semiconductor) and the sign of magnetic susceptibility.

The optical properties of the compounds of the VIIIA-group elements were reported, in particular, by Kjekshus et al.[7.372,7.373] ($FeSi_2$ reflectivity); O'Shaughnessy and Smith[7.374] ($PtSb_2$ IR absorption); Balberg and Pinch[7.375] (Fe_2O_3 optical absorption); Lutz et al.[7.376] ($FeAs_2$ IR absorption); Kaner et al.[7.362] (RuP_2 optical absorption); Hulliger[7.377] (RuP_2, $RuAs_2$, $RuSb_2$, OsP_2, $OsAs_2$, and $OsSb_2$ diffuse reflectance) and Reference 7.378 (RhP_2, $RhAs_2$, $RhSb_2$, IrP_2, and $IrAs_2$ diffuse reflectance); Odile et al.[7.364] (NiP_2 optical absorption); Lutz and Willich[7.379] (PtP_2 infrared absorption); and Damon et al.[7.380] ($PtSb_2$ optical absorption). The optical properties of the VIIIA-VB_3 skutterudite-type compounds were reported by Lutz et al.[7.381-7.383]

7.8. SEMICONDUCTOR SOLID SOLUTIONS

7.8.1. Alloys of Elemental Semiconductors

Among the solid solutions of the elemental semiconductors, the alloys of silicon and germanium are of particular interest as the most currently used materials for thermoelectric energy conversion at elevated temperatures. Stöhr and Klemm[7.385] and Johnson and Christian[7.386] showed that the lattice constant of the Si_xGe_{1-x} alloys is a linear function of x (Vegard's rule):

$$a \text{ (nm)} = 0.56579 - 0.02269x \quad \text{(at 297 K)}$$

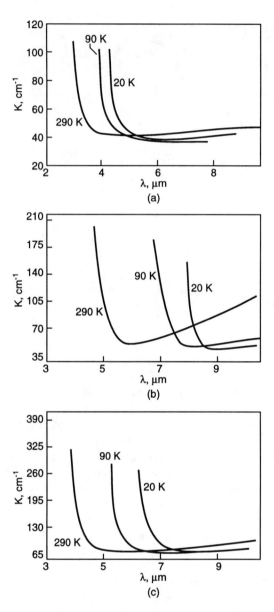

Figure 7.59 Absorption coefficient for PbS (a), PbSe (b), and PbTe (c) vs. wavelength at 20 K, 90 K, and 290 K. (After A.F. Gibson, W.D. Lowson, and T.S. Moss, *Proc. Phys. Soc.*, A64, 1054, 1951. With permission.)

Both elemental components, as well as their alloys, are indirect semiconductors. In accordance with Krishnamurti et al.,[7.384] the calculated dependence of the energy gap on composition of the Si_xGe_{1-x} alloys may be expressed (in eV) as

$$E_g(\Gamma \rightarrow X) = 0.8941 + 0.0421x + 0.1691x^2$$

and

$$E_g(\Gamma \rightarrow L) = 0.7596 + 1.0860x + 0.3306x^2$$

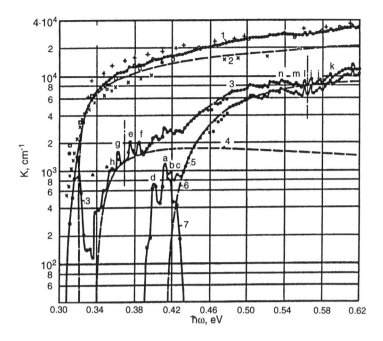

Figure 7.60 Compilation of the data for the PbTe absorption coefficient from different sources. (1) Experimental data, (2) calculated data for the absorption coefficient due to the direct allowed transition between the nearest band extrema at the L-point, (3) difference between the data in 1 and 2, (4) calculated contribution due to an indirect transition from the second valence band extremum (VB[Σ]) to the L-minimum, (5) difference between the data in 1 and the sum of the data in 2 and 4, (6) calculated contribution due to an indirect transition between a valence band Λ-extremum and the L-minimum of the conduction band. The structures a to d are assumed to be caused by phonon-assisted transitions between bound energy levels of lattice defects. The two symmetrical sets of peaks, e to h and i to n, are assumed to correspond to transitions from the same level to two different levels related to different conduction band extrema and separated by the energy 195 meV. The peaks e, f, i, j, and k include the emission of an optical phonon, while the peaks g, h, l, m, and n include the absorption of the same phonon. (After T.R. Globus, B.L. Gelmont, K.I. Geiman, V.A. Kondrashov, and A.V. Matveenko, *Soviet Phys. JETP*, 53, 1000, 1981. With permission.)

Table 7.22 Availability of IV-VI Compounds

Compound	Purity (%)	Form	Price (US$ per gram) From	To
GeS	99.99	Powder	24.68	28.00
GeSe	99.99	Lumps	30.40	41.10
SnS	99.5	8-Mesh powder	3.72	7.54
SnSe	99.999	3- to 12- mm pieces	7.40	10.20
SnTe	99.999	Lumps	2.94	3.68
PbS	99.995	Crystals	12.90	16.90
PbSe	99.999	3- to 6-mm lumps	1.72	2.24
PbTe	99.999	3- to 12-mm pieces	4.85	5.76

From Johnson-Matthey, 1995.

The room temperature mobility of both electrons and holes in the alloys passes through a smooth minimum of about 170 cm²/V·sec and 130 cm²/V·sec, respectively, near the composition $Si_{0.5}Ge_{0.5}$.[7.387,7.388] A review of the properties of Si_xGe_{1-x} alloys is given by Herman et al.[7.389] (see also References 7.390 and 7.391). The thermal conductivity and Seebeck effect in the alloys have been reported, in particular, by Toxen,[7.392] Steele and Rosi,[7.393] Klemens,[7.402]

Table 7.23 V_2-VI_3 Compounds. Molecular Weight, Crystal Structure and Lattice Parameters, and Density

Compound	Molecular weight	Mean atomic weight	Mean atomic number	Crystal structure and space group	a (nm)	b (nm)	c (nm)	β°	Density (g/cm³)
As_2O_3, arsenolite	197.84	39.57	18	Cubic, O_h^7-Fd3m, Z = 80	1.10745[7.1]				3.89[7.1]
As_2O_3, claudetite				Monoclinic C_{2h}^5-P2$_1$/n, Z = 20	0.525	1.290[7.1]	0.453	93.9	4.23[7.1]
As_2S_3, orpiment	246.02	49.20	22.8	C_{2h}^5-P2$_1$/n, Z = 20	1.1475	0.9577[7.1]	0.4256	90.68	3.46[7.2]
As_2Se_3	386.72	77.34	33.6	C_{2h}^5-P2$_1$/n, Z = 20	1.2053	0.9890[7.1]	0.4277	90.47	4.75[7.2]
As_2Te_3	532.64	106.53	44.4	Monoclinic C_{2h}^3-C2/m, Z = 20	1.4339	0.4006[7.1]	0.9873	95.0	6.50[7.2]
Sb_2O_3, valentinite	291.50	58.30	25.2	Orthorhombic D_h^{10}-Pccn	0.492	1.246[7.1]	0.542		5.6[7.1]
Sb_2S_3, stibnite	339.68	67.94	30	Orthorhombic D_{2h}^{16}-Pnam or C_{2v}^9-Pna2$_1$	1.1299	1.1310	0.3839		4.60[7.175]
Sb_2Se_3	480.38	96.08	40.8	D_{2h}^{16}-Pnam	1.162	1.177	0.3962		5.81[7.2]
Sb_2Te_3	626.30	125.26	51.6	Trigonal D_{3d}^5-R$\bar{3}$m, Z = 15	0.4264[a] 0.425[a] 1.0426[b 7.136]		3.0458[a] 3.03[a]	23.52[b 7.136]	6.505[7.1] 6.49[7.136]
α-Bi_2O_3	465.96	93.19	38	C_{2h}^5-P2$_1$/c, Z = 20	0.5848	0.8166	0.7510	113.0	8.9[7.2]
β-Bi_2O_3				Tetragonal D_{2d}^4-P$\bar{4}$2$_1$c, Z = 20	0.7743		0.5631		8.929[7.1]
Bi_2S_3	514.16	102.83	42.8	D_{2h}^{16}-Pnam, Z = 20	1.1150	1.1300[7.2]	0.3981		
Bi_2Se_3	654.84	130.97	53.6	D_{3d}^5-R$\bar{3}$m	0.4138[a] 0.414[a] 0.9841[b]		2.864[a 7.1] 2.87a	24.27[b 7.1]	7.51[7.136]
Bi_2Te_3	800.76	160.15	64.4	D_{3d}^5-R$\bar{3}$m	0.43835[a] 1.0473[b]		3.0487[a 7.2]	24.17[b]	7.73[7.136]

[a] Hexagonal axes.
[b] Rhombohedral axes.

Table 7.24 V$_2$-VI$_3$ Compounds. Thermal and Thermochemical Properties

Compound	Melting point (K)	Boiling point (K)	Enthalpy (kJ/mol) of Formation	Enthalpy (kJ/mol) of Fusion	Heat capacity (RT) (J/mol·K)	Debye temperature (K)	Comments
As$_2$O$_3$, arsenolite	551 [7.1] 547 [7.2] under pressure	733 [7.2]			96.08 [7.1]		$C_p = 35.04 + 0.2348\,T$ between 273 K and 548 K for both allotropes [7.1]
As$_2$O$_3$, claudetite	586 [7.2]						
As$_2$S$_3$	585 [7.1] 583 [7.2]	980 [7.2]	−169.0 [7.2]		116.3 [7.2] 116.7 [7.1]		$C_p = 105.72 + 0.365\,T$ between 298 K and 585 K [7.1]
As$_2$Se$_3$	650 [7.1] 533 [7.2]				121.6 [7.1]		$C_p = 95.88 + 0.858\,T$ between 298 K and 650 K [7.1]
As$_2$Te$_3$	648 [7.1] 621 [7.2]				127.93 [7.1]		$C = 135.28 + 0.044\,T - 1.86 \cdot 10^6 T^{-2}$ between 298 K and 648 K [7.1]
Sb$_2$O$_3$, senarmonite	843 (trans.) [7.2] 928 [7.1]	1698 [7.2]			101.45 [7.1]		$C_p = 79.97 + 0.07195\,T$ [7.1]
Sb$_2$O$_3$, valentinite							
Sb$_2$S$_3$	823 [7.1]				120.1 [7.1] 12.98 (80 K) [7.1]	310 (80 K) [7.1]	$C_p = 101.91 + 0.0606\,T$ between 298 K and 823 K [7.1]
Sb$_2$Se$_3$	884 [7.2] 885 [7.1]				127.0 [7.1] 16.33 (80 K) [7.1]	240 (80 K) [7.1]	$C_p = 123.93 + 0.01026\,T$ between 290 K and 888 K [7.1]
Sb$_2$Te$_3$	893 [7.2]		−117.2 [7.136]	98.95 [7.136]	128.8 [7.1] 20.52 (80 K)	160 (80 K) [7.1]	
Bi$_2$O$_3$	1090 [7.1]	2163 [7.2]	−573.9 [7.2]		113.5 [7.2]		
Bi$_2$S$_3$	1123 [7.2] 1036 [7.1]		−143.2 [7.2]		122.9 [7.1] 122.2 [7.2]		
Bi$_2$Se$_3$	979 [7.1] 983 (decomp.) [7.2]		−139.95 [7.136]		14.24 (80 K) [7.1] 124.3 [7.1]	182 (0 K) [7.1]	$C_p = 114.55 + 0.0277\,T$ between 298 K and 1036 K [7.1] $C_p = 118.61 + 0.01926\,T$ between 298 K and 995 K [7.1]
Bi$_2$Te$_3$	858 [7.1] 853 [7.2]		−102.09 [7.136]	118.62 [7.136]	124.7 [7.1]	164.9 (0 K) [7.1] 155 [7.4]	$C_p = 108.06 + 0.0553\,T$ [7.1] between 293 K and 850 K [7.1]

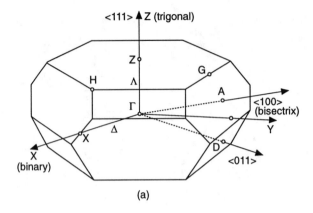
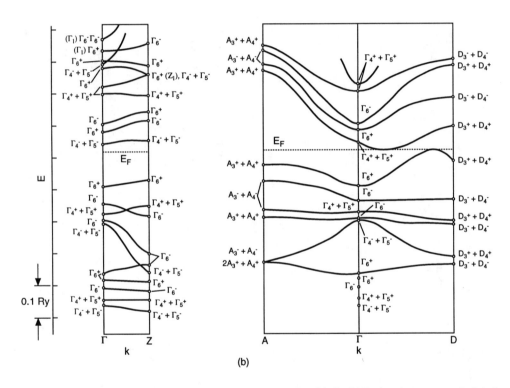

Figure 7.61 (a) Brillouin zone for a tetradymite-structure crystal, (b) the Bi_2Te_3 band structure calculated using the empirical pseudopotential method (spin-orbit interaction is included). (After F. Borghese and E. Donato, *Nuovo Cimento*, 53B, 283, 1968. With permission.)

and the Ioffes[7.403] (at temperatures below 300 K); the high-temperature (up to 1200 K) thermal conductivity was measured by Abeles et al.[7.404] The specific heat of the alloys is a linear function of the alloy composition.

Another group of binary alloys with semiconductor properties is the continuous row of solid solutions in the system Se-Te.[7.55] The lattice parameter of the alloys is an almost linear function of the component concentration. The alloys are synthesized in the form of both poly- and monocrystals. The electrical conductivity and Hall coefficient in a wide range of temperature (4 K to 500 K) was measured by A. Nussbaum (see in Reference 7.246).

Some information regarding the binary alloys with semiconductor properties is included in Chapter 3 of this book.

Table 7.25 V$_2$-VI$_3$ Compounds. Energy Band Parameters

Compound	Energy gap (eV) Direct 0 K	Energy gap (eV) Direct 300 K	Energy gap (eV) Indirect 300 K	Energy gap (eV) Thermal	dE_g/dT (10^{-4} eV/K)	dE_g/dp (10^{-6} eV/bar)	Spin-orbit splitting, meV for band Conduction	Spin-orbit splitting, meV for band Valence	Effective mass in m_o units Electrons	Effective mass in m_o units Holes
As$_2$S$_3$		2.68[7.1]		2.78 (0 K)[7.1]	−7 (77–290 K)[7.1]	−14 (300–400 K)[7.1]				
As$_2$Se$_3$	2.15 (10 K)[7.1]	1.85[7.1]		2.01 (0 K)[7.1]	−7.9 (80–274 K)[7.1]	−17 (300 K)[7.1]	15[7.1]	40[7.1]	0.36[7.1]	0.50[7.1]
As$_2$Te$_3$				0.9[7.290]						
Sb$_2$S$_3$		1.88[7.1] 1.63[7.291]			−9[7.1]	−7[7.1]				
Sb$_2$Se$_3$		≃1	1.11[7.1]	1.08–1.32 (0 K)[7.1]	−1.36[7.1]	−14[7.1]				
Sb$_2$Te$_3$		0.21–0.28[7.1]		0.3[7.136]					0.3 (m_{ds})[7.1]	
Bi$_2$O$_3$	3.1 (77 K)[7.1]	2.8[7.1]		1.6 (0 K)[7.1]	−9[7.1]					
Bi$_2$S$_3$	1.45 (77 K)[7.1]	1.3[7.1]		1.0 (0 K)[7.1]						
Bi$_2$Se$_3$	0.160 (77 K)[7.1]	0.35[7.136]		0.160[7.1]	−2[7.1]				0.155[7.1]	0.125[7.1]
Bi$_2$Te$_3$			0.13–0.14[7.1]	0.15–0.16[7.1] 0.21[7.136]	−1.5[7.1]	−2[7.1]			0.27–0.32[7.1]	0.35–0.46[7.1]

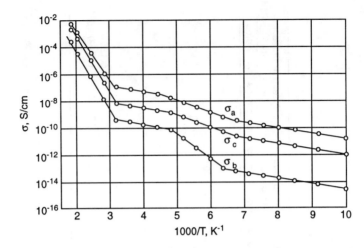

Figure 7.62 Temperature dependence of the Sb_2Te_3 electrical conductivity along the a, b, and c axes. (After B. Roy, B.R. Chakraborty, R. Bhattacharya, and A.K. Dutta, *Solid State Commun.*, 25, 937, 1978. With permission.)

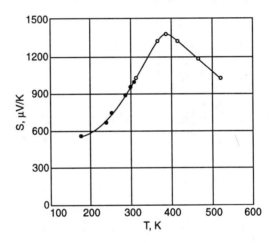

Figure 7.63 Temperature dependence of the Sb_2Se_3 Seebeck coefficient. (After B.R. Chakraborty, B. Roy, R. Bhattacharya, and A.K. Dutta, *J. Phys. Chem. Solids*, 41, 93, 1980. With permission.)

7.8.2. III-V/III-V Semiconductor Alloys

The III-V/III-V alloys belong to the alloy systems with common anions or common cations (ternary alloys), or to the systems with simultaneous substitution of both cations and anions (quaternary alloys). The information regarding the limits of existence of solid solutions in the systems is gathered in Table 7.33.

The problem of formation and utilization of the III-V/III-V solid solutions was raised even before the properties of all possible III-V compounds were examined, because some of the applications of the materials needed combinations of properties intermediate to the properties of individual III-V phases. By now, III-V/III-V solid solutions have become virtually the most important materials of electronic industry, especially for the semiconductor laser and optoelectronic devices industries (see, e.g., References 7.399 to 7.401).

At least in the first approximation, the solid solutions' lattice parameters follow Vegard's rule. The dependence of the lattice parameters on composition is displayed in Figure 7.74

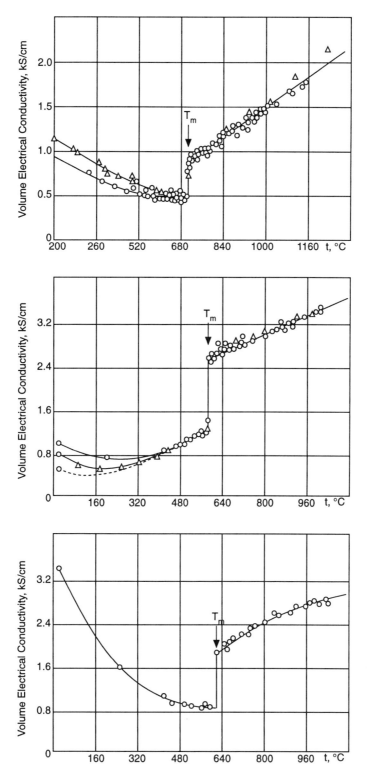

Figure 7.64 Temperature dependence of the electrical conductivity of Bi_2Se_3 (a), Bi_2Te_3 (b), and Sb_2Te_3 in the solid and liquid states. (After V.M. Glazov, S.N. Chizhevskaya, and N.N. Glagoleva, *Liquid Semiconductors,* Nauka Publishing House, Moscow, 1967. With permission.)

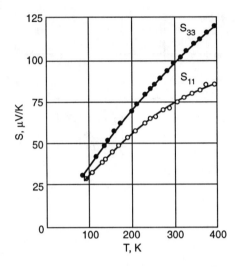

Figure 7.65 Temperature dependence of the Sb_2Te_3 Seebeck coefficient across (S_{11}) and along (S_{33}) the c axis. (After M. Stordeur and W. Heiliger, *Phys. Stat. Sol.*, B78, K103, 1976. With permission.)

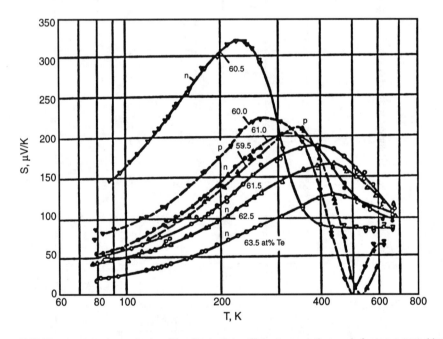

Figure 7.66 Temperature dependence of the Seebeck coefficient across the c axis for two p-type (dashed lines) and five n-type single crystals of Bi_2Te_3 with various deviations from stoichiometric composition. (After C.H. Champness and A.L. Kipling, *Canad. J. Phys.*, 44, 769, 1966. With permission.)

(adapted from Reference 7.398). The correlation of the lattice parameters of the alloys of various III-V semiconductors with their energy gap is shown in Figure 7.75 (adapted from Reference 7.413). The analytical expressions for $E_g = E_g(x)$ are collected in Table 7.34.

The areas of existence of solid solutions with sphalerite structure in the sections GaP-GaAs-InAs-InP, GaP-GaSb-InSb-InP, and GaAs-GaSb-InSb-InAs of the quaternary systems Ga-In-P-As, Ga-In-P-Sb, and Ga-In-As-Sb, respectively, are shown in Figure 7.76 (adapted

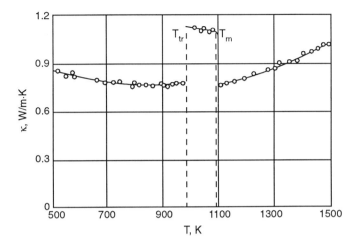

Figure 7.67 Temperature dependence of the Bi_2O_3 thermal conductivity measured on a polycrystalline sample. T_{tr} — temperature of phase transition in the solid state; T_m — melting point. (After A. Krost, in Landolt-Börnstein, *New Series*, Vol. III/17f, O. Madelung, Ed., Springer-Verlag, New York, 1984, p. 528. With permission.)

Table 7.26 V_2VI_3 Compounds. Magnetic Susceptibility and Dielectric Permittivity

Compound	Molar magnetic susceptibility (10^{-5} SI units/gram-mol)	Dielectric permittivity (dielectric constant)	
		Static	Optical
As_2O_3	−51.8		
As_2S_3		12.1 (E∥a)	8.8 (E∥a)
		10.7 (E∥c)	7.0 (E∥c)
		5.9 (E∥b)	5.7 (E∥b)
As_2Se_3		12.4 (E∥c)	8.8 (E∥c)
		13.9 (E∥a)	10.5 (E∥a)
Sb_2O_3		12.8 (≈2 MHz)	
Sb_2S_3	−105.3 [7.297]	180 (E∥c)	9.5 (E∥c)
		15 (E⊥c)	7.2 (E⊥c) (at 320 K)
Sb_2Se_3	−103.8 [7.298]	133 (E∥c)	15.1 (E∥c)
		≈110 (10 GHz)	13.7 (E⊥c)
Sb_2Te_3	−97.1	36.5 (80 K) (E∥c)	32.5 (E∥c)
		168 (80 K) (E⊥c)	51 (E⊥c)
Bi_2O_3	−104.3 [7.312]	112 (1 kHz) [7.2]	
Bi_2S_3	−154.6	120 (E∥c)	13 (E∥c)
		38 (E⊥c)	9 (E⊥c)
Bi_2Se_3		113 (E⊥c)	29 (E⊥c)
Bi_2Te_3	≈−450 (B∥c)	75 (15 K) (E∥c)	50 (E∥c)
	≈−190 (B⊥c)	290 (15 K) (E⊥c)	85 (E⊥c)

From Landolt-Börnstein, New Series III/17e and III/17g, O. Madelung, M. Schulz, and H. Weiss, Eds., Springer-Verlag, New York, 1982–1984.

from Reference 7.398). Back in 1958, Köster and Ulrich[7.414] suggested the existence of wide areas of solubility in the system AlSb-GaSb-InSb. This suggestion proved to be correct, and not only for this system, as one can see in Tables 7.33 and 7.34.

The hardness of $Al_xGa_{1-x}As$ and $GaAs_xSb_{1-x}$ alloys was studied by Swaminathan et al.;[7.415] the compositional dependence of the Knoop microhardness anisotropy of $Ga_xIn_{1-x}As_yP_{1-y}$ alloys was reported in Reference 7.416. Other works devoted to study of the III-V/III-V alloy

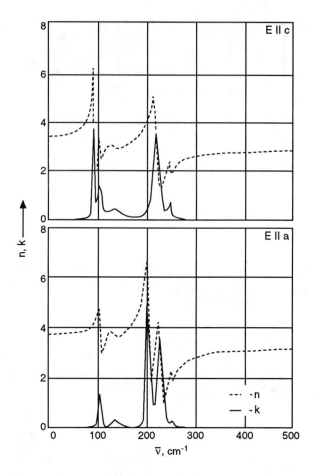

Figure 7.68 Index of refraction, n, and extinction coefficient, K, for As_2Se_3 vs. wavenumber. (After R. Zallen, M. Slade, and A.T. Ward, *Phys. Rev.*, B3, 4257, 1971. With permission.)

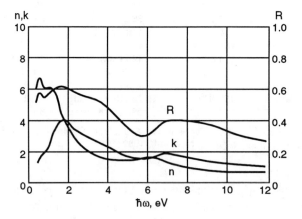

Figure 7.69 Experimental magnitudes of reflectivity, R, and the calculated index of refraction, n, and extinction coefficient, K, (Kramer-Kronig analysis) for Sb_2Te_3 vs. photon energy at 300 K (the electric field vector is perpendicular to the trigonal axis). (After W. Richter, A. Krost, U. Novack, and E. Anastassakiss, *Zs. Phys.*, B49, 191, 1982. With permission.)

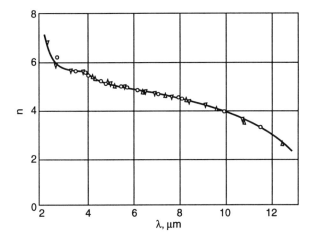

Figure 7.70 Spectral distribution of the index of refraction at 300 K calculated from reflectivity and optical transmission measurements on Bi_2Se_3 samples with the thickness from 0.7 μm to 5.9 μm and the electric field vector perpendicular to the c axis. (After H. Köhler and C.R. Becker, *Phys. Stat. Sol.*, B61, 533, 1974. With permission.)

of these alloys was reported by Pietsch and Marlow.[7.419] A detailed description of thermal, elastic, and electrical properties of the $Al_xGa_{1-x}As$ alloys is compiled in Reference 7.421. The electrical and thermal transport properties of the $GaAs_{1-x}P_x$ alloys at 77 K and 300 K were studied by Takeda et al.[7.420] The thermal, electrical, and optical properties of the $Ga_{1-x}In_xAs$ alloys were published by Abrahams et al.[7.422] The information regarding electron mobility in these alloys and thermal conductivity in them was reported by Chan et al.[7.423] and Abeles,[7.424] respectively. The thermal conductivity of the $InAs_xSb_{1-x}$ was reported in Reference 7.425. The information regarding the electrical transport and optical reflectance for $Ga_xIn_{1-x}As_ySb_{1-y}$ is presented, in particular, in Reference 7.426. Kowalsky et al.[7.427] published the results of their measurement of the optical absorption coefficient and index of refraction of the alloys in the system $In_xGa_{1-x}P_yAs_{1-y}$.

As was mentioned earlier, modern electronics and optoelectronics widely utilize the III-V/III-V alloys. It is interesting to note that the very first laser, based on semiconducting solid solutions, was built using a GaP-GaAs alloy.

7.8.3. II-VI/II-VI Solid Solutions

Substitutional solid solutions of the II-VI/II-VI type are similar, in many features, to the binary compounds forming them, and therefore present a substantial interest for applications in optical devices. Some of them are known as minerals (e.g., onofrite, HgS_xSe_{1-x}).

In principle, it is possible to have thirty-two II-VI/II-VI chalcogenide alloy systems, nine with cation substitution, nine with anion substitution, and fourteen alloy systems with substitution of both cations and anions. From these systems, at least 13 are continuous rows of solid solutions (Table 7.35). In the first approximation, the lattice constants of the solutions follow Vegard's rule.

Zinc selenide dissolves in solid state up to 70 mol% of CdSe; solubility of ZnS in ZnTe and ZnTe in ZnS does not exceed 5 to 10 mol%; the solid solutions $CdS_{1-x}Te_x$ ($x \leq 0.2$) have sphalerite structure, while the solid solutions $CdTe_{1-x}S_x$ ($x \leq 0.25$) crystallize in wurtzite structure. There is a miscibility gap with $0.43 < x < 0.60$ in the alloy system $Cd_{1-x}Hg_xS$; the alloys with $x < 0.43$ have wurtzite structure, while the phases with $x > 0.60$ crystallize in the sphalerite lattice. There is a small miscibility gap between $x = 0.25$ and $x = 0.30$ in the alloy

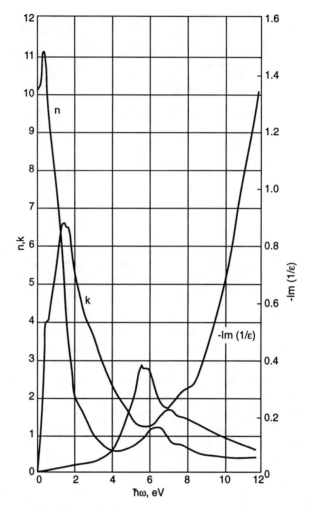

Figure 7.71 Index of refraction, n, extinction coefficient, K, and energy loss function, $-\text{Im}(1/\varepsilon)$, vs. photon energy for Bi_2Te_3 (with $E \perp C$). (After D.L. Greenaway and G. Harbeke, *J. Phys. Chem. Solids*, 26, 1585, 1965. With permission.)

Table 7.27 V_2-VI_3 Compounds. Commercial Availability

Compound	Purity (%)	Form	Price (US$ per gram)	
			From	To
As_2O_3	99.999	100-mesh powder	1.78	24.90
As_2S_3	99.999	325-mesh powder	5.26	6.60
As_2Se_3	99.999	1- to 6-mm pieces	6.80	69.30
Sb_2O_3	99.999	Powder	1.44	11.58
Sb_2S_3	99.999	Lump	3.12	4.20
Sb_2Se_3	99.999	3- to 12-mm pieces	9.76	12.60
Sb_2Te_3	99.999	3- to 12-mm lump	5.96	7.50
Bi_2O_3	99.9998	20-mesh powder	0.48	0.86
Bi_2S_3	99.9	200-mesh powder	0.76	1.06
Bi_2Se_3	99.99	Lump	2.73	3.40
Bi_2Te_3	99.999	Pieces under 6 mm	3.88	5.30

From Johnson-Matthey, 1995–1996.

Table 7.28 Compounds Containing Group VIIIA Elements. Basic Chemical and Crystallographic Parameters

Compound	Molecular weight	Mean atomic weight	Mean atomic number	Crystal structure and space group	Lattice parameters a (nm)	b (nm)	c (nm)	β°	Density (300 K) (g/cm³)	Ref.
FeSi	83.93	41.97	20						6.1	7.2
FeSi$_2$	112.02	37.34	18	Orthorhombic, D_{2h}^{18}-Cmca	0.9863	0.7791	0.7833		4.74	7.1, 7.2
Fe$_2$P	142.66	47.55	22.3						6.8	7.2
FeAs	130.77	65.38	29.5						7.85	7.2
FeAs$_2$	205.69	68.56	30.7	Orthorhombic, D_{2h}^{12}-Pnnm	0.5300	0.5981	0.2882		7.477	7.2
FeAs$_3$	280.61	70.15	31.3	Cubic, T_h^5-Im3	0.81814				6.7158	7.2
FeSb$_2$	299.35	99.8	42.7							
FeO	71.84	35.92	17						6.0	7.2
Fe$_3$O$_4$	231.53	33.08	15.7	Cubic, O_h^7-Fd3m	0.83940				5.2003	7.2
Fe$_2$O$_3$	159.69	31.94	15.2	Hexagonal-R, D_{3d}^6-R$\bar{3}$c	0.50329		1.37492		5.2749	7.2
FeS	87.91	43.96	21	Hexagonal, D_{6h}^4-P6$_3$/mmc	0.3446		0.5877		4.7	7.2
FeS$_2$ pyrite	119.98	39.99	19.3	Cubic, T_h^6-Pa3	0.54175 0.54188				5.0116	7.1, 7.2, 7.332
FeS$_2$ marcasite				D_{2h}^{12}-Pnnm	0.4443	0.5423	0.33876		4.8813	7.2
Fe$_3$S$_4$	295.78	42.25	20.3	O_h^7-Fd3m	0.9876				4.079	7.2
FeSe	134.81	67.41	30						6.7	7.2
FeSe$_2$	213.77	71.26	31.3	D_{2h}^{12}-Pnnm	0.4801	0.5778	0.3587		7.134	7.2
FeTe	183.45	91.73	39							
FeTe$_2$	311.05	103.68	43.3	D_{2h}^{12}-Pnnm	0.5265	0.6265	0.3869		8.094	7.2
Co$_2$Si	145.95	48.65	22.7	Orthorhombic	0.4918	0.3737	0.7109		6.8	7.333 7.2
CoSi	87.02	43.5	20.5							
CoSi$_2$	115.10	38.37	18.3						4.9	7.2
Co$_2$P	148.84	49.61	23						6.4	7.2
CoP$_2$	120.88	40.29	19	Monoclinic, C_{2h}^5-P2$_1$/c	0.5610	0.5591	0.5643	116.82	5.08	7.1
CoP$_3$	151.85	37.96	18							

Table 7.28 (continued) Compounds Containing Group VIIIA Elements. Basic Chemical and Crystallographic Parameters

Compound	Molecular weight	Mean atomic weight	Mean atomic number	Crystal structure and space group	a (nm)	b (nm)	c (nm)	β°	Density (300 K) (g/cm³)	Ref.
CoAs	133.86	66.93	30		0.5918	0.5876	0.5964	116.45	8.22	7.2
CoAs$_2$	208.78	69.59	31	C_{2h}^5-P2$_1$/c	0.5049	0.5872	0.3127	90.45	7.479	7.1 7.2
CoAs$_3$	286.70	71.67	31.5	T_h^5-Im3	0.82060				6.7298	7.2
CoSb$_2$	302.43	100.81	43	C_{2h}^5-P2$_1$/c	0.6508	0.6388	0.6543	117.66	8.34	7.1
CoSb$_3$	424.18	106.05	45	T_h^5-Im3						7.3
CoS$_2$	123.07	41.02	19.7	T_h^6-Pa3	0.55345				4.8213	7.2
Co$_3$S$_4$	305.04	43.58	20.7	O_h^7-Fd3m	0.9876				4.079	7.3
CoSe$_2$	216.85	72.28	31.7	T_h^6-Pa3	0.58588				7.1618	7.2
NiSi	86.78	43.39	21	Orthorhombic, D_{2h}^{16}-Pnma	0.5233	0.3258	0.5695		4.83	7.333 7.2
NiSi$_2$	114.86	38.29	18.7							
Ni$_2$Si	145.47	48.5	23.3	D_{2h}^{16}-Pnma	0.499	0.372	0.701		7.40	7.333
Ni$_2$P	148.36	49.5	23.7						7.33	7.2
NiP$_2$	120.64	40.21	19.3	Monoclinic, C_{2h}^6-P2/c	0.6366	0.5615	0.6072	126.2		7.1
NiAs	133.62	66.81	30.5							
NiAs$_2$ rammelsbergite	208.53	69.51	31.3	D_{2h}^{12}-Pnnm	0.4757	0.5797	0.3542		7.091	7.2
NiAs$_2$ pararammelsbergite				Orthorhombic	0.575	0.582	1.1428		7.244	7.2
NiAs$_3$	283.45	70.86	31.8	T_h^5-Im3	0.8330				6.4286	7.2
NiSb	180.45	90.23	39.5						8.75	7.2
NiS	90.76	45.38	22	Hexagonal, D_{3d}^5-P$\bar{3}$m	0.9616		0.3152		5.3743	7.2
Ni$_3$S$_2$	240.21	48.04	23.2	Hexagonal-R, D_3^7-R32	0.5746		0.7134		5.867	7.2
Ni$_3$S$_4$	304.34	43.48	21.1	O_h^7-Fd3m	0.9480				4.7458	7.2
NiS$_2$	122.81	40.94	20	T_h^6-Pa3	0.56873				4.4350	7.2
NiSe	137.65	68.83	31						7.20	7.2
NiSe$_2$	216.61	72.20	32	T_h^6-Pa3	0.59604				6.7948	7.2
NiTe$_2$	313.89	104.63	44	Hexagonal, D_{3d}^3-P$\bar{3}$m1	0.3869		0.5308		7.575	7.2

NONADAMANTINE & VARIABLE-COMPOSITION SEMICONDUCTOR PHASES

Compound				Structure	a	b	c	angle		Ref
RuAl$_2$	155.03	51.68	23.3	Orthorhombic, D_{2h}^{24}-Fddd	0.4715	0.8015	0.8715			7.1
					0.799	0.471	0.876			7.326
RuGa$_2$	240.51	80.17	35.3	D_{2h}^{24}-Fddd	0.4749	0.8184	0.8696			7.1
					0.8184	0.479	0.8696			7.330
RuSi$_2$	157.24	52.41	24	D_{2h}^{18}-Cmca						7.1
Ru$_2$Si$_3$	286.40	57.28	26	Tetragonal	1.1075		0.8954		6.96	7.331
				Orthorhombic	1.1057	0.8934	0.5533		9.23	7.1, 7.334
Ru$_2$Ge$_3$	419.91	83.98	36.8	Pbcn	1.1463	0.9238	0.5716		9.23	7.334
RuP$_2$	163.02	54.34	24.7	D_{2h}^{12}-Pnnm	0.5117	0.5892	0.2871		6.256	7.1
RuAs$_2$	250.91	83.64	36.7	D_{2h}^{12}-Pnnm	0.5428	0.6183	0.2969		8.364	7.1
RuAs$_3$	325.83	81.46	35.8							
RuSb$_2$	344.57	114.86	48.7	D_{2h}^{12}-Pnnm	0.5951	0.6674	0.3179		9.063	7.1
RuSe$_2$	258.99	86.33	37.3	T_h^6-Pa3	0.5934					7.1
RuTe$_2$	356.27	118.76	49.3	T_h^6-Pa3	0.639					7.1
				D_{2h}^{12}-Pnnm	0.5271 (at 77 K)	0.6387 (at 77 K)	0.4083 (at 77 K)			7.1
RhSi	130.99	65.50	29.5	D_{2h}^{16}-Pnma	0.55310	0.3063	0.6362			7.335
RhP$_2$	164.85	54.95	25	C_{2h}^5-P2$_1$/c	0.57417	0.57951	0.58389	112.911	6.12	7.1
RhAs$_2$	252.75	84.25	37	C_{2h}^5-P2$_1$/c	0.60629	0.60816	0.61498	114.707	8.15	7.1
RhSb$_2$	346.41	115.47	49	Same	0.66156	0.65596	0.66858	116.821	8.89	7.1
—RhBi$_2$	520.87	173.62	70.3	Same	0.69207	0.67945	0.69613	117.735	11.86	7.1
RhS$_3$	199.09	49.77	23.3	T_h^6-Pa3	0.558				6.40	7.1
				Trigonal distortion	0.558			>90		7.1
RhSe$_2$	260.83	86.94	37.7	D_{2h}^{16}-Pnam	2.091					7.1
RhSe$_3$	339.79	84.95	36.8	Trigonal distortion	0.5962	0.595	0.3709	90.73		7.1
PdS	138.48	69.24	31	Tetragonal, C_{4h}^2-P4$_2$/m	0.6429		0.6608		6.6	7.1
									6.7	7.2
PdS$_2$	170.54	56.85	26	Orthorhombic, D_{2h}^{15}-Pbca	0.546	0.554	0.753		4.7–4.8	7.1
PdSe	185.38	92.69	40	C_{4h}^2-P4$_2$/m	0.673		0.691			7.1
PdSe$_2$	264.37	88.12	38	D_{2h}^{15}-Pbca	0.574	0.587	0.769			7.1
OsSi$_2$	246.37	82.12	34.7	D_{2h}^{18}-Cmca (Z = 16)	1.01496	0.81168	0.82230			7.1

Table 7.28 (continued) Compounds Containing Group VIIIA Elements. Basic Chemical and Crystallographic Parameters

Compound	Molecular weight	Mean atomic weight	Mean atomic number	Crystal structure and space group	a (nm)	b (nm)	c (nm)	β°	Density (300 K) (g/cm³)	Ref.
Os_2Si_3	464.56	92.93	38.8	Orthorhombic, D_{2h}^{14}-Pbcn	1.1158	0.8962	0.5579			7.1
					1.1124	0.8932	0.5570		11.15	7.334
				Hexagonal close packed	0.27333	0.43191				7.2
Os_2Ge_3	598.17	119.63	34.4	D_{2h}^{14}-Pbcn	1.1544	0.9281	0.5783		12.82	7.1
OsP_2	252.15	84.05	35.3	D_{2h}^{12}-Pmmn	0.51012	0.59022	0.29183		9.531	7.1
OsP_4 (RT)	314.10	62.82	27.2	C_{2h}^{5}-$P2_1/c$	0.4694	0.4683	0.7096	80.43	6.78	7.1
OsP_4 (HT)				Triclinic, C_i^1-$P\bar{1}$	0.7540	0.7153	0.4718	100.38 (α) 90.34 (β) 111.20 (γ)	6.73	7.1
$OsAs_2$	340.04	113.35	47.3	D_{2h}^{12}-Pnnm	0.54115	0.61900	0.30127		11.191	7.1
$OsSb_2$	433.70	144.57	59.3	Same	0.59411	0.66873	0.32109		11.291	7.1
OsS_2	254.32	84.77	36	T_h^6-Pa3	0.5619				9.47	7.1
$OsSe_2$	348.12	116.04	48	Same	0.5945					7.1
$OsTe_2$	445.4	148.5	60	Same	0.6397					7.1
				D_{2h}^{12}-Pnnm	0.52804	0.64018	0.40481			7.1
Ir_3Ga_5	925.26	115.66	48.3	D_{2d}^{8}-$P\bar{4}n2$ (Z = 4)	0.5823		1.420			7.1
Ir_4Ge_5	1131.83	125.76	52	D_{2d}^{2}-$P\bar{4}c2$ (Z = 4)	0.56123		1.8361			7.1
IrP_2	254.17	84.72	35.7	D_{2h}^{5}-$P2_1/c$ (Z = 4)	0.5746	0.5791	0.5851	111.60	9.32	7.1
IrP_3	285.14	71.29	30.5	T_h^5-Im3	0.80151				7.36	7.1
$IrAs_2$	342.06	114.02	47.7	C_{2h}^{5}-$P2_1/c$ (Z = 4)	0.60549	0.60717	0.61587	113.197	10.86	7.1
$IrAs_3$	416.98	104.25	44	T_h^5-Im3	0.84673				9.12	7.1
$IrSb_2$	435.72	145.24	59.7	C_{2h}^{5}-$P2_1/c$ (Z = 4)	0.65945	0.65492	0.66951	115.158	10.98	7.1
$IrSb_3$	557.47	139.37	57.5	T_h^5-Im3	0.92533				9.35 9.32	7.1 7.336
$IrBi_2$ (?)	610.18	203.39	80.3	C_{2h}^{5}-$P2_1/c$	0.70	0.69	0.71	117	13.3	7.1
IrS_2	256.34	85.45	36.3	D_{2h}^{16}-Pnam	1.978	0.5624	0.3565		8.43	7.1
IrS_3	288.4	72.1	31.3	Trigonal distortion	0.55			>90 (α)		7.1

Compound				Space group						
IrSe$_2$	350.14	116.71	48.3	D_{2h}^{16}-Pnam	2.094	0.593	0.374			7.1
IrSe$_3$	429.10	107.28	44.8	Trigonal	0.596					7.1
PtP$_2$	257.03	85.68	36	T_h^6-Pa3	0.56956			>90 (α)	9.238	7.1
PtAs$_2$	344.92	114.97	48	Same	0.59665				10.79	7.1
PtSb$_2$	438.58	146.19	60	Same	0.6428				10.964	7.1
PtS	227.14	113.57	47	Tetragonal, D_{4h}^9-P4$_2$/mmc	0.3469		0.6109		10.04	7.1
PtS$_{2.06}$	253.35	85.30		Trigonal, D_{3d}^3-P$\bar{3}$m1	0.3542		0.5043			7.1
PtS$_2$	259.21	86.40	36.7						7.85	7.2
PtSe$_2$	353.0	117.67	48.7	D_{3d}^3-P$\bar{3}$m1	0.3724		0.5062			7.1

Table 7.29 Compounds Containing Group VIIIA Elements. Thermochemical, Thermal, and Thermal Transport Properties and Hardness (Reference 7.2 Unless Indicated Otherwise)

Compound	Standard enthalpy of formation (kJ/mol)	Heat capacity at 298.15 K (J/mol·K)	Thermal conductivity at 300 K (mW/cm·K)	Debye temperature (K)	Melting or decomposition temperature (K)	Hardness (GPa)	300-K volume expansion coefficient (10^{-5} 1/K)
$FeSi_2$	-70.3[7.346]		40[7.337]	630 (0 K)	1533[7.340] 1493		
$FeAs_2$					1287[7.1]		
$FeSb_2$				327[7.342]	1021[7.343]	2.00	
FeO	-272.0				1650		
Fe_2O_3	-824.2	109.3			1838		
Fe_3O_4	-1118.4	143.4			1870		
FeS	-100.0	50.5					
FeS_2	-178.2	62.2			1016[7.332]		
$CoSi_2$	-102.9[7.346]				1599[7.347]		4.2[7.1]
$CoAs_2$					1215		
$CoAs_3$					1204[7.1]		
$CoSb_2$					1132[7.1]		
$CoSb_3$			90 ± 10[7.336] ≈103[7.337] 89.1[7.338]	307[7.338]	1146[7.337]		
CoO	-237.9	55.2					
Co_3O_4	-891.0	123.4					
CoS	-82.8						
Co_2S_3	-147.3						
$NiSi_2$	-86.6[7.346]		40[7.337]		≈1270[7.347]		
$NiAs_2$					1125[7.344]		3.0[7.344]
$NiSb_2$					892[7.344]		5.3[7.344]
Ni_2O_3	-489.5						
NiS	-82.0	47.1				≈5.5[7.1]	
$RuAl_2$					≈2020[7.326]		

NONADAMANTINE & VARIABLE-COMPOSITION SEMICONDUCTOR PHASES

Compound							
Ru$_2$Si	−66.5 [7.348]						
RuSi	−133.1 [7.348]						
Ru$_2$Si$_3$	−334.72 [7.348]		20 [7.349]	435 (1.5–5 K) [7.325]	2070 [7.339]		
Ru$_2$Ge$_3$					1770 [7.325]		
Ru$_2$Sn					≈1500 [7.325]		
RuSb$_2$					≈1600 [7.1]	4.20 [7.1]	
RuO$_2$	−305.0						
RuO$_4$	−239.3						
RhSb$_2$			≈100 [7.336]		≈1373 [7.1]		3.7 [7.344]
RhSb$_3$					≈1170 [7.1]		2.3 [7.344]
RhBi$_2$							
Rh$_2$O$_3$	−343.0						
PdO	−85.4						
PdS	−75.0						
Os$_2$Si$_3$	−517.8 [7.348]						
OsSi$_2$	−308.8 [7.348]						
OsSb$_2$						5.60 [7.1]	
OsO$_4$	−394.1						
IrAs$_2$							2.4 [7.344]
IrSb$_2$			95 [7.336]		>1373 [7.1]		2.4 [7.344]
IrSb$_3$			89.1 [7.338]	303 [7.338]	≈1170 [7.1]		
IrO$_2$	−274.1						
IrS$_2$	−138.0						
Ir$_2$S$_3$	−234.0						
PtSi			260 [7.1]		1502		
PtP$_2$							
PtAs$_2$			400 [7.345]	≈325 [7.345]	>1533 [7.1]		
PtSb$_2$			390 [7.1]		1499 [7.1]		
PtS	−81.6	43.4					
PtS$_2$	−108.8	65.9					

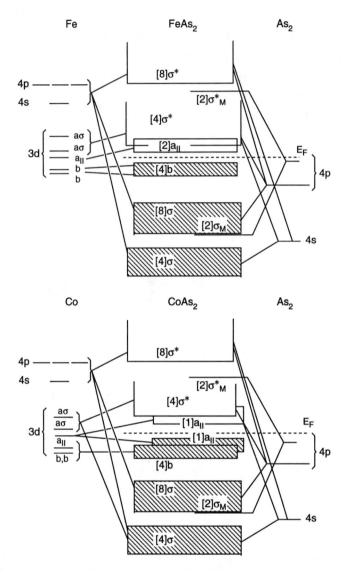

Figure 7.72 Energy level scheme for valence electrons in loellingite (a) and in an arsenopyrite-type phase with normal d^5 configuration (b). (After J.B. Goodenough, *J. Solid State Chem.*, 6, 144, 1972. With permission.)

system $Cd_{1-x}Hg_xSe$; the phases with $x < 0.25$ are the wurtzite-type solid solutions, and the alloys with $x > 0.30$ have the sphalerite structure. There is a similar distribution of structures in the alloy system $CdSe_{1-x}Te_x$ with a miscibility gap between $x = 0.4$ and $x = 0.6$.[7.397]

The energy gap of solid solutions can be approximated by a parabolic polynomial

$$E_g(x) = E_g(x = 0) + ax + bx^2$$

Tyagai et al.[7.445] performed an analysis of the nonlinearity parameter, b, in this equation for $Zn_xCd_{1-x}S$, $Zn_xCd_{1-x}Se$, $Zn_xCd_{1-x}Te$, CdS_xSe_{1-x}, CdS_xTe_{1-x}, $CdSe_xTe_{1-x}$, using the results of measurements of electroreflectance. Solid solutions in the quaternary (or quasiternary) systems $ZnS_xSe_yTe_{1-x-y}$ and $CdS_xSe_yTe_{1-x-y}$ were investigated, in particular, by Burlakov et al.[7.447]

Among the most studied ternary II-VI/II-VI alloys are solid solutions $Hg_{1-x}Cd_xTe$. The possibility to make infrared detectors for the atmosphere infrared transparency window

Table 7.30 Compounds Containing Group VIIIA Elements. Band Structure Parameters

Compound	Energy gap (eV) Optical (300 K) Direct	Energy gap (eV) Optical (300 K) Indirect	Energy gap (eV) Thermal	dE_g/dT (meV/K)	Density of states (in 10^{22} cm^{-3})	Effective mass in the units of m_o Electrons	Effective mass in the units of m_o Holes
FeSi$_2$			0.85–1.0 [7.1]	−0.45 (700–1200 K) [7.1]	4.9–7.46 [7.1]		
FeP$_2$			0.37–0.4 [7.1]				
FeAs$_2$			0.2–0.22 [7.1]				
FeSb$_2$			≈0.17 [7.1] 0.02 [7.353]				
FeS$_2$ pyrite	≥0.95 [7.332] 0.95 [7.360]		0.52 [7.1]				
FeSe			0.14 [7.1]				
FeSe$_2$			0.31 [7.1]				
FeTe$_2$			0.92 [7.1]				
CoP$_2$			>0.02 [7.363]				
CoP$_3$	0.45 [7.1]						
CoAs$_2$			≈0.35 [7.1]				
CoAs$_3$	<0.4 [7.1]		≈0.25 [7.1]				
CoSb$_2$			0.17 [7.1]				
CoSb$_3$	<0.4 [7.365]		0.55 [7.337] 0.49 [7.1]				
NiP$_2$	0.73 [7.364]						
NiAs$_2$ (rammel.)	≈0.05 [7.1]						
NiSb$_2$			Metal [7.1]				
NiS$_2$	0.27 [7.1]		0.64 [7.1] (>380 K)	−0.4 [7.1]			
NiS$_{1.99}$	0.370 [7.359]						
RuGa$_2$	0.85 [7.1]		0.42 ± 0.4 [7.326]				
Ru$_2$Si$_3$ (HT)			0.44 [7.348]				
Ru$_2$Si$_3$ (LT)			≈0.7 [7.1]				
Ru$_2$Ge$_3$ (HT)			0.34 [7.349]				
Ru$_2$Ge$_3$ (LT)			0.52 [7.1]				
RuP$_2$	0.8 [7.362]						
RuAs$_2$	≈0.8 [7.1]						
RuSb$_2$			>0.3 [7.1]				
RuS$_2$	1.8 [7.1]						
RuSe$_2$	1.0 [7.3]		≥0.6 [7.1]				
RuTe$_2$			0.25 [7.1]				
RhP$_2$	≈1.0 [7.1]						
RhAs$_2$	1.15 [7.1]						
RhAs$_3$	>0.1 [7.1]						
Rh$_2$S$_3$			0.8 [7.1]				
RhS$_3$	>1.5 [7.1]						
RhSe$_2$	0.6 [7.1]						
RhSe$_3$	0.7 [7.1]						
PdP$_2$			0.6–0.7 [7.1]				
PdS			0.5 [7.1]				
PdS$_2$			0.7–0.8 [7.1]				
PdSe			0.2 [7.1]				
PdSe$_2$			0.4 [7.1]				
OsSi$_2$	≈2 [7.1]						
OsP$_2$	2.1 [7.1]						
OsAs$_2$	0.9 [7.1]						
OsSb$_2$			>0.3 [7.1]				
OsS$_2$	2.0 [7.1]						
OsTe$_2$			>0.2 [7.1]				
IrP$_2$	1.1 [7.362]						
IrAs$_2$	≈1 [7.1]						

Table 7.30 (continued) Compounds Containing Group VIIIA Elements. Band Structure Parameters

Compound	Energy gap (eV)			dE_g/dT (meV/K)	Density of states (in 10^{22} cm^{-3})	Effective mass in the units of m_0	
	Optical (300 K)		Thermal			Electrons	Holes
	Direct	Indirect					
IrSb$_3$			1.18[7.336]			1.51[7.336]	0.17[7.336]
IrS$_2$	0.9[7.1]						
IrS$_3$	2.0[7.1]						
IrSe$_2$	≈1.0[7.1]						
IrSe$_3$			≥0.45[7.1]				
PtP$_2$			≥0.6[7.1]				
PtAs$_2$			0.8[7.1]				
PtSb$_2$	>0.4[7.1]	≈0.10[7.1]	0.11[7.1]			1.6[7.1]	0.2[7.1]
PtS	0.8[7.1]		0.64[7.1]				
PtS$_{2.06}$		0.95[7.1]	0.4[7.1]	−0.37[7.1]			
PtSe$_2$			0.1[7.1]				

(8 μm to 12 μm) and closeness of the binary constituents' lattice constants (a_{HgTe}/a_{CdTe} is between 0.99694 and 0.9966, from different sources) caused this "child of the cold war" to be widely used in the night vision instruments, rocket plume detecting systems, missile guiding systems, as well as in weather satellite sensors and in some medical devices (see, e.g., Reference 7.428 and 7.429). The direct energy gap of the Hg$_{1-x}$Cd$_x$Te alloys (transition $\Gamma_{8v} \to \Gamma_{6c}$) was expressed by Weileer et al.[7.430] for temperatures between 24 K and 91 K and $0.175 \leq x \leq 0.269$ as

$$E_g, eV = -0.31 + 1.88x + 5 \cdot 10^{-4} T(1 - 2x) \quad (7.1)$$

while Hansen et al.[7.431,7.432] suggest for $0 \leq x \leq 0.6$ and $4 K \leq T \leq 300 K$

$$E_g, eV = -0.302 + 1.93x + 5.35 \cdot 10^{-4} T(1 - 2x) - 0.180x^2 + 0.332x^3 \quad (7.2)$$

An IEE Data Review publication[7.433] contains the following expression for $0 \leq x \leq 1$ and $T > 70$ K:

$$E_g, eV = -0.313 + 1.787x + 0.444x^2 - 1.237x^3 + 0.932x^4$$
$$+ \left(6.67 \cdot 10^{-4} - 1.714 \cdot 10^{-3} x + 7.6 \cdot 10^{-4} x^2\right) T \quad (7.3)$$

The Hg$_{1-x}$Cd$_x$Te alloys have relatively high electron mobility. Giriat et al.[7.434] observed a mobility at low temperatures exceeding $5 \cdot 10^5$ cm^2/V·sec, substantially higher than in both CdTe and HgTe. Patterson et al.[7.435] showed that the electron mobility in the compound Hg$_{0.807}$Cd$_{0.193}$Te may be close to $5.3 \cdot 10^5$ cm^2/V·sec at 60 K.

The data on the concentration dependence of the II-VI/II-VI solid solutions' energy gap are gathered in Table 7.36.

The effective mass of electrons in Zn$_{1-x}$Cd$_x$S at Γ_{7c} point is a parabolic function of x, while the hole effective mass is virtually a linear function of x [7.448] (the maximum deviation from linearity does not exceed 5%). A composition nonlinearity of the electron effective mass has been also observed on Hg$_{1-x}$Cd$_x$Te[7.430] (see, also, Reference 7.449) and may be related to the compositional nonlinearity of the energy gap.

The electrical transport properties of the II-VI/II-VI alloys were studied and reviewed in References 7.397 and 7.449 to 7.454. The optical properties of the alloys were reported, in particular, in References 7.455 and 7.456 (Zn$_{1-x}$Cd$_x$S and ZnS$_x$Se$_{1-x}$), Reference 7.457 (Zn$_{1-x}$Cd$_x$Te), and References 7.458 and 7.462 (Hg$_x$Cd$_{1-x}$Te).

Table 7.31 Group VIIIA Elements' Binary Compounds with the Atoms of Third, Fourth, Fifth, and Sixth Groups. Electrical and Thermoelectric Properties at 300 K

Compound	Electrical conductivity (S/cm)	Mobility (cm²/V·sec) Electrons	Mobility (cm²/V·sec) Holes	Carrier concentration (10^{18} cm^{-3})	Seebeck coefficient (10^{-6} V/K)	Comments
FeSi$_2$		0.26	2–4			
FeP$_2$				≈0.1 (n, 5 K)		
FeAs$_2$	20–500			≈50 (n, 60 K)	−200	
FeSb$_2$	91 [7.370]				30	
FeS	10^{-3}				10	
FeS$_2$	1–10 [7.332]	≤360 [7.332]		0.005–0.5 [7.332]		
	0.57	230			−500	Synthetic
	16.7	100				Natural
FeSe$_2$	1.0				62	
FeTe$_2$	66.7				64	
	100 [7.370]					Marcasite
CoP$_2$	175					
CoP$_3$	1800				≈50	
CoAs$_2$					−200	
CoAs$_3$	3200				≈±100	
CoSb$_2$	400				≈140	
CoSb$_3$	1200 [7.337]		≤3445 [7.337]		≤233 [7.337]	
	7100				−400 (max.)	
NiP$_2$	2.4–11.1			0.87 [7.363]	392 [7.363]	
NiS$_2$	1.0				311	
RuAl$_2$	≈300					
RuGa$_2$	5 [7.326]					
Ru$_2$Ge$_3$				0.7 (n, 1.5 K)		
RuP$_2$	3.8–11.1			0.74	−350	
RuAs$_2$					−350	
RuSb$_2$	≈1.0				−250	
RuSe$_2$					−120	
RuTe$_2$					−200	
RhAs$_3$					70	
RhSb$_2$	≈500				30	
RhSb$_3$					60	
Rh$_2$S$_3$					−300	
RhS$_3$					−400	
RhSe$_2$					−80	
RhSe$_3$					−30	
PdP$_2$					≈−100	
PdS					−250	
PdS$_2$	0.01				−240	
PdSe					−120	
PdSe$_2$	1.0				500	
OsSb$_2$	0.0067				−200	
OsSe$_2$					−80	
OsTe$_2$					−100	
IrP$_2$				6.9	250	
IrP$_3$					200	
IrAs$_3$					150	
IrSb$_3$					200	
	≈2500 [7.336]		1000 [7.336]	≈20 [7.336]	≈170 [7.336]	
	≈4 [7.336]	40 [7.336]		≈800 [7.336]	≈−300 [7.336]	
IrS$_2$					250	
IrSe$_2$					300	
IrSe$_3$					−150	
PtP$_2$					−90 to 180	
PtAs$_2$					−200 to 310	
PtSb$_2$	500 (250 K)	420	830	7.7 (n$_i$)	80	
		3700 (77 K)	7300 (77 K)			

Table 7.31 (continued) Group VIIIA Elements' Binary Compounds with the Atoms of Third, Fourth, Fifth, and Sixth Groups. Electrical and Thermoelectric Properties at 300 K

Compound	Electrical conductivity (S/cm)	Mobility (cm²/V·sec)		Carrier concentration (10^{18} cm^{-3})	Seebeck coefficient (10^{-6} V/K)	Comments
		Electrons	Holes			
PtS	0.01				50	
PtS$_{2.062}$	50 ($\perp$c)				500 ($\perp$c)	
PtSe$_2$					40	

From Landolt-Börnstein, *New Series*, III/17e and III/17g, O. Madelung, M. Schulz, and H. Weiss, Eds., Springer-Verlag, New York, 1982–1984.

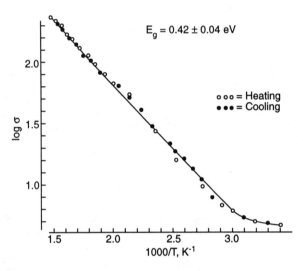

Figure 7.73 Temperature dependence of the electrical conductivity (σ in S/cm) for polycrystalline RuGa$_2$ (a heating-cooling cycle). (After J. Evers, G. Oehlinger, and H. Meyer, *Mater. Res. Bull.*, 19, 1177, 1984. With permission.)

7.8.4. Solid Solutions of II-VI and III-V Compounds

The quaternary alloys (II-VI)$_x$(III-V)$_{1-x}$ are of interest for some practical applications because they combine high energy gap of II-VI compounds with high carrier mobility in III-V compounds. The first system, in which continuous solid solutions have been discovered by Goryunova and Fedorova,[7.459] is (GaAs)$_x$(ZnSe)$_{1-x}$. Harsy and Bertoti[7.460] studied electroluminescence and photoluminescence in the crystals (ZnS)$_x$(GaP)$_{1-x}$; Voitsekhovskii et al.[7.461] investigated optical and electrical properties of (ZnTe)$_x$(InAs)$_{1-x}$ and (CdTe)$_x$(InAs)$_{1-x}$ alloys. By now, there is information regarding the properties of about thirty II-VI/III-V systems. Brief information regarding them is compiled in Table 7.37.

Among the other ternary and quaternary alloy systems, which present interest from both theoretical and practical points of view, the IV-VI/IV-VI and II-VI/IV-VI alloys occupy an important place.[7.260,7.501-7.505] The scope of this book does not provide an opportunity to discuss these and other heterovalent ternary and quaternary systems in detail.

7.8.5. Organic Semiconductors

Organic solids may be insulators, semiconductors, or metals. The majority of them are insulators, but the number of metallic and semiconducting organic compounds is quite substantial and is growing, first of all, as a result of the studying of the doping processes and

Table 7.32 Thatogenides, Pnictogenides, and Chalcogenides of Group VIIIA Elements. Room Temperature Magnetic Susceptibility, Dielectric Constant, and Index of Refraction (Reference 7.1 If Not Indicated Otherwise)

Compound	Molar magnetic susceptibility (10^{-6} emu·cm^3/mol)	Dielectric constant	Index of refraction
$FeSi_2$	−4.48	61.6 (static) 27.6 (optical)	≈5.5 (1 μm)[7.371]
FeP_2	−16.5		
$FeAs_2$	−14.0		
$FeSb_2$	−522 [7.370]		
$Fe_{0.94}S$	Antiferrimagnetic (T_N = 600 K)		
FeS_2	−24.0 [7.332]	19.8 [7.332] (optical)	
$FeSe$	Ferrimagnetic (T_c = 445 K)		
$FeTe_2$	−243 [7.370]		
CoP_3	−14		
$CoAs_2$	−16		
$CoAs_3$	−30		
$CoSb_2$	−4		
$CoSb_3$	−47		
NiP_2	−19.0		
$NiAs_2$	≈−15		
$NiSb_2$	−30		
NiS_2	Ferrimagnetic (T_c = 30 K) Antiferrimagnetic (T_N ≈ 50 K) Paramagnetic (above 50 K)		
Ru_2Si_3	−148.9		
Ru_2Ge_3	−180.6 (900 K)		
Ru_2Sn_3	−195.4 (400 K)		
RuP_2	−44		
$RuAs_2$	−73		
$RuSb_2$	−82		
RuS_2	<0		
$RuSe_2$	<0		
$RuTe_2$	<0		
RhP_2	−37		
RhP_3	−33.6		
$RhAs_2$	−104		
$RhAs_3$	−70		
$RhSb_2$	−70		
$RhSb_3$	−95		
$RhBi_2$	−161		
RhS	<0		
$RhSe_3$	<0		
PdP_2	−6		
PdS	<0		
PdS_2	<0		
$PdSe_2$	<0		
OsP_2	−74		
$OsAs_2$	−100		
$OsSb_2$	−94		
OsS_2	<0		
$OsSe_2$	<0		
$OsTe_2$	<0		
IrP_2	−61		
$IrAs_2$	−93		
$IrAs_3$	−105.5		
$IrSb_2$	−70		
$IrSb_3$	−145		
IrS_2	<0		

Table 7.32 (continued) Thatogenides, Pnictogenides, and Chalcogenides of Group VIIIA Elements. Room Temperature Magnetic Susceptibility, Dielectric Constant, and Index of Refraction (Reference 7.1 If Not Indicated Otherwise)

Compound	Molar magnetic susceptibility (10^{-6} emu·cm^3/mol)	Dielectric constant	Index of refraction
IrSe$_2$	<0		
IrSe$_3$	<0		
PtP$_2$	−64		
PtAs$_2$	−48		
PtSb$_2$	−56	32 (infrared)	5.5 (16 µm)
		31.4 (25.7 GHz)	
PtS	<0		
PtS2	<0		

Table 7.33 III-V/III-V Systems. Information Regarding Solid Solutions

Alloy system	Character of chemical interaction in the solid state	Solubility (mol%)	Ref.
Ternary Phases — Common Anion			
BP-AlP	Very small solubility near binary constituents	<1	7.397
BP-GaP	Same	<1	7.397
BAs-GaAs	Substantial solubility	Unknown	7.397
AlN-GaN	Complete solubility	100	7.397
AlP-GaP	AlP dissolves in GaP	≤50	7.397
AlP-InP	Complete solubility	100	7.405
AlAs-GaAs	Same	100	7.397
AlAs-InAs	Same	100	7.397
AlSb-GaSb	Same	100	7.397
AlSb-InSb	Same	100	7.397
GaP-InP	Same	100	7.397
GaAs-InAs	Same	100	7.397
GaSb-InSb	Same	100	7.397
Ternary Phases — Common Cation			
BP-BAs	Substantial solubility	Unknown	7.397
AlP-AlAs	Small solubility of AlP in AlAs	≤4	7.394
GaP-GaAs	Solubility of GaP in GaAs	≤45	7.395
GaAs-GaSb	Solubility of GaAs in GaSb	≤20	7.396
	Solubility of GaSb in GaAs	≤15	7.396
InP-InAs	Complete solubility	100	7.397
InAs-InSb	Same	100	7.397
InSb-InBi	Solubility of InBi in InSb	≤3	7.395
Quaternary Phases — Substitution of both Cations and Anions			
AlAs-GaSb	Limited solubility	Unknown	7.395
GaP-InAs	Complete solubility	100	7.397
GaP-InSb	Limited solubility of each of the binary compounds in another	<5	7.397
GaAs-InSb	Same	<15	7.397
InP-GaAs	Complete solubility	100	7.397
InP-GaSb	Limited solubility of each of the binary compounds in another	<20	7.397
InAs-GaSb	Same	<15	7.398
Quaternary Phases — Substitution of One Cation by Two			
AlP-GaP-InP	Substantial solubility		7.395
AlAs-GaAs-InAs	Same		7.395
AlSb-GaSb-InSb	Same		7.395

NONADAMANTINE & VARIABLE-COMPOSITION SEMICONDUCTOR PHASES

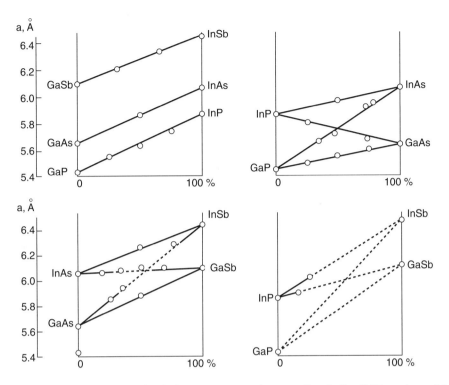

Figure 7.74 Correlation between the lattice parameter and composition in the III-V-based quasibinary systems. (After E.K. Müller and J.L. Richards, *J. Appl. Phys.*, 35, 1233, 1964. With permission.)

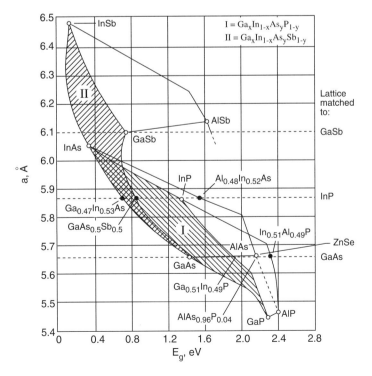

Figure 7.75 Correlation between the room temperature value of the lattice parameter and energy gap for III-V compounds and their alloys. (After Landolt-Börnstein, *New Series,* O. Madelung, Ed., Vol. III/22a, Springer-Verlag, 1987, p. 343. With permission.)

Table 7.34 III-V/III-V Semiconductor Alloy Systems. Concentration Dependence of the Energy Gap

Alloy system	Analytical expression for the energy gap vs. concentration	Comments	Ref.
	III$_x$-III$_{1-x}$-V Phases		
Al$_x$Ga$_{1-x}$P	$E_{g,\,ind} = 2.28 + 0.16x$	RT	7.395
	$E_{g,\,ind} = 2.34 + 0.011x + 0.179x^2$	6 K	7.405
Al$_x$Ga$_{1-x}$As	$E_g\,(\Gamma_{8v} - \Gamma_{6c}) = 1.420 + 1.087x + 0.438x^2$	RT	7.406
	$E_g\,(\Gamma_{8v} - L_{6c}) = 1.705 + 0.695x$	RT	7.406
	$E_g\,(\Gamma_{8v} - X_{6c}) = 1.905 + 0.10x + 0.16x^2$	RT	7.406
	$E_{g,\,dir} = 1.424 + 1.247x$	x < 0.45	7.401
	$E_{g,\,dir} = 1.424 + 1.247x + 1.147(x - 0.45)^2$	x > 0.45	7.401
Al$_x$Ga$_{1-x}$Sb	$E_g\,(\Gamma_{8v} - \Gamma_{6c}) = 0.73 + 1.10x + 0.47x^2$	300 K	7.407
	$E_g\,(\Gamma_{8v} - X_{6c}) = 1.05 + 0.56x$	300 K	7.407
	$E_g\,(\Gamma_{8v} - \Gamma_{6v}) = 0.813 + 1.097x + 0.40x^2$	4.2 K	7.408
	$E_{g,\,dir} = 0.726 + 1.129x + 0.368x^2$	300 K	7.401
Al$_x$In$_{1-x}$P	$E_g\,(\Gamma_v - \Gamma_c) = 1.34 + 2.23x$	300 K	7.395
	$E_g\,(\Gamma_v - X_c) = 2.24 + 0.18x$	300 K	7.395
	$E_{g,\,dir} = 1.351 + 2.23x$	300 K	7.401
Al$_x$In$_{1-x}$As	$E_g\,(\Gamma_v - _c) = 0.37 + 1.91x + 0.74x^2$	300 K	7.395
	$E_g\,(\Gamma_v - X_c) = 1.8 + 0.4x$	300 K	7.395
	$E_{g,\,dir} = 0.360 + 2.012x + 0.698x^2$	300 K	7.401
Al$_x$In$_{1-x}$Sb	$E_{g,\,dir} = 0.172 + 1.621x + 0.43x^2$	300 K	7.401
Ga$_x$In$_{1-x}$P	$E_{g,\,dir} = 1.351 + 0.643x + 0.786x^2$	300 K	7.401, 7.409
Ga$_x$In$_{1-x}$As	$E_g\,(\Gamma_v - \Gamma_c) = 0.324 + 0.7x + 0.4x^2$	300 K	7.410
	$E_g\,(\Gamma_v - \Gamma_c) = 0.422 + 0.7x + 0.4x^2$	2 K	7.410
	$E_{g,\,dir} = 0.360 + 1.064x$	300 K	7.401
Ga$_x$In$_{1-x}$Sb	$E_{g,\,dir} = 0.172 + 0.139x + 0.415x^2$	300 K	7.401
	$E_{g,\,dir} = 0.235 + 0.1653x + 0.413x^2$	0 K	7.395
	III-V$_x$-V$_{1-x}$ Phases		
GaP$_x$As$_{1-x}$	$E_{g,\,dir} = 1.424 + 1.150x + 0.176x^2$	300 K	7.401
	$E_g\,(\Gamma_v - \Gamma_c) = 1.515 + 1.172x + 0.186x^2$	2 K	7.395
	$E_g\,(\Gamma_v - X_c) = 1.9715 + 0.144x + 0.211x^2$	2 K	7.395
GaAs$_x$Sb$_{1-x}$	$E_{g,\,dir} = 0.726 - 0.502x + 1.2x^2$	300 K	7.401
	$E_{g,\,dir} = 0.73 - 0.5x + 1.2x^2$	300 K	7.411
InP$_x$As$_{1-x}$	$E_g\,(\Gamma_v - \Gamma_c) = 0.356 + 0.675x + 0.32x^2$	300 K	7.395
	$E_{g,\,dir} = 0.360 + 0.891x + 0.101x^2$	300 K	7.401
	$E_g\,(\Gamma_v - \Gamma_c) = 0.414 + 0.64x + 0.36x^2$	77 K	7.395
InAs$_x$Sb$_{1-x}$	$E_{g,\,dir} = 0.18 - 0.41x + 0.58x^2$	300 K	7.401
	III$_x$-III$_{1-x}$-V$_y$-V$_{1-y}$ Phases		
Ga$_x$In$_{1-x}$As$_y$P$_{1-y}$	$E_{g,\,dir} = 1.35 + 0.668x - 1.068y + 0.758x^2 + 0.078y^2 - 0.069xy - 0.322x^2y + 0.03xy^2$	300 K	7.395, 7.412
	III$_x$-III$_y$-III$_{1-x-y}$V Phases		
Al$_x$Ga$_y$In$_{1-x-y}$As	$E_g = 0.36 + 2.093x + 0.629y + 0.577x^2 + 0.436y^2 + 1.013xy - 2.0xy(1 - x - y)$	300 K	7.395
Al$_x$Ga$_y$In$_{1-x-y}$Sb	$E_{g,\,dir} = 0.095 + 1.76x + 0.28y + 0.345(x^2 + y^2) + 0.085(1 - x - y)^2 + (23 - 28y)xy(1 - x - y)$	300 K	7.413

complex formation in organics. Some of the organic metals have even shown superconductive behavior (see, e.g., References 7.506 and 7.507).

Comprehensive information relevant to the history and directions of investigation of organic semiconductors is contained in References 7.508 to 7.515. The results of the first

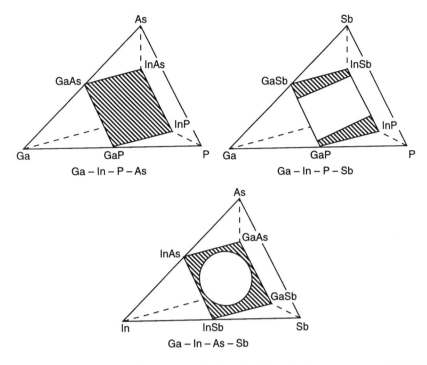

Figure 7.76 Areas of existence of the phases with the sphalerite structure in some quaternary systems (shaded). (After E.K. Müller and J.L. Richards, *J. Appl. Phys.*, 35, 1233, 1964. With permission.)

Table 7.35 Ternary and Quaternary II-VI/II-VI Alloys with Continuous Solubility in Solid State (Reference 7.397 Unless Indicated Otherwise)

		Lattice parameter at room temperature (nm)	
Alloy	Crystal structure	a	c
Single Anion			
$Zn_{1-x}Cd_xS$	Wurtzite	$0.3823 + 0.312x$	$0.62565 + 0.0228x$
		$0.38230 + 0.03124x^{7.1}$	$0.62565 + 0.04555x^{7.1}$
$Zn_{1-x}Cd_xTe$	Sphalerite	$0.6101 + 0.0381x$	
$Zn_{1-x}Hg_xS$	Same	$0.54145 + 0.04265x$	
$Zn_{1-x}Hg_xSe$	Same	$0.5653 + 0.0431x$	
$Zn_{1-x}Hg_xTe$	Same	$0.6101 + 0.0361x$	
$Cd_{1-x}Hg_xTe$	Same	$0.64822 + 0.00199x$	
Single Cation			
$ZnS_{1-x}Se_x$	Sphalerite	$0.54154 + 0.02385x$	
		$0.54093 + 0.02617x^{7.1}$	
$ZnSe_{1-x}Te_x$	Same	$0.5653 + 0.0448x$	
$CdS_{1-x}Se_x$	Wurtzite	$0.41348 + 0.01652x$	$0.67490 + 0.0742x$
$HgS_{1-x}Se_x$	Sphalerite	$0.5841 + 0.0243x$	
$HgS_{1-x}Te_x$	Same	$0.58410 + 0.06213x$	
$HgSe_{1-x}Te_x$	Same	$0.6084 + 0.0378x$	
Cation/Anion Substitution			
$(ZnSe)_{1-x}(CdTe)_x$	Sphalerite	$0.5653 + 0.8292x$	

Table 7.36 II-VI/II-VI Alloys. Energy Gap (eV) Vs. Component Concentration[a]

Alloy system	Analytical expression for energy gap	Comments	Ref.
\multicolumn{4}{c}{Common Anion}			

Alloy system	Analytical expression for energy gap	Comments	Ref.
$Zn_{1-x}Cd_xS$	$E_{g,dir} = 3.78 - 1.78x + 0.61x^2$	300 K	7.3, 7.438
$Zn_{1-x}Cd_{1-x}Se$	$E_{g,dir} = 1.90 + 0.64x + 0.26x^2$	300 K	7.445
$Zn_{1-x}Cd_xTe$	$E_{g,dir} = 2.28 - 1.27x + 0.60x^2$	300 K	7.439
$Zn_{1-x}Hg_xSe$	$E_{g,dir} = 2.76 - 2.97x$	$x < 0.93$, 77 K	7.440
$Cd_xHg_{1-x}Se$	$E_g(x,T) = -0.209(1 - 7.172x - 2.174x^2)$ $+ 7.37 \cdot 10^{-4}(1 - 1.277x - 0.151x^2)T$ $+ 2.001 \cdot 10^{-9}(1 + 23.45x - 599.4x^2)T^2$	5–300 K, $1 \cdot 10^{16} < n < 9 \cdot 10^{16}$ cm^{-3}, $0.15 < x < 0.68$	7.440
$Cd_xHg_{1-x}Te$	$E_g(\Gamma_{8v} - \Gamma_{6c}) = -0.302 + 1.93x$ $+ 5.34 \cdot 10^{-4}(1 - 2x)T - 0.810x^2 + 0.832x^3$	$0 \leq x \leq 1$	7.1, 7.441, 7.442
$Cd_{1-x}Zn_xTe$	$E_{g,dir} = 1.510 + 0.606x + 0.139x^2$	300 K	7.1
	$E_{g,dir} = 1.598 + 0.614x + 0.166x^2$	12 K	7.1

Common Cation

Alloy system	Analytical expression for energy gap	Comments	Ref.
$ZnS_{1-x}Se_x$	$E_{g,dir} = 2.721 + 0.352(1 - x) + 0.630(1 - x)^2$	300 K, film	7.1, 7.443
	$E_{g,dir} = 2.67 + 0.75(1 - x) + 0.25(1 - x)^2$	300 K, powder samples	7.444

Substitution of Both Anions and Cations

Alloy system	Analytical expression for energy gap	Comments	Ref.
$(ZnSe)_x(CdTe)_{1-x}$	$E_g = 1.474 + 0.0404x + 0.0470x^2 + 1.0886x^3$		7.1, 7.446

[a] For data for the II-VI/II-VI systems with miscibility gaps, see Reference 7.1, Vol. III/22a.

systematic electrical measurements, the purpose of which was the search for possible applications of metallophthalocyanines, were published by Eley and Parfitt.[7.516] Semiconducting properties of free phthalocyanine and the phthalocyanines of Mg, Cu, Zn, Ni, Co, and Pt were reported in References 7.517 to 7.521. The electrical conductivity vs. temperature for phthalocyanine of nickel and Cl-phthalocyanines of Ti, Zr, Hf, Pd, and Pt was measured by Nasirdinov et al.[7.522] To date, metallophthalocyanines of more than 70 metals have been synthesized (see Reference 7.513, p. 74). The optical and electrical properties of phthalocyanine semiconductors, first of all copper phthalocyanine, were reported, in particular, in References 7.523 to 7.525. The semiconductor properties of Sc-, Zr-, and Co-polyphthalocyanines were investigated in Reference 7.526. Interest in organic high-conductivity materials has been growing, especially after the discovery of the polyacetylene semiconductor synthesis techniques (by the Shirakawa group[7.527] and by Edwards et al.[7.528,7.529]). These methods can be applied to various high-molecular systems.

Two groups of linear aromatic hydrocarbons: (1) acenes, with the general formula $C_{4n+2}H_{2n+4}$, and phenyls, $C_{6n}H_{4n+2}$, (n is the number of benzene rings in a molecule) and their derivatives are of substantial interest as materials for optical and electrooptical applications in the UV and visible parts of the spectrum; and (2) a large group of compounds consisting of two or more parts bonded by a complete or partial charge transfer between them; these compounds possess high electrical conductivity of semiconducting or metallic character.

The area of application of conducting polymers is continuously expanding. In particular, polysulfurnitride,[7.530] polyacetylene, and polypyrrole[7.535,7.536] can and are used as materials for ohmic contacts, Schottky barriers, and corrosion protection on n-type "classical" semiconductors. Electrically conductive plastics are used for electromagnetic shielding of the field-sensitive electronic elements.[7.537] Table 7.38 contains a list of conducting polymers and the molecular entities used as the (initially insulating) host parts or electrically active groups for formation of organic solids with semiconductor properties.

Selected information regarding the thermal, mechanical, electrical, and optical properties of organic semiconductors is compiled in Tables 7.39 and 7.40.

Table 7.37 II-VI/III-V Alloys. Limits of Solubility and Energy Gap

Alloy system	Limits of solubility	Energy gap (eV) vs. composition	Ref.
$(ZnS)_x(AlP)_{1-x}$	<1 mol% AlP in ZnS		7.397, 7.463
$(ZnS)_x(GaP)_{1-x}$	Wide range, sphalerite structure	$E_{g,dir} = 2.76 - 2.83x + 3.74x^2$	7.397, 7.460, 7.464–7.467
$(ZnS)_x(InP)_{1-x}$			7.468
$(ZnS)_x(GaAs)_{1-x}$	Wide range, sphalerite structure	$E_{g,dir} = 1.45 - 0.05x + 2.30x^2$	7.468–7.470
$(ZnSe)_x(GaP)_{1-x}$	Wide range, sphalerite structure	$E_{g,dir} = 2.78 - 2.53x + 2.33x^2$	7.397, 7.465, 7.466, 7.471–7.473
$(ZnSe)_x(InP)_{1-x}$			7.468
$(ZnSe)_x(GaAs)_{1-x}$	Continuous solid solutions, sphalerite structure	$E_{g,dir} = 1.43 - 0.09x + 1.26x^2$	7.397, 7.465, 7.470, 7.473–7.475
$(ZnTe)_x(GaP)_{1-x}$			7.468
$(ZnTe)_x(GaAs)_{1-x}$	Continuous solid solutions	$E_{g,dir} = 1.45 - 0.59x + 1.38x^2$	7.470, 7.473
$(ZnTe)_x(AlSb)_{1-x}$			7.468
$(CdS)_x(GaP)_{1-x}$			7.468
$(CdS)_x(InP)_{1-x}$	Continuous solid solutions, wurtzite structure	$E_{g,dir} = 1.36 - 1.30x + 2.43x^2$	7.397, 7.468, 7.473
$(CdS)_x(GaAs)_{1-x}$			7.397, 7.468, 7.473
$(CdS)_x(InAs)_{1-x}$	Up to 20 mol% of CdS dissolves in InAs		7.397, 7.476, 7.477
$(CdSe)_x(GaP)_{1-x}$			7.468
$(CdSe)_x(InP)_{1-x}$	Continuous solid solutions, sphalerite structure		7.397, 7,478
$(CdSe)_x(GaAs)_{1-x}$			7.468, 7.479
$(CdSe)_x(InAs)_{1-x}$			7.376, 7.480
$(CdTe)_x(InP)_{1-x}$	Solubility up to 1 mol%		7.478
$(CdTe)_x(GaAs)_{1-x}$	Limited solubility		7.481, 7.482
$(CdTe)_x(InAs)_{1-x}$	Solubility up to 30 mol% of InAs in CdTe and up to 35 mol% of CdTe in InAs	$E_{g,dir} = 0.36 - 1.18x + 2.42x^2$	7.397, 7.461, 7.473, 7.484–7.489, 7.500
$(CdTe)_x(AlSb)_{1-x}$	Continuous solubility, sphalerite structure	$E_{g,dir} = 2.20 - 0.63x - 0.02x^2$	7.397, 7.473, 7.490, 7.491
$(CdTe)_x(InSb)_{1-x}$	Solubility up to 5 mol% of CdTe in InSb		7.397, 7.473, 7.487, 7.492–7.495
$(HgTe)_x(InAs)_{1-x}$	Continuous solubility, sphalerite structure		7.397, 7.496
$(HgTe)_x(GaSb)_{1-x}$	Limited solubility of HgTe in GaSb		7.397, 7.499
$(HgTe)_x(InSb)_{1-x}$	Solubility up to 42 mol% of HgTe in InSb	$E_{g,dir} = 0.18 - 0.60x + 0.12x^2$	7.497, 7.498

The technology of preparation and applications of organic semiconductors is quite advanced, while up to now there has not been a satisfactory theoretical model for their electrical conductivity or reliable empirical correlations between their structure and electrical properties. Two methods of controlling the polymer electrical conductivity, namely, pyrolysis[7.561,7.562] and (polyacetylene) oxidation,[7.538] offer wide opportunities for their use in numerous applications (as briefly mentioned before) in some electronic devices including remotely readable indicators of temperature, humidity, radiation, and chemical release; as materials for electrostatic charge dissipation and electromagnetic shielding of sensitive electronic circuitry, in particular, in computers; as the materials for electronic gas analysis devices and as the polymeric electrode materials for rechargeable electric batteries (polypyrrole-lithium cells — Allied-Signal, Morristown, NJ, USA/BASF, Germany). The number of known

Table 7.38 Semiconducting Polymers and Donors/Acceptors Forming Organic Semiconductor Charge-Transfer Molecules (References 7.1, Vol. III/17, 7.512, and 7.513 and Ones Cited Below)

Substance	Structure	Characteristic parameters	Ref.
Polymers			
Polyacetylene $(CH)_x$		RT conductivity 10^3–10^5 S/cm	7.538, 7.541
Trans-		Rhombohedral, a = 0.446 nm, b = 0.768 nm, c = 0.438 nm	
Cis-		Monoclinic, a = 0.424 nm, b = 0.732 nm, c = 0.246 nm, β = 91.5°	
Polyaniline $(C_{16}H_5N)_x$		0.1–1 S/cm	7.542
Polyazulene $(C_{10}H_8)_x$		0.1–1 S/cm	7.543
Polyfuran $(C_4H_2O)_x$		10^2 S/cm (max)	7.544
Poly(p-phenylene) $(C_6H_4)_x$		Up to 10^3 S/cm	7.545
		Up to 10 cm²/V·sec	7.546
Poly(p-phenylene sulfide) (PPS), $(C_6H_4S)_x$		10–100 S/cm	7.547–7.549
Poly(p-phenylene vinylene) (PPV)		10^2–10^3 S/cm	7.550–7.552

Compound	Structure	Conductivity	Ref.
Polymetallophthalocyanines (PPcM)	[macrocyclic M-phthalocyanine structure]	10^{-4}–10^0 S/cm	7.553
Polypyrrole		10^2–10^3 S/cm	7.554
		10^1–10^2 S/cm	7.555
		10^{-3} S/cm	7.556
Polysulfurnitride $(SN)_x$		10^4 S/cm (superconductor below 0.26 K)	7.558
Polythiophene $(C_4H_2S)_x$ (polythiofuran)		10^1–10^2 S/cm	7.557

Molecular Entities Forming the Charge-Transport Bonded Semiconductors

Compound	Structure	Property
p-Chloranil	[tetrachloro-1,4-benzoquinone structure]	$E^a_{1/2} = +0.01$ eV

Table 7.38 (continued) Semiconducting Polymers and Donors/Acceptors Forming Organic Semiconductor Charge-Transfer Molecules (References 7.1, Vol. III/17, 7.512, and 7.513 and Ones Cited Below)

Substance	Structure	Characteristic parameters	Ref.
Tetrathiotetracene (TTT)		$E_{1/2} = +0.08$ eV	
Tetracyanoquinodimethane (TCNQ)		$E_{1/2} = +0.13$ eV	7.512
Tetracyanoethylene (TCNE)		$E_{1/2} = +0.15$ eV	
11,11,12,12-Tetracyano-naphtho-2,6-quinodimethane (TNAP)		$E_{1/2} = +0.21$ eV	7.512

Tetrafulvalenes

Tetramethylthiofulvalene (TMTTF) — X — sulfur, R — methyl — $E_{1/2} = +0.27$ eV

Tetrathiofulvalene (TTF) — X — sulfur, R — hydrogen — $E_{1/2} = +0.33$ eV

Tetraselenofulvalene (TSeF) — X — selenium, R — hydrogen — $E_{1/2} = +0.48$ eV

2,3-Dichloro-5,6-dicyano benzoquinone (DDQ) — $E_{1/2} = 0.51$ eV 7.512

Tetraethyl-triphenylene diamine (TETD) — $E_{1/2} = +0.68$ eV

N-Ethylcarbazole — $E_{1/2} = +1.25$ eV

Table 7.38 (continued) Semiconducting Polymers and Donors/Acceptors Forming Organic Semiconductor Charge-Transfer Molecules (References 7.1, Vol. III/17, 7.512, and 7.513 and Ones Cited Below)

Substance	Structure	Characteristic parameters	Ref.
Hexamethylbenzene (HMB)		$E_{1/2} = +1.57$ eV	
Tetramethyl-p-phenylene-diamine (TMPD)		$E_{1/2} > 0$	
N-Methylacridinium (NMAd)		$E_{1/2} = -0.41$ eV	
Trinitrofluorenone (TNF)		$E_{1/2} = -0.45$ eV	

Tetracyanobenzene (TCNB) $E_{1/2} = -0.71$ eV

N-Methylquinolinium (NMQn) $E_{1/2} = -0.86$ eV

N-Methylphenazinium (NMP)

N-Methylpiridinium (NMPy) $E_{1/2} = -1.28$ eV

Pyromelitic dianhydride (PMDA)

Triethylammonium (TEA) $E_{1/2} = -2.8$ eV

[a] $E_{1/2}$ denotes the half-wave oxidation/reduction potential relative to saturated calomel electrode.

Table 7.39 Chemical, Thermochemical, Thermal, and Mechanical Properties and Crystal Structure of Organic Semiconductors (from Reference 7.1, Vol. III/17, Unless Indicated Otherwise)

Compound	Melting point (K)	RT density (g/cm³)	Enthalpy (kJ/mol) of Fusion	Enthalpy (kJ/mol) of Sublimation	Crystal structure and space group	Lattice parameter a (nm)	Lattice parameter b (nm)	Lattice parameter c (nm)	β°	Elastic moduli (GPa)
Anthracene, $C_{14}H_{10}$	489	1.28	28.86	100.5 (392.3 K)	Monoclinic, C_{2h}^5-$P2_1/a$, $Z = 2$	0.8562	0.6038	1.118	124.42	11.6 (c_{22})
Benzene, C_6H_6	278.6	0.8765 [7.2]			Orthorhombic, D_{2h}^{15}-$Pbca$, $Z = 4$	0.7460	0.9666	0.703		6.14 (c_{11}) 6.56 (c_{22}) 5.83 (c_{33}) (250 K)
Biphenyl, $C_{12}H_{10}$	344	1.19								0.9 (c_{22})
Dibenzothiophene, $C_{12}H_8S$ (Dibenzothiophene, diphenylenesulfide)	372–373 372.65 [7.2]				C_{2h}^5-$P2_1/a$, $Z = 2$ Monoclinic, C_{2h}^5-$P2_1/c$, $Z = 4$	0.812 0.867	0.563 0.600	0.951 1.870	95.1 113.9	
1,4-Dibromonaphthalene, $C_{10}H_6Br_2$	356	2.037			C_{2h}^5-$P2_1/a$, $Z = 8$	2.7320	1.6417	0.4048	91.95	
Alpha-9,10-dichloroanthracene, $C_{14}H_8Cl_2$	483.5	1.525			C_{2h}^5-$P2_1/a$	0.704	1.793	0.863	102.93	
1,4-Diiodobenzene, $C_6H_4I_2$	404–405	2.79			D_{2h}^{15}-$Pbca$, $Z = 4$	1.7008	0.7321	0.5949		
Durene, $C_{10}H_{14}$	352.4	1.03			C_{2h}^5-$P2_1/a$, $Z = 2$	1.157	0.577	0.703	113.3	
Iodoform, CHI_3	396	4.01			Hexagonal, C_6^6-$P6_3$, $Z = 2$	0.6818		0.7524		
9-Methylanthracene, $C_{15}H_{12}$	354.7	10471 [7.2]			C_{2h}^5-$P2_1/c$, $Z = 4$	0.8920	1.4641	0.8078	96.47	
Naphthalene, $C_{10}H_8$	353.45	1.152	18.78	72.4 (287 K)	C_{2h}^5-$P2_1/a$	0.82606	0.59872	0.86816	122.671	
Alpha-perylene, $C_{20}H_{12}$	550–552	1.322	31.87	126	C_{2h}^5-$P2_1/a$, $Z = 4$	1.128	1.083	1.026	100.55	
Alpha-phenazine, $C_{12}H_8N_2$	449–450	1.34			C_{2h}^5-$P2_1/a$	1.322	0.5061	0.7088	109.22	
Phenothiazine, $C_{12}H_9NS$	459–462	1.352			Orthorhombic, D_{2h}^{16}-$Pnma$, $Z = 4$	0.7916	2.0974	0.5894		

Substance	(1)	(2)	(3)	(4)	Crystal system, space group	a	b	c	Angles (°)
Beta-phthalocyanine, $C_{32}H_{18}N_8$					C_{2h}^5-$P2_1/a$, $Z = 2$	1.985	0.472	1.48	122.25
Ryrene, $C_{16}H_{10}$	429	1.44			C_{2h}^5-$P2_1/a$, $Z = 4$	1.3649	0.9256	0.8470	100.28
trans-Stilbene, $C_{14}H_{12}$	397–399	1.27	$0.9707^{7.2}$	97.5	C_{2h}^5-$P2_1/a$, $Z = 4$	1.2381	0.5723	1.5571	114.11
p-Terphenyl, $C_{18}H_{14}$	487.0	1.234			C_{2h}^5-$P2_1/a$, $Z = 2$	0.8119	0.5615	1.3618	92.07
Tetracene, $C_{18}H_{12}$	630	1.24		113	Triclinic, C_i^1-$P\bar{1}$	0.790	0.603	1.353	100.3 (α) 113.2 (β) 86.3 (γ)
Tetracyanoethylene, C_6N_4 (TCNE)	471–473	1.31			Monoclinic, C_{2h}^5-$P2_1/n$, $Z = 2$	0.751	0.621	0.700	97.17
7,7,8,8-Tetracyanoquinodimethane, $C_{12}H_4N_4$ (TCNQ)		1.315			Monoclinic, C_{2h}^6-$C2/c$, $Z = 4$	0.8906	0.7060	1.6395	98.54
(Tetramethyltetraselenofulvalene)$_2$:hexafluorophosphate, $(C_{10}H_{12}Se_4)_2$:PF_6, (TMTSeF)$_2$:PF_6					C_i^1-$P\bar{1}$, $Z = 1$	0.7297	0.7711	1.3522	83.39 (α) 86.27 (β) 71.01 (γ)
(Perylene)$_2$:(PF$_6$)$_{1.1}$ × 0.8(CH$_2$Cl$_2$), $C_{40}H_{24}$:(PF$_6$)$_{1.1}$ × 0.8(CH$_2$Cl$_2$)					Orthorhombic, D_{2h}^{12}-$Pnmn$, $Z = 1$	0.4285	1.2915	1.4033	
(Tetrathiotetracene)$_2$:I_3, (TTT)$_2$:I_3, $(C_{18}H_8S_4)_2$:I_3					Orthorhombic, D_{2h}^{18}-$Abam$, $Z = 2$	1.8394	0.4962	1.8319	
Tetrathiofulvalene:bromine, TTF:Br$_{0.7}$, $C_6H_4S_4$:Br$_{0.7}$					Monoclinic, C_{2h}^3-$C2/m$, $Z = 4$	1.5617	1.5627	0.3572	91.23
Potassium:tetracyanoquinodimethane, K:TCNQ, K:$C_{12}H_4N_4$					C_{2h}^5-$P2_1/n$, $Z = 8$	0.70835	1.7773	1.7859	
Tetrathiofulvalene:tetracyanoquinodimethane, TTF:TCNQ, $C_6H_4S_4$:$C_{12}H_4N_4$					C_{2h}^5-$P2_1/c$	1.2298	0.3819	1.8468	104.46
Tetrathiofulvalene:tetrachloro-p-benzoquinone, TTF:chloranil, $C_6H_4S_4$:$C_6Cl_4O_2$					C_{2h}^5-$P2_1/n$, $Z = 2$	0.7411	0.7621	1.4571	99.20

Table 7.40 Organic Semiconductors. Electronic and Optical Properties (from References 7.1, Vol. III/17, and 7.510 to 7.513 and the References Indicated Below). Room Temperature

Substance	Dark conductivity (S/cm)	Energy gap (eV)	Mobility (cm^2/V·sec) Electrons	Mobility (cm^2/V·sec) Holes	Dielectric constant	Index of refraction
Antracene	<10^{-15}	3.88–4.1	1.74 (max)	2.07 (max)	2.90	1.55 (589 nm)
Benzene		7	2 (250 K)			
Naphthalene		4.9–5.1	0.64 (max)	1.50 (max)	3.43 (max)	1.945 (546 nm) (max)
Perylene		3.10	5.53 (max)	87.4 (max) (60 K)		
Phenothiazine			1.1 (max)			1.96 (589 nm) (max)
Phenazine			2.45	0.02		1.95 (589 nm) (max)
Phthalocyanine, PcH$_2$	10^{-7}	2	1.2 (373 K)	1.1 (373 K)		
PcNi	$6·10^{-10}$ [7.522]	1.35 [7.522]				
Cl-PcPt	$5·10^{-10}$ [7.522]	1.60 [7.522]				
Pyrene				0.50 (max)	3.80 (max)	
Salycilal-N-alkylimine-Cu	$6·10^{-15}$ [7.563]	1.62 [7.563]				
trans-Stilbene				1.4		
p-Terphenyl			1.2 (max)	0.80		
Tetracene		3.40		0.85		
Tetracyanoethylene				0.26 (max)		
7,7,8,8-Tetracyanoquino-dimethane			0.65			
(TMTSF)$_2$:PF$_6$	10^5 (14 K)		10^4–10^5 (4.2 K)			
(Perylene)$_2$:(PF$_6$)$_{1.1}$ × 0.8(CH$_2$Cl$_2$)	900			0.91		
(TTT)$_2$:I$_3$	10^3					
	3000 (40 K)					
TTF:Br$_{0.7}$	100–500	2.00				
K:TCNQ	$3·10^{-4}$–$2·10^{-2}$ (at 9.3 GHz)			1.1		
TTF:TCNQ	400 ± 100	≈0.5	300–450 (total, 58 K)			
	10^4 (at 58 K)					
TTF:chloranil	$8·10^{-4}$	≈0.6				

organic semiconductors is very large. Many organic dyes and pigments (including chlorophyll, carotene, and the blood pigments) are semiconductors. Organic semiconductors, similar to their inorganic counterparts, are catalytically active. Some electrical and optical properties of a selected group of these materials are also gathered in Tables 7.38 and 7.40.

Among the organic semiconductors which currently attract particular attention as materials for application in electronic devices are some oligomers of thiophene, such as dimethylquaterthienyl, terthienyl, and hexathienyl[7.564] that possess record-high charge mobility; semiconductor/superconductors based on tetrathiofulvalene, such as $(ET)_2MHg(SCN)_4$ quasi-1D and quasi-2D conducting systems (M is for K, Tl, Rb, or NH_4);[7.565] and some polymers including acrylonitrile-butadiene-styrene (ABS) complex systems, such as $FeCl_3$-ABS and Fe^{+3}-ABS.[7.566] In the field of both development of organic semiconductors and their application in electronic industry, the next five to ten years promise to be very fruitful.

REFERENCES

7.1. Landolt-Börnstein, *New Series*, Vol. III/17e and III/17g, O. Madelung, M. Schulz, and H. Weiss, Eds., Springer-Verlag, New York, 1982–1984.

7.2. D. R. Lide, Ed., *Handbook of Chemistry and Physics*, 76th ed., CRC Press, Boca Raton, FL, 1995–1996.

7.3. O. Madelung, Ed., *Semiconductors Other than Group IV Elements and III-V Compounds*, in *Data in Science and Technology*, R. Poerschke, Ed., Springer-Verlag, New York, 1992.

7.4. K. W. Böer, *Survey of Semiconductor Physics*, Van Nostrand Reinhold, New York, 1990.

7.5. E. W. Saker and F. A. Cunnel, *Research (London)*, 7, 114, 1954.

7.6. L. Pincherle and J. Radcliffe, *Adv. Phys. (Philos. Mag. Suppl.)*, 5, 271, 1956.

7.7. E. Mooser and W. P. Pearson, *Phys. Rev.*, 101, 492, 1956.

7.8. W. E. Spicer, A. H. Sommer, and J. C. White, *Phys. Rev.*, 115, 57, 1959.

7.9. G. Kienast and J. Verma, *Z. Anorg. Allg. Chem.*, 310, 143, 1961.

7.10. H. Overhof, J. Knecht, R. Fischer, and F. Hensel, *J. Phys.*, F8, 1607, 1978.

7.11. J. Knecht, R. Fischer, H. Overhof, and F. Hensel, *J. Chem. Soc. Chem. Commun.*, p. 905, 1978.

7.12. G. K. Wertheim, R. L. Cohen, G. Crecelius, K. W. West, and J. H. Wernick, *Phys. Rev.*, B20, 860, 1979.

7.13. G. K. Wertheim, C. W. Bates, and D. N. E. Buchanan, *Solid State Commun.*, 30, 473, 1979.

7.14. F. Wooten and G. A. Condas, *Phys. Rev.*, 131, 657, 1963.

7.15. R. F. Hall and H. C. Wright, *Br. J. Appl. Phys.*, 18, 33, 1967.

7.16. J. Korringa, *Physica*, 13, 392, 1947; W. Kohn and N. Rostoker, *Phys. Rev.*, 94, 1411, 1954.

7.17. A. Hasegawa and M. Watanabe, *J. Phys.*, F7, 75, 1977.

7.18. S.-H. Wei and H. Krakauer, *Phys. Rev. Lett.*, 55, 1200, 1985.

7.19. C. Norris and L. Wallden, *Phys. Stat. Sol.*, A2, 381, 1970.

7.20. P. Görlich, *Z. Phys.*, 101, 335, 1936.

7.21. E. Zintl and G. Brauer, *Z. Phys. Chem.*, B20, 245, 1933.

7.22. D. T. Cromer, *Acta Crystallogr.*, 12, 36 and 41, 1959.

7.23. E. Zintl and W. Dullenkopf, *Z. Phys. Chem.*, B16, 183, 1932.

7.24. E. Busman and S. Lohmeyer, *Z. Anorg. Allg. Chem.*, 312, 53, 1961.

7.25. H. G. von Schnering, W. Höhle, and G. Krogull, *Z. Naturforsch.*, 34B, 1678, 1979.

7.26. J. P. Suchet, *Chemical Physics of Semiconductors*, Van Nostrand, London, 1965.

7.27. R. Suhrman and C. Kangro, *Naturwissenschaften*, 40, 137, 1953.

7.28. Y. A. Ugai and T. N. Vigutova, *Sov. Phys. Solid State*, 1, 1635, 1960.

7.29. C. Kunze, *Ann. Phys.*, 6, 89, 1960.

7.30. R. P. Elliot, *Constitution of Binary Alloys*, 1st Suppl., McGraw-Hill, New York, 1965.

7.31. K. Miyake, *J. Appl. Phys.*, 31, 76, 1960.

7.32. U. Heilig, *Phys. Stat. Sol.*, 2, K296, 1962.

7.33. H. Miyazawa, K. Noga, S. Chikazumi, and A. Kobayashi, *J. Phys. Soc. Jpn.*, 7, 647, 1952.

7.34. H. Miyazawa and S. Fukuhara, *J. Phys. Soc. Jpn.*, 7, 645, 1952.

7.35. S. Pastorello, *Gazz. Chim. Ital.*, 60, 493, 1930.

7.36. S. Pastorello, *Gazz. Chim. Ital.,* 61, 47, 1931.
7.37. W. B. Pearson, *Handbook of Lattice Spacings and Structures of Metals and Alloys,* Pergamon Press, Elmsford, NY, 1958.
7.38. R. E. Tate and F. W. Schonfeld, *Trans. AIME,* 215, 296, 1959.
7.39. R. W. G. Wyckoff, *Crystal Structures,* Interscience, New York, 1965.
7.40. G. Brauer and E. Zintl, *Z. Phys. Chem.,* B37, 323, 1937.
7.41. G. Gnutzmann, F. W. Dorn, and W. Klemm, *Z. Anorg. Allg. Chem.,* 309, 210, 1961.
7.42. A. A. Dowman, F. H. Jones, and A. H. Beck, *J. Phys.,* D8, 69, 1975.
7.43. A. H. Sommer and W. H. McCarroll, *J. Appl. Phys.,* 37, 174, 1966.
7.44. N. N. Zhuravlev, V. A. Smirnov, and T. A. Mingazin, *Sov. Phys. Crystallogr.,* 5, 124, 1960.
7.45. K. H. Jack and M. M. Wachtel, *Proc. R. Soc. London,* A239, 46, 1957.
7.46. C. W. Bates, D. Das Gupta, L. Galan, and D. N. E. Buchanan, *Thin Solid Films,* 69, 175, 1980.
7.47. H. Krebs, *Acta Crystallogr.,* 9, 95, 1956.
7.48. A. A. Mostovskii, V. A. Chaldyshev, V. P. Kiselev, and A. I. Klimin, *Izv. Akad. Nauk SSSR Ser. Fiz.,* 38, 195, 1974; *Bull. Acad. Sci. U.S.S.R. Phys. Ser.,* 38, 10, 1974.
7.49. A. A. Mostovskii, V. A. Chaldyshev, G. F. Karavaev, A. I. Klimin, and I. N. Ponomarenko, *Izv. Akad. Nauk SSSR Ser. Fiz.,* 40, 2490, 1976; *Bull. Acad. Sci. U.S.S.R. Phys. Ser.,* 40, 36, 1976.
7.50. R. Gobrecht, *Phys. Stat. Sol.,* 13, 429, 1966.
7.51. R. Gobrecht, *Ann. Phys.,* 20, 262, 1967.
7.52. S. Imamura, *J. Phys. Soc. Jpn.,* 14, 1491, 1959.
7.53. A. H. Sommer, *Photoemissive Materials,* John Wiley & Sons, New York, 1968.
7.54. W. Wepper and R. A. Huggins, *J. Sol. State Chem.,* 22, 297, 1977.
7.55. J. A. Skunk, *Constitution of Binary Alloys,* 2nd Suppl., McGraw-Hill, New York, 1969.
7.56. W. E. Spicer, *Phys. Rev.,* 112, 114, 1958.
7.57. A. Ebina and T. Takahashi, *Phys. Rev.,* B7, 4712, 1973.
7.58. F. W. Dorn and W. Klemm, *Z. Anorg. Allg. Chem.,* 309, 189, 1961.
7.59. C. Ghosh and B. P. Varma, *J. Appl. Phys.,* 49, 4549, 1978.
7.60. W. J. Harper and W. J. Choike, *J. Appl. Phys.,* 27, 1358, 1956.
7.61. A. H. Sommer and W. E. Spicer, *J. Appl. Phys.,* 32, 1036, 1961.
7.62. D. G. Fisher, *J. Appl. Phys.,* 47, 3471, 1976.
7.63. E. A. Taft and H. R. Philipp, *Phys. Rev.,* 115, 1583, 1959.
7.64. N. D. Mogulis, F. G. Borsyak, and B. T. Dyatlovitskaya, *Izv. Akad. Nauk SSSR,* 12, 126, 1948.
7.65. G. Wallis, *Ann. Phys.,* 17, 401, 1956.
7.66. K. Hirschberg and K. Deutscher, *Phys. Stat. Sol.,* 26, 527, 1968.
7.67. J. Königsberger and J. Weiss, *Ann. Phys. Leipzig,* 35, 1, 1911.
7.68. M. Faraday, *Exp. Res. Electr.,* 1, 432 and 1340, 1839.
7.69. F. Braun, *Poggendorff Ann. Phys. Chem.,* 3, 556, 1874.
7.70. A. Schuster, *Philos. Mag.,* 48, 251, 1874.
7.71. L. O. Grondal, *Phys. Rev.,* 27, 813, 1926.
7.72. V. P. Zhuze and B. V. Kurchatov, *Phys. Z. Sov. Union,* 2, 453, 1933.
7.73. J. Benar, *Oxydation des Metaux,* Gauthier, Villars, 1964.
7.74. J. Bloem, *Philips Res. Rep.,* 13, 167, 1958.
7.75. Gmelins Handb. Anorg. Chemie. Kupfer, Teil B1, 1958; Silber, Teil B1, 1971; Silber Teil B3, 1973.
7.76. M. H. Manghnani, W. S. Brower, and H. S. Parker, *Phys. Stat. Sol.,* A25, 69, 1974.
7.77. M. M. Beg and S. M. Shapiro, *Phys. Rev.,* B13, 1728, 1976.
7.78. G. K. White, *J. Phys.,* C11, 2171, 1978.
7.79. R. E. Gerkin and K. S. Pitzer, *J. Am. Chem. Soc.,* 84, 2662, 1962.
7.80. V. F. Drobny and D. L. Pulfrey, *Thin Solid Films,* 61, 89, 1979.
7.81. L. Kleinman and K. Mednick, *Phys. Rev.,* B21, 1549, 1980.
7.82. S. Nikitine, J. B. Grun, and M. Sieskind, *J. Phys. Chem. Sol.,* 17, 292, 1961.
7.83. G. Mönch, *Z. Phys.,* 78, 728, 1932.
7.84. H. K. Henisch, *Rectifying Semiconductor Contacts,* Oxford University Press, New York, 1957.
7.85. W. Feldman, *Phys. Rev.,* 64, 113, 1943.
7.86. G. F. G. Garlick, Ref. see in 7.246, chap. 11, Sect. 9.
7.87. J. B. Grun, M. Sieskind, and S. Nikitine, *J. Phys. Radium,* 22, 176, 1961.

7.88. J. W. Hodby, T. E. Jenkins, T. E. Schwab, C. Tamura, and D. Trivich, *J. Phys.*, C9, 1429, 1976.
7.89. P. Hanscombe, *Thin Solid Films*, 51, L25, 1978.
7.90. H. Dünwald and C. Wagner, *Z. Phys. Chem.*, 22, 208, 1933.
7.91. J. Hallberg and R. C. Hanson, *Phys. Stat. Sol.*, 42, 305, 1970.
7.92. C. Noguet, M. Tapiero, and J. P. Zielinger, *Phys. Stat. Sol.*, 24, 565, 1974.
7.93. M. O'Keeffe and F. S. Stone, *J. Phys. Chem. Solids*, 23, 261, 1962.
7.94. K. F. Young and H. P. R. Frederikse, *J. Phys. Chem. Ref. Data*, 2, 313, 1973.
7.95. C. Noguet, C. Schwab, C. Sennett, M. Sieskind, and C. Viel, *J. Phys. (Paris)*, 26, 317, 1965.
7.96. M. O'Keeffe, *J. Chem. Phys.*, 39, 1789, 1963.
7.97. P. Dawson, M. M. Hargreave, and G. R. Wilkinson, *J. Phys. Chem. Solids*, 34, 2201, 1973.
7.98. H. Wieder and A. W. Czanderna, *J. Appl. Phys.*, 37, 184, 1966.
7.99. E. B. Bakulin, M. M. Bredov, E. G. Ostroumova, and V. V. Shcherbinina, *Sov. Phys. Solid State*, 19, 1307, 1977.
7.100. B. Prévot, C. Carabatos, and M. Sieskind, *Phys. Stat. Sol.*, A10, 455, 1972.
7.101. B. K. Bairamov, B. S. Predtechenskii, L. S. Starostina, and Z. M. Khashkhozhev, *Sov. Phys. Solid State*, 17, 2030, 1975.
7.102. M. Wautelet, A. Roos, R. Lazzeroni, F. Hanus, and G. Lambin, *J. Mater. Res.*, 8, 327, 1993.
7.103. S. Nikitine, *Exitons. Optical Properties of Solids*, Plenum Press, New York, 1969.
7.104. F. I. Kreingold and B. S. Kulinkin, *Fiz. Tekh. Poluprovodnikov*, 4, 2353, 1970.
7.105. F. I. Kreingold, B. S. Kulinkin, and V. I. Tsurikov, *Sov. Phys. Solid State*, 20, 2191, 1978.
7.106. A. S. Povarennykh, *Crystal Chemistry. Classification of Minerals*, Vol. 1, Plenum Press, New York, 1972.
7.107. V. V. Gorbachev and I. M. Putilin, *Phys. Stat. Sol.*, A16, 553, 1973.
7.108. P. Junod, *Helv. Phys. Acta*, 32, 567 and 601, 1959.
7.109. T. Sakuma, K. Jide, K. Honma, and H. Okasaki, *J. Phys. Soc. Jpn.*, 43, 538, 1977.
7.110. N. E. Pingitore, B. F. Ponce, M. P. Eastman, F. Moreno, and C. Podpora, *J. Mater. Res.*, 7, 2219, 1992.
7.111. A. J. Frueh, *Am. Miner.*, 46, 654, 1961.
7.112. P. Kubaschewski and J. Nölting, *Ber. Bunsenges. Phys. Chem.*, 77, 74, 1973.
7.113. P. N. Walsh, E. W. Art, and D. White, *J. Phys. Chem.*, 66, 1546, 1962.
7.114. P. Junod, H. Hediger, B. Kilchör, and B. Wullschleger, *Philos. Mag.*, 36, 941, 1977.
7.115. N. Nakayama, *Jpn. J. Appl. Phys.*, 8, 450, 1969.
7.116. B. J. Mulder, *Phys. Stat. Sol.*, A15, 409, 1973; A18, 633, 1973.
7.117. A. A. Voskanyan, P. N. Inglizian, S. P. Lalykin, I. A. Plyutto, and Y. M. Shevchenko, *Sov. Phys. Semicond.*, 12, 2096, 1978.
7.118. V. M. Glazov and A. S. Burkhanov, *Sov. Phys. Semicond.*, 7, 1401, 1973.
7.119. O. P. Astakhov and V. V. Lobankov, *Teplofizika Vysokikh Temp.*, 10, 107, 1972.
7.120. R. Daiven, *J. Appl. Phys.*, 41, 5034, 1970.
7.121. A. Goswami and B. V. Rao, *Indian J. Phys.*, 50, 50, 1976.
7.122. O. P. Astakhov, L. I. Berger, K. Dovletov, and K. Tashliev, *Izv. Akad. Nauk Turkmen. SSR Ser. Fiz. Tekh. Nauk*, 4, 108, 1973.
7.123. H. Rickert, *Festkörperprobleme*, 6, 85, 1967.
7.124. H. P. Geserich and W. Suppanz, *Phys. Stat. Sol.*, 35, 381, 1969.
7.125. H. H. Dörner, H. P. Geserich, and H. Rickert, *Phys. Stat. Sol.*, 37, K85, 1970.
7.126. J. Appel, *Z. Naturforsch.*, 10A, 530, 1955.
7.127. C. Wood, C. Harrap, and W. M. Kane, *Phys. Rev.*, 121, 978, 1961.
7.128. O. P. Astakhov, L. I. Berger, K. Dovletov, and K. Tashliev, *Izv. Akad. Nauk Turkmen. SSR Ser. Fiz. Tekh. Nauk*, 5, 101, 1973.
7.129. S. Duchemin and F. Guastavino, *Solid State Commun.*, 26, 187, 1978.
7.130. N. S. Kurnakov and N. S. Stepanov, *Zh. Russ. Fiz. Khim. Obshchestva (J. Russ. Phys. Chem. Soc.)*, 34, 520, 1902; 37, 668, 1905.
7.131. E. A. Owen and G. D. Preston, *Proc. Phys. Soc. London*, 36, 541, 1924.
7.132. A. Sacklowski, *Ann. Phys.*, 77, 264, 1925.
7.133. J. Friauf, *J. Am. Chem. Soc.*, 48, 1906, 1926.
7.134. Alfa-Aesar, Johnson Matthey Catalog, 1995–1996.
7.135. W. D. Robertson and H. H. Uhlig, *Trans. AIMME*, 180, 345, 1949.

7.136. V. M. Glazov, S. N. Chizhevskaya, and N. N. Glagoleva, *Zhidkiye Poluprovodniki (Liquid Semiconductors)*, Nauka, Moscow, 1967.
7.137. J. M. Eldridge, E. Miller, and K. L. Komarek, *Trans. Met. Soc. AIME*, 239, 775, 1967.
7.138. E. Anastassakis and C. H. Perry, *Phys. Rev.*, B4, 1251, 1971.
7.139. B. C. Gerstein, F. J. Jelineck, M. Habenschuss, W. D. Shickell, J. R. Mullay, and P. L. Chung, *J. Chem. Phys.*, 47, 2109, 1967.
7.140. B. C. Gerstein, P. L. Chung, and B. C. Danielson, *J. Phys. Chem. Solids*, 27, 1161, 1966.
7.141. F. J. Jelineck, W. D. Shickell, and B. C. Gerstein, *J. Phys. Chem. Solids*, 28, 267, 1967.
7.142. R. G. Schwartz, H. Shanks, and B. C. Gerstein, *J. Solid State Chem.*, 3, 533, 1971.
7.143. U. Winkler, *Helv. Phys. Acta*, 28, 633, 1955.
7.144. W. B. Witten, P. L. Chung, and G. C. Danielson, *J. Phys. Chem. Solids*, 26, 49, 1965.
7.145. Y. I. Dutchak, V. P. Yarmolyuk, and Y. I. Fedyshchin, *Inorg. Mater.*, 11, 1047, 1975.
7.146. P. H. Chung, W. B. Witten, and G. C. Danielson, *J. Phys. Chem. Solids*, 26, 1753, 1965.
7.147. L. C. Davis, W. B. Witten, and G. C. Danielson, *J. Phys. Chem. Solids*, 28, 439, 1967.
7.148. F. Aymerich and G. Mula, *Phys. Stat. Sol.*, B42, 697, 1970.
7.149. J. P. Van Dyke and F. Hermann, *Phys. Rev.*, B2, 1644, 1970.
7.150. R. F. Blunt, H. P. R. Frederikse, and W. R. Hosler, *Phys. Rev.*, 100, 663, 1955.
7.151. M. W. Heller and G. C. Danielson, *J. Phys. Chem. Solids*, 23, 601, 1962.
7.152. G. Busch and U. Winkler, *Physica*, 20, 1067, 1954.
7.153. R. G. Morris, R. D. Redin, and G. C. Danielson, *Phys. Rev.*, 109, 1909, 1958.
7.154. G. Busch and M. Schneider, *Physica*, 20, 1084, 1954.
7.155. H. P. R. Frederikse, W. R. Hosler, and D. E. Roberts, *Phys. Rev.*, 103, 67, 1956.
7.156. R. D. Redin, R. G. Morris, and G. C. Danielson, *Phys. Rev.*, 109, 1916, 1958.
7.157. M. W. Heller and G. C. Danielson, *J. Phys. Chem. Solids*, 23, 601, 1962.
7.158. L. I. Berger, *Sov. Phys. Proc. VUZ*, No. 2, 64, 1962.
7.159. P. W. Li, S. N. Lee, and G. C. Danielson, *Phys. Rev.*, B6, 442, 1972.
7.160. J. J. Martin, *J. Phys. Chem. Solids*, 33, 1139, 1972.
7.161. H. R. Shanks, *J. Cryst. Growth*, 23, 190, 1974.
7.162. R. Geick, W. J. Habel, and C. H. Perry, *Phys. Rev.*, 148, 824, 1966.
7.163. G. A. Stringer and R. J. Higgins, *Phys. Rev.*, B3, 506, 1971.
7.164. L. L. Korenblit and A. P. Kolesnikov, *Sov. Phys. Tech. Phys.*, 1, 924, 1956.
7.165. D. McWilliams and D. W. Lynch, *Phys. Rev.*, 130, 2248, 1963.
7.166. E. Anastassakis and E. Burstein, *Solid State Commun.*, 9, 1525, 1971.
7.167. W. S. Scouler, *Phys. Rev.*, 178, 1353, 1969.
7.168. J. C. Bose, U.S. Patent 755,840, 1904.
7.169. L. Eisenmann, *Ann. Phys.*, 38, 121, 1940.
7.170. H. Hintenberg, *Z. Phys.*, 119, 1, 1942.
7.171. R. A. Smith, *Adv. Phys. (Philos. Mag. Suppl.)*, 2, 321, 1953.
7.172. N. Kh. Abrikosov and L. E. Glukhikh, *Zh. Neorgan. Khim.*, 8, 1792, 1963.
7.173. R. F. Brebrick, *Phys. Chem. Solids*, 24, 27, 1963.
7.174. L. E. Shelimova and N. Kh. Abrikosov, *Zh. Neorg. Khim.*, 9, 1879, 1964.
7.175. K. C. Mills, *Thermodynamic Data for Inorganic Sulfides, Selenides and Tellurides*, Butterworth & Co. Ltd. (U.K.), 1974.
7.176. V. M. Zhdanov, *Zh. Phys. Khim.*, 45, 1357, 1971.
7.177. A. P. Lambros, D. Gerleas, and N. A. Economou, *J. Phys. Chem. Solids*, 35, 537, 1974.
7.178. H. S. Belson and B. Houston, *J. Appl. Phys.*, 41, 422, 1970.
7.179. S. I. Novikova and N. Kh. Abrikosov, *Sov. Phys. Solid State*, 5, 1397, 1963.
7.180. K. G. Subhandra and D. B. Sirdeshmukh, *Pramana*, 10, 357, 1978.
7.181. Y. S. Touloukian, R. K. Kirby, R. E. Taylor, and T. Y. R. Lee, *Thermophysical Properties of Matter. Thermal Expansion — Nonmetallic Solids*, Vol. 13, Plenum Press, New York, 1977.
7.182. S. I. Novikova and N. Kh. Abrikosov, *Sov. Phys. Solid State*, 19, 683, 1977.
7.183. N. Kh. Abrikosov, O. G. Karpinskii, L. E. Shelimova, and M. A. Korzhuev, *Inorg. Mater.*, 13, 2160, 1977.
7.184. S. I. Novikova, *Fiz. Tverd. Tela*, 5, 1913, 1963.
7.185. P. Leroux-Hugon, X. X. Nguyen, and M. Rodot, *J. Phys.*, 28, Cl-73, 1967.
7.186. L. I. Berger and I. A. Dobrushina, *Trudy IREA (Proc. IREA)*, 26, 302, 1964.

7.187. N. Kh. Abrikosov, V. F. Bankina, L. V. Poretskaya, L. E. Shelimova, and E. V. Skudnova, *Semiconducting II-VI, IV-VI and V-VI Compounds,* Plenum Press, New York, 1969.
7.188. A. J. Miller, G. A. Saunders, and Y. K. Yogurtcu, *J. Phys.,* C14, 1569, 1981.
7.189. T. Grandke and L. Ley, *Phys. Rev.,* B16, 832, 1977.
7.190. A. W. Parke and G. P. Srivastava, *Phys. Stat. Sol.,* B101, K31, 1980.
7.191. W. Albers, C. Haas, and F. Van Der Maesen, *J. Phys. Chem. Solids,* 15, 306, 1960.
7.192. P. Nicolic and S. Vujatovic, Proc. 12th Int. Conf. Phys. Semicond., Stuttgart, 1974, 331.
7.193. R. Car, G. Ciucci, and L. Quartapelle, *Phys. Stat. Sol.,* B86, 471, 1978.
7.194. R. Tsu, W. E. Howard, and L. Esaki, *Phys. Rev.,* 172, 779, 1968.
7.195. M. L. Cohen, Y. Tung, and P. B. Allen, *J. Phys.,* 29, C4-163, 1968.
7.196. F. Herman, R. C. Kortum, I. B. Ortenburger, and J. P. Van Dyke, *J. Phys.,* 29, C4-62, 1968.
7.197. Y. Tung and M. L. Cohen, *Phys. Rev.,* 180, 823, 1969.
7.198. S. K. Bahl and K. L. Chopra, (a) *J. Appl. Phys.,* 40, 4940, 1969; (b) *J. Appl. Phys.,* 41, 2169, 1970.
7.199. M. Baleva, *Phys. Stat. Sol.,* 99, 341, 1980.
7.200. J. R. Burke, R. S. Allgaier, B. B. Houston, J. Babiskin, and P. G. Siebenmann, *Phys. Rev Lett.,* 14, 360, 1965.
7.201. J. R. Burke, B. Houston, and H. T. Savage, *J. Phys. Soc. Jpn.,* 21, 384, 1966.
7.202. S. Rabii, *Phys. Rev.,* 182, 821, 1969.
7.203. Y. W. Tsang and M. L. Cohen, *Phys. Rev.,* B3, 1254, 1971.
7.204. H. T. Savage, B. Houston, and J. R. Burke, *Phys. Rev.,* B6, 2292, 1972.
7.205. P. C. Kemeny and M. Cardona, *J. Phys.,* C9, 1361, 1976.
7.206. J. C. Melvin and D. C. Hendry, *J. Phys.,* C12, 3003, 1979.
7.207. S. Katayama, *Phys. Rev.,* B22, 336, 1980.
7.208. G. Nimtz and B. Schlicht, *Narrow Gap Semiconductors. Springer Tracts in Modern Physics,* Vol. 98, Springer-Verlag, Berlin, 1983.
7.209. H. Preier, *Appl. Phys.,* 20, 189, 1979.
7.210. M. L. Cohen and J. R. Chelikowsky, *Electronic Structure and Optical Properties of Semiconductors,* Springer-Verlag, New York, 1989.
7.211. P. J. T. Eggenkamp, H. J. M. Swagten, K. Kopinga, and W. J. M. deJonge, *Phys. Rev.,* B48, 3770, 1993.
7.212. K. Murase, S. Sugai, T. Higuchi, S. Takaoka, T. Fukunaga, and H. Tawamura, 14th Int. Conf. Phys. Semicond., Edinburgh, September 1978; *Inst. Phys. Conf. Ser.,* 43, 437, 1979.
7.213. J. van den Dries and R. Lieth, *Phys. Stat. Sol.,* A5, K171, 1971.
7.214. A. Stanchev and C. Vodenicharov, *Thin Solid Films,* 38, 67, 1976.
7.215. T. Yabumoto, *J. Phys. Soc. Jpn.,* 13, 559, 1957.
7.216. A. S. Okhotin, A. N. Krestovnikov, A. A. Aivazov, and A. S. Pushkarskii, *Phys. Stat. Sol.,* 31, 485, 1969.
7.217. R. Klann, T. Höfer, R. Buhleier, T. Elsaesser, and J. W. Tomm, *J. Appl. Phys.,* 77, 277, 1995.
7.218. D. S. Kiriakos, O. Vassalides, and N. M. Economou, 14th Int. Conf. Phys. Semicond., Edinburgh, September 1978; *Inst. Phys. Conf. Ser.,* 43, 501, 1979.
7.219. M. A. Korzhuev, L. E. Shelimova, and N. Kh. Abrikosov, *Sov. Phys. Semicond.,* 11, 171, 1977.
7.220. B. B. Anisimov, A. A. Gabedava, S. Z. Dzhamagidze, I. I. Elfimova, Y. A. Maltsev, and R. R. Shvangiradze, *Inorg. Mater.,* 14, 1417, 1978.
7.221. R. A. Hein, J. W. Gibson, R. Mazelsky, R. C. Miller, and J. K. Huhn, *Phys. Rev. Lett.,* 12, 320, 1964.
7.222. N. Kolomoets, E. Lev, and L. Sysoeva, *Sov. Phys. Semicond.,* 5, 2101, 1964; 6, 551, 1964.
7.223. W. Albers, C. Haas, H. J. Vink, and J. D. Wasscher, *J. Appl. Phys.,* 32, 2220, 1961.
7.224. T. Yabumoto, *J. Phys. Soc. Jpn.,* 13, 972, 1958.
7.225. J. Umeda, *J. Phys. Soc. Jpn.,* 16, 124, 1961.
7.226. H. Maier and D. R. Daniel, *J. Electron. Mater.,* 6, 693, 1977.
7.227. S. Asanabe, *J. Phys. Soc. Jpn.,* 14, 281, 1959.
7.228. P. B. Allen and M. L. Cohen, *Phys. Rev.,* 177, 704, 1969.
7.229. K. L. I. Kobayashi, Y. Kato, Y. Katayama, and K. F. Komatsubara, *Solid State Commun.,* 17, 875, 1975.
7.230. R. C. Keezer, D. L. Bowman, and J. H. Becker, *J. Appl. Phys.,* 39, 2062, 1968.

7.231. A. F. Ioffe and A. R. Regel, *Prog. in Semicond. (London)*, 4, 237, 1960.
7.232. A. V. Ioffe and A. F. Ioffe, *Fiz. Tverd. Tela*, 2, 5, 1960.
7.233. D. H. Damon, C. R. Martin, and R. C. Miller, *J. Appl. Phys.*, 34, 3083, 1963.
7.234. R. Allgaier and W. Scanton, *Phys. Rev.*, 111, 1029, 1958.
7.235. U. Schlichting and K. H. Gobrecht, *J. Phys. Chem. Solids*, 34, 753, 1973.
7.236. D. L. Rode, *Phys. Stat. Sol.*, B53, 244, 1972.
7.237. B. Schlicht, R. Dornhaus, G. Nimtz, L. D. Haas, and T. Jacobus, *Solid State Electron.*, 21, 1481, 1978.
7.238. J. D. Wiley, W. J. Buckel, and R. L. Schmidt, *Phys. Rev.*, B13, 2489, 1976.
7.239. (a) H. R. Chandrasekhar and U. Zwick, *Solid State Commun.*, 18, 1509, 1976; (b) D. I. Siapkas, D. S. Kyriakos, and N. A. Economou, *Solid State Commun.*, 19, 765, 1976.
7.240. A. F. Gibson, W. D. Lawson, and T. S. Moss, *Proc. Phys. Soc.*, A64, 1054, 1951.
7.241. H. Chandrasekhar, R. G. Humphrey, U. Zwick, and M. Cardona, *Phys. Rev.*, B15, 2177, 1977.
7.242. R. B. Schoolar, H. R. Riedl, and J. R. Dixon, *Solid State Commun.*, 4, 423, 1966.
7.243. A. E. Eunos, *J. Opt. Soc. Am.*, 52, 261, 1962.
7.244. R. Dalven, *Solid State Physics*, Vol. 28, H. Ehrenreich, F. Seitz, and D. Turnbull, Eds., Academic Press, New York, 1973, 179.
7.245. J. N. Zemel, J. D. Jensen, and R. B. Schoolar, *Phys. Rev.*, A140, 330, 1965.
7.246. W. E. Tenant, *Solid State Commun.*, 20, 613, 1976.
7.247. R. A. Smith, *Semiconductors*, Cambridge University Press, New York, 1959.
7.248. N. Suzuki, K. Sawai, and S. Adachi, *J. Appl. Phys.*, 77, 1249, 1995.
7.249. I. Gregora, B. Velicky, and M. Zavetova, *J. Phys. Chem. Solids*, 37, 785, 1976.
7.250. I. A. Drabkin, T. B. Zhukova, I. V. Nelson, and L. M. Sysoeva, *Inorg. Mater.*, 15, 1194, 1979.
7.251. R. F. Bis and J. R. Dixon, *Phys. Rev.*, B2, 1004, 1970.
7.252. R. B. Schoolar and J. R. Dixon, *J. Opt. Soc. Am.*, 58, 119, 1968.
7.253. D. G. Avery, *Proc. Phys. Soc.*, B64, 1087, 1951.
7.254. A. F. Gibson, *Proc. Phys. Soc.*, B65, 378, 1952.
7.255. B. Spranger, U. Schiessl, A. Lambrecht, H. Böttner, and M. Tacke, *Appl. Phys. Lett.*, 53, 2582, 1988.
7.256. N. Piccioli, J. M. Benson, and M. Balkanski, *J. Phys. Chem. Solids*, 35, 971, 1974.
7.257. J. R. Lowney and S. D. Senturia, *J. Appl. Phys.*, 47, 1773, 1976.
7.258. T. R. Globus, B. L. Gelmont, K. I. Geiman, V. A. Kondrashov, and A. V. Matveenko, *Sov. Phys. JETP*, 53, 1000, 1981.
7.259. A. N. Veis, R. F. Kuteinikov, S. A. Kuznetsov, and Y. I. Ukhanov, *Sov. Phys. Semicond.*, 10, 1320, 1976.
7.260. H. Preier, *Semicond. Sci. Technol.*, 5, S12, 1990.
7.261. W. W. Scanlon, *J. Phys. Chem. Solids*, 8, 423, 1959.
7.262. Y. V. Maltsev, I. K. Smirnov, Y. I. Ukhanov, and A. N. Veis, *J. Phys. (Paris)*, 29, 99, 1968.
7.263. I. A. Drabkin, L. Y. Morozovskii, I. V. Nelson, and Y. I. Ravich, *Sov. Phys. Semicond.*, 6, 1156, 1972.
7.264. A. N. Veis, V. I. Kaidanov, Y. I. Ravich, L. A. Ryabtseva, and Y. I. Ukhanov, *Sov. Phys. Semicond.*, 10, 62, 1976.
7.265. O. Ziep and D. Gensow, *Phys. Stat. Sol.*, B96, 359, 1979.
7.266. B. L. Gelmont, T. R. Globus, and A. V. Matveenko, *Solid State Commun.*, 38, 931, 1981.
7.267. F. Rössler, *Z. Anorg. Allgem. Chem.*, 9, 546, 1895.
7.268. D. Harker, *Z. Kristal.*, 89, 175, 1934.
7.269. C. Frondell, *Am. Miner.*, 24, 185, 1939.
7.270. P. W. Lange, *Naturwissenschaften*, 27, 133, 1939.
7.271. E. Dönges, *Z. Anorg. Chem.*, 265, 56, 1951.
7.272. K. Schubert and H. Fricke, *Z. Metallk.*, 44, 457, 1953.
7.273. S. A. Semiletov, *Proc. Inst. Kristallogr.*, 10, 76, 1954; *Kristallografia*, 1, 403, 1956.
7.274. J. R. Wiese and L. Muldawer, *Phys. Chem. Solids*, 15, 13, 1960.
7.275. I. Teramoto and S. Takayanagi, *J. Phys. Chem. Solids*, 19, 124, 1961.
7.276. K. A. Becker, K. Plieth, and I. N. Stranski, *Progress in Inorganic Chemistry*, Vol. 4, F. A. Cotten, Ed., John Wiley & Sons, New York, 1962, 1.
7.277. V. G. Kuznetsov and K. K. Palkina, *Zh. Neorg. Khim.*, 8, 1204, 1963.

7.278. S. Nakajima, *J. Phys. Chem. Solids,* 24, 479, 1963.
7.279. G. Malmros, *Acta Chem. Scand.,* 24, 479, 1963.
7.280. H. B. Krause, *Acta Crystallogr.,* A31, 66, 1975.
7.281. J. W. Medernach and R. L. Snyder, *J. Am. Ceram. Soc.,* 61, 494, 1978.
7.282. T. J. Wieting and M. Schlüter, *Electrons and Phonons in Layered Crystal Structures,* Reidel Publishing, Dordrecht, Holland, 1979.
7.283. L. V. Poretskaya, N. Kh. Abrikosov, and V. M. Glazov, *Zh. Neorg. Chim.,* 8, 1196, 1963.
7.284. A. Kozakov, H. Neumann, and G. Leonhardt, *Phys. Lett.,* 62A, 95, 1977.
7.285. H. L. Althaus, G. Weiser, and S. Nagel, *Phys. Stat. Sol.,* B87, 117, 1978.
7.286. G. Weiser, *Proc. Int. Conf. Phys. Selenium and Tellurium, Königstein, 1979,* E. Gerlach and P. Grosse, Eds., Springer-Verlag, New York, 1979.
7.287. A. G. Khasabov and I. Y. Nikiforov, *Sov. Phys. Crystallogr.,* 16, 28, 1971.
7.288. L. P. Caywood and G. R. Miller, *Phys. Rev.,* B2, 3209, 1970.
7.289. F. Borghese and E. Donato, *Nuovo Cimento,* 53B, 283, 1968.
7.290. N. S. Platakis, *J. Non Crystal Solids,* 24, 365, 1977.
7.291. K. A. Verkhovskaya, I. P. Grigas, and V. M. Fridkin, *Sov. Phys. Solid State,* 10, 1583, 1969.
7.292. H. J. Goldsmid, *J. Electron.,* 1, 218, 1955.
7.293. A. R. Regel, *Ukr. Fiz. Zh.,* 7, 833, 1962.
7.294. V. I. Fedorov and I. Y. Davydov, *High Temp.,* 16, 654, 1978.
7.295. L. B. Schein, *Phys. Rev.,* B15, 1024, 1977.
7.296. M. Abkowitz, D. F. Blossey, and A. I. Lakatos, *Phys. Rev.,* B12, 3400, 1975.
7.297. B. Roy, B. R. Chakraborty, R. Bhattachariya, and A. K. Dutta, *Solid State Commun.,* 25, 937, 1978.
7.298. B. R. Chakraborty, B. Roy, R. Bhattachariya, and A. K. Dutta, *J. Phys. Chem. Solids,* 41, 913, 1980.
7.299. W. Eichler and G. Simon, *Phys. Stat. Sol.,* B86, K85, 1978.
7.300. H. Schwartz, G. Björck, and O. Beckmann, *Solid State Commun.,* 5, 905, 1967.
7.301. H. Köhler and A. Fabricius, *Phys. Stat. Sol.,* B71, 487, 1975.
7.302. G. R. Hyde, H. A. Beale, and I. L. Spain, *J. Phys. Chem. Solids,* 35, 1719, 1974.
7.303. J. A. Woollam, H. A. Beale, and I. L. Spain, *Phys. Lett.,* 41A, 319, 1972.
7.304. M. Stordeur and W. Kühnberger, *Phys. Stat. Sol.,* B69, 377, 1975.
7.305. C. H. Champness and A. L. Kipling, *Can. J. Phys.,* 44, 769, 1966.
7.306. M. K. Zhitinskaya, V. I. Kaidanov, and V. P. Kondratev, *Sov. Phys. Semicond.,* 10, 1300, 1976.
7.307. H. Süssmann, *Wissensch. Z. Univ. Halle,* 12, 65, 1973.
7.308. A. V. Ioffe, V. G. Kuznetsov, and K. K. Palkina, *Zh. Neorg. Khim. (Russ. J. Inorg. Chem.),* 8, 1113, 1963.
7.309. I. A. Smirnov, M. I. Shadrichev, and V. A. Kutasov, *Sov. Phys. Solid State,* 11, 2681, 1970.
7.310. O. Sälzer and H. Nieke, *Ann. Phys. (Lpz.),* 15, 192, 1965.
7.311. I. R. Drabble, *Progress in Semiconductors,* Vol. 7, A. F. Gibson and R. E. Burgess, Eds., John Wiley & Sons, New York, 1963.
7.312. Tables de Constantes Sélectionneés. Diamagnétisme et Paramagnétisme, G. Foëx, Paris, 1957.
7.313. D. Treacy and P. C. Taylor, *Phys. Rev.,* B11, 2941, 1975.
7.314. R. Zallen, M. L. Slade, and A. T. Ward, *Phys. Rev.,* B3, 4257, 1971.
7.315. B. P. Grigas and V. I. Valyukenas, *Sov. Phys. Semicond.,* 12, 1420, 1978.
7.316. W. Richter, A. Krost, U. Nowak, and E. Annastassakis, *Z. Phys.,* B49, 191, 1982.
7.317. W. P. Doyle, *J. Phys. Chem. Solids,* 4, 144, 1958.
7.318. D. M. Mattox and L. Gildart, *J. Phys. Chem. Solids,* 18, 215, 1961.
7.319. L. Gildart, J. M. Kline, and D. M. Mattox, *J. Phys. Chem. Solids,* 18, 286, 1961.
7.320. D. L. Greenway and G. Harbeke, *J. Phys. Chem. Solids,* 26, 1585, 1965.
7.321. H. Imaizumi, H. Yamaguti, H. Kaibe, and I. Nishida, Proc. 7th Int. Conf. on Thermoelectric Energy Conversion, Arlington, VA, 1988, 141.
7.322. H. Kaibe, M. Sakata, Y. Isoda, and I. Nishida, *J. Jpn. Inst. Metals,* 53, 958, 1989.
7.323. H. Wada, T. Sato, K. Takahashi, and N. Nakastukasa, *J. Mater. Res.,* 5, 1052, 1990.
7.324. H. Wada, M. Watanabe, J. Morimoto, and T. Miyakawa, *J. Mater. Res.,* 6, 1711, 1991.
7.325. C. P. Susz, J. Miller, K. Yvon, and E. Parté, *J. Less Common Met.,* 71, P1, 1980.
7.326. J. Evers, G. Oehlinger, and H. Meyer, *Mater. Res. Bull.,* 19, 1177, 1984.

7.237. A. Ennaoui, S. Fiechter, H. Goslowsky, and H. Tributsch, *J. Electrochem. Soc.*, 132, 1579, 1985.
7.328. A. Ennaoui, S. Fiechter, W. Jaegermann, and H. Tributsch, *J. Electrochem. Soc.*, 133, 98, 1986.
7.329. C. B. Vinning and C. E. Allevato, Proc. 10th Int. Conf. Thermoelectrics, Cardiff, U.K., 1991, 167.
7.330. W. Jeitschko, H. Holleck, H. Nowotny, and F. Benesovsky, *Monatsh. Chem.*, 94, 838, 1963.
7.331. F. M. d'Heurle and P. Gas, *J. Mater. Res.*, 1, 205, 1986.
7.332. S. Fiechter, M. Birkholz, A. Hartmann, P. Dulski, M. Giersig, H. Tributsch, and R. J. D. Tilley, *J. Mater. Res.*, 7, 1829, 1992.
7.333. F. d'Heurle, C. S. Petersson, J. E. E. Baglin, S. J. LaPlaca, and C. Y. Wong, *J. Appl. Phys.*, 55, 4208, 1984.
7.334. D. J. Poutcharovsky and E. Parthé, *Acta Crystallogr.*, B30, 2692, 1974.
7.335. S. Petersson, R. Anderson, J. Baglin, J. Dempsey, W. Hammer, F. d'Heurle, and S. LaPlaca, *J. Appl. Phys.*, 51, 373, 1980.
7.336. T. Caillat, A. Borshchevsky, and J.-P. Fleurial, AIP Conf. Proc. 316, 13th Int. Conf. Thermoelectrics, Kansas City, MO, B. Mathiprakasam and P. Heenan, Eds., AIP, New York, 1995, 31.
7.337. T. Caillat, A. Borshchevsky, and J.-P. Fleurial, AIP Conf. Proc. 316, 13th Int. Conf. Thermoelectrics, Kansas City, MO, B. Mathiprakasam and P. Heenan, Eds., AIP, New York, 1995, 58.
7.338. A. Borshchevsky, J.-P. Fleurial, E. Allevato, and T. Caillat, AIP Conf. Proc. 316, 13th Int. Conf. Thermoelectrics, Kansas City, MO, B. Mathiprakasam and P. Heenan, Eds., AIP, New York, 1995, 3.
7.339. D. J. Poutcharovsky, K. Yvon, and E. Parté, *J. Less Common Met.*, 40, 139, 1975.
7.340. I. Nishida, *Phys. Rev.*, B17, 2710, 1973.
7.341. Y. Dusausoy, J. Protas, R. Wandji, and B. Roques, *Acta Crystallogr.*, B27, 1209, 1971.
7.342. J. Steger and E. Kostiner, *J. Solid State Chem.*, 5, 131, 1972.
7.343. F. Gronvold, A. J. Highe, and E. F. Westrum, *J. Chem. Thermodyn.*, 9, 773, 1977.
7.344. A. Kjekshus and T. Rakke, *Acta Chem. Scand.*, A31, 517, 1977.
7.345. J. Kundrotas and A. Dargys, *Litov. Fizich. Sbornik (Lituanian Phys. Proc.)*, 22, 74, 1982.
7.346. S. S. Lau, J. W. Mayer, and K. N. Tu, *J. Appl. Phys.*, 49, 4005, 1978.
7.347. K. N. Tu, G. Ottaviani, R. D. Thompson, and J. W. Mayer, *J. Appl. Phys.*, 53, 4406, 1982.
7.348. C. S. Petersson, J. E. E. Baglin, J. J. Dempsey, F. M. d'Heurle, and S. J. LaPlaca, *J. Appl. Phys.*, 53, 4886, 1982.
7.349. J.-P. Fleurial and A. Borshchevsky, Proc. 12th Int. Conf. Thermoelectrics, K. Matsuura, Ed., Yokohama, 1993, 236.
7.350. W. Jeitschko, *Acta Crystallogr.*, B33, 2347, 1977.
7.351. U. Birkholz, A. Frühauf, and J. Schelm, Proc. 10th Int. Conf. Phys. Semicond., Cambridge, MA, 1970, 311.
7.352. C. Blaauw, F. van der Woude, and G. A. Sawatski, *J. Phys.*, C6, 2371, 1973.
7.353. A. K. L. Fan, G. H. Rosenthal, H. L. McKinzie, and A. Wold, *J. Solid State Chem.*, 5, 136, 1972.
7.354. J. B. Goodenough, *Solid State Chem.*, 5, 144, 1972.
7.355. L. D. Dudkin and N. Kh. Abrikosov, *Zh. Neorg. Khim.*, 1, 2096, 1956.
7.356. C. M. Pleass and R. D. Heyding, *Can. J. Chem.*, 40, 590, 1962.
7.357. T. Caillat, A. Borshchevsky, and J.-P. Fleurial, Proc. 27th Intersoc. Energy Conversion Eng. Conf., San Diego, CA, 1992, 499.
7.358. L. F. Mattheiss, *Phys. Rev.*, B10, 995, 1974.
7.359. R. L. Kautz, M. S. Dresselhaus, D. Adler, and A. Linz, *Phys. Rev.*, B6, 2078, 1972.
7.360. R. Schieck, A. Hartmann, S. Fiechter, R. Könenkamp, and H. Wetzel, *J. Mater. Res.*, 5, 1567, 1990.
7.361. K. Mason and G. Müller-Vogt, *J. Cryst. Growth*, 63, 34, 1983.
7.362. R. Kaner, C. A. Castro, R. P. Gruska, and A. Wold, *Mater. Res. Bull.*, 12, 1143, 1977.
7.363. P. C. Donohue, *Mater. Res. Bull.*, 7, 943, 1972.
7.364. J. P. Odile, S. Soled, C. A. Castro, and A. Wold, *Inorg. Chem.*, 17, 283, 1978.
7.365. J. Ackerman and A. Wold, *J. Phys. Chem. Solids*, 38, 1013, 1977.
7.366. W. D. Johnston, R. C. Miller, and D. H. Damon, *J. Less Common Met.*, 8, 272, 1965.
7.367. A. Dargys and J. Kundrotas, *J. Phys. Chem. Solids*, 44, 261, 1983.
7.368. D. H. Damon, R. C. Miller, and A. Sagar, *Phys. Rev.*, A138, 636, 1965.

7.369. J.-P. Fleurial, T. Caillat, and A. Borshchevsky, AIP Conf. Proc. 316, 13th Int. Conf. Thermoelectronics, Kansas City, MO, 1994, B. Mathiprakasam and P. Heenan, Eds., AIP Press, New York, 1995, 40.
7.370. G. Yamaguchi, M. Shimada, and M. Koizumi, *J. Solid State Chem.*, 34, 241, 1980.
7.371. U. Birholz, H. Finkenrath, J. Naegele, and N. Uhle, *Phys. Stat. Sol.*, 30, K81, 1968.
7.372. E. Bjerkelund and A. Kjekshus, *Acta Chem. Scand.*, 24, 3317, 1970.
7.373. G. Brostigen and A. Kjekshus, *Acta Chem. Scand.*, 24, 2993, 1970.
7.374. J. O'Shaughnessy and C. Smith, *Solid State Commun.*, 8, 481, 1970.
7.375. I. Balberg and H. L. Pinch, *J. Mag. Magn. Mater.*, 7, 12, 1978.
7.376. H. D. Lutz, G. Schneider, and G. Kliche, *Phys. Chem. Minerals*, 9, 109, 1983.
7.377. F. Hulliger, *Nature (London)*, 198, 1081, 1963.
7.378. F. Hulliger, *Phys. Lett.*, 4, 282, 1963.
7.379. H. D. Lutz and P. Willich, *Z. Anorg. Allg. Chem.*, 428, 199, 1977.
7.380. D. H. Damon, R. C. Miller, and P. R. Emtage, *Phys. Rev.*, B5, 2175, 1972.
7.381. H. D. Lutz and G. Kliche, *Z. Anorg. Allg. Chem.*, 480, 105, 1981.
7.382. H. D. Lutz and G. Kliche, *J. Solid State Chem.*, 40, 64, 1981.
7.383. H. D. Lutz and G. Kliche, *Phys. Stat. Sol.*, B112, 549, 1982.
7.384. S. Krishnamurti, A. Sher, and A. Chen, *Appl. Phys. Lett.*, 47, 160, 1985.
7.385. H. Stöhr and W. Klemm, *Z. Anorg. Chem.*, 241, 305, 1939.
7.386. E. R. Johnson and S. M. Christian, *Phys. Rev.*, 95, 560, 1954.
7.387. M. Glickman, *Phys. Rev.*, 100, 1146, 1955; 111, 125, 1958.
7.388. G. Busch and O. Vogt, *Helv. Phys. Acta*, 33, 437, 1960.
7.389. F. Herman, M. Glickman, and R. A. Parmenter, *Prog. Semicond.*, 2, 1, 1957.
7.390. R. Braunstein, A. R. Moore, and F. Herman, *Phys. Rev.*, 109, 695, 1958.
7.391. E. F. Kustov, E. A. Melnikov, A. A. Suchenkov, A. I. Levadnyi, and V. A. Filikov, *Sov. Phys. Semicond.*, 17, 481, 1983.
7.392. A. M. Toxen, *Phys. Rev.*, 122, 450, 1961.
7.393. M. C. Steele and F. D. Rosi, *J. Appl. Phys.*, 29, 1517, 1958.
7.394. N. Kobayashi and T. Fukui, *J. Cryst. Growth*, 67, 513, 1984.
7.395. O. Madelung, Ed., *Semiconductors. Group IV Elements and III-V Compounds;* Series, *Data in Science and Technology,* R. Poerschke, Ed., Springer-Verlag, New York, 1991.
7.396. M. J. Cherng, G. B. Stringfellow, and R. M. Cohen, *Appl. Phys. Lett.*, 44, 677, 1984.
7.397. N. A. Goryunova, *Slozhnyye Almazopodobnyye Poluprovodniki (Multinary Diamond-Like Semiconductors),* Soviet Radio, Moscow, 1968.
7.398. E. K. Müller and J. L. Richards, *J. Appl. Phys.*, 35, 1233, 1964.
7.399. M. I. Nathan, *Proc. IEEE*, 54, 927, 1966.
7.400. H. Kressel, *Handbook of Semiconductors,* Vol. 4, C. Hilsum, Ed., North-Holland, Amsterdam, 1981.
7.401. H. C. Casey and M. B. Panish, *Heterostructure Lasers,* Academic Press, New York, 1978.
7.402. P. G. Klemens, *Phys. Rev.*, 119, 507, 1960.
7.403. A. V. Ioffe and A. F. Ioffe, *Izv. Akad. Nauk SSSR Ser. Fiz.*, 20, 65, 1956.
7.404. B. Abeles, D. S. Beers, G. D. Cody, and J. P. Dismukes, *Phys. Rev.*, 125, 44, 1962.
7.405. A. Onton, R. J. Chicotka, and M. R. Lorenz, *J. Cryst. Growth*, 27, 166, 1974.
7.406. A. K. Saxena, *Phys. Stat. Sol.*, B105, 777, 1981.
7.407. C. Alibert, A. Joullié, A. M. Joullié, and C. Ance, *Phys. Rev.*, B27, 4946, 1983.
7.408. Y. F. Biryulin, S. P. Vul, V. V. Chaldyshev, and Y. V. Shmartzev, *Sov. Phys. Semicond.*, 17, 65, 1983.
7.409. M. Bugajski, A. M. Kontkiewicz, and H. Mariette, *Phys. Rev.*, B28, 7105, 1983.
7.410. Y. F. Biryulin, N. V. Ganina, M. G. Milvidskii, V. V. Chaldyshev, and Y. V. Shmartzev, *Sov. Phys. Semicond.*, 17, 68, 1983.
7.411. R. L. Nahory, M. A. Pollack, J. C. DeWinter, and K. M. Williams, *J. Appl. Phys.*, 48, 513, 1977.
7.412. E. Kuphal, *J. Cryst. Growth*, 67, 441, 1984.
7.413. K. Zbitnew and J. C. Woolley, *J. Appl. Phys.*, 52, 6611, 1981.
7.414. W. Köster and W. Ulrich, *Z. Metallk.*, 49, 365, 1958.
7.415. V. Swaminathan, W. R. Wagner, and P. J. Anthony, *J. Electrochem. Soc.*, 130, 2468, 1983.
7.416. D. Y. Watts and A. F. W. Willoughby, *Mater. Lett.*, 2, 355, 1984.

7.417. V. N. Vigdorovich and A. Y. Nashelskii, *Sov. Powder Metallurgy Metal Ceramics,* 2, 123, 1963.
7.418. N. N. Sirota and L. A. Makovetskaya, *Sov. Phys. Doklady,* 4, 1342, 1960.
7.419. U. Pietsch and D. Marlow, *Phys. Stat. Sol.,* A93, 143, 1986.
7.420. K. Takeda, N. Matsumoto, A. Taguchi, H. Taki, E. Ohta, and M. Sakata, *Phys. Rev.,* B32, 1101, 1985.
7.421. S. Adachi, *J. Appl. Phys.,* 58, R1, 1985; 53, 8755 and 5867, 1985.
7.422. M. S. Abrahams, R. Braunstein, and F. D. Rosi, *J. Phys. Chem. Solids,* 10, 204, 1959.
7.423. K. T. Chan, L. D. Zhu, and J. M. Ballantyne, *Appl. Phys. Lett.,* 47, 44, 1985.
7.424. B. Abeles, *Phys. Rev.,* 131, 1906, 1963.
7.425. L. I. Berger, I. K. Shchukina, and A. E. Balanevskaya, *Trudy IREA (Proc. IREA),* 29, 246, 1966; *Chem. Abstr.,* 67, 57821f, 1967.
7.426. C. Pickering, *J. Electron. Mater.,* 15, 51, 1986.
7.427. W. Kowalsky, H.-H. Wehmann, F. Fiedler, and A. Schlachetski, *Phys. Stat. Sol.,* A77, K75, 1983.
7.428. R. A. Reynolds, *J. Vac. Sci. Technol.,* A7, 269, 1989.
7.429. D. G. Seiler and C. L. Littler, Eds., *Narrow Gap Semiconductors and Related Materials,* Adam Higler Publishing, New York, 1990.
7.430. M. H. Weileer, R. L. Aggarwal, and B. Lax, *Phys. Rev.,* B16, 3606, 1977.
7.431. G. L. Hansen, J. L. Schmit, and T. N. Casselman, *J. Appl. Phys.,* 53, 7099, 1983.
7.432. G. L. Hansen and J. L. Schmit, *J. Appl. Phys.,* 54, 1639, 1983.
7.433. J. Bruce and P. Capper, Eds., EMIS Data Reviews Series No. 3, INSPEC. IEE, New York, 1987, 107.
7.434. W. Giriat, Z. Dziuba, R. Galazka, L. Sosnowski, and I. Zakrzewski, Proc. Int. Congr. Phys. Semicond., Paris, 1964, 1251.
7.435. J. D. Patterson, W. A. Cobba, and S. L. Lehoczky, *J. Mater. Res.,* 7, 2211, 1992.
7.436. S. Roland, A. Lasbley, A. Seyni, R. Grander, and R. Triboulet, *Rev. Phys. Appl.,* 24, 795, 1989.
7.437. J. A. Valdivia and G. E. Barberis, *J. Phys. Chem. Solids,* 56, 1141, 1955.
7.438. L. G. Suslina, A. G. Plyukhin, O. Goede, and D. Hennig, *Phys. Stat. Sol.,* B94, K185, 1979.
7.439. K. Saito, A. Ebina, and T. Takahashi, *Solid State Commun.,* 11, 841, 1972.
7.440. N. P. Gavaleshko, W. Dobrowolski, M. Baj, L. Dmowski, T. Dietl, and V. V. Khomyak, *Physics of Narrow Gap Semiconductors,* J. Rauluszkiewicz, M. Gorska, and E. Kaczmarek, Eds., Elsevier, New York, 1978, 331.
7.441. G. L. Hansen, J. L. Schmit, and T. N. Casselman, *J. Appl. Phys.,* 53, 7099, 1982.
7.442. J. Chu, S. Xu, and D. Tang, *Appl. Phys. Lett.,* 43, 1064, 1983.
7.443. L. Soonckindt, D. Etienne, I. P. Marchand, and L. Lassabatére, *Surf. Sci.,* 86, 378, 1979.
7.444. L. I. Berger and V. M. Petrov, *Trudy IREA (Proc. IREA),* 31, 424, 1969.
7.445. V. A. Tyagai, O. V. Snitko, V. N. Bondarenko, N. I. Vitrikhovskii, V. B. Popov, and A. N. Krasiko, *Fiz. Tverd. Tela,* 16, 1373, 1974.
7.446. P. Chandrasekharam, D. Raja Reddy, and B. K. Reddy, *Phys. Stat. Sol.,* B119, K5, 1983.
7.447. V. M. Burlakov, A. P. Litvinchuk, V. N. Pyrkov, G. G. Tarasov, and N. I. Vitrikhovskii, *Phys. Stat. Sol.,* B128, 389, 1985.
7.448. M. S. Brodin, M. V. Kurik, and N. I. Vitrikhovskii, Proc. Int. Conf. Phys. Semicond., Kyoto, 1966; *J. Phys. Soc. Jpn. Suppl.,* 21, 127, 1966.
7.449. R. Dornhaus and G. Nimtz, *Springer Tracts in Modern Physics,* Vol. 78, 1976; Vol. 98, 1983.
7.450. R. K. Willardson and A. C. Beer, Eds., *Semiconductors and Semimetals,* Vol. 18, Academic Press, New York, 1981.
7.451. R. Mach, P. Frögel, L. G. Suslina, A. G. Aleshkin, J. Maege, and G. Voight, *Phys. Stat. Sol.,* B109, 607, 1982.
7.452. A. Sher, D. Eger, and A. Zemel, *Appl. Phys. Lett.,* 46, 59, 1985.
7.453. F. L. Madarasz and F. Szmulowicz, *J. Appl. Phys.,* 58, 2770, 1985.
7.454. J. P. Stadler, G. Nimtz, and G. Remenyi, *Solid State Commun.,* 57, 459, 1986.
7.455. E. A. Davis, R. E. Drews, and E. L. Lind, *Solid State Commun.,* 5, 573, 1967.
7.456. L. G. Suslina, A. G. Areshkin, V. G. Melekhin, and D. L. Fedorov, *JETP Lett.,* 39, 54, 1984.
7.457. D. J. Olego, J. P. Faurie, S. Srivanathan, and P. M. Raccah, *Appl. Phys. Lett.,* 47, 1172, 1985.
7.458. T. E. Korsak, N. P. Sysoeva, B. M. Ayupov, V. V. Antonov, A. V. Voitsekhovskii, and E. F. Titova, *Sov. Phys. Semicond.,* 19, 222, 1985.
7.459. N. A. Goryunova and N. N. Fedorova, *Sov. Phys. Solid State,* 1, 344, 1959.

7.460. M. Harsy and I. Bertoti, *Phys. Stat. Sol.,* 11, K135, 1965.
7.461. A. V. Voitsekhovskii, F. P. Kesamanly, V. K. Mityurev, and Y. V. Rud, *Ukrain. Fiz. Zh.,* 10, 1349, 1965.
7.462. O. P. Gorodnichii, I. N. Gorbatyuk, I. M. Rachenko, and R. M. Vinetskii, *Phys. Stat. Sol.,* B130, K169, 1985.
7.463. A. Addamiano, *J. Electrochem. Soc.,* 107, 1006, 1960.
7.464. I. Bertoti, E. Barta, J. Schanda, and P. Sviszt, Proc. Int. Conf. Luminescence, Budapest, 1966, 1102.
7.465. W. M. Yum, *J. Appl. Phys.,* 40, 2617, 1969.
7.466. H. Sonomura, T. Uragaki, and T. Miyauchi, *Jpn. J. Appl. Phys.,* 12, 968, 1973.
7.467. A. Shiutani and S. Minagawa, *J. Phys. Chem. Solids,* 34, 911, 1973.
7.468. W. M. Yim, J. P. Dismukes, and H. Kressel, *RCA Rev.,* 31, 662, 1970.
7.469. R. Hill, *Thin Solid Films,* 27, L1, 1975.
7.470. R. Hill, *Thin Solid Films,* 34, 395, 1976.
7.471. A. V. Voitsekhovskii, V. S. Manzhara, and T. P. Stetsenko, *Ukrain. Fiz. Zh.,* 24, 1055, 1979.
7.472. M. Glickman, A. Catalano, and A. Wold, *Solid State Commun.,* 49, 799, 1984.
7.473. M. Glickman and W. D. Kraeft, *Solid State Electron.,* 28, 151, 1985.
7.474. S. M. Ku and L. J. Bodi, *J. Phys. Chem. Solids,* 29, 2077, 1968.
7.475. N. A. Borisov, V. M. Lakeenkov, M. G. Milvidskii, and O. V. Pelevin, *Sov. Phys. Semicond.,* 10, 948, 1976.
7.476. A. V. Voitsekhovskii and V. P. Drobyazko, *Ukrain. Fiz. Zh.,* 12, 461, 1967.
7.477. A. V. Voitsekhovskii, V. P. Drobyazko, V. K. Mityurev, and V. P. Vasilenko, *Inorg. Mater.,* 4, 1681, 1968.
7.478. N. A. Goryunova and V. I. Sokolova, *Izv. Moldav. Filiala Akad. Nauk SSSR,* No. 3(69), 31, 1960.
7.479. A. V. Voitsekhovskii and A. D. Pashun, *Ukrain. Phys. Zh.,* 18, 1558, 1973.
7.480. V. P. Drobyazko, *Sov. Phys. Semicond.,* 12, 354, 1978.
7.481. V. A. Anishchenko, A. V. Voitsekhovskii, and V. E. Sharenko, *Ukrain. Fiz. Zh.,* 18, 1735, 1973.
7.482. V. A. Anishchenko, A. V. Voitsekhovskii, and A. D. Pashun, *Sov. Phys. Semicond.,* 13, 850, 1979.
7.483. A. V. Voitsekhovskii, Thesis, State Polytechnic Institute, Kiev, 1965.
7.484. A. D. Stuckes and R. P. Chasmar, *J. Phys. Chem. Solids,* 25, 469, 1964.
7.485. A. V. Voitsekhovskii and V. P. Drobyazko, *Inorg. Mater.,* 3, 1976, 1967.
7.486. N. A. Semikolenova and E. N. Khabarov, *Izv. VUZ Fiz. (U.S.S.R.),* 16, 76 and 82, 1973; *Sov. Phys. J. (U.S.A.),* 16, 801 and 807, 1973.
7.487. G. I. Buzevich, L. A. Skorobogatova, and E. N. Khabarov, *Sov. Phys. Semicond.,* 7, 1389, 1974.
7.488. V. N. Morozov, A. L. Petrov, and L. A. Skorobogatova, *Sov. Phys. Semicond.,* 11, 1192, 1977.
7.489. A. L. Petrov and G. A. Kuzmina, *Sov. Phys. Semicond.,* 14, 604, 1980.
7.490. G. A. Kuzmina and E. N. Khabarov, Proc. 24th Sci. Conf. Leningr. Inst. of Constr. Engineers, Sect. Physics, 1966, 21.
7.491. G. A. Kuzmina, M. V. Kurik, V. S. Manzhara, and E. N. Khabarov, *Sov. Phys. Semicond.,* 6, 1579, 1973.
7.492. E. Khabarov, A. Inyutkina, E. Kolosov, L. Osnach, V. Khabarova, and P. Sharavskii, *Izv. Akad. Nauk SSSR Fizika,* 28, 1010, 1964.
7.493. L. A. Skorobogatova and E. N. Khabarov, *Sov. Phys. Semicond.,* 8, 257, 1974.
7.494. L. A. Skorobogatova, A. L. Petrov, and G. A. Kuzmina, *Sov. Phys. Semicond.,* 13, 480, 1979.
7.495. A. L. Petrov, *Sov. Phys. Semicond.,* 15, 339, 1981.
7.496. L. A. Osnach, *Research in Mathematical and Experimental Physics,* Leningrad Institute of Construction Engineers, 1965.
7.497. L. I. Berger, 5th Int. Conf. Vapor Growth and Epitaxy, Abstracts, Coronado, CA, 1981, 331.
7.498. L. I. Berger, Quaternary Narrow-Band Semiconductors $(HgTe)_x(InSb)_{1-x}$ for Far-Infrared Detectors, SDSU and Office of Naval Research, Grant No. N00014-83-K-0588, Final Report, 1986.
7.499. V. P. Ambros and I. I. Burdiyan, Proc. State Pedag. Inst., Tiraspol, 1966, 12.
7.500. V. N. Morozov and V. G. Chernov, *Inorg. Mater.,* 15, 1036, 1979.
7.501. G. S. Oleinik, P. A. Mizetskii, A. I. Nizkova, L. A. Polivtsev, and I. A. Ryadnina, *Izv. Akad. Nauk SSSR Neorg. Mater.,* 19, 1799, 1983; *Inorg. Mater.,* 19, 1587, 1983.
7.502. M. Jain, H. K. Sehgal, and A. V. R. Warrier, *Phys. Stat. Sol.,* A77, 67, 1983.
7.503. B. A. Akimov, N. B. Brandt, L. I. Ryabova, D. R. Khokhlov, S. M. Chudinov, and O. B. Yatsenko, *Sov. Phys. JETP Lett.,* 31, 279, 1980.

7.504. S. Takaoka, T. Itoga, and K. Murase, *Jpn. J. Appl. Phys.*, 23, 216, 1984.
7.505. H. Preier, M. Bleicher, W. Riedel, H. Pfeiffer, and H. Maier, *Appl. Phys.*, 12, 277, 1977.
7.506. T. Ishiguro and K. Yamayi, *Organic Superconductors. Series in Solid State Sciences,* Springer-Verlag, Berlin, 1990.
7.507. A. E. Kovalev, S. I. Pestoskii, A. Gilevskii, and N. D. Kushch, *JETP Lett.*, 59, 560, 1994.
7.508. H. Akamatu, H. Inokuchi, and Y. Matsunada, *Nature*, 173, 168, 1954.
7.509. H. Meier, *Die Photochemie der Organischen Farbstoffe (Photochemistry of Organic Dyes),* Springer-Verlag, New York, 1963.
7.510. J. E. Katon, Ed., *Organic Semiconducting Polymers,* Marcel Dekker, New York, 1968.
7.511. N. B. Hannay, Ed., *Treeatse on Solid State Chemistry,* Vol. 3, *Crystalline and Non-Crystalline Solids,* Plenum Press, New York, 1976.
7.512. C. L. Braun, Organic Semiconductors, in *Handbook of Semiconductors,* Vol. 3, S. P. Keller, Ed., North-Holland, Amsterdam, 1980.
7.513. J. Simon and J. J. André, *Molecular Semiconductors,* Springer-Verlag, New York, 1985.
7.514. O. Inganäs and I. Lundström, *Synth. Metals,* 10, 5, 1984/1985.
7.515. *Survey of Semiconductor Physics,* Vol. 1, Van Nostrand Reinhold, New York, 1990, 927.
7.516. D. D. Eley and G. D. Parfitt, *Trans. Faraday Soc.*, 51, 1529, 1955.
7.517. P. E. Fielding and F. Gutman, *J. Chem. Phys.*, 26, 411, 1957.
7.518. W. Feldmayer and G. Wolf, *J. Electrochem. Soc.*, 105, 141, 1958.
7.519. A. T. Vartanyan and I. A. Karpovich, *Zh. Fiz. Khim.*, 32, 178, 1958.
7.520. A. T. Vartanyan and L. D. Rozenshtein, *Izv. Akad. Nauk SSSR Ser. Fiz.*, 29, 428, 1961.
7.521. C. Hamman and J. Storbeck, *Naturwissenschaften,* 50, 327, 1963.
7.522. S. D. Nasirdinov, E. A. Shugam, L. I. Berger, V. E. Plyushchev, and L. P. Shklover, *Zh. Fiz. Khim.*, 40, 741, 1966.
7.523. A. Eptsein and B. S. Wildi, *J. Chem. Phys.*, 32, 324, 1960.
7.524. E. L. Frankevich, L. I. Busheva, E. I. Balabanov, and L. G. Cherkashina, *Vysokomolekul. Soedin.*, 5, 1684, 1963.
7.525. H. Meier, W. Albrecht, and E. Zimmerhackl, *Polymer Bull.*, 13, 43, 1985.
7.526. S. D. Nasirdinov, E. A. Shugam, L. I. Berger, L. P. Shklover, and M. Z. Gurevich, *Zh. Fiz. Khim.*, 40, 2614, 1966.
7.527. T. Ito, H. Shirakawa, and S. Ikeda, *J. Polym. Sci. Polym. Chem.*, 12, 11, 1974.
7.528. J. H. Edwards and W. J. Feast, *Polymer,* 21, 595, 1980.
7.529. J. H. Edwards, W. J. Feast, and D. C. Bott, *Polymer,* 25, 395, 1984.
7.530. K. Yoshino, S. Ura, S. Sasa, K. Kaneto, and Y. Inuishi, *Jpn. J. Appl. Phys.*, 21, L507, 1982.
7.531. P. M. Grant, T. Tani, W. D. Gill, M. Krounbi, and T. C. Clarke, *J. Appl. Phys.*, 52, 869, 1981.
7.532. B. R. Weinberger, M. Akhtar, and S. C. Gau, *Synth. Met.*, 4, 187, 1982.
7.533. J. Tsukamoto, H. Ohigashi, K. Matsumura, and A. Takahashi, *Synth. Met.*, 4, 177, 1982.
7.534. D. L. Peebles, J. S. Murday, D. C. Weber, and J. Milliken, *J. Phys. (Paris) Colloq.*, 44, C3-591, 1983.
7.535. K. K. Kanazawa, A. F. Diaz, W. D. Gill, P. M. Grant, G. B. Street, and G. P. Gardini, *Synth. Met.*, 1, 329, 1979/1980.
7.536. O. Inganas and I. Lundstrom, *Synth. Met.*, 10, 5, 1984/1985.
7.537. E. A. Hassan, *Plastic Technol.*, 6, 67, 1981.
7.538. C. K. Chiang, M. A. Druy, S. C. Gau, A. J. Heeger, E. J. Louis, A. G. MacDiarmid, and Y. W. Park, *J. Am. Chem. Soc.*, 100, 1013, 1987.
7.539. C. R. Fincher, Jr., C.-E. Chen, A. J. Heeger, A. G. MacDiarmid, and J. B. Hastings, *Phys. Rev. Lett.*, 48, 100, 1982.
7.540. J. C. Chien, F. E. Karasz, and K. Shimamura, *J. Polym. Sci. Polym. Lett. Ed.*, 20, 97, 1982.
7.541. N. Basescu, Z.-X. Liu, D. Moses, A. J. Heeger, H. Naarmann, and N. Theophilou, *Nature,* 327, 403, 1987.
7.542. A. G. MacDiarmid, M.-C. Chiang, M. Halpern, W. S. Huang, J. R. Krawczuk, R. J. Mammone, S. L. Mu, N. L. Somasiri, and W. Wu, *Polym. Prepr. Am. Chem. Soc. Div. Polym. Chem.*, 25, 248, 1984.
7.543. J. Bargon, S. Mohmand, and R. J. Waltman, *Mol. Cryst. Liq. Cryst.*, 93, 279, 1983.
7.544. G. Tourillon and F. Garnier, *J. Electroanal. Chem.*, 135, 173, 1982.

7.545. D. M. Ivory, G. G. Miller, J. M. Sova, L. W. Shacklette, R. R. Chance, and R. H. Baughman, *J. Chem. Phys.*, 71, 1506, 1979.
7.546. S. W. Shacklette, H. Eckhardt, R. R. Chance, G. G. Miller, D. M. Ivory, and R. H. Baughman, *J. Chem. Phys.*, 73, 4098, 1980.
7.547. J. F. Rabolt, T. C. Clarke, K. K. Kanazawa, J. R. Reynolds, and G. B. Street, *J. Chem. Soc. Chem. Commun.*, p. 347, 1980.
7.548. R. R. Chance, L. W. Shacklette, G. G. Miller, D. M. Ivory, J. M. Sowa, R. L. Elsenbaumer, and R. H. Baughman, *J. Chem. Soc. Chem. Commun.*, p. 348, 1980.
7.549. T. C. Clarke, K. K. Kanazawa, V. Y. Lee, J. F. Rabolt, and G. B. Street, *J. Polym. Sci. Polym. Phys.*, 20, 117, 1982.
7.550. I. Murase, T. Ohnishi, T. Noguchi, and M. Hirooka, *Polym. Commun.*, 25, 327, 1984.
7.551. G. E. Wnek, J. C. W. Chien, F. E. Karasz, and C. P. Lillya, *Polym. Commun.*, 20, 1443, 1979.
7.552. D. R. Gagnon, J. D. Capistran, F. E. Karasz, and R. W. Lenz, *Polym. Bull.*, 12, 293, 1984.
7.553. H. Meier, W. Albrecht, and E. Zimmerhackl, *Polym. Bull.*, 13, 43, 1985.
7.554. A. F. Diaz, K. K. Kanazawa, and G. P. Gardini, *J. Chem. Soc. Chem. Commun.*, p. 635, 1979.
7.555. G. B. Street, T. C. Clarke, M. Kroubni, K. K. Kanazawa, V. Lee, P. Pfluger, J. C. Scott, and G. Weiser, *Mol. Cryst. Liq. Cryst.*, 83, 253, 1982.
7.556. A. Watanabe, M. Tanaka, and J. Tanaka, *Bull. Chem. Soc. Jpn.*, 54, 2278, 1981.
7.557. G. Kossmehl and G. Chatzitheodorou, *Makromol. Chem. Rapid. Commun.*, 2, 551, 1981.
7.558. M. M. Labes, P. Love, and L. F. Nickols, *Chem. Rev.*, 79, 1, 1979.
7.559. E. A. Silinsh, *Organic Molecular Crystals,* Springer-Verlag, New York, 1980.
7.560. A. I. Belkind and V. V. Grechov, *Phys. Stat. Sol.*, A26, 377, 1974.
7.561. T. R. Walton, J. R. Griffith, and J. P. Reardon, *J. Appl. Polym. Sci.*, 30, 2921, 1985.
7.562. T. M. Keller, *J. Polym. Sci.*, C24, 211, 1986.
7.563. E. A. Shugam, L. I. Berger, E. G. Rukhadze, and G. V. Panova, *Zh. Phys. Khim.*, 39, 481, 1965.
7.564. T. Siegrist, R. M. Fleming, R. C. Haddon, R. A. Laudise, A. J. Lovinger, H. E. Katz, P. Bridenbaum, and D. D. Davis, *J. Mater. Res.*, 10, 2170, 1955.
7.565. V. N. Topnikov, S. I. Pesotskii, and V. N. Laukhin, *JETP Lett.*, 59, 374, 1994.
7.566. N. A. Bakr and M. I. Abdel-hamid, *J. Mater. Res.*, 10, 2653, 1995.

CHAPTER 8

Semiconductor Devices

Semiconductors are used as the basic materials of the majority of today's electronic, optoelectronic, and thermoelectronic devices. The main consumer of the electronic devices now is the computer industry; introduction of personal computers and continuously growing demand for them and related electronic systems have increased the world annual production of devices such as transistors to 10^{16} to 10^{17} units. This astronomical demand has been followed by introduction of many very impressive technological and circuit innovations which caused the price of a single semiconductor device to decrease from about $1.00 in 1969 to $0.000001 to $0.0000005 in 1994, a million plus decrease.

Simultaneously with the development of a great variety of transistor-based devices and integrated circuits, we see continuous progress taking place in the development of the technology of manufacturing of light-emitting diodes and semiconductor lasers, solar energy conversion devices, thermoelectric converters, and thermoelectric heat pumps.

8.1. p-n JUNCTION AND SCHOTTKY BARRIER

Contact of a metal and a semiconductor or the contact of two semiconductors with different conductivity types is the basic element of the majority of electronic devices designed for rectification, amplification, switching, electro-optic (opto-electronic) conversion, and other operations in electronic and electrical circuits.

8.1.1. p-n Junction

A p-n junction is (1) an electrical contact between an n-type and a p-type semiconductor or (2) a border inside of a piece of an initially intrinsic semiconductor crystal that was created by the diffusion of donor impurities through one surface, and acceptor impurities through the opposite surface of the crystal. Of course, the interface between two pieces of semiconductors has a disadvantage related to the additional electrical resistance caused by the microscopic irregularities on the contacting surfaces: a mechanical contact usually occupies but a few percent of the total interface area. One of the solutions to this problem is the so-called epitaxial growth of, say, n-type material on a carefully chosen and well-prepared surface of a p-type material from either the liquid or vapor phase. In this process, the crystal lattice of the properly chosen deposited material (almost) perfectly matches the crystal lattice of the substrate and the contact electrical resistance is small. Formation of a p-n junction in the bulk of the same crystal virtually excludes the additional resistance, but the transition from n- to p-part of the crystal may not be abrupt.

There are quite a few methods which may be utilized for p-n junction fabrication. The grown junction technique was a method most commonly used in the beginning period of the development of semiconductor devices. The crystal was grown from the melt containing dopants of one kind (say, donors). At a certain moment in the process, an acceptor impurity was added in the melt in the amount higher than the concentration of the initial dopant. The part of the crystal which grows after this moment has the conductivity of the opposite sign. This method has been widely used for junction fabrication in silicon; the commonly used dopants are phosphorus (donor) and boron (acceptor).

Another process, which has been commonly used in manufacturing of germanium-based diodes and transistors, consists of melting a pellet of indium (T_m = 429.8 K) on the surface of an n-Ge crystal at about 430 K to 435 K; liquid indium dissolves an amount of germanium, forming a dip in the germanium crystal filled with the Sn-Ge alloy. With the following temperature decrease, the solubility of Ge in In decreases, and the germanium atoms from the melt are deposited on the walls of the dip, forming a regrown area in the initial crystal that has the same atom arrangement as the rest of the crystal, but with p-type conductivity originated from the presence of indium atoms dissolved in the regrown germanium. The only condition to fabricate a p-n junction is the donor concentration in the initial germanium crystal, N_D, to be lower than the indium atom (acceptor) concentration, N_A, in the regrown volume of germanium. This process is similar to the liquid phase epitaxy, LPE (from Greek επι — on, upon; and ταξη — order, succession, regularity), with the difference that in LPE the initial crystal is not dissolved but is serving as a substrate for a liquid solution containing atoms of the substrate. At a proper temperature, these atoms, being captured by the dangling bonds of the atoms on the surface of the substrate, form the new layer of the substrate which, in its turn, provides dangling bonds for capturing the atoms of the next layer, and so on. The growing layers have atom arrangement exactly the same as in the underlying layer. The same method can be used for growth of the layers of one semiconductor on the surface of another semiconductor. The conditions of successful growth are, first of all, the lattice match between the materials of the substrate and the growing layer and similarity of the nature of the chemical bond in them.

A p-n junction may be grown in a semiconductor crystal as a result of diffusion of a dopant into it. One of the most common methods of the diffused p-n junction formation, which is widely used now in integrated circuits fabrication, is diffusion from the gas phase. If a surface of a semiconductor plate (wafer) is exposed at elevated temperature to the dopant gas, the atoms of the gas may be captured by the surface, and then they may migrate in the direction of the negative gradient of their concentration, i.e., into the bulk of the plate. At high-enough temperature, some of the host atoms are leaving their equilibrium lattice sites and the vacancies they left may be occupied by the dopant atoms. If the initial plate contains, say, donor impurities, and the diffusing atoms are acceptors, the diffused acceptor concentration may be high enough to compensate the donors ($N_A > N_D$) up to a certain depth from the plate surface, at which $N_A = N_D$. This is the location of the p-n junction.

The disadvantage of this technology, namely, the high temperature of the host plate, is eliminated in the ion implantation method of the p-n junction formation in which a beam of chosen doping ions, accelerated to kinetic energies high enough (up to several MeV) to penetrate into a semiconductor plate to a desirable depth, is focused at a selected spot of the plate surface. The ions penetrate into the crystal lattice forming a p-n junction at that depth with very high accuracy. Unfortunately, the bombarding ions create lattice damage that ought to be healed by annealing, but the annealing temperatures are relatively moderate. The annealing creates difficulties related to the possible dissociation of the host material if it is a compound; this problem may be solved by the proper incapsulation of the annealed part or by annealing locally only the damaged spots using a well-focused laser beam. For reviews of the p-n junction technology, see References 8.1 to 8.6.

SEMICONDUCTOR DEVICES

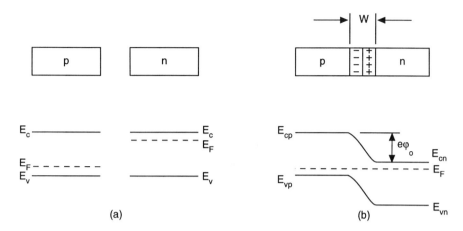

Figure 8.1 Schematic presentation of the energy bands in semiconductors (E_c — the conduction band bottom, E_v — the valence band top, E_F — Fermi level). (a) Energy gaps of a p-type and an n-type extrinsic semiconductor; (b) band diagram for a p-n junction in equilibrium.

8.1.2. p-n Junction Diode

An electronic device utilizing electronic properties of p-n junction is called a p-n junction diode. It consists of a semiconductor crystal with a p-n junction in it and two so-called ohmic metal-semiconductor contacts separated by the junction and serving as the connection of the diode to the external electronic circuits. A DC voltage applied between the contacts is called a bias. To understand the work principles of a p-n junction diode, let us consider first the distribution of electrons and holes in and around the junction at zero bias and thermal equilibrium (see, e.g., References 8.7 and 8.8). If the n-type and p-type parts of the semiconductor on both sides from the p-n junction were mechanically separated, their energy gap would look as in Figure 8.1a. The Fermi level may be considered, as was mentioned in Chapter 2, as the electron chemical potential. When the two parts are connected, the Fermi levels on both sides from the junction must be equal (like the water levels in interconnected containers). This requirement causes the deformation of the lower and upper borders of the energy gap in the area of the junction: the minimum energy of conducting electrons at the p-side of the junction is greater than in the n-side, i.e., a potential barrier is formed (Figure 8.1b) with the height

$$\varphi = \frac{k_B T}{e} \ln \frac{N_A N_D}{n_i^2} \tag{8.1}$$

where k_B is Boltzmann's constant, e is the electron charge, and n_i is the intrinsic charge concentration.

The origin of the barrier can be understood from the consideration of the processes that take place during the junction formation. From the moment of (electric) contact of two parts of the junction, the electrons from the n-part start to diffuse into the p-part and holes from the p-part will move in the opposite direction, because the electron concentration in the n-side, n_n, is much greater than that on the p-side, n_p, and the hole concentration on the p-side, p_p, is much greater than that on the n-side, p_n ($n_n \gg n_p$, $p_p \gg p_n$). When electrons leave the n-area, a positive space charge is formed by the positively charged ionized donors left uncompensated by the electrons near the junction. On the p-side of the junction, a negative space charge is formed by the ionized acceptor atoms, whose charge is not compensated as a result of diffusion of the holes through the junction. The migrated electrons and holes form an electric field

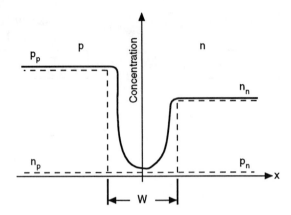

Figure 8.2 Schematic presentation of the carrier concentration distribution in a p-n junction in equilibrium.

with direction from the n-part to the p-part, with its intensity growing with migration of the electrons and holes until it stops the charges from moving through the junction; adjacent to the junction is a layer of the high electrical resistivity (barrier) filled with ionized atoms of donors and acceptors.

The real distribution of the free carrier concentration in and around the depletion area (solid line in Figure 8.2) can be replaced without a noticeable decrease in accuracy by the distribution shown by the dash lines. Using the Poisson equation, it is possible to calculate the total depth of the space charge area penetration into the semiconductor crystal from the p-n junction (depleted area, transition area) at equilibrium, i.e., at $V_{bias} = 0$:

$$W_o = \left(\frac{\varepsilon_s \varepsilon_o}{e} \cdot \frac{N_A + N_D}{N_A N_D} \right)^{1/2} \qquad (8.2)$$

where $\varepsilon_s \varepsilon_o$ is the dielectric constant of the semiconductor.

Equation 8.2 is derived with the assumption of a sharp border between the parts containing donor and acceptor impurities, i.e., abrupt replacement of donors by acceptors. For a gradual transition from n-section to p-section of the crystal, with a linear dependence of the impurity concentration on the distance, $d(N_D - N_A)/dx = G = $ constant, Equation 8.2 can be written in the form:[8,10]

$$W_o = (12 \varepsilon_s \varepsilon_o \varphi / eG)^{1/3} \qquad (8.3)$$

In this case, the assumption is $N_{An} = N_{Dp} = 0$ (donors are only in the n-part and acceptors are only in the p-part), $G = (N_A + N_D)/W_o$.

When an external voltage is applied to the p-n junction, the height of the potential barrier changes and the transition is not in equilibrium. If the voltage, V, increases the barrier height as a result of application of the negative potential to the p-section (V < 0), the flux of the majority carriers through the junction decreases up to zero, while the flux of the minority carriers is not changing (they never have a barrier) (see Figure 8.2). The minority carrier currents depend on the rate of the thermal generation of the electron-hole pairs. Per unit volume, this rate is equal to p_n/τ_p in the n-section and to n_p/τ_n in the p-section (τ_n and τ_p are the lifetimes of the pairs in the n- and p-section, respectively). These pairs migrate toward the barrier and are divided by its field which creates an electric current through the junction. Only the minority charges that were formed at the distance from the junction equal or smaller than diffusion length, L_p and L_n, can reach and pass the barrier; the rest of them will recombine before and will not participate in the current. The diffusion length is equal to $L_n = (D_n\tau_n)^{1/2}$ for electrons in p-section and $L_p = (D_p\tau_p)^{1/2}$ for holes in n-section; D is the diffusion coefficient. At V < 0, the current virtually does not depend on V and is equal to

$$I_s = eA\left(p_n L_p/\tau_p + n_p L_n/\tau_n\right) \quad (8.4)$$

where A is the area of p-n junction. This current is called the saturation current.

When V > 0 (the "plus" of the outer current source is connected to p-section), the height of the barrier decreases and the majority charges can pass (inject) through the junction to its other side, where they become the minority and recombine at the distance L_n or L_p from the barrier. Their motion through the p-n junction creates a current in the outer circuit. The concentration of the injected carries is proportional to $\exp(eV/k_BT)$; the total current is equal to

$$I = I_s\left[\exp(eV/k_BT) - 1\right] \quad (8.5)$$

When V is positive and $\exp(eV/k_BT) \gg 1$, the current through the p-n junction diode is an exponential function of V; when V < 0, even at relatively low absolute magnitudes of V, $\exp(eV/k_BT) \ll 1$ and current is independent of the voltage drop between the n- and p-parts of the diode. Equation 8.5 is called the current-voltage characteristic of a p-n diode.

Equation 8.4 is valid if the distance, t, between the p-n transition area and metal (base) electrodes is greater than the diffusion length, L_p and/or L_n. If one or both bulk semiconductor parts of the p-n junction diode are thin (t < L), a part of the injected carriers reaches the base electrode(s) and recombines there. Then the magnitude of τ in Equation 8.4 has to be replaced by τ_{eff} which is equal to

$$\tau_{eff} = \frac{1+(sL/D)t/L}{sL/D+tL}\tau \quad (8.6)$$

where s is the recombination rate.

Presence of the high-resistance depletion area between two conductive parts in a p-n junction diode makes it similar to a capacitor with the capacitance per unit area, C'_d, equal to

$$C'_d = \left[\frac{\varepsilon\varepsilon_o eN_A N_D}{(N_A+N_D)(\varphi-V)}\right]^{1/2} \quad (8.7)$$

Equation 8.7 is the base for a method of experimental determination of the diffusion voltage, φ, by plotting $1/C^2$ vs. bias voltage, V.

When an AC voltage is applied to a p-n junction diode simultaneously with a bias, the diode behaves as a complex resistance with the capacitive reactance (the alternative voltage lags the alternative current).

8.1.3. Metal-Semiconductor Rectifying Contact. Schottky Barrier Diode

A contact of a metal and a semiconductor can also be used to fabricate a diode (Schottky diode). When a metal with the work function $e\varphi_m$ is brought in contact with an n-type semiconductor, the work function of which is $e\varphi_s$ ($e\varphi_m > e\varphi_s$), the electrons from the semiconductor side near the junction transfer to the metal until the Fermi levels of both metal and semiconductor have equalized. The electron departure increases the potential of the semiconductor (it is charged positively by the positive donor ions left uncompensated by the migrated electrons). As a result, a depletion area of the width W is formed in the semiconductor (there is no change in the electron distribution on the metal side of the contact because of a very high concentration of free electrons in the metal). The positive charge of the donors matches the excessive negative charge on the metal. If a metal is in contact with a p-type semiconductor and $e\varphi_m < e\varphi_s$, the electrons migrate through the junction from the metal to semiconductor. This forms in the semiconductor a layer of the width W depleted of the free charges but filled with (nondrifting) negatively charged acceptor ions. This ion layer forms a positively charged area on the metal side of the junction.

The migration of the charges forms an electric field in the depleted area, which will increase until an equilibrium is reached when this electric field prevents the charges from diffusing through the junction. The equilibrium potential drop across the depleted area is equal to $V_o = \varphi_m - \varphi_s = \varphi_b$. The barrier thus formed is called the Schottky barrier, and the reader understands that the application of an external voltage across the barrier in one or another direction will either expand or contract the width of the depletion area, decreasing or increasing the electrical resistance of the junction, i.e., the contact will be rectifying. The electric current through the junction area A obeys the formula similar to that for a p-n junction diode:

$$I = I_o \left[\exp(eV/k_B T) - 1 \right] \quad (8.8)$$

where $I_o = \dfrac{\alpha A e m_n^* (k_B T)^2}{2\pi^2 \hbar^3} \exp(-e\varphi_b / k_B T)$ is an empirical constant (Crowell and Sze[8.9] suggest $\alpha \simeq 0.5$).

The basic applications of the Schottky diodes and p-n junction diodes are similar. Some information pertinent to fabrication of Schottky barriers on different semiconductors is presented in Table 8.1. The use of Schottky barriers on GaAs is partially related to the difficulty of producing good insulating layers between metal electrodes and GaAs for the gate preparation on the field-effect transistors — the devices currently having the greatest demand by the electronic industry. One of possible solutions is deposition of silicon on GaAs crystal by, e.g., chemical vapor deposition technique with the following: (1) thermal oxidation of the deposited Si and (2) deposition of the metal electrode on the oxide layer.

Table 8.1 Fabrication of Schottky Barriers on Various Semiconductors

Semiconductor	Surface preparation	Schottky metal (with barrier height, eV)	Substrate temperature (°C)
n-ZnS	1% Br in CH_3OH	Ag, Au, Cr	100
CdS		Pt (0.88), Au (0.73–0.83)	
p-$Hg_{1-x}Cd_x$Te		Al, Cr	
p-PbS	Annealing in vacuum (170°C, 30 min)	In, Pb	25
p-PbSe	Same	In, Pb	25
p-$PbS_{1-x}Se_x$	Same	In, Pb	25
p-PbTe	Etching in HBr/Br_2 solution	Cu, Zn, In, Pb	Cooled
p-$Pb_{1-x}Sn_x$Se $0.05 < x < 0.70$	Annealing in vacuum (170°C, 30 min)	Pb	25
p-$Pb_{1-x}Sn_x$Te $x = 0.21$	Etching in HBr/Br_2 solution	In, Pb	25

Mainly from References 8.15 and 8.18 to 8.20.

SEMICONDUCTOR DEVICES

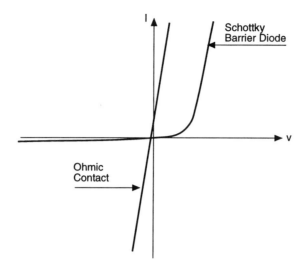

Figure 8.3 Current-voltage characteristics of an ohmic contact and a Schottky barrier diode.

8.2. OHMIC CONTACT

Virtually no electronic or electro-optic device can be fabricated without so-called ohmic contacts, i.e., contacts of semiconductors with metals in which current is a linear function of the applied voltage (Ohm's law) that does not depend on the current direction. The current-voltage characteristics of a Schottky barrier and an ohmic contact are shown in Figure 8.3. Another two quite common requirements for an ohmic contact are (1) its low electrical resistance and (2) its reasonably low dependence on temperature.

The ways to fabricate a satisfactory ohmic contact to an n-type semiconductor are (1) to use a metal or alloy that meets the requirement $\varphi_m < \varphi_s$ or (2) to introduce a high amount of donor impurities in the thin layer of the semiconductor adjacent to the contact with the metal. The first method is more difficult because only a very few metallic materials which possess properties adequate for their applications have lower work function than the commonly used semiconductors. The second method is more practical. If in a heavily doped area in close proximity to the contact, the donor concentration, n⁺, is high enough for the thickness, W, of the depletion area to become so small that electrons cross the barrier by tunneling, the barrier becomes equally passable for them and the resistance in both directions of the current remains the same. The correlation between the specific contact conductance, g_c, defined as

$$g_c = \frac{1}{r_c} = \frac{1}{A}\frac{dI}{dV}\bigg|_{V \to 0}$$

and the material parameters, such as the dielectric constant, effective mass of majority charges, and $\varphi_b = \varphi_m - \varphi_s$, is discussed in References 8.10, p. 304 ff and 8.11, p. 210 ff.

Currently, the semiconductors most commonly used by the electronic devices industry are silicon and gallium arsenide. Table 8.2 contains information regarding the electrode materials for silicon including some technological parameters.

A brief qualitative description of the criteria which have to be used for the choice of proper material and processes for Schottky and ohmic contact fabrication for silicon and GaAs is given in Reference 8.15, p. 998 ff. Fabrication of ohmic electrodes on GaAs has a few specific difficulties related to decomposition of GaAs at elevated temperatures, the large barrier height at the contact of GaAs with a majority of metals (Table 8.3), and high specific

Table 8.2 Metals and Alloys Used as Electrode Material for Silicon-Based Devices. Physical and Physicochemical Properties

Material	Melting point (K)	Coefficient of thermal expansion (10^{-6} K^{-1})	Barrier height to silicon (eV) n-type	Barrier height to silicon (eV) p-type	Electrical conductivity (10^4 S/cm)	Reacts with Si at (K)	Stable on Si up to (K)	Sintering temperature (K)
La	993	4.9 ± 0.1	0.4	0.7	1.8			
Y	1796	9.3 (25–1000°C)	0.4	0.7	1.4–1.5			
Ti	1941	8.1	0.50	0.61	1.8			
Cr	2180	6.2	0.61		5.3–7.9			
W	3695	30	0.67	0.45	17.7			
Mo	2893	5	0.68	0.42	6.7–17			
Al	933	23	0.72	0.58	33–37	~520	~520	
Au	1337.33	14.2	0.8	0.35	44–45			
Pt	2041.55	8.9	0.9		8.6–9.3			
MoSi$_2$	2253	8.25	0.55		1–2.5	798	>1270	1273
CrSi$_2$			0.57			723		
TaSi$_2$	~2470	8.8–10.7	0.6		2.0–2.6	923	≥1270	1273
TiSi$_2$	1813	12.5	0.6		6.3–7.7	823	≥1220	1173
TiSi						773		
NbSi$_2$			0.62			923		
CoSi						648–773		
CoSi$_2$	1593	10.14	0.65		5.6–10	623–773	≤1223	1173
WSi$_2$	2438	6.25–7.9	0.65		1.4–3.3	923	≥1270	1273
Ni$_2$Si						473–623		
NiSi$_2$	1226	12.06	0.7		~2.0	≥1020	≤1120	1173
Pd$_2$Si			0.75					
PdSi						1123		
PdSi$_2$					2.0–3.3	373–573		673
PtSi$_2$	1502						≤1220	873–1073
PtSi			0.84		2.9–3.6	573		
Pt$_2$Si	1671				2.9–3.3			
VSi$_2$						873		
MnSi						1073		
MnSi$_2$						673–773		
FeSi$_2$						823		
FeSi						723–823		
ZrSi$_2$					2.5–2.9	973		
RhSi						623–698		
HfSi$_2$					2.0–2.2	1023		1173
HfSi						823–973		
IrSi$_2$						1273		
IrSi$_{1.7}$						773–1273		
IrSi						673–773		
TiB$_2$					10–17	>870	>870	
TiC	3530				~1	723–773	723	
TiN	3223				0.66–2.5	723–773	723	
ZrN	3253				1.0–5.0	723–773	723	
HfN	3270				1.0–3.3	723–773	723	
TaC	4258				~1.0			
TaN	3360				0.5	723–773	723	
NbN	2573				~2	723–773	723	

Mainly from References 8.10 to 8.15.

electrical resistance of the contacts (the metallization technique used to produce ohmic contacts on GaAs[8.15,8.16] is presented in Table 8.4). The same difficulties have to be overcome with the other III-V wide-energy gap semiconductors. Table 8.5 contains some information regarding the ohmic contact technology for III-V, III-III-V, and III-V-V compounds.

Table 8.3 Metal-GaAs Barrier Height (in eV) for Some Metals and Alloys at 300 K

Metal (alloy)	n-GaAs	p-GaAs	Metal (alloy)	n-GaAs	p-GaAs
Cu	0.82		Ti	0.82	
Ag	0.93	0.44	Ti–Ag	0.80	
Au	0.95	0.48	Sb	0.86	
Au-Ga	0.71–0.75		Bi	0.89	
Au-Ge	0.27–0.35		Cr	0.77	
Au-Sn	0.72		W	0.77	
Al	0.80	0.63	Ni	0.83	
In	0.82		Pt	0.94	0.48
			Pt-Ni	0.95	

From References 8.14 and 8.15.

Table 8.4 GaAs. Ohmic Contact Fabrication Technique

Metal or alloy	Technique	Temperature (°C)
n-GaAs		
0.75Ag-0.25In	Sinter	500
0.90Ag-0.05In-0.05Ge	Alloy	600
0.98Ag-0.02Sn	Same	550–650
1Ag-2Sn	Same	
0.88Au-0.12Ge	Alloy with Au overlayer	450
0.88Au-0.12Ge	Alloy with Ni overlayer	480
0.88Au-0.12Ge	Alloy with In overlayer	
0.94Au-0.06Si	Alloy	425
In/Ni	Ni plated on In	
Pd-Ge	Sinter (2 hr)	500
Sn-Ni	Ni plated on Sn and alloyed	
0.96Sn-0.04Sb	Alloy	300–400
p-GaAs		
0.75Ag-0.25In	Alloy	500
0.94Au-0.06Si	Evaporation	550
0.80Au-0.20Sn	Alloy	450
Sn-In-Au	Alloy with Au overlayer	300
0.90Au-0.10Te	Laser alloy	RT
0.98Au-0.02Te	Alloy	500
0.98Au-0.02Te	Evaporation	350–700
In	Melt	300
In-Al	Alloy	320
0.90In-0.10Au	Alloy	550
0.8Ag-0.1In-0.1Zn	Alloy	600
0.9Ag-0.1Zn	Same	450
0.99Au-0.01Be		
0.99Au-0.01Zn	Electroless, evaporation	600
0.8Au-0.2In	Preform	
Semiinsulating GaAs		
In-Zn		

Mainly from References 8.12, 8.15, and 8.16.

Table 8.5 III-V, III-III-V, and III-V-V Compounds. Ohmic Contact Technology

Compound	Conductivity type	Contact material	Technique	Alloy temperature (°C)
AlN	p, $\sigma = 10^{-11}$ to 10^{-13} S/cm	Si	Preform	1500–1800
	Same	Al, Al-In	Same	
	Same	Mo, W	Sputter	1000
AlP	n	Ga-Ag	Preform	500–1000
AlAs	n, p	In-Te	Same	150
	n, p	Au	Same	160
	n, p	Au-Ge	Same	700
	n	Au-Sn	Same	
GaN	n, semiinsulating	Al-In	Same	
GaP	p	0.99Au-0.01Zn	Preform, evaporation	700
	p	Au-Ge	Preform	
GaP	n	0.62Au-0.38Sn	Preform	360
	n	0.98Au-0.02Ge	Evaporation	700
GaSb	p	In	Preform	500
	n	In	Same	
InP	p	In	Same	350–600
	n	In, In-Te	Same	350–600
	n	Ag-Sn	Preform, evaporation	600
InAs	n	In	Preform	
	n	0.99Sn-0.01Te	Same	
InSb	n	In	Same	
	n	0.99Sn-0.01Te	Same	
$Al_xGa_{1-x}P$	n	Sn	Same	
$Al_xGa_{1-x}As$	p	Au-In	Electroplate	400–450
	p	Au-Zn	Evaporation	
	p	Al	Same	500
	n	Au-Ge-Ni	Same	500
	n	Au-Sn	Same	450–485
			Electroless	450
	n	Au-Si	Evaporation	
$Ga_xIn_{1-x}As$ x = 0.17 and 0.20	n	Sn, Pt	Preform	
$Ga_xIn_{1-x}Sb$	n	Sn-Te	Evaporation	
$GaAs_{1-x}P_x$	p	Au-Zn	Same	500
	p	Al	Same	500
	n	Au-Ge-Ni	Same	450
$GaAs_{1-x}P_x$	n	Au-Sn	Evaporation	450
$InAs_xSb_{1-x}$	n	In-Te	Preform	

From References 8.12 and 8.17.

8.3. p-n-JUNCTION AND SCHOTTKY-BARRIER TWO-ELECTRODE DEVICES

Depending on particular details related to the choice of the diode materials (the energy gap, concentration and distribution of the donors and acceptors, mobility of the current carriers, etc.) and the diode design (the junction geometry, breakdown voltage, exposition to radiation, etc.), the diodes perform different roles in electronic circuits, such as rectification of an alternating current (rectifier), stabilization of voltage (breakdown diode), gigahertz oscillation generation (tunnel diode), conversion of the light energy into electric energy (solar cell/photodiode, avalanche photodiode) and vice versa (light-emitting diode, semiconductor laser), and LC circuit tuning (varactor diode, etc.). The geometry of the active parts of a diode, their basic electrical and dielectric properties (capacitance, resistance, and inductance)

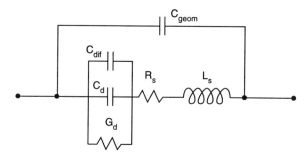

Figure 8.4 Equivalent circuit of a p-n junction. C_d, C_{diff}, and C_{geom} are the depletion, diffusion, and geometric capacitances, respectively; G_d is the differential conductance; R_s is the series resistance which consists of the resistance of the neutral semiconductor areas outside of the p-n junction area and the contact resistances; L_s is a parasitic inductance. (After M. Shur, *Physics of Semiconductor Devices*, Prentice-Hall, Englewood Cliffs, NJ, 1990. With permission.)

and their dependence on the applied voltage determine the impedance of a p-n junction diode. The equivalent circuit of a diode is shown in Figure 8.4. Here $C_d = AC'_d$ is the depletion capacitance (Equation 8.7), A is the p-n junction area, C_{diff} is the diffusion capacitance,

$$C_{diff} = j_o \tau_p A \sqrt{\frac{2k_B T}{e}}$$

(j_o is the current density), R_d is the depletion area resistance,

$$R_d = \frac{k_B T}{e} \bigg/ j_o A,$$

R_s is the series resistance of nonactive parts of the diode (semiconductor outside of the depletion area and the ohmic contacts), C_{geom} is the geometric capacitance, $C_{geom} = \varepsilon \varepsilon_o A/L$ (L is the distance between the ohmic contacts or its equivalent). At a large forward bias, $C_{diff} \gg C_d$ and $C_{diff} \gg C_{geom}$ and the measured diode impedance, Z_{meas}, may be expressed (Reference 8.11, p. 174) as

$$Z_{meas} = R_s + R_d/(1 + \omega C_{diff} R_d) + i\omega L_s \qquad (8.9)$$

An important characteristic of a diode is its DC and dynamic rectification ratios, which can be derived using Equation 8.5. As was mentioned before, in forward bias, $\exp(eV_f/k_B T) \gg 1$ and the diode DC resistance, $R_{dcf} \equiv \frac{V_f}{I_f}$, is

$$R_{dcf} = \frac{V_f}{I_s} \exp(-eV_f/k_B T) \qquad (8.10)$$

while the dynamic resistance

$$R_{df} \equiv \frac{dV_f}{dI_f} = \frac{k_B T}{eI_s} \exp(-eV_f/k_B T) = \frac{k_B T}{eI_f} \qquad (8.11)$$

In reverse bias, $\exp(eV/k_B T) \ll 1$, the reverse DC resistance is

$$R_{dcr} = V_f/I_s \qquad (8.12)$$

and the dynamic resistance is

$$R_{dr} = \frac{k_B T}{eI_s} \exp(eV_r/k_B T) \qquad (8.13)$$

The DC rectification ratio, R_{dcr}/R_{dcf}, is

$$\frac{R_{dcr}}{R_{dcf}} = \exp(eV_f/k_B T) \qquad (8.14)$$

The dynamic rectification ratio is

$$\frac{R_{dr}}{R_{df}} = \frac{I_f}{I_s} \exp(eV_r/k_B T) \qquad (8.15)$$

8.3.1. Varistor and Varactor

Dependence of the dynamic resistance, dV/dI, of a p-n junction diode on applied bias for small AC signals makes it a variable resistor (varistor). Dependence of the p-n junction capacitance on bias (Equation 8.7) permits use of a diode for electrically controlled tuning in LC circuits. Diodes used for this purpose are called varactors (variable reactance operators). The more abrupt the doping profile at the p-n junction, the higher is the sensitivity of the varactor to the bias change, dC/dV_{bias}.[8.22] The quality factor of a varactor is a function of frequency and is equal to

$$Q \simeq \frac{\omega(C_{diff} + C_d)}{1 + \omega^2(C_{diff} + C_d)^2 R_s R_d} \qquad (8.16)$$

for the equivalent circuit shown in Figure 8.4 if $C_{geom} \ll C_{diff} + C_d$ and $\omega L_s \ll R_s$. Equation 8.16 yields a maximum (Reference 8.15, p. 1106):

$$Q_{max} = (R_d/4R_s)^{1/2}$$

at

$$\omega = C^{-1}(R_d R_s)^{-1/2}$$

with

$$C = \frac{\varepsilon_s \varepsilon_o A}{W}$$

The information regarding applications of varistors and varactors in electrical and electronic circuits is presented, in particular, in Reference 8.23.

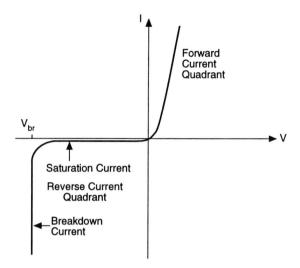

Figure 8.5 Full current-voltage characteristic of a p-n junction.

8.3.2. Zener and Avalanche Diodes

In accordance with Equation 8.5, the current through a diode is independent of the applied voltage at $V < 0$ (reverse bias) and $\exp(eV/k_BT) \ll 1$. Increase of the reverse bias results in increase of electron kinetic energy. At certain voltage, V_{br} (breakdown voltage), impact ionization takes place and the free electrons resulting from this process will also be accelerated and, in their turn, will ionize the neutral atoms, producing more free electrons. This avalanche-like process decreases the differential resistance of the diode, dV/dI, to just a few ohms. The current-voltage characteristic of a p-n junction diode with the complete reverse breakdown branch is shown in Figure 8.5.

The avalanche process described above is utilized in the diodes called avalanche diodes. A convenient characteristic of the avalanche breakdown diode is the electron multiplication, M_n, which can be expressed as a function of the probability, p_i, of an ionizing collision of an electron with an atom in the crystal lattice:

$$M_n \equiv \frac{n_{out}}{n_{in}} = \frac{1}{1-p_i} = \frac{1}{1-(V/V_{br})^\alpha}$$

where $3 < \alpha < 6$, depending on the material of the junction.

Another breakdown process — Zener effect[8.21] — is connected to distortion of the covalent bonds under the action of the strong electric field in the depletion area. If the field is strong enough, some of the bonds will be broken, thus forming the electron-hole pairs which will move in the opposite directions from the depletion area. In the energy band diagram, this process corresponds to a transition of a valence electron from the valence to the conduction band. Penetration of an electron through the energy gap is called tunneling. The avalanche breakdown takes place in the lower electric fields in comparison with Zener breakdown; to make the Zener effect dominating, the material of the diode has to be heavily doped.

Zener diode's active part is an abrupt p⁺-n⁺ barrier in a semiconductor. The high concentration of dopants prevents the depletion area from substantial widening under the reverse bias. When reverse bias passes through a certain critical magnitude (200 to 500 mV), the entire depletion area is filled by the electrons migrating from the p- to n-part of the barrier,[8.21] and any attempt to increase the voltage drop across the barrier is virtually unsuccessful — the differential resistance of the barrier is very small. Current in a circuit containing a

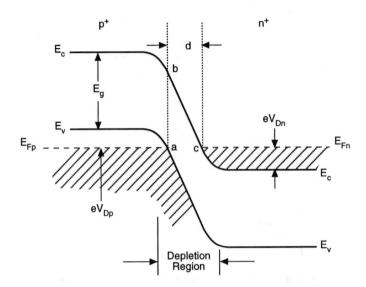

Figure 8.6 Band diagram of a tunnel diode in equilibrium. The tunneling barrier, abc, is triangular. (After K.W. Böer, *Survey of Semiconductor Physics*, Vol. 2, Van Nostrand Reinhold, New York, 1992. With permission.)

breakdown diode may increase to very large magnitudes and destroy the diode if the circuit does not have in it a resistor that limits the current through the diode.

8.3.3. Tunnel Diode

A junction of very heavily (up to 10^{20} cm^{-3}) doped n- and p-type semiconductors with the very narrow equilibrium depletion area (about 5 nm to 15 nm) is used in a device called a tunnel diode. The tunneling effect with a reverse bias in this diode is the Zener effect, and the diode can be used as described in the previous section. Application of the forward bias leads to an effect, which was first described by Esaki[8.24] (Esaki received the Nobel prize in 1973 for his work on this effect); it consists of presence of an area in the forward-bias part of the current-voltage characteristic with the negative differential resistance.

In a relatively pure semiconductor, the impurity atoms are usually distributed throughout its volume so far from each other that interaction between them is negligibly small and an energy level of an impurity atom may be the same for all atoms of this kind in the crystal. A continuous increase of the concentration of particular impurity atoms (say, donors) leads to such a close distribution of them that the donor energy states, in accordance with Pauli's principle, form a band of levels which may overlap the bottom of the conduction band. If this process increases the free electron concentration to the level higher than the effective density of states, N_c, the Fermi level will move from the energy gap to the conduction band. In the case of an acceptor impurity, this process results in dislocation of the Fermi level into the valence band. The described materials are called degenerated n-type and p-type semiconductors, respectively. In the former, the energy states between the conduction band bottom, E_c (Figure 8.6, adapted from Reference 8.15), and up to the Fermi level, E_F, are almost completely filled with electrons; in the latter, the energy states between the Fermi level and the top of the valence band, E_v, are essentially completely filled with holes.

The current-voltage characteristic is shown in Figure 8.7 (adapted from Reference 8.15). The solid curve results from superposition of the diffusion (thermal) current and the tunneling current. The shape of the I-V curve can be explained in terms of energy bands. From Figure 8.6, it is seen that the conduction electrons in the n-part of the diode are in the energy states already occupied by electrons in the valence band of the p-part; for the Fermi level to

SEMICONDUCTOR DEVICES

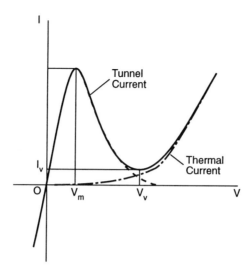

Figure 8.7 Tunnel diode. Current-voltage characteristic. (After K.W. Böer, *Survey of Semiconductor Physics*, Vol. 2, Van Nostrand Reinhold, New York, 1992. With permission.)

be the same on both sides, the bands ought to overlap on the energy scale. If a forward or reverse bias, V, is applied, the Fermi levels on both sides of the barrier are not even ($E_{Fp} - E_{Fn} = eV$). Formation of different parts of the I-V characteristic (Figure 8.7) is schematically shown in Figure 8.8 (adapted from Reference 8.6). If a negative bias is applied (Figure 8.8b), $E_{Fp} > E_{Fn}$ and the electrons in the valence band of the p-part are in the states which are not occupied in the conduction band of the n-part; therefore, they tunnel from the p- to n-part through the barrier, and the greater is the absolute magnitude of the bias, the greater is the (negative) current. When the magnitude of the bias decreases, the Fermi levels approach each other and the number of available states in the n-part conduction band decreases, which is followed by the current decrease until $E_{Fp} = E_{Fn}$ and I = 0 (Figure 8.8a).

Biasing the diode forward initially creates a situation in which the states of conduction electrons in the n-part and holes in the valence band of the p-part match each other and the forward current increases as the bias increases (Figure 8.8c). This is followed by the decrease of the overlapping states and, in spite of the bias increase, the forward current decreases, but instead of reaching zero magnitude, it passes through a minimum (valley) at a certain voltage and starts to increase exponentially with the bias, as in a common p-n junction diode with the relatively low magnitudes of N_A and N_D (Figure 8.8e).

The part of the I-V curve in Figure 8.7 between the current peak, I_p, and the valley minimum, I_v, presents a particular interest. In these limits, the dynamic resistance of the tunnel diode, dV/dI, is negative. This effect is useful in a number of applications. The tunnel diode itself can be used, e.g., to make a voltage divider which can actually be an amplifier. As an example, consider a circuit presented in Figure 8.9 (adapted from Reference 8.25). The circuit analysis gives

$$V_{out} = \frac{R}{R + r_{diode}} \left(V_{sig} + V_{batt} \right) \quad (8.17)$$

from whence

$$\Delta V_{out} = \frac{R}{R + r_{diode}} \Delta V_{sig} \quad (8.18)$$

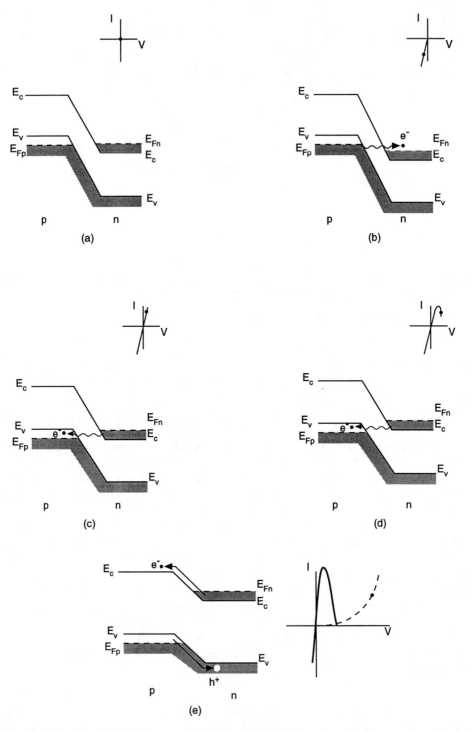

Figure 8.8 Tunnel diode band diagrams and I-V characteristics for various biasing conditions. (a) Zero bias; (b) reverse bias; (c) forward bias up to V_m; (d) forward bias between V_m and V_n (negative differential resistance part, see Figure 8.7); (e) forward bias beyond the tunnel current region–thermal (diffusion) current area. (After B.G. Streetman, *Solid State Electronic Devices*, 2nd ed., Prentice-Hall, Englewood Cliffs, NJ, 1980. With permission.)

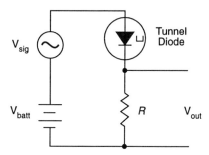

Figure 8.9 Schematics of a tunnel diode/resistor amplifier. (After P. Horowitz and W. Hill, *The Art of Electronics*, Cambridge University Press, New York, 1980. With permission.)

If V_{batt} is chosen in such a way that it is in the limits of the negative resistance part of the diode's I-V characteristic, r_{diode} is negative, and if its magnitude is close to R, $\Delta V_{out} \gg \Delta V_{sig}$.

Currently, the tunnel diodes per se are not widely used; the negative resistance barriers are utilized as a part of many and more sophisticated semiconductor devices.

8.3.4. Photodiode (Solar Cell)

A photodiode is a p-n junction diode in which energy of photons may be converted into the energy of an electric field (or electric current) by means of the photovoltaic effect. The photovoltaic effect consists of the appearance of a forward voltage across a p-n junction of a properly illuminated diode.

A typical design of a photodiode is shown in Figure 8.10 (adapted from Reference 8.26). The photons entering the diode's p-part (in this case) generate a certain number of electron-hole pairs in the depletion area if the photon energy is greater than the semiconductor energy gap. The electric field of the depletion area separates the pairs, pushing the electrons to the n-part, where they occupy the lowest available energy positions in the conduction band, and, consequently, pushing the holes to the highest available energy states in the valence band on the p-side. Hence, there is an electric current through the depletion area proportional to the rate of generation of the electron-hole pairs, g_{op}, by the incident photons. This current (additional to one formed by the bias voltage source) is also proportional to the hole diffusion length, L_p, and the electron charge (the current density, $j_o \equiv j_{optical} \equiv j_L = eg_{op}L_p$). The greater is g_{op}, the greater is the (reverse) additional current through the depletion area. Depending on g_{op}, the I-V characteristic of the photodiode will cross the current axis at different points (Figure 8.11, adapted from Reference 8.6). It is understandable that the photodiode will be in equilibrium ($j = 0$) when the voltage across the diode is of a certain magnitude (V_{oc} is the open circuit voltage).

The current-voltage characteristic of a photodiode can be presented in the form

$$j = j_o \left[\exp(eV/k_B T) - 1 \right] - eg_{op}L_p \tag{8.19}$$

where j_o is the reverse current density in the dark, i.e., when electron-hole pairs are generated only by the lattice thermal motion.

For the visible and near-infrared range, a relatively high (but lower than breakdown) reverse bias is usually applied to the photodiode to decrease the carrier transit time and the diode capacitance. The general figures of merit of a photodiode are (1) the quantum efficiency, which is equal to the number of generated electron-hole pairs per incident photon:

$$\eta_q = \frac{j_p \cdot hf}{eP_{opt}} \tag{8.20}$$

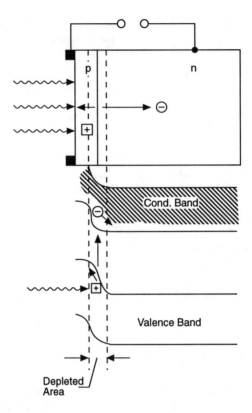

Figure 8.10 Solar cell; the p-region is about 1 μm thick. (After R.E. Hummel, *Electronic Properties of Materials,* Springer-Verlag, New York, 1985. With permission.)

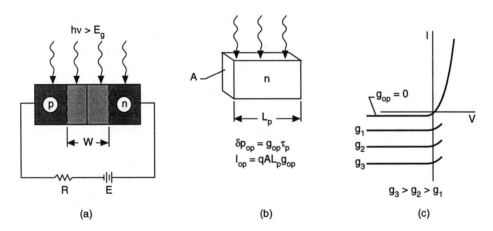

Figure 8.11 Optical generation of carriers in a p-n junction. (a) Absorption of photons by the device; (b) current I_{op} resulting from electron-hole pair generation within a diffusion length of the junction on the n-side; (c) the I-V characteristic of an illuminated p-n junction. (After B.G. Streetman, *Solid State Electronic Devices,* 2nd ed., Prentice-Hall, Englewood Cliffs, NJ, 1980. With permission.)

where j_p is density of the photogenerated current resulting from absorption of incident optical power, P_{opt}, with the photon energy hf; (2) the responsivity, which is equal to the generated photocurrent per unit of the optical power:

SEMICONDUCTOR DEVICES

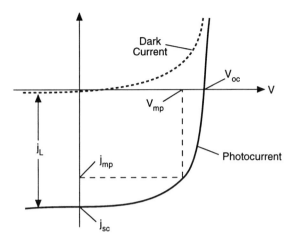

Figure 8.12 Current density-voltage characteristic of a photodiode. V_{mp} and j_{mp} — the maximum power parameters.

$$R = \frac{j_p}{P_{opt}} = \frac{\eta_q e}{hf} \quad (8.21)$$

and (3) the device noise that depends on a few parameters described, e.g., in Reference 8.12.

If power is extracted from a photodiode, the fourth quadrant of the I-V characteristic (Figure 8.12, solid curve) is used. The important points of the curve here are V_{oc} — the open-circuit voltage ($j = 0$), j_{sc} — short-circuit current ($V = 0$, the only resistance in the circuit is the resistance of the diode), and the maximum power point ($P_{max} = j_{mp} V_{mp}$). The V_{oc}, j_{sc}, j_{mp}, and V_{mp} can be estimated using Equation 8.19. If $j = 0$, circuit is open and

$$j_o \left[\exp(eV_{oc}/k_B T) - 1\right] = j_L$$

or

$$j_L + j_o = j_o \exp(eV_{oc}/k_B T)$$

and

$$\ln\left(\frac{j_L}{j_o} + 1\right) = \frac{eV_{oc}}{k_B T}$$

Usually, $j_L/j_o \gg 1$; therefore,

$$V_{oc} = \frac{k_B T}{e} \ln\left(\frac{j_L}{j_o}\right) \quad (8.22)$$

or

$$V_{oc} = \frac{k_B T}{e} \ln\left(\frac{eg_o L_p}{j_o}\right) \quad (8.23)$$

At V = 0 (short circuit), Equation 8.19 gives

$$j_{sc} = j_L = eg_o L_p \tag{8.24}$$

The diffusion length, L_p, is related to the hole mobility, μ_p, and lifetime, τ_p, by expression

$$L_p^2 \simeq \mu_p \tau_p k_B T/e \tag{8.25}$$

Hence, the experimentally measurable magnitudes of V_{oc} and j_{sc} may be used for estimation of the carrier diffusion length, lifetime, and mobility.

The maximum power, P_{max}, point's coordinates, j_{mp} and V_{mp} (Figure 8.12), may be determined using expressions

$$P = jV$$

and

$$dP = jdV + Vdj = 0 \tag{8.26}$$

It is followed from Equation 8.26 that j_{mp} and V_{mp} magnitudes lie at the point of the I-V curve with the slope equal to –1, i.e.,

$$\frac{dj(j_{mp}, V_{mp})}{dV} \cdot \frac{V_{mp}}{j_{mp}} = -1 \tag{8.27}$$

The power conversion efficiency of a photodiode is the ratio of P_{max} and the total light or solar energy flux approaching the photodiode aperture:

$$\eta_c = \frac{P_{max}}{P_{inc}} = \frac{j_{mp} V_{mp}}{P_{inc}} \tag{8.28}$$

A traditional technical characteristic of a solar cell is a parameter, called the fill-factor, physical meaning of which is understandable from the following expression:

$$FF = \frac{P_{max}}{j_{sc} V_{oc}} = \frac{j_{mp} V_{mp}}{j_{sc} V_{oc}} \tag{8.29}$$

A photodiode that is operated in a high reverse bias mode, close to its breakdown voltage, is called the avalanche photodiode. The electron-hole pairs, which were generated by the light impinging on the photodiode, are accelerated to high velocities; and, passing through the depletion area, they generate secondary electron-hole pairs that, in their turn, being accelerated, generate more pairs. The result is a photocurrent gain between 10^2 and 10^3. These diodes are used for low light power applications and for very high frequencies (up to 10^{10} Hz), e.g., the fiber-optic communication systems.

A photodiode which is operated in the forward bias mode serves as a generator of light. This effect, called injection electroluminescence, requires application of semiconductors with a direct energy gap. This prevents the heat release of the crystal lattice resulting from the

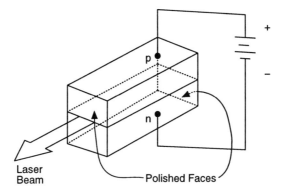

Figure 8.13 Schematic setup of a semiconductor laser.

indirect recombination, although proper doping of the material may correct this handicap (see, e.g., Reference 8.30). There is a wide range of semiconductor materials which are and can be used for these light-emitting diodes (LED); for example, zinc sulfide is used for generation of the ultraviolet photons, InSb is used for the infrared part of the spectrum, and between them there are numerous compounds and variable composition phases (solid solutions) useful for the visible part of the spectrum (the spectral distribution of semiconductors, which present interest for use in LEDs, is analyzed, in particular, by Bergh and Dean[8.28]).

Besides direct applications of the photodiode as a light signal detector/light (solar energy) converter and a light-emitting device, an important application of it is in the device called optoisolator or optocoupler. It consists of an LED and a photodetecting diode immersed into a light-transmitting packaging medium. The input signal enters the LED; its light output enters the photodiode which, in turn, forms the output electric signal. The optocoupler isolates electrically the LED and photodetector electrical circuits and prevents the noise transfer through it. The current transfer ratio, I_2/I_1, is usually of the order of 10^{-3}, but in a high-gain optoisolator with a phototransistor as the photodetector this ratio can be increased to ten.[8.28]

Semiconductor laser in its most simple form (homojunction laser) is a p^+-n^+ junction diode made of material of a proper energy gap; essentially it is a combination of an LED with two Fabry-Pérot geometry's naturally cleaved or highly polished mirrors at the opposite side of the junction (Figure 8.13). The semiconductor laser differs in a few features from the solid-state and gas laser,[8.11] although it also emits a narrow-band monochromatic light beam. In a semiconductor laser, the quantum transitions occur between the bands, while in the other lasers, transitions are related to the discrete energy levels; the spectral and spatial characteristics of the laser strongly depend on the junction medium properties.

The advantages of semiconductor lasers are (1) their small size (about 0.1 mm long); (2) simplicity of the laser circuitry: the beam is produced by passing the forward current through the diode; (3) small laser dimensions and very short photon lifetimes permit modulation of the laser beam at very high frequencies; (4) low driving voltage (0.5 to 2 V) and high lasing efficiency; and (5) the capability to be a part of common integrated circuit structures.[8.15] A disadvantage of the semiconductor laser for some applications is a considerably large divergence of the laser beam related to the short distance between the reflecting surfaces.

The small size of the semiconductor lasers and the simplicity of the beam intensity modulation have made them the most applicable light sources for fiber-optic communication systems. Semiconductor variable-composition phases of III-V, II-VI, and IV-VI types that are either used or can be used in (heterostructure) semiconductor lasers of particular wavelength are shown in Figure 8.14 together with their working temperature (adapted from Reference 8.29). Various semiconductor compounds of interest for applications either individually or in combinations (in heterojunction devices) are presented in Figure 8.15 (adapted from Reference 8.28).

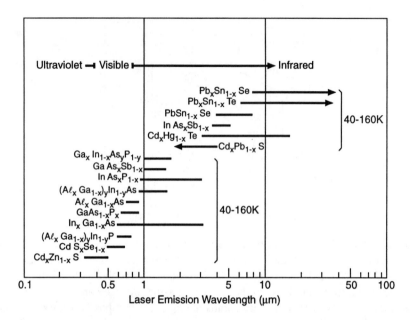

Figure 8.14 II-VI/II-VI, III-V/III-V, and IV-VI/IV-VI semiconductor phases and the emission wavelength range for heterostructure lasers made on their basis. (After H.C. Casey and M.B. Panich, *Heterostructure Lasers,* Academic Press, New York, 1978. With permission.)

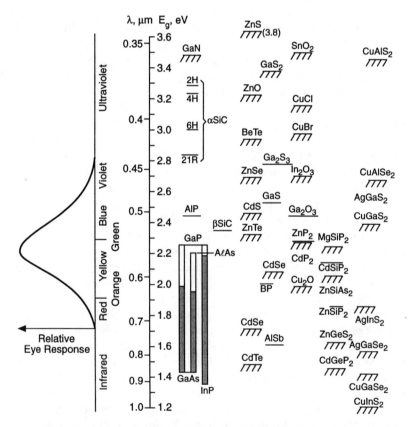

Figure 8.15 Semiconductor compounds and ternary alloys with bandgaps of interest for light-emitting diodes. The relative eye response is presented at the left with color bands identified. (After A.A. Bergh and P.J. Dean, *Light Emitting Diodes,* Clarendon Press, 1976. With permission.)

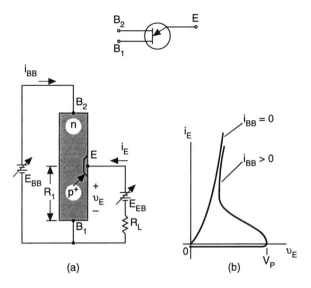

Figure 8.16 Schematic presentation of the biasing circuit (a) and I-V characteristic of a unijunction transistor (b). (After B.G. Streetman, *Solid State Electronic Devices,* 2nd ed., Prentice-Hall, Englewood Cliffs, NJ, 1980. With permission.)

8.3.5. Unijunction Transistor

The unijunction transistor is a single-p-n-junction device which operates by the volume conductivity modulation. In the process of development of semiconductor device fabrication technology, the name of this device changed from filamentary transistor[8.30] to double-base diode,[8.31] to unijunction transistor (UJT).[8.32] The common use of UJT is as a triggering device for power switches, such as thyristors.

The schematic diagram of a UJT and its current-voltage characteristic are shown in Figure 8.16 (adapted from Reference 8.6). The device consists of a p⁺-n junction in a relatively long, high-resistivity n-type bar (or layer). The p⁺-part is called the emitter; the two ohmic base contacts, B_1 and B_2, connect the n-part to the current source E_{BB}. When current between the base contacts, i_{BB}, is equal to zero, the current i_E is described by the usual I-V characteristic of a forward-biased diode. If $i_{BB} > 0$, it creates a voltage drop in the n-part below the emitter, $v_{R_1} = i_{BB}R_1$. Up until the voltage drop in this part, v_E, formed by the source E_{EB}, is not greater than v_{R_1}, the junction voltage is negative, and only a small saturation current passes through it. If v_E becomes slightly greater than v_{R_1}, the holes are injected into the n-area. It decreases R_1, and this is followed by the decrease of v_{R_1} and increase of the current j_{BB} until the I-V characteristic virtually coincides with a characteristic of the unbiased junction. The required turn-on (peak point) voltage, V_p, can be controlled by variation of E_{BB}.

8.4. MULTITERMINAL (MULTIBARRIER) DEVICES

All modern semiconductor devices are essentially combinations of p-n junctions, Schottky barriers, and n⁺-n (p⁺-p) transitions connected in different sequences to each other, and furnished with the ohmic metal-semiconductor contacts for connection to the outer circuit.

From the time of discovery by Bardeen, Brattain, and Shockley[8.33-8.35] of the first active semiconductor device called, the transistor (acronym for transfer resistor), solid-state electronics has progressed to a point that it modified not only economical, but even social and political structures of the human kind.

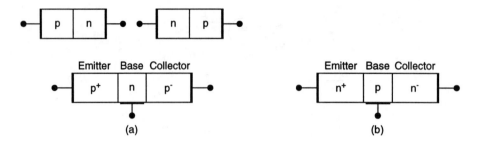

Figure 8.17 Bipolar junction transistor as a combination of two back-to-back p-n junctions.

The multiterminal transistor-type devices can be divided into two groups:

1. Bipolar devices in which both electrons and holes transfer through the junctions changing their parameters in desirable directions, and
2. Unipolar devices operating with only one carrier type

The description of transistor will be started from the bipolar devices, since historically they were first.

8.4.1. Bipolar Devices

8.4.1.1. Bipolar Junction Transistor

A bipolar junction transistor, BJT, may be considered as two diodes connected back-to-back to each other; this makes either n-p-n BJT or p-n-p BJT (see Figure 8.17). It is more accurate to name them either n⁺-p-n⁻ or p⁺-n-p⁻ BJTs because one of the outer parts of the structure is more heavily doped and the other outer part is less heavily doped in comparison with the part in the middle. A typical dopant distribution (see, e.g., in Reference 8.11) is 10^{20} cm^{-3}–10^{17} cm^{-3}–10^{15} cm^{-3}.

In the transistor design, it is important to have the distance between the emitter and collector parts (base width, W_b) smaller than the diffusion length of the base minority carriers (holes in p-n-p, and electrons in n-p-n transistors). The band diagram of a p-n-p BJT in equilibrium is shown in Figure 8.18a. When a forward bias, V_{EB}, is applied between emitter and base and a (larger) reverse bias, V_{BC}, is applied between the base and collector, the band diagram looks like the one shown in Figure 8.18b (the reader understands that in the case of an n-p-n BJT, each of two diagrams has to be turned "upside down"). There are three different ways to connect a transistor to a circuit for its use as either a switch or an amplifier. They are called (1) common-base configuration, (2) common-emitter configuration, and (3) common-collector configuration and are shown in Figure 8.19 (adapted from Reference 8.15). The forward bias V_{EB} creates the current I_E which consists mainly of holes (p⁺ condition). A large number of holes that enter the n-part diffuse toward the collector part of the transistor (Figure 8.19a). If a reverse bias V_{BC} is applied to the p-n junction between the base and collector, the holes entering the depletion area of this junction will be captured by the electric field of the area and accelerated into the p-part. As the result of this, the reverse current through the right p-n junction will be increased by the magnitude proportional to V_{EB}. The current I_B is much smaller than I_C; therefore, $I_C \simeq I_E$. It is possible to show that the common-base configuration current gain, $\alpha \equiv dI_C/dI_E$, can be expressed as

$$\alpha = 1 - (W_B/L_p)^2/2 \simeq 1 \quad (8.30)$$

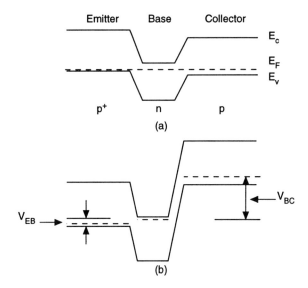

Figure 8.18 Energy band diagram of a p-n-p BJT (a) without and (b) with bias.

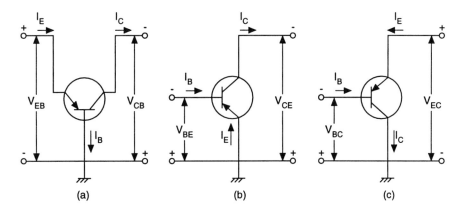

Figure 8.19 Common-base (A), common-emitter (B), and common-collector (C) circuit configurations for a bipolar junction transistor.

Here, W_B is the width of the base minus the parts of it occupied by the depletion areas on the borders with the emitter and collector, and L_p is the diffusion length of holes in the base. The experimentally evaluated magnitudes of α are on the level of 0.998 to 0.999.

The voltage and power gains by a BJT are controlled not only by its own properties, but depend also on the outer circuit parameters; two of them are the incremental dynamic resistance of the emitter junction, $R_E = dV_{BE}/dI_E$ and the load resistance in the collector part of the circuit. The voltage gain is equal to

$$G_V = \Delta V_{out}/\Delta V_{in} = (\Delta I_C/\Delta I_E)(R_{out}/R_E) \tag{8.31}$$

In Equation 8.31, the ratio $\Delta I_C/\Delta I_E = \alpha \approx 1$, and R_{out} can be as high as the dynamic resistance of the collector, $R_C = dV_{CB}/dI_C$. From the plot I_C vs. V_{CB} (Figure 8.20) one can see that R_C has very high magnitudes and G_V can reach a magnitude of up to 10^4.

A correlation between $\Delta V_{in} = \Delta V_{EB}$ and $\Delta V_{out} = \Delta V_{CB}$ is shown in Figure 8.20. It can be shown that the power gain, G_P is also essentially equal to the dynamic resistance ratio:

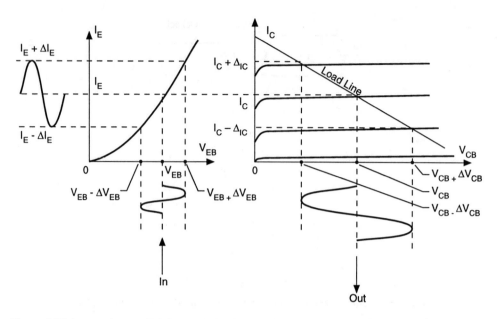

Figure 8.20 Input and output I-V characteristics for a transistor with a common-base circuit configuration.

$$G_P \equiv \Delta P_{out}/\Delta P_{in} = (\Delta I_C/\Delta I_E)^2 (R_{out}/R_E) \qquad (8.32)$$

The choice of a circuit configuration for a BJT depends on parameters of the other elements that form the circuit. In the common emitter configuration, the input current is equal to the base current I_B. In this case, the current gain, β_{ce}, can be expressed via the current gain α in the common-base configuration considered above as

$$\beta_{ce} \equiv \Delta I_C/\Delta I_B = \frac{\alpha}{1-\alpha} \qquad (8.33)$$

and because $\alpha \simeq 1$ and $1 - \alpha > 0$, β_{ce} is high — up to 400 to 600. The output resistance in the common-emitter configuration is approximately equal to the product of the collector differential resistance and $(1 - \alpha)$; the input resistance is equal to $-dV_{BE}/dI_B$. The configuration with common collector is characterized by the high input resistance and low output resistance. The voltage gain $V_{EC}/V_{BC} < 1$, but the current and power gains are high.

Depending on the direction of the emitter-base and collector-base biases, there are four modes of BJT operation presented in Table 8.6. The active mode is one that has been used in the above description of the BJT function. The saturation and cutoff modes can be utilized when a BJT is used as a switch (the saturation mode is for the "ON" state and the cutoff mode is for the "OFF" state). As is shown in a quantitative consideration in Reference 8.11, p. 250, when a BJT operates in the inverse mode, the transistor gain is very low: more holes are injected into the collector region than electrons into the base insofar as the dopant content in the base is higher. The low gain makes this mode of operation unimportant for use in electronic circuits, but it can be efficiently used for evaluation of some transistor parameters (see, e.g., Reference 8.36).

8.4.1.2. Thyristor

The thyristor (acronym for Greek θυρα — door, gate, doorway; and *resistor*) is a four-layer p^+-n-p^+-n^+ structure with three or four electrodes which can be switched from a

Table 8.6 The Modes of BJT Operation

Operation mode	Emitter-base bias	Collector-base bias
Active mode	Forward	Reverse
Saturation mode	Forward	Forward
Cutoff mode	Reverse	Reverse
Inverse mode	Reverse	Forward

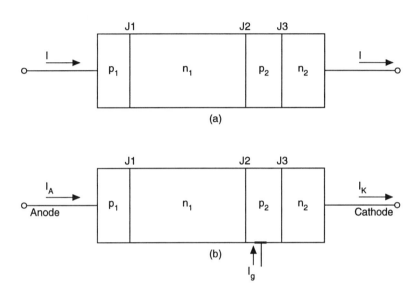

Figure 8.21 Structure of (a) two-terminal p-n-p-n diode (Shockley diode) and (b) three-terminal thyristor (I_g — gate current).

high-impedance "OFF" state to a low-impedance "ON" state. To analyze the thyristor principle of operation, consider first a four-layer diode.[8.37] It consists of three p-n junctions connected in series in the order p-n/n-p/p-n (Figure 8.21a). If a voltage is applied to the structure in such a way that p_1-part is under a negative potential relative to n_2-part, the junctions J1 and J3 are under a reverse bias and the diode's current-voltage characteristic is one of a regular p-n junction for $V < 0$. For a good diode, a breakdown voltage for this bias is very high — up to several kilovolts. With a bias of the opposite polarity (shown in Figure 8.21), J1 and J3 junctions are under a forward bias, whereas J2 junction is under a reverse bias, and until the voltage reaches a certain magnitude (critical peak forward voltage, V_p), the p-n-p-n diode's I-V characteristic has a typical reverse-bias shape. At the voltage V_p, the minority carriers, which are injected by the outer p-n junctions (J1 and J3), pass through J2 barrier and enter into the n_1- and p_2-parts. These uncompensated charges decrease the height of the potential barriers of J1 and J3 junctions; this increases the minority carrier injection through these junctions, which causes further current increase and so on. This positive feedback is responsible for the part of the diode's I-V characteristic with negative resistance (Figure 8.22a). With this feedback, the diode remains in the forward conducting state until the applied voltage becomes lower than a relatively small magnitude called the holding voltage V_h.

A thyristor differs from a p-n-p-n diode by the existence of an additional ohmic contact (gate electrode, Figure 8.21b) at the p_2-region or n_1-region (or both). If a voltage is applied between p_2-region and the cathode in such a way that a current I_g enters the p-region, the critical peak voltage (switching voltage, V_p) can be controlled by the I_g magnitude (Figure 8.22b, adapted from Reference 8.15). Increase of the I_g magnitude results in decrease of the peak voltage V_p, and when V_p becomes equal to the anode-cathode voltage drop, V_o, the thyristor transfers into the "ON" state. After turning the thyristor in the conducting state,

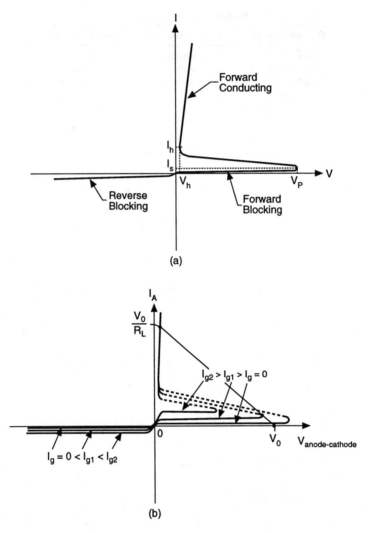

Figure 8.22 Typical I-V characteristics of (a) a p-n-p-n diode and (b) a thyristor. V_p — forward breaking voltage, I_s — switching current, V_h and I_h — holding voltage and current, R_L — load resistance, V_o — voltage applied to the load-thyristor circuit.

the triggering circuit carrying the I_g current can be disconnected. It is clear from the description of the thyristor operation that it can be turned from the "ON" to the "OFF" position only by decreasing the anode-cathode voltage drop to the magnitude $V_o < V_h$ (see Figure 8.22).

8.4.2. Unipolar Devices

Unipolar devices are ones in which currents contain carriers of only one type. The operation of these devices is mainly based on the field effect, i.e., on displacement of the majority charges in a semiconductor crystal by an external electric field resulting in redistribution of electrical resistivity in the volume of the crystal. A device in which the field effect is used to control the current passing through it is called the field effect transistor (FET). Besides the unipolarity, the FETs differ from the BJTs by the controlling entity (the voltage in the former and the current in the latter).

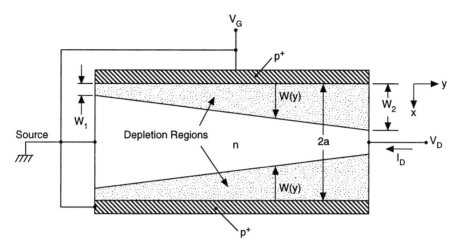

Figure 8.23 Schematics of a positively biased active channel. V_D — bias, $W(y)$ — depletion layer width. (After S.M. Sze, *Semiconductor Devices, Physics and Technology*, John Wiley & Sons, 1985. With permission.)

Similar to the BJT, the FET contains three terminals, two of which, called "source" and "drain", serve to pass the working (controlled) current through the bulk of the device, while the third terminal (gate) is used to form an electric field in the device that serves to control the electrical properties of the medium between the source and the drain.

The concept and first design of the FET were proposed by Lilienfeld[8.38] and Heil[8.39] long before the invention of bipolar transistors, but the first unipolar devices of commercial value became available only in the late 1960s. A similar device, the metal-insulator-semiconductor diode (MIS diode), was proposed by Moll[8.40] and by Pfann and Garrett.[8.41]

8.4.2.1. Field-Effect Transistors

Depending on their construction, FETs can be divided into three groups: the junction FETs (JFET), the metal-semiconductor FETs (MESFET), and the metal-oxide-semiconductor FETs (MOSFET).

JFET is a voltage-controlled resistor. Its schematic structures are shown in Figures 8.23 and 8.24 (from References 8.12 and 8.15). The two p^+-n junctions are separated by the n-type layer of the thickness 2a. When a reverse voltage drop V_g is applied to both gates (relative to the source), it forms two high-resistivity depletion regions of the thickness W in the n-layer. This decreases the cross section of the part of the channel through which electrons may pass from the source to the drain; the resistance of this narrowed part (active channel) is, therefore, increased. If a voltage V_D is applied to the drain relative to the source, the current I_D will flow toward the source. When I_D is small, the geometry of the depletion area is the same as in the absence of a current. But, with the I_D increase, the boundary between the channel and depletion area becomes more and more slanted (W is increasing along the direction of the conduction electron drift). This is a result of the increase of $V(y)$ with y, and, therefore, the greater is y, the greater is the magnitude of the total (reverse) bias applied to the p^+-n junctions. Increase of V_D and, therefore, increase of I_D result in decrease of W_1 and increase of W_2. At a certain current $I_D = I_p$ and voltage $V_D = V_p$ (subscript p is for pinch-off), the depletion areas merge at the drain end of the channel and the current I_D remains equal to I_p at any magnitude of $V_D > V_p$. It is understandable that the greater is $|V_g|$, the smaller are magnitudes of the pinch-off current and voltage. A family of the I-V characteristics of a JFET is shown in Figure 8.25 (adapted from Reference 8.15).

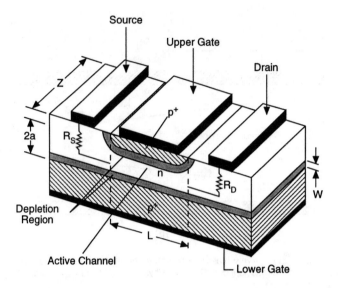

Figure 8.24 Schematics of a junction field-effect transistor. R_S and R_D represent the resistance of the n-region outside of the active channel. (After K.W. Böer, *Survey of Semiconductor Physics*, Vol. 2, Van Nostrand Reinhold, New York, 1992. With permission.)

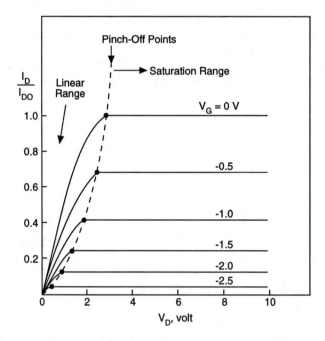

Figure 8.25 Family of the I-V characteristics of a JFET (V_G — gate voltage). (After K.W. Böer, *Survey of Semiconductor Physics*, Vol. 2, Van Nostrand Reinhold, New York, 1992. With permission.)

In accordance with Shockley[8.42] analysis (see, also, Reference 8.15), the magnitudes of I_p and V_p at $V_g = 0$ can be expressed as the functions of the device geometry and the properties of the channel material:

$$I_{po} = \frac{e^2 N_D^2 a^3 Z \mu_n}{2\varepsilon\varepsilon_o L} \qquad (8.34)$$

SEMICONDUCTOR DEVICES

and

$$V_{po} = \frac{eN_D a^2}{2\varepsilon\varepsilon_o} \tag{8.35}$$

The linear part of the current-voltage characteristic of a JFET with $V_D \ll V_p$ can be approximated as

$$I_D \simeq \frac{I_p}{V_p}\left[1 - \left(\frac{V_T - V_g}{V_p}\right)^{1/2}\right] V_D \tag{8.36}$$

where

$$V_T = \frac{k_B T}{e} \ln\left(\frac{N_A N_D}{n_i^2}\right)$$

The concept of the MESFET was suggested by Mead.[8.43] This device has a Schottky barrier instead of a p-n junction; the lower part of the channel is not commonly used in the device operation. The MESFET operation principle and current-voltage characteristic are similar to those of the JFET, although the MESFET has an important advantage. Since its design is not connected to the dopant diffusion, the channel in the MESFET can be made so thin that the Schottky barrier will keep the channel blocked even without any bias applied to the gate. Therefore, the MESFET can be fabricated either as a "normally-on" or a "normally-off" device. The technological simplicity of Schottky barriers allows fabrication of MESFETs of very small dimensions and high dimensional precision. These devices are made using GaAs, InP, and other III-V compounds with much higher carrier mobility than that in silicon; they have found a relatively wide application in high-speed digital and microwave circuits.[8.44,8.45] The schematic structure of the normally-off and normally-on MESFET is shown in Figure 8.26 (adapted from Reference 8.11).

The metal-oxide-semiconductor field-effect transistor (MOSFET), based on silicon, is the most widely used semiconductor device of the modern electronic industry. The gate in a JFET is a p-n junction; in a MESFET it is a metal-semiconductor Schottky barrier. In both, there is a leakage current which is equal to just a few nanoamperes at ordinary temperatures, but it increases substantially at elevated temperatures. If the gate is forward biased, relative to either source or drain, the gate barrier decreases its impedance and a large current may pass through the gate; this current can destabilize the FET parameters and is difficult to consider in the process of FET design.

In silicon MOSFET, the gate is electrically separated from the conducting channel by a thin layer of vitreous silicon dioxide grown onto the channel by oxidation of the silicon surface. In MOSFETs, the gate may have either of the polarities without a current; this simplifies the circuit design.

The schematics of a silicon MOSFET is shown in Figure 8.27 (adapted from Reference 8.15). The areas under the source and drain electrodes have the n-type conductivity with a high donor atom concentration. The inversion layer shown in the figure is not a part of the transistor construction, but is a result of application of a sufficient positive potential to the gate electrode. The layer is formed by the electrons that are capacitively pulled by the gate electrode's electric field from the bulk of the p-layer, leaving behind them a high-volume-resistivity depletion area. The inversion layer serves as a conducting bridge between the two

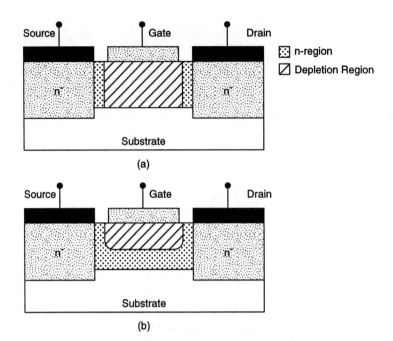

Figure 8.26 Depletion regions in normally-off (a) and normally-on (b) MESFETs at zero gate-to-source and drain-to-source voltages. (After M. Shur, *Physics of Semiconductor Devices*, Prentice-Hall, Englewood Cliffs, NJ, 1990. With permission.)

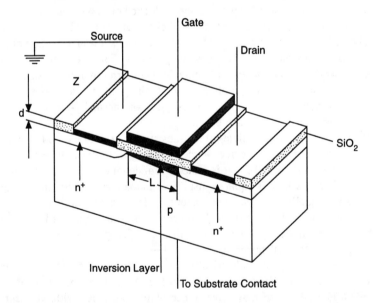

Figure 8.27 Schematics of a MOSFET with the p-type base n+-doped for source and drain. A contact is attached directly to the substrate for threshold voltage adjustment. (After K.W. Böer, *Survey of Semiconductor Physics*, Vol. 2, Van Nostrand Reinhold, New York, 1992. With permission.)

n+-areas, which are otherwise separated by two p-n+ junctions. The depletion area is very helpful in construction of integrating circuits (ICs): it separates electrically a MOSFET from other devices fabricated on the same IC surface.

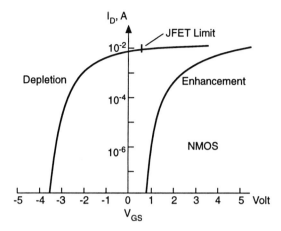

Figure 8.28 Dependence of drain current on gate-source voltage at a fixed value of drain voltage for the depletion-type (left curve) and enhancement-type (right curve) MOSFETs. (After P. Horovitz and W. Hill, *The Art of Electronics*, Cambridge University Press, New York, 1980. With permission.)

The current-voltage characteristics of MOSFETs are similar to those of other FETs considered here. An important parameter for a MOSFET is the minimum gate voltage (threshold voltage, V_T) which is sufficient for formation of an n-channel in a MOSFET with the p-type substrate or a p-channel, if the substrate is of n-type. The common acronyms for these transistors are NMOS and PMOS, respectively.

Similar to MESFETs, the MOSFETs can be either normally-off or normally-on FETs, depending on their design. The normally-off MOSFET is called the enhancement-type device (the gate voltage has to be applied to enhance the device conductance). The normally-on MOSFET is called the depletion-type device (the gate voltage has to be applied to deplete the area between the two n⁺- (or p⁺-) sectors under the source and drain electrodes). The dependence of the drain current on the gate-source voltage at a fixed value of the drain voltage for a depletion-type NMOS and an enhancement-type NMOS is shown in Figure 8.28 (adapted from Reference 8.25). In contrast with JFET, the forward bias in MOSFETs may be much higher without a risk of device destruction.

The analytical expression for the MOSFET I-V characteristic[8.15] is similar to that of the MESFET:

$$I_D = \left(\frac{Z}{L}\right)\mu_n C_o \left\{ \left(V_g - 2V_T - \frac{V_{DS}}{2}\right)V_{DS} - \frac{2}{3C_o}(2\varepsilon\varepsilon_o e N_A)^{1/2}\left[(V_{DS}+2V_T)^{3/2} - (2V_T)^{3/2}\right]\right\} \quad (8.37)$$

where C_o is the gate SiO$_2$ layer's capacitance per unit area and

$$V_T = \frac{k_B T}{e} \ln\left(\frac{N_A}{n_i}\right).$$

For Z and L, see Figure 8.27.

MOSFETs are sometimes known by different acronyms: IGFET (insulated-gate FET), MISFET (metal-insulator-semiconductor FET), or MOST (metal-oxide-semiconductor transistor). Comprehensive information regarding the technology of MESFETs fabrication and their use in electronic circuits is compiled in References 8.2, 8.5, 8.10 to 8.12, 8.15, and 8.46 to 8.50.

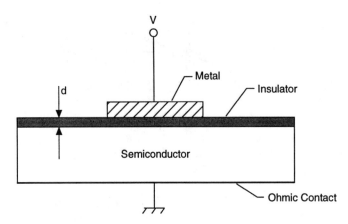

Figure 8.29 Schematics of a metal-insulator-semiconductor diode.

8.4.2.2. Metal-Insulator-Semiconductor Diode and Charge-Coupled Device

The metal-insulator-semiconductor diode (MIS diode) is essentially a variable capacitor (see Figure 8.29, adapted from Reference 8.12), as was first suggested by the authors of References 8.40 and 8.41. Most commonly, MIS diodes are based on silicon with SiO_2 as the insulator, although the relatively low carrier mobility in silicon in comparison with, say, III-V compounds stimulates the search for different materials to be utilized in MOS diodes and different geometry, including that of the Schottky diode.

The MIS diode shown in Figure 8.29 is a capacitor in which the upper metal layer and the semiconductor connected by the ohmic contact to the bottom metal layer are (parallel) electrodes; the oxide layer of the thickness d and a depletion semiconductor layer (if any) adjacent to the oxide layer serve as the insulator. Therefore, the total capacitance of the MIS diode per unit area is

$$C = \frac{C_i C_d}{C_i + C_d} \tag{8.38}$$

where $C_i = \varepsilon_i \varepsilon_o / d$ is the unit area capacitance of the oxide layer, $\varepsilon_i \varepsilon_o$ is the oxide dielectric constant, $C_d = \varepsilon_s \varepsilon_o / w$ is the depletion layer capacitance per unit area, $\varepsilon_s \varepsilon_o$ is the semiconductor dielectric constant, and W is the thickness of the depletion layer.

For a simplified MIS diode (ideal MIS diode) model, in which (1) both metal and semiconductor have the same work function magnitude at zero bias, (2) at any bias, the only charges in the diode are those in the semiconductor and in the metal electrode adjacent to the oxide layer (with equal concentrations per unit area and opposite signs), and (3) the DC resistivity of the (combined) insulator is infinitely high, the dependence of the diode capacitance on bias is shown in Figure 8.30 (for a diode with a p-type semiconductor). Under any negative bias, holes are accumulated at the surface of the semiconductor adjacent to the oxide layer, and the diode unit area capacitance is equal to C_i. At $V > 0$, a depletion layer of the width W appears in the semiconductor and the total capacitance is $C = \left(C_i^{-1} + C_d^{-1}\right)^{-1}$ and is decreasing with the rise of the bias. The capacitance reaches its minimum at $V = V_{threshold}$, when the growing bias forms an inversion layer with the n-type conductivity. With inversion, C becomes independent of voltage (at $V < V_{breakdown}$) for all measurement frequencies with the exception of the low ones (below, say, 10^2 Hz). At these frequencies, the diode capacitance starts to increase until it reaches the C_i magnitude and again becomes independent of the bias. For explanation of this effect, consider the process of the capacitance measurement

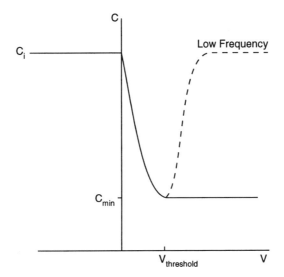

Figure 8.30 Capacitance-voltage characteristic for an ideal n-channel (p-substrate) MOS capacitor.

which is connected with application to the (biased) diode of a usually small alternating voltage. At high frequencies, the small increments of voltage happen in periods of time too short for thermally generated electron-hole pairs to influence the processes which control the depletion area width, W. If, however, the frequency is low, the electrons from the electron-hole pairs, generated during the relatively long period of voltage oscillation, will be able to move under the action of the bias voltage to the semiconductor surface adjacent to the oxide layer and, as a result, to increase the diode capacitance up to C_i, as is shown by the dashed line in Figure 8.30.

The charge-coupled device (CCD) belongs to the family of the charge-transfer devices (see Reference 52) and was first proposed by Boyle and Smith.[8.53] It may be used for moving an electric charge along a predetermined pass in the limits of the device, being controlled by properly chosen electric pulses (clock pulses) applied to a sequence of MIS capacitors forming the device. The basic feature of a CCD is a series of closely located metal electrodes which form an array of MIS (or MOS) capacitors on a silicon substrate; the substrate is separated from the electrodes by an SiO_2 layer. A CCD can be made on other semiconductors, such as GaAs, using the Schottky barrier capacitors (see, e.g., References 8.48 and 8.54). The schematics of a silicon-based CCD (after Reference 8.54) and a GaAs-based Schottky barrier CCD (adapted from Reference 8.55) are shown in Figure 8.31. CCDs have found application in various logic functions, in imaging, and in signal processing.

In the described CCD, the minority-carrier charge packets are transferred along the border of the semiconductor and oxide or the semiconductor surface with the Schottky barriers. The surface of semiconductor contains the charge traps related to so-called interface states; their presence decreases the magnitude of the transferred charge. Boyle and Smith[8.56] suggested a CCD in which the charge packets are moving along a channel lying beneath the surface. This device, called the buried-channel CCD (BCCD), is almost completely free of the interface trapping. For details regarding the physics and technology of the BCCD, see References 8.57 and 8.58. A similar improvement of the MOSFET (buried-channel MOSFET) is described by Huang and Taylor.[8.59]

8.5. CONCLUSION

Specific intrinsic properties of semiconductor materials, among them high sensitivity to temperature, electric and magnetic fields and radiation, the low (in comparison with metals)

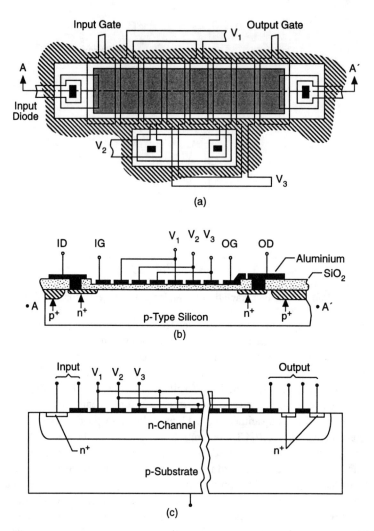

Figure 8.31 Charge-coupled devices. (a) Layout of a silicon-based CCD; (b) its cross-sectional diagram; (c) schematics of a CCD based on formation of Schottky barriers in a GaAs n-channel. (After C.K. Kim, Charge-Coupled Devices and Systems, M.J. Howes and D.V. Morgan, Eds., John Wiley & Sons, 1979 (a and b) and D.K. Schroder, *Metal-Semiconductor Schottky Barrier Junctions and their Applications*, B.L. Sharma, Ed., Plenum Press, 1984 (c).. With permission.)

intrinsic concentration of conduction electrons and their high mobility, and high sensitivity of their electrical conductivity to external mechanical pressure, are the basis of their application in electronic devices.

Devices with one p-n junction can be used for generation of electrical oscillations, rectification of the alternating currents, parametric amplification, frequency conversion, radiation detection, light radiation, etc. Devices with two p-n junctions are used mainly for generation and amplification of electric signals. Devices with three p-n junctions, e.g., thyristors, are used as amplifiers and switches. Multijunction devices are used in the signal processing and in logic circuits.

Semiconductor thermoelements are utilized for conversion of thermal energy into electric energy. Absence of moving parts in the thermoelectric generators makes them very reliable and able to operate unattended for many years (as an example, the Voyager spacecrafts traveling to the outer parts of the Solar system, where solar radiation is too weak to be used as the source for electric energy generation, are equipped with the thermoelectric generators using the heat released by radioactive isotopes and are able to function for up to 20 years).

Semiconductor sensors are used in measuring instruments to measure temperature, radiation level, electric and magnetic field parameters, mechanical pressure, etc. Development of the communication systems is virtually impossible without utilization of light detectors and light-emitting devices.

Quite a few books reviewing the achievements in the field of semiconductor device research and development are published consistently. But the continuously growing needs of industry and a gigantic application potential possessed by semiconductor materials make the information compiled in these books substantially outdated in a 2- to 3-year period, and the only way to keep pace is to follow the information reported in the current periodicals.

Their variety notwithstanding, the semiconductor electronic devices consist of a few elements, which have been described in this chapter, and their various combinations and modifications. Familiarity with these elements opens the door for in-depth study of the material relevant to a particular semiconductor device or integrated circuit. Thus, the information that has been presented in this chapter may be considered as a useful overview of the topic of semiconductor devices, added to which are the fundamental references that can be used for more detailed comprehension of the semiconductor device subject.

REFERENCES

8.1. D. H. Navon, *Electronic Materials and Devices*, Houghton Mifflin, Boston, 1975.
8.2. A. S. Grove, *Physics and Technology of Semiconductor Devices*, John Wiley & Sons, New York, 1967.
8.3. F. Beltram, F. Capasso, and S. Sen, *Electronic Materials*, J. R. Chelikovsky and A. Franciosi, Eds., Springer-Verlag, New York, 1991, 233.
8.4. J. M. Poate, *Electronic Materials*, J. R. Chelikovsky and A. Franciosi, Eds., Springer-Verlag, New York, 1991, 307.
8.5. E. S. Yang, *Fundamentals of Semiconductor Devices*, McGraw-Hill, New York, 1978.
8.6. B. G. Streetman, *Solid State Electronic Devices*, Prentice-Hall, Englewood Cliffs, NJ, 1980.
8.7. W. Shockley, *Bell Syst. Tech. J.*, 28, 435, 1949.
8.8. J. Bardeen, *Handbook of Physics*, 2nd ed., E. U. Condon and H. Odishaw, Eds., McGraw-Hill, New York, 1967, p. 8-52.
8.9. C. R. Crowell and S. M. Sze, *Solid State Electron.*, 9, 695, 1966.
8.10. S. M. Sze, *Physics of Semiconductor Devices*, 2nd ed., John Wiley & Sons, New York, 1981.
8.11. M. Shur, *Physics of Semiconductor Devices. Prentice Hall Series in Solid State Physical Electronics*, N. Holonyak, Jr., Ed., Prentice-Hall, Englewood Cliffs, NJ, 1990.
8.12. S. M. Sze, *Semiconductor Devices. Physics and Technology*, John Wiley & Sons, New York, 1985.
8.13. S. P. Murarka and M. C. Peckerar, *Electronic Materials Science and Technology*, Academic Press, New York, 1989.
8.14. W. R. Runyan and K. E. Bean, *Semiconductor Integrated Circuit Processing Technology*, Addison-Wesley, Reading, MA, 1990.
8.15. K. W. Böer, *Survey of Semiconductor Physics, Vol. 2, Barriers, Junctions, Surfaces and Devices*, Van Nostrand Reinhold, New York, 1992.
8.16. R. E. Williams, *Gallium Arsenide Processing Techniques*, Artech House, Dedham, MA, 1984.
8.17. V. L. Rideout, *Solid State Electron.*, 18, 541, 1975.
8.18. S. C. Gupta and H. Preier, *Metal-Semiconductor Schottky Barrier Junctions and Their Applications*, B. L. Sharma, Ed., Plenum Press, New York, 1984.
8.19. L. J. Brillson, *Thin Solid Films*, 89, 461, 1982; *Surf. Sci. Rep.*, 2, 123, 1982.
8.20. E. H. Rhoderick and R. H. Williams, *Metal-Semiconducting Contacts*, Oxford Scientific Publishing, Oxford, 1988.
8.21. C. Zener, *Phys. Rev.*, 71, 323, 1947.

8.22. M. H. Norwood and E. Shatz, *Proc. IEEE,* 56, 788, 1968.
8.23. M. J. Howes and D. V. Morgan, *Variable Impedance Devices,* John Wiley & Sons, New York, 1978.
8.24. L. Esaki, *Phys. Rev.,* 109, 603, 1958.
8.25. P. Horowitz and W. Hill, *The Art of Electronics,* Cambridge University Press, New York, 1980.
8.26. R. E. Hummel, *Electronic Properties of Materials,* Springer-Verlag, New York, 1985.
8.27. H. Holonyak, J. C. Campbell, M. H. Lee, J. D. Verdeyen, W. L. Johnson, M. G. Crawford, and D. Finn, *J. Appl. Phys.,* 44, 5517, 1973.
8.28. A. A. Bergh and P. J. Dean, *Light-Emitting Diodes,* Clarendon Press, Oxford, 1976 (cited in S. M. Sze, *Semiconductor Devices. Physics and Technology,* John Wiley & Sons, New York, 1985.
8.29. H. C. Casey and M. B. Panish, *Heterostructure Lasers,* Academic Press, New York, 1978.
8.30. W. Shockley, G. L. Pearson, and J. R. Haynes, *Bell Syst. Tech. J.,* 28, 344, 1949.
8.31. I. A. Lesk and V. P. Mathis, *IRE Conv. Rec.,* 6, 2, 1953.
8.32. V. A. Bluhm and T. P. Sylvan, *Solid State Design,* 5, 26, 1964.
8.33. J. Bardeen and W. H. Brittain, *Phys. Rev.,* 74, 130, 1948.
8.34. W. Shockley, *Bell Syst. Tech. J.,* 28, 435, 1949.
8.35. W. Shockley, M. Sparks, and G. K. Teal, *Phys. Rev.,* 83, 151, 1951.
8.36. ASTM Standards: (a) F528-91 "Test Method for Measurement of Common-Emitter DC Gain of Junction Transistors"; (b) F570-90 "Test Methods for Transistor Collector-Emitter Saturation Voltage"; and (c) F769-91 "Methods for Measuring Transistor and Diode Leakage Currents".
8.37. J. L. Moll, M. Tanenbaum, J. M. Goldey, and N. Holonyak, *Proc. IRE,* 44, 1174, 1956.
8.38. J. E. Lilienfeld, U.S. Patent No. 1,745,175, 1930.
8.39. O. Heil, Br. Patent No. 439,457, 1935.
8.40. J. L. Moll, *Wescon Conv. Rec.,* No. 3, 32, 1959.
8.41. W. G. Pfann and C. G. B. Garrett, *Proc. IRE,* 47, 2011, 1959.
8.42. W. Shockley, *Proc. IRE,* 40, 1365, 1952.
8.43. C. A. Mead, *Proc. IEEE,* 54, 307, 1966.
8.44. L. Johnston and W. T. Lynch, *IEDM Techn. Digest,* IEEE Publishing, Los Angeles, 1986, 252.
8.45. B. J. Van Zeghbrock, W. Patrick, H. Meier, P. Vettiger, and P. Wolf, *IEDM Techn. Digest,* IEEE Publishing, Los Angeles, 1986, 832.
8.46. E. H. Nicolian and J. R. Brews, *MOS Physics and Technology,* John Wiley & Sons, New York, 1982.
8.47. A. D. Milnes, Ed., *MOS Devices. Design and Manufacture,* Edinburgh University Press, 1983.
8.48. D. K. Schroeder, *Advanced MOS Devices,* Addison-Wesley, Reading, MA, 1987.
8.49. G. Rabbat, Ed., *Handbook of Advanced Semiconductor Technology and Computer Systems,* Van Nostrand Reinhold, New York, 1988.
8.50. Y. P. Tsividis, *Operation and Modeling of the MOS Transistor,* McGraw-Hill, New York, 1987.
8.51. A. S. Grove, E. H. Snow, B. E. Deal, and C. T. Sah, *J. Appl. Phys.,* 35, 2458, 1964.
8.52. C. H. Séquin and M. F. Tompsett, *Charge Transfer Devices,* Academic Press, New York, 1975.
8.53. W. S. Boyle and G. E. Smith, *Bell Syst. Tech. J.,* 49, 587, 1970.
8.54. D. K. Schroder, *Metal-Semiconductor Schottky Barrier Junctions and Their Applications,* B. L. Sharma, Ed., Plenum Press, New York, 1984, 225.
8.55. C. K. Kim, *Charge-Coupled Devices and Systems,* M. J. Howes and D. V. Morgan, Eds., John Wiley & Sons, New York, 1979, 1.
8.56. W. S. Boyle and G. E. Smith, U.S. Patent No. 3,792,322, 1974.
8.57. R. H. Walden, R. H. Krambeck, R. J. Strain, J. McKenna, N. L. Schryer, and G. E. Smith, *Bell Syst. Tech. J.,* 51, 1635, 1972.
8.58. D. J. Burt, Proc. Int. Conf. Technol. Applic. CCD, Edinburgh University, 1974, 1.
8.59. J. S. T. Huang and G. W. Taylor, *IEEE Trans.,* ED-22, 995, 1975.

Index

A

Absolute dilatometers, 33, see also Dilatometers
Absorption coefficient, see also Optical properties
  aluminum antimonide crystals, 131
  cadmium selenide crystals, 203, 205
  copper bromide crystals, 216–217
  diamond crystals, 48, 50
  germanium crystals, 73, 74, 78, 79
  gray tin crystals, 83
  I-III-VI_2 compounds, 254, 258
  I-IV_2-V_3 compounds, 245
  I_3-VI_4 compounds, 268–269, 272
  IA-IB compounds, 299, 302
  $(IA)_3$-VB compounds, 301, 312
  II-IV-V_2 compounds, 240, 245
  III-V/III-V semiconductor alloys, 367
  IV-VI compounds, 331, 333, 335–336, 354, 356, 357
  silver and copper chalcogenides, 310, 323, 328, 329
  silver and copper oxides, 306–307, 320
Acceptors, see also Donors
  boron nitride crystals, 116
  aluminum nitride crystals, 123
  diamond crystals, 44, 47
  electrical conductivity, 19
  gallium antimonide crystals, 150
  gallium phosphide crystals, 136
  VIIIA-containing elements, 351
  photoconductivity, 26
  p-n junction, 412–414
  Schottky barrier diode, 416
Acetylene, 55, see also Synthesis
Adamantine semiconductors, see Ternary adamantine semiconductors
Allotropes
  aluminum antimonide crystals, 128
  aluminum nitride, 123
  aluminum phosphide, 125
  boron crystals, 40–41
  boron nitride crystals, 112
  boron phosphide, 116, 118
  cadmium selenide crystals, 201–202
  cadmium sulfide crystals, 198
  cadmium telluride crystals, 203
  carbon, 43, see also Diamond
  copper bromide crystals, 216
  copper chloride crystals, 214
  copper iodide crystals, 217
  gallium arsenide crystals, 138
  gallium phosphide crystals, 134
  III_2-VI_3 compounds, 275
  indium arsenide crystals, 160–161
  indium phosphide crystals, 155–156, 158
  lead oxide, 327
  mercury selenide crystals, 210
  mercury sulfide crystals, 207–208
  mercury telluride crystals, 211
  phosphorus, 84, 85
  selenium, 86
  silicon, 56–58
    carbide crystals, 105, 108
  silver iodide crystals, 219
  zinc oxide crystals, 183
  zinc selenide crystals, 189
  zinc sulfide crystals, 186
  zinc telluride crystals, 193
Alloys
  elemental semiconductors, 355–357, 360
  I_2-IV-VI_3 compounds, 263–265
  II-VI and III-V compounds, 380, 384, 386–387
  II-VI/II-VI compounds, 367, 376, 378, 380
  III-V/III-V compounds, 362, 364–365, 367, 382, 383, 385
  ternary and photodiode applications, 431, 432
Alpha-rhombohedral boron, see Rhombohedral structure
Alpha-tetragonal boron, see Tetragonal structure
Aluminum, 185, see also Dopants/doping; Zinc oxide
Aluminum antimonide, 128–131, see also Binary IV-IV and III-V semiconductors
Aluminum arsenide, 125–128, see also Binary IV-IV and III-V semiconductors
Aluminum nitride, 123–124, see also Binary IV-IV and III-V semiconductors
Aluminum phosphide, 124–126, see also Binary IV-IV and III-V semiconductors
Amorphous state, 86, 87, 231, 271, see also Selenium; Group II-IV-V_2; Group IV_3-V_4
Amplification effect, 2
Annealing, 412, see also p-n Junction
Antifluorite structure (group II_2-IV compounds)
  band structure, 315, 332–333

crystal structure, 313, 330
electric transport and electrical properties, 316, 334, 335
optical properties, 317, 336
synthesis, 316–317
thermochemical, thermal, elastic properties, 315, 331, 336
Antimonides, 230, see also Group II-IV-V_2
Antimony, 1, 82, 325–326, see also Individual entries
Antimony triselenide, 344, 347, 362, 364, 365, see also Group V_2-VI_3
Antimony trisulfide, 339, 342, 348, see also Group V_2-VI_3
Antimony tritelluride, 339, 344, 345, 348, 366, see also Group V_2-VI_3
Antiphase domain boundaries (APBs), 144–145, see also Gallium arsenide
APBs, see Antiphase domain boundaries
Applications
  aluminum nitride crystals, 123–124
  boron nitride crystals, 116
  diamond crystals, 43
  elemental boron, 42–43
  gallium antimonide crystals, 152
  gallium arsenide crystals, 143
  gallium phosphide crystals, 137
  germanium, 66
  I-III-VI_2 compounds, 251
  $(IA)_3$-VB compounds, 297, 301–302
  II-VI/II-VI solid solutions, 376, 378
  III-V/III-V semiconductor alloys, 362, 367
  IV_3-V_4 compounds, 273
  IV-VI compounds, 317, 336, 338
  indium antimonide crystals, 170–171
  indium arsenide crystals, 165, 166
  mercury sulfide crystals, 209
  organic semiconductors, 386
  semiconductor devices, 446
    Schottky barrier and p-n junction diodes, 416
    varistors and varactors, 422
  silicon, 56
    carbide crystals, 105, 108, 110
  tellurium crystals, 91
  zinc oxide crystals, 185, 186
  zinc selenide crystals, 192–193
  zinc sulfide crystals, 187
APW, see Augmented plane wave
Aromatic hydrocarbons, 386, see also Organic semiconductors
Arsenic, 1, see also Individual entries
Arsenic triselenide, 339, 347, 348, 365, 366, see also Group V_2-VI_3
Arsenic trisulfide, 339, 365, see also Group V_2-VI_3
Arsenic tritelluride, 339, 347, see also Group V_2-VI_3
Arsenides, 230, see also Group II-IV-V_2
Arsenolite, 339, 348, see also Group V_2-VI_3
Arsenopyrite, 353, 376, see also Group VIIIA
Auger recombination, 31
Augmented plane wave (APW), 296, see also Group IA-IB

Avalanche diodes, 423–424
Avalanche photodiode, 430, see also Multiterminal (multibarrier) devices; Photodiode

B

Band diagram, 434, 435, see also Bipolar junction transistors
Band gap, 31, 41
Band structure
  aluminum antimonide crystals, 129, 130
  aluminum arsenide crystals, 127
  aluminum nitride crystals, 123, 124
  aluminum phosphide crystals, 125, 126
  boron arsenide crystals, 121
  boron nitride crystals, 113, 114, 116, 117
  boron phosphide crystals, 119, 120
  cadmium selenide crystals, 202, 203
  cadmium sulfide crystals, 199, 200
  cadmium telluride crystals, 205, 207
  copper bromide crystals, 216
  copper chloride crystals, 214, 215
  copper iodide crystals, 218, 219
  diamond crystals, 52, 53
  gallium antimonide crystals, 149, 150
  gallium arsenide crystals, 139, 142
  gallium nitride crystals, 131–132
  gallium phosphide crystals, 136
  germanium crystals, 69–70, 75
    -silicon alloys, 93
  gray selenium crystals, 87
  gray tin crystals, 76, 79
  I-III-VI_2 compounds, 250–254
  I_2-IV-VI_3 compounds, 256
  I_3-VI_4 compounds, 267
  IA-IB compounds, 296, 298
  $(IA)_3$-VB compounds, 299, 304–308
  II_2-IV compounds, 315, 332–333
  II-IV-V_2 compounds, 238–240
  IV-VI compounds, 319–320, 339, 340, 343
  V_2-VI_3 compounds, 339, 360
  VIIIA-containing elements, 353
  indium antimonide crystals, 169
  indium arsenide crystals, 162, 164
  indium nitride crystals, 153–154
  mercury telluride crystals, 211, 212
  phosphorus crystals, 84, 85
  silicon crystals, 56, 57
    carbide crystals, 108, 109
  silver and copper oxides, 303, 305, 316, 317
  silver iodide crystals, 220, 221
  zinc oxide crystals, 184
  zinc selenide crystals, 191
  zinc sulfide crystals, 187, 188
  zinc telluride crystals, 196
Barrier(s)
  height in metal-gallium arsenide semiconductors, 418, 419
  p-n junction diode, 413–415
  thyristors, 437
  Zener diodes, 423

INDEX

BCCD, see Buried-channel charge-coupled device
Beryllium, 116, see also Boron nitride; Dopants/doping
Beta-rhombohedral boron, see Rhomohedral structure
Beta-tetragonal boron, see Tetragonal structure
Bias, see also Forward bias; Reverse bias
    metal-insulator-semiconductor diodes, 444
    p-n junction diode, 413, 415
    two-electrode devices, 421–423
        tunnel diode conditions, 424–426
Binary II-VI and I-VII tetrahedral semiconductors
    I-VII compounds
        copper bromide, 216–217
        copper chloride, 214–216
        copper fluoride, 214
        copper iodide, 217–219
        silver iodide, 219–222
    II-VI compounds
        cadmium selenide, 201–203
        cadmium sulfide, 198–201
        cadmium telluride, 203–207
        mercury selenide, 210
        mercury sulfide, 207–210
        mercury telluride, 210–213
        zinc oxide, 183–186
        zinc selenide, 187, 189–194
        zinc sulfide, 186–188
        zinc telluride, 193, 195–198
Binary IV-IV and III-V semiconductors
    IIIB-VB compounds
        aluminum antimonide, 128–131
        aluminum arsenide, 125–128
        aluminum nitride, 123–124
        aluminum phosphide, 124–126
        boron arsenide, 121–122
        boron nitride, 112–117
        boron phosphide, 116, 118–122
        crystal structure, 110–112
        gallium antimonide, 145, 147–152
        gallium arsenide, 138–146
        gallium nitride, 131–134
        gallium phosphide, 134–137
        indium antimonide, 165–171
        indium arsenide, 160–165
        indium nitride, 152–155
        indium phosphide, 155–160
    IVB-IVB compounds, 105–110
Binary inorganic semiconductors, 7–11
Binary valence compounds, 3, 4
    ternary analogs, see Ternary valence compounds
Bipolar devices, 433–438, see also Multiterminal (multibarrier) devices
Bipolar junction transistor (BJT), 434–436, see also Multiterminal (multibarrier) devices
Bismite, 348, see also Group V_2-VI_3
Bismuth, 1, 80–81, see also Dopants/doping; Gray tin
Bismuth selenide, 339, 345, 363, see also Group V_2-VI_3
Bismuth sesquioxide, 345, 347, 365, see also Group V_2-VI_3
Bismuth triselenide, 349, 367, see also Group V_2-VI_3
Bismuth trisulfide, 348, see also Group V_2-VI_3
Bismuth tritelluride, 349, 368, see also Group V_2-VI_3

BJT, see Bipolar junction transistor
Black phosphorus, 85, 86, see also Phosphorus
Bonding strength, 38, see also Boron
Borazone, 116, see also Boron nitride
Boron, see also Elemental semiconductors
    chemical interactions, 91, 92
    impurities and diamond properties, 43, 44
    p-n junction fabrication for semiconducting devices, 412
    properties, 37–43
Boron arsenide, 121–122, see also Binary IV-IV and III-V semiconductors
Boron carbides, see Boron, chemical interactions
Boron nitride, 112–116, see also Binary IV-IV and III-V semiconductors
Boron phosphide, 116, 118–121, see also Binary IV-IV and III-V semiconductors
Bravais structure, 339, see also Group V_2-VI_3
Breaking strength, 139, 148
Bridgman method, see also Synthesis
    cadmium telluride crystals, 206
    copper chloride crystals, 215
    indium antimonide crystals, 170
    mercury selenide crystals, 210
    mercury telluride crystals, 213
Bridgman-Stockbarger method, 199, 203, see also Synthesis
Brillouin zone, see also Band structure
    aluminum antimonide crystals, 129
    boron arsenide crystals, 121
    boron nitride crystals, 113, 114
    boron phosphide crystals, 119
    cadmium selenide crystals, 202
    cadmium sulfide crystals, 199
    diamond crystals, 52, 53
    gallium antimonide crystals, 149
    germanium crystals, 69
    IA-IB compounds, 296, 297
    $(IA)_3$-VB compounds, 299, 304
    II_2-IV compounds, 315
    II-IV-V_2 compounds, 235, 238–239
    IV-VI compounds, 319, 323, 339–342
    V_2-VI_3 compounds, 339, 360, 361
    indium antimonide crystals, 169
    phosphorus crystals, 84
    zinc oxide crystals, 184
Brinell hardness, 87, see also Gray selenium; Hardness
Bulk modulus, see also Elastic properties
    aluminum antimonide crystals, 128
    copper chloride crystals, 214
    diamond crystals, 48, 51
    I-III-VI_2 compounds, 247, 250
    I-IV_2-V_3 compounds, 247
    II-IV-V_2 compounds, 235, 237
    IV-VI compounds, 338
    indium arsenide crystals, 161
    indium phosphide crystals, 158
    mercury sulfide crystals, 208
    silver and copper oxides, 315
Buried-channel charge-coupled device (BCCD), 445

Buried-channel metal-oxide semiconductor field effect transistor (Buried-channel MOSFET), 445
Buried-channel MOSFET, see Buried-channel metal-oxide semiconductor field effect transistor
Burstein shift, 323, see also Binary II-VI and I-VIII tetrahedral semiconductors

C

Cadmium selenide, 201–203, see also Binary II-VI and I-VIII tetrahedral semiconductors
Cadmium sulfide, 198–201, see also Binary II-VI and I-VIII tetrahedral semiconductors
Cadmium telluride, 203–207, see also Binary II-VI and I-VIII tetrahedral semiconductors
Calorimetry method, 33, see also Specific heat
Capacitance-voltage characteristic, 445
Capacitor, 444, see also Field effect transistors
Carbon, 39, 55, 91, 92, see also Diamond
Carrier free path time, 23, see also Electrical conductivity
Carrier mobility, see Mobility
Cauchy relations, 48, see also Diamond
CCD, see Charge-coupled devices
Cesium antimonide, 301, 311, 312, see also Group $(IA)_3$-VB
Cesium auride, 296, 298, 299, see also Group IA-IB
Cesium chloride, 296
Chalcogenides
  properties of VIIIA-containing elements, 353, 354, 381–382
  silver and copper semiconductor properties
    chemical properties and crystal structure, 307, 321
    commercial availability, 313, 330
    electric transport, 312, 324
    electrical conductivity, 312, 325
    energy gap, 307, 309–311
    optical properties, 310, 323, 326–329
    Seebeck effect, 312, 326
    thermal, thermodynamic, mechanical properties, 307, 322
Chalcopyrite structure
  I-III-VI_2 compounds, 246, 247, 249
  II-IV-V_2 compounds, 230, 235
  sphalerite structure comparison, 235–239
Charge-coupled devices (CCD), 445, 446, see also Multiterminal (multibarrier) devices
Chemical bonds, 15–18
Chemical inertness, 119, see also Boron phosphide
Chemical interactions, 91–94, see also Elemental semiconductors
Chemical methods, 42, see also Synthesis
Chemical transport, 187, see also Synthesis; Zinc sulfide
Chemical vapor deposition (CVD)
  boron nitride crystals, 113
  boron phosphide crystals, 120
  cadmium sulfide crystals, 199
  copper chloride crystals, 216
  defect ternary compounds, 284
  diamond crystals, 55, 56
  IA-IB compounds, 295–296
  mercury selenide crystals, 210
  zinc selenide crystals, 192, 195
Cinnabar structure, 208, 211, see also Mercury sulfide; Mercury telluride
Circuit configurations, 434–436, see also Bipolar junction transistors; Multiterminal (multibarrier) devices
Clausius equation, 198, see also Cadmium sulfide
Cobalt, 145, see also Gallium arsenide
Coefficient of thermal expansion
  aluminum antimonide crystals, 128
  aluminum arsenide crystals, 125–126
  aluminum nitride crystals, 123
  boron nitride crystals, 114–116
  boron phosphide crystals, 119
  cadmium selenide crystals, 202
  cadmium sulfide crystals, 198
  cadmium telluride crystals, 204, 206
  copper chloride crystals, 214
  copper iodide crystals, 217, 218
  diamond crystals, 46
  gallium antimonide crystals, 145, 147
  gallium arsenide crystals, 138, 139
  gallium nitride crystals, 131, 132
  gallium phosphide crystals, 134, 135
  germanium crystals, 66, 69
  gray selenium crystals, 87
  gray tin crystals, 79, 80
  I-III-VI_2 compounds, 247–249
  I-IV_2-V_3 compounds, 244, 247
  I_2-IV-VI_3 compounds, 256, 259
  I_3-VI_4 compounds, 266, 267
  II_2-IV compounds, 315, 331
  II-IV-V_2 compounds, 231, 234, 235
  III-V/III-V semiconductor alloys, 365, 367
  IV_3-V_4 compounds, 274
  IV-VI compounds, 338
  VIIIA-containing elements, 374–375
  indium arsenide crystals, 161, 162
  indium phosphide crystals, 155–157
  measurement in semiconductors, 33–34
  mercury selenide crystals, 210
  mercury telluride crystals, 211, 212
  silicon crystals, 57–59
    carbide crystals, 106, 107
  silver and copper oxides, 315
  silver iodide crystals, 220
  tellurium, 89
  zinc oxide crystals, 183
  zinc selenide crystals, 189, 190
  zinc sulfide crystals, 186
  zinc telluride crystals, 193, 195
Collector, 434–436, see also Bipolar junction transistors; Multiterminal (multibarrier) devices
Combustion flame synthesis, 56, see also Diamond; Synthesis
Combustion process, 107–108, see also Silicon carbide; Synthesis
Comparator dilatometers, 33, see also Dilatometers
Composition nonlinearity, 378, see also Group II-VI/II-VI, solid solutions

Computers, 2, 411
Concentration triangle, 242, 246, 256, 258, see also Group I_2-IV-VI_3; Group I-IV_2-V_3
Condensed matter, 16, see also Gas phase
Conduction band, see also Individual entries
    diagram of tunnel diode, 424
    electrical conductivity in semiconductor crystals, 19, 20
    photodiodes, 427, 428
Conduction electrons, see also Electrons, mobility
    diamond crystals, 52
    generation and n-type impurities in semiconductor crystals, 18
    germanium crystals, 69
    semiconductor crystals, 15–17
Copper bromide, 216–217, see also Binary II-VI and I-VIII tetrahedral semiconductors
Copper chloride, 214–216, see also Binary II-VI and I-VIII tetrahedral semiconductors
Copper fluoride, 214, see also Binary II-VI and I-VIII tetrahedral semiconductors
Copper iodide, 217–219, see also Binary II-VI and I-VIII tetrahedral semiconductors
Copper-germanium phosphide, see Group I-IV_2-V_3 compounds
Corrosion, 121, see also Boron arsenide
Cost/availability
    aluminum antimonide, 131
    aluminum phosphide, 125
    boron nitride, 116
    boron phosphide, 121
    cadmium selenide, 203
    cadmium sulfide, 199
    cadmium telluride, 207
    copper chloride, 216
    copper iodide, 218
    elemental boron, 42
    gallium arsenide, 145
    gallium phosphide, 137
    I-III-VI_2 compounds, 256
    II_2-IV compounds, 317
    IV-VI compounds, 338, 357
    V_2-VI_3 compounds, 351, 368
    indium antimonide, 171
    indium phosphide, 160
    mercury selenide, 210
    mercury sulfide, 210
    mercury telluride, 213
    natural diamonds, 43
    phosphorus crystals, 86
    silicon carbide, 110
    silver and copper chalcogenides, 313, 330
    silver and copper oxides, 307
    silver iodide, 222
    zinc oxide, 186
    zinc selenide, 193
    zinc sulfide, 187
    zinc telluride, 198
Covalent bonding, 15, 231, 234, see also Group II-IV-V_2
Crystal growth, see Synthesis
Crystal structure
    boron crystals, 37
        boron phosphide, 119
    modifications, 39
    cuprite, 315
    defect ternary compounds, 282, 283, 287
    gray selenium, 86–87
    gray tin, 74
    I-III-VI_2 compounds, 247, 249
    I_2-IV-VI_3 compounds, 256, 259
    I_3-VI_4 compounds, 266
    IA-IB compounds, 295, 296
    $(IA)_3$-VB compounds, 299, 303
    II_2-IV compounds, 313, 330
    II-IV-V_2 compounds, 231–233
    IIIB-VB compounds, 110, 112
    III_2-VI_3 compounds, 275–277
    IV_3-V_4 compounds, 274
    IV-VI compounds, 318, 337
    V_2-VI_3 compounds, 339, 358
    VIIIA-containing elements, 369–373
    mathematical method of composition prediction, 2–3
    organic semiconductors, 394–395
    p-n junction in semiconductor devices, 411
    silicon carbide crystals, 105
    silver and copper chalcogenides, 321
    silver and copper oxides, 314
Current carriers
    germanium crystals, 70
    mobility in semiconductor crystals, 16
    p-n junction diode, 414–415
    silicon crystals, 60, 64
Current-voltage characteristic
    field effect transistors, 441
    ohmic contact and Schottky barrier diode, 417
    photodiode, 427, 429
    p-n junction diode, 415, 423
    tunnel diodes, 425
    unijunction transistors, 433
CVD, see Chemical vapor deposition
Cyclotron resonance, 69, see also Germanium
Czochralski technique, see also Synthesis
    aluminum antimonide crystals, 131
    copper chloride crystals, 216
    gallium antimonide crystals, 150
    gallium arsenide crystals, 138, 143
    indium antimonide crystals, 170
    silicon crystals, 64, 68

D

Dark conductivity, 26, 396
DC jet method, 54, see also Diamond; Synthesis
Debye characteristics
    aluminum antimonide crystals, 128, 129
    aluminum arsenide crystals, 126
    aluminum nitride crystals, 123
    boron phosphide crystals, 119
    cadmium selenide crystals, 202
    cadmium telluride crystals, 204
    copper bromide crystals, 216
    copper chloride crystals, 214
    copper iodide crystals, 218
    gallium antimonide crystals, 148
    gallium arsenide crystals, 138, 141

gallium phosphide crystals, 134
germanium crystals, 69–71
I-III-VI$_2$ compounds, 249
I$_2$-IV-VI$_3$ compounds, 256, 260
I$_3$-VI$_4$ compounds, 266, 267
II-IV-V$_2$ compounds, 235, 236
III$_2$-IV-VI compounds, 289
IV-VI compounds, 338
V$_2$-VI$_3$ compounds, 359
VIIIA-containing elements, 374–375
indium antimonide crystals, 168
indium arsenide crystals, 161, 163
indium nitride crystals, 153
indium phosphide crystals, 156, 158
mercury selenide crystals, 210
mercury telluride crystals, 211
silicon crystals, 58, 61
silver and copper chalcogenides, 321
silver and copper oxides, 315
silver iodide crystals, 220
zinc oxide crystals, 183
zinc selenide crystals, 189
zinc sulfide crystals, 186
zinc telluride crystals, 195
Decomposition, 125, 186, 231, 235, see also Melting point
Defect adamantine semiconductors, see Ternary adamantine semiconductors
Defect ternary compounds, 281–286, 288
Defects, 16, 143, 353
Deformation, 138
Degenerated n- and p-type semiconductors, 424
Dember effect, 28, see also Photoconductivity
Density
　aluminum antimonide crystals, 128
　aluminum nitride crystals, 123
　aluminum phosphide crystals, 125
　boron phosphide crystals, 119
　cadmium selenide crystals, 202
　cadmium sulfide crystals, 198
　cadmium telluride crystals, 204
　copper iodide crystals, 217
　defect ternary compounds, 285–286
　gallium antimonide crystals, 145
　gallium arsenide crystals, 139–140
　gallium nitride crystals, 131
　gallium phosphide crystals, 134
　germanium crystals, 66
　gray selenium crystals, 87
　gray tin crystals, 79
　I-III-VI$_2$ compounds, 246–248
　I$_2$-IV-VI$_3$ compounds, 256, 259
　IA-IB compounds, 296, 300
　(IA)$_3$-VB compounds, 299, 308
　II$_2$-IV compounds, 330
　II-IV-V$_2$ compounds, 231, 235
　III$_2$-VI$_3$ compounds, 275, 279
　IV-VI compounds, 337
　VIIIA-containing elements, 369–373, 377–378
　indium antimonide crystals, 165
　indium arsenide crystals, 161
　indium nitride crystals, 152

　mercury sulfide crystals, 208
　mercury telluride crystals, 211
　organic semiconductors, 394–395
　phosphorus crystals, 85
　silicon carbide crystals, 107
　silicon crystals, 58, 60
　silver and copper chalcogenides, 321
　silver and copper oxides, 314
　silver iodide crystals, 219
　tellurium, 89
　zinc oxide crystals, 183
　zinc sulfide crystals, 186
　zinc telluride crystals, 193
Diamond
　elemental semiconductor, 43–56
　-like dopants and gray tin crystals, 80–81
Diamond-like materials, 2–3, 131, see also Aluminum antimonide
Dielectric constant
　aluminum antimonide crystals, 131
　aluminum arsenide crystals, 127
　aluminum nitride crystals, 123–124
　aluminum phosphide crystals, 125
　boron nitride crystals, 115, 116
　boron phosphide crystals, 120
　cadmium selenide crystals, 202
　cadmium sulfide crystals, 199
　cadmium telluride crystals, 206
　copper bromide crystals, 216
　copper chloride crystals, 215
　copper iodide crystals, 218, 220
　defect ternary compounds, 288
　gallium antimonide crystals, 151, 154
　gallium arsenide crystals, 147
　gallium nitride crystals, 133
　gallium phosphide crystals, 136–137
　germanium crystals, 73–74, 78, 79
　gray selenium crystals, 87
　I-III-VI$_2$ compounds, 251, 256
　II$_2$-IV compounds, 316, 336
　II-IV-V$_2$ compounds, 240, 244
　III$_2$-VI$_3$ compounds, 281, 282
　IV$_3$-V$_4$ compounds, 274
　IV-VI compounds, 331, 335, 349
　V$_2$-VI$_3$ compounds, 347, 365
　VIIIA-containing elements, 355, 381–382
　indium antimonide crystals, 170, 171
　indium arsenide crystals, 163, 165
　indium nitride crystals, 154
　indium phosphide crystals, 160, 161
　mercury selenide crystals, 210
　mercury sulfide crystals, 209
　mercury telluride crystals, 213
　organic semiconductors, 395–396
　phosphorus crystals, 84
　silicon crystals, 61, 66, 67
　　carbide crystals, 108
　silver and copper chalcogenides, 313, 326
　silver and copper oxides, 306, 319
　silver iodide crystals, 222
　zinc oxide crystals, 185
　zinc selenide crystals, 192

zinc sulfide crystals, 187
zinc telluride crystals, 197
Diffusion coefficient, 74, see also Germanium
Diffusion length, 31, 415, 430, see also Semiconductor devices
Diffusive reflectivity, see also Reflectivity
 I-IV$_2$-V$_3$ compounds, 242, 244, 246
 I$_3$-VI$_4$ compounds, 268–269, 272, 273
Dilatometers, 134, 247, see also Gallium phosphide; Group I-III-VI$_2$
Dislocation formation, 143–144, see also Gallium arsenide
Donors, see also Acceptors
 boron nitride crystals, 116
 aluminum nitride crystals, 123
 diamond crystals, 44, 47
 electrical conductivity, 18, 21
 gallium antimonide crystals, 150
 gallium phosphide crystals, 136
 VIIIA-containing elements, 351
 ohmic contact in semiconductors, 417
 photoconductivity, 26
 p-n junction, 412–414
 Schottky barrier diode, 416
Dopants/doping
 boron nitride crystals, 116
 electrical conductivity of aluminum nitride crystals, 123
 electrical transport in aluminum antimonide crystals, 129
 gallium arsenide crystals
  hardening, 138–139
  synthesis, 144, 145
 gray selenium crystals, 87
 gray tin crystals, 80–81
 nitrogen and mobility in silicon carbide crystals, 108
 p-n junction, 412, 422
 sodium and resistivity in (IA)$_3$-VB compounds, 299, 301, 310
 sulfur and hardness of indium phosphide crystals, 158, 159
 Tunnel diodes, 424
 Zener diodes, 423
 zinc oxide crystals, 185
Drift mobility, see also Mobility
 aluminum nitride crystals, 123
 gallium arsenide crystals, 142, 144
 gallium nitride crystals, 133
 gallium phosphide crystals, 136
 indium antimonide crystals, 170
 indium nitride crystals, 155
 silicon crystals, 60, 64
 zinc sulfide crystals, 187
Ductility, 15
Dynamic resistance, 434, 435, see also Semiconductor devises

E

Effective mass, see also Electrical conductivity
 aluminum antimonide crystals, 129
 aluminum arsenide crystals, 127
 aluminum nitride crystals, 123
 boron phosphide crystals, 119
 cadmium selenide crystals, 202
 cadmium sulfide crystals, 199
 cadmium telluride crystals, 205
 copper bromide crystals, 216
 copper chloride crystals, 214
 defect ternary compounds, 288
 electrical conductivity, 23
 gallium antimonide crystals, 149
 gallium arsenide crystals, 139–140
 gallium nitride crystals, 132
 gallium phosphide crystals, 136
 germanium crystals, 70, 76
 gray tin crystals, 80
 I-III-VI$_2$ compounds, 250, 254
 I$_2$-IV-VI$_3$ compounds, 256, 258, 261
 I$_3$-VI$_4$ compounds, 270
 II$_2$-IV compounds, 333
 II-IV-V$_2$ compounds, 239, 242
 II-VI/II-VI solid solutions, 378
 IV-VI compounds, 343
 V$_2$-VI$_3$ compounds, 339, 361
 VIIIA-containing elements, 377–378
 indium antimonide crystals, 169
 indium arsenide crystals, 162
 indium nitride crystals, 154
 indium phosphide crystals, 158–159
 mercury selenide crystals, 210
 mercury telluride crystals, 211
 phosphorus crystals, 84
 silicon crystals, 60, 64
  carbide crystals, 108, 109
 silver and copper chalcogenides, 307, 309, 311
 silver and copper oxides, 305
 tellurium crystals, 90
 zinc oxide crystals, 184
 zinc selenide crystals, 191
 zinc sulfide crystals, 187
 zinc telluride crystals, 196
 semiconductor crystals, 18
Einstein's equation, 27–28, see also Photoconductivity
Elastic properties
 aluminum antimonide crystals, 128
 aluminum arsenide crystals, 126–127
 aluminum phosphide crystals, 125
 boron nitride crystals, 116
 boron phosphide crystals, 119
 cadmium selenide crystals, 202
 cadmium sulfide crystals, 199
 cadmium telluride crystals, 204
 copper bromide crystals, 216
 copper chloride crystals, 214
 copper iodide crystals, 218
 diamond crystals, 48, 51
 gallium antimonide crystals, 148, 149
 gallium arsenide crystals, 138–139, 141
 gallium nitride crystals, 131
 gallium phosphide crystals, 134, 135
 germanium crystals, 69, 72–75
 gray tin crystals, 79
 I-III-VI$_2$ compounds, 247, 250

I-IV$_2$-V$_3$ compounds, 244, 247
I$_2$-IV-VI$_3$ compounds, 256, 260
II$_2$-IV compounds, 315, 331
IV-VI compounds, 338
indium antimonide crystals, 168, 169
indium arsenide crystals, 161–162
indium nitride crystals, 153
indium phosphide crystals, 156, 158
mercury sulfide crystals, 208
mercury telluride crystals, 211
organic semiconductors, 394–395
silicon crystals, 59, 62
    carbide crystals, 108
silver and copper oxides, 315
silver iodide crystals, 220
zinc oxide crystals, 183
zinc selenide crystals, 190
zinc sulfide crystals, 186
zinc telluride crystals, 195
Elasticity modulus, 42, see also Boron
Electric transport, see also Electrical conductivity; Mobility
    aluminum arsenide crystals, 127
    aluminum antimonide crystals, 129
    boron crystals, 41
    copper chloride crystals, 214–215
    copper iodide crystals, 218
    diamond crystals, 44
    gallium antimonide crystals, 149
    germanium crystals, 70–71
    I$_3$-VI$_4$ compounds, 270
    IA-IB compounds, 296
    II$_2$-IV compounds, 315–316
    II-VI/II-VI solid solutions, 378
    III-V/III-V semiconductor alloys, 367
    V$_2$-VI$_3$ compounds, 339
    VIIIA-containing elements, 354
    indium nitride crystals, 154
    mercury sulfide crystals, 208
    mercury telluride crystals, 212
    silicon carbide crystals, 108
    silver and copper chalcogenides, 312, 324
    silver iodide crystals, 220
    tellurium crystals, 90
    zinc oxide crystals, 185
    zinc selenide crystals, 191, 193
    zinc sulfide crystals, 187, 189
    zinc telluride crystals, 196
Electrical conductivity
    aluminum antimonide crystals, 129, 131
    aluminum nitride crystals, 123
    aluminum phosphide crystals, 125
    boron crystals, 41
        -carbon alloys, 91
        nitride crystals, 116, 118
        phosphide crystals, 119, 121
    cadmium sulfide crystals, 199
    cadmium telluride crystals, 206
    diamond, 43
    elemental semiconducting alloys, 360
    gallium arsenide crystals, 140–142, 145
    gallium nitride crystals, 133
    germanium crystals, 71–72
    -silicon alloys, 93
    gray tin crystals, 76, 80, 81
    I-III-VI$_2$ compounds, 247
        annealing, 250, 255
    I-IV$_2$-V$_3$ compounds, 244, 247
    I$_2$-IV-VI$_3$ compounds, 260, 262–264
    I$_3$-VI$_4$ compounds, 267–269
    IA-IB compounds, 297
    (IA)$_3$-VB compounds, 308–310
    II$_2$-IV compounds, 315–316, 334, 335
    II-IV-V$_2$ compounds, 239, 242
    III$_2$-VI$_3$ compounds, 280, 281
    IV-VI compounds, 323, 325, 330–331, 348
    V$_2$-VI$_3$ compounds, 339, 344, 345, 347
    VIIIA-containing elements, 354, 379–380
    indium arsenide crystals, 163
    mercury sulfide crystals, 208
    mercury telluride crystals, 213
    metals vs. nonmetals, 1
    organic semiconductors, 386
    phosphorus crystals, 84
    semiconductor crystals, 17–24
        conduction electrons, 15–16
        measurement, 30
    silicon crystals, 57, 59, 60, 65
    silver and copper chalcogenides, 307, 312, 325, 328
    solid and liquid selenium, 87, 89
    solid electrolytes, 1
    zinc sulfide crystals, 187, 189
    zinc telluride crystals, 196, 197
Electrical resistivity, see Resistivity
Electrode materials, 417–419, see also Semiconductor devices
Electroluminescence, 137, 380, see also Binary II-VI and I-VIII tetrahedral semiconductors; Gallium phosphide
Electrolytes, solid, 1
Electromagnetic radiation, 29, see also p-n Junction
Electron effective mass tensors, 22–23, see also Electrical conductivity
Electronegativity, 231, 234, 295, see also Group IA-IB; Group II-IV-V$_2$
Electronic charge transport, 1
Electronic materials, 1
Electronic semiconductors, 16
Electrons, see also Holes
    acceleration calculation, 22
    behavior in solid materials, 16
    bipolar junction transistors, 434–436
    dissipation, 16
    distribution in p-n junction diode, 413–415
    emission coefficient, 300, see also Group IA-IB
    energy in diamond crystals, 52
    germanium crystals, 70, 76
    lifetime calculation, 31
    metal-insulator-semiconductor diodes, 444–445
    mobility
        aluminum antimonide crystals, 129
        aluminum arsenide crystals, 127
        aluminum phosphide crystals, 125
        aluminum nitride crystals, 123

INDEX 457

    boron phosphide crystals, 119, 121
    cadmium selenide crystals, 202
    cadmium sulfide crystals, 199
    cadmium telluride crystals, 205–206
    elemental semiconducting alloys, 357
    gallium antimonide crystals, 149–151
    gallium arsenide crystals, 140–142
    gallium nitride crystals, 132–133
    gallium phosphide crystals, 136
    germanium-silicon alloys, 94
    gray tin crystals, 75, 80, 82
    I_2-IV-VI_3 compounds, 264, 265
    II_2-IV compounds, 334
    II-IV-V_2 compounds, 235, 237–239, 242
    II-VI/II-VI solid solutions, 378
    III-V/III-V semiconductor alloys, 367
    IV-VI compounds, 323, 327, 331
    V_2-VI_3 compounds, 339, 344
    VIIIA-containing elements, 379–380
    indium antimonide crystals, 169–170
    indium arsenide crystals, 163
    indium nitride crystals, 155–156
    indium phosphide crystals, 159–160
    mercury telluride crystals, 212
    organic semiconductors, 395–396
    photoconductivity in semiconductor crystals, 28
    selenium crystals, 86
    silicon crystals, 59, 108, 111
    silver and copper chalcogenides, 324
    silver and copper oxides, 305–306
    tellurium crystals, 90
    zinc oxide crystals, 185
    zinc selenide crystals, 191
    zinc sulfide crystals, 187
    zinc telluride crystals, 196
  photodiodes, 427, 428, 430
  recombination in semiconductor crystals, 25–26
Elemental semiconductors
  boron, 37–43
  chemical interactions, 91–94
  diamond, 43–56
  germanium, 65–74, 77
  gray tin, 74–76, 78–81
  Periodic table groups III, IV, V, VI, 6–7
  phosphorus (group V), 81–86
  selenium, 86–88
  silicon, 56–65
  tellurium, 88–91
Emitter, 433–436, see also Semiconductor devices
Energy band concept, 16
Energy gap
  aluminum antimonide crystals, 129, 130
  aluminum arsenide crystals, 127
  aluminum nitride crystals, 123
  aluminum phosphide crystals, 125
  boron alloys, 92
  boron arsenide crystals, 121, 122
  boron nitride crystals, 113, 116, 117
  boron phosphide crystals, 119
  cadmium selenide crystals, 202
  cadmium sulfide crystals, 199
  cadmium telluride crystals, 205

  copper bromide crystals, 216
  copper chloride crystals, 214
  copper iodide crystals, 218
  defect ternary compounds, 288
  diamond crystals, 43, 52
  gallium antimonide crystals, 149
  gallium arsenide crystals, 138, 139, 142
  gallium nitride crystals, 131–133
  gallium phosphide crystals, 134, 136
  germanium crystals, 70, 76
    -silicon alloys, 93
  gray arsenic, 81
  gray selenium crystals, 87
  gray tin crystals, 75, 80, 81, 84
  I-III-VI_2 compounds, 247, 250, 254
  I-IV_2-V_3 compounds, 242, 244, 247
  I_2-IV-VI_3 compounds, 256, 261
  I-V binary systems, 297
  I_3-VI_4 compounds, 267, 268
  IA-IB compounds, 296, 297, 300
  $(IA)_3$-VB compounds, 308
  II-IV-V_2 compounds, 237–240
  II-VI/II-VI solid solutions, 376, 378, 386
  II-VI and III-V alloys, 380, 387
  II_2-IV and II_2-IV compounds, 333
  III-V/III-V semiconductor alloys, 364, 383, 384, 386
  III_2-VI_3 compounds, 280
  IV-VI compounds, 319, 320, 323, 336, 343, 356
  V_2-VI_3 compounds, 339, 361
  VIIIA-containing elements, 377–378
  indium antimonide crystals, 169
  indium arsenide crystals, 162–163
  indium phosphide crystals, 158
  mercury selenide crystals, 210
  mercury sulfide crystals, 208
  mercury telluride crystals, 211–212
  monoclinic selenium, 86
  organic semiconductors, 395–396
  phosphorus crystals, 84
  photodiodes, 430–431
  p-n junction diode, 413
  semiconductor crystals, 18–19, 27
  silicon crystals, 56, 57
    carbide crystals, 105, 108, 110
  silver and copper chalcogenides, 307, 309–311
  silver and copper oxides, 303, 305
  silver iodide crystals, 220
  tellurium crystals, 90, 91
  zinc oxide crystals, 184
  zinc selenide crystals, 190–191
  zinc sulfide crystals, 186–187
  zinc telluride crystals, 195–196
Energy loss function, 350, 368, see also Group V_2-VI_3
Energy spectrum, 16, see also Gas phase
Enhancement-type devices, 443, see also Field effect transistors
Enthalpy of formation
  I-III-V_2 compounds, 249
  I-III-VI_2 compounds, 247
  II_2-IV compounds, 331
  II-IV-V_2 compounds, 230
  III_2-VI_3 compounds, 275, 279

IV-VI compounds, 338
V$_2$-VI$_3$ compounds, 359
VIIIA-containing elements, 374–375
organic semiconductors, 394–395
silver and copper chalcogenides, 321
silver and copper oxides, 315
Enthalpy of fusion
II$_2$-IV compounds, 331
IV-VI compounds, 338
V$_2$-VI$_3$ compounds, 359
selenium, 87
silver and copper chalcogenides, 321
Epitaxial layers
aluminum arsenide crystals, 127
boron phosphide crystals, 118, 120–121
gallium arsenide crystals, 142, 143
gallium nitride crystals, 133
gray tin crystals, 80–81, 84
indium nitride crystals, 152
zinc oxide crystals, 183
zinc selenide crystals, 191, 192
Equienergy points, 23, see also Electrical conductivity
Equilibrium transition temperature, 75–76, see also Gray tin
Errors, 33, 34, see also Coefficient of thermal expansion; Heat capacity
Etching, 66, see also Germanium
Ethane, 55, see also Diamond; Synthesis
Eutectic formation, 145, see also Gallium antimonide
Evaporation, 42, see also Boron; Synthesis
Extinction coefficient, see also Optical properties
cadmium selenide crystals, 203, 205
germanium crystals, 73, 74, 78, 79
gray tin crystals, 83
(IA)$_3$-VB compounds, 301, 311, 312
II$_2$-IV compounds, 316, 336
V$_2$-VI$_3$ compounds, 348, 366, 368
indium arsenide crystals, 165
silver and copper oxides, 306
Extraction, 66, see also Germanium
Extrinsic conductivity, see also Intrinsic conductivity
aluminum phosphide crystals, 125
boron nitride crystals, 116
gallium phosphide crystals, 136
IA-IB compounds, 297
(IA)$_3$-VB compounds, 301, 316
semiconductor crystals, 26

F

Fabrication of semiconductor devices, 412, 417–419, see also Ohmic contact; p-n Junction
Fermi energy
p-n junction diode, 413
Schottky barrier diode, 416
thermoelectric properties of semiconductor crystals, 24–25
tunnel diodes, 424, 425
Fiber-optic communication, 431, see also Multiterminal (multibarrier) devices
Field effect transistor, 439–443, see also Multiterminal (multibarrier) devices
Filaments, 42, see also Boron
Fill-factor, 430, see also Multiterminal (multibarrier) devices
First-principles method, 190, 191, see also Zinc selenide
Flash evaporation technique, 254, see also Group I-III-VI$_2$
Float-zone method, 64, see also Silicon; Synthesis
Formation schemes, 229, see also Ternary adamantine semiconductors
Forward bias, 28, 430, 434, 437, see also Bias; Multiterminal (multibarrier) devices
Four-point method, 29, see also Resistivity
Free charges, 27, 28, see also Photoconductivity; p-n Junction
Free state, 16
Free valence electrons, see Conduction electrons
Fused silica dilatometers, 34, see also Dilatometers

G

Galenite crystals, 2
Galenite-type compounds, see Group IV-VI
Gallium, 69, see also Germanium
Gallium antimonide, 145, 147–152, see also Binary IV-IV and III-V semiconductor
Gallium arsenide, 138–145, 417, 418, see also Binary IV-IV and III-V semiconductor; Ohmic contact
Gallium nitride, 131–134, see also Binary IV-IV and III-V semiconductor
Gallium phosphide, 134–137, see also Binary IV-IV and III-V semiconductor
Gas phase
composition of II-IV-V$_2$ compounds, 231
diffusion method for p-n junction fabrication, 412
quantum system, 16
synthesis of diamond crystals, 54, 55
Gate metallization, 143, see also Gallium arsenide
Generation/recombination, 31–32, 93, see also Recombination
Germanium
chemical interactions, 92–94
dopant
gray tin crystals, 80–81
hardness in gallium arsenide crystals, 138
early use, 2
elemental semiconductor properties, 65–74
alloys, 355–357, 360
heat capacity, 58, 60
p-n junction fabrication, 412
Germanium selenide, 319, 323, 331, 344, 350, see also Group IV-VI
Germanium sulfide, 323, 331, 344, 349, see also Group IV-VI
Germanium telluride, 320, 323, 325, 331, 342, 350, see also Group IV-VI
Glass transformation temperature, see Temperature, glass transformation
Gold, 87, see also Dopants/doping; p-Type conductivity
Graphite, 46, 128, see also Aluminum arsenide; Diamond; Synthesis

INDEX

Gray arsenic, 81–82, see also Arsenic
Gray selenium, 86–87, see also Selenium
Gray tin, 2, 74–76, 78–81, see also Elemental semiconductors
Greenockite, see Wurtzite structure
Group I_2-IV-VI_3, 256, 258, 260–265
Group I_3-VI_4, 266–271
Group IA-IB, 295, 297, 300
Group $(IA)_3$-VB compounds, 297, 299, 301–302, 304, 308
Group I-III-VI_2, 246–256
Group I-IV_2-V_3, 242, 244–246
Group I-V binary systems, see Nonadamantine semiconductors, I-V binary systems
Group I-VI
  copper and silver chalcogenides
    chemical properties and crystal structure, 307, 321
    electric transport, 312, 324
    electrical conductivity, 312, 325
    energy gap, 307, 309–311, 323
    optical properties, 313, 328
    thermal, thermodynamic, mechanical properties, 307, 322
    thermoelectric properties, 312, 326
  copper and silver oxides, 303, 305–307, 314
Group I-VII compounds, 2, see also Binary II-VI and I-VIII tetrahedral semiconductors
Group II-III_2-VI_4, 282–286, 288, see also Defect ternary compounds
Group II_2-IV
  band structure, 315, 332–333
  crystal structure, 313, 330
  electic transport and electrical properties, 316, 334, 335
  optical properties, 317, 336
  synthesis, 316–317
  thermochemical, thermal, elastic properties, 315, 331, 336
Group II-IV-V_2, 230–243
Group II-VI, 2, see also Binary II-VI and I-VIII tetrahedral semiconductors
Group II-VI/II-VI, 431, 432
  solid solutions, 367, 376, 378, 380
Group II-VI and III-V alloys, 380
Group $(II-VI)_x(III_2-VI_3)$, 283, 284, see also Defect ternary compounds
Group IIIB compounds, 37–43, see also Binary IV-IV and III-V semiconductors
Group III-III-V, 420
Group III_2-IV-VI, 284, 287, 289
Group III-V, 2, 420, see also Binary IV-IV and III-V semiconductors
Group III-V-V, 420
Group III-V/III-V alloys, 362, 364–365, 367, 382, 383, 385
Group III_2-VI_3, 273, 275–281
Group IV-IV compounds, see Binary IV-IV and III-V semiconductors
Group IV_3-V_4, 271–274
Group IV-VI
  applications, 336, 338, 357
  crystal structure and chemical parameters, 318, 337
  electrical conductivity and resistivity, 323, 325–327, 329–331, 344, 346–348
  electronic properties and band structure, 318–320, 323, 339–343
  intrinsic current, 327, 329, 347
  magnetic and optical properties, 331, 333, 335–336, 349–357
  mobility, 327, 345
  natural minerals, 317
  thermal conductivity, 323, 344
  thermochemical, thermal, mechanical properties, 318, 338
Group IV-VI/IV-VI, 431, 432
Group V compounds, 81–86, see also Elemental semiconductors
Group V_2-VI_3
  band structure, 339, 360, 361
  chemical properties and crystal structure, 339, 358
  commercial availability, 351, 368
  electrical conductivity, 339, 342, 344–345, 347, 362, 363
  magnetic properties, 347, 365
  natural minerals, 338–339
  optical properties, 348, 350, 366–368
  Seebeck effect, 345, 347, 364
  thermal and thermochemical properties, 339, 359
  thermal conductivity, 347, 365
Group VII elements, 87, see also p-Type conductivity
Group VIIIA
  band structure, 354, 377–378
  chemical properties and crystal structure, 351, 369–375
  electronic properties, 351, 353
  electrical and thermoelectric properties, 354–355, 379–380
  energy gap, 353–354, 376
  optical properties, 355, 381–382
  thermal and thermochemical properties, 351, 377–378
Grown junction method, 412, see also p-n Junction; Synthesis
Grünesian constant, 156, see also Indium phosphide

H

Hall effect
  aluminum phosphide crystals, 125
  boron crystals, 41
  cadmium selenide crystals, 202, 204
  cadmium sulfide crystals, 199, 201
  cadmium telluride crystals, 205–206, 208, 209
  diamond crystals, 44, 47
  elemental semiconducting alloys, 360
  gallium antimonide crystals, 150, 152
  gallium arsenide crystals, 142, 143
  gallium nitride crystals, 133
  gallium phosphide crystals, 136
  germanium crystals, 70
  gray tin crystals, 76, 80
  I_2-IV-VI_3 compounds, 260, 262
  I_3-VI_4 compounds, 267–270
  IA-IB compounds, 296

II$_2$-IV compounds, 334
IV-VI compounds, 323, 325, 327, 345, 346
indium antimonide crystals, 170
indium arsenide crystals, 162
indium nitride crystals, 154–155
indium phosphide crystals, 160
measurement in semiconductor crystals, 30–31
silicon crystals, 57, 60
silver and copper chalcogenides, 311
tellurium crystals, 90
zinc oxide crystals, 185
zinc sulfide crystals, 187, 189
zinc telluride crystals, 196
Halogens, 66, see also Germanium
Hardness, see also Knoop microhardness; Mechanical properties; Mohs microhardness; Vickers microhardness
   aluminum antimonide crystals, 128
   aluminum arsenide crystals, 126
   aluminum nitride crystals, 123
   boron arsenide crystals, 121
   boron crystals, 42
   boron phosphide crystals, 119
   cadmium selenide crystals, 202
   cadmium telluride crystals, 204
   copper bromide crystals, 216
   diamond crystals, 43
   I-III-VI$_2$ compounds, 247, 250
   I$_2$-IV-VI$_3$ compounds, 256, 260
   II-IV-V$_2$ compounds, 235, 237
   VIIIA-containing elements, 374–375
   semiconductor crystals and bonding types, 15
   zinc telluride crystals, 195
Harmonic oscillations, 138, see also Gallium arsenide
Hawleyite structure, 198, see also Cadmium sulfide
Heat capacity
   aluminum antimonide crystals, 128, 129
   boron nitride crystals, 112
   boron phosphide crystals, 119
   gallium antimonide crystals, 148
   gallium phosphide crystals, 134
   germanium crystals, 69, 70
   gray selenium crystals, 87
   II$_2$-IV compounds, 315, 331
   II-IV-V$_2$ compounds, 235, 236
   IV-VI compounds, 338
   V$_2$-VI$_3$ compounds, 359
   VIIIA-containing elements, 374–375
   indium antimonide crystals, 166, 168
   indium arsenide crystals, 161, 163
   indium nitride crystals, 152–153
   indium phosphide crystals, 156, 158
   silicon crystals, 58, 60
   silver and copper chalcogenides, 321
   silver and copper oxides, 315
Heat of atomization
   defect ternary compounds, 285–286
   I-III-VI$_2$ compounds, 247, 249
   I$_2$-IV-VI$_3$ compounds, 256, 260
   II-IV-V$_2$ compounds, 235, 236
Heat of formation, see Enthalpy of formation
Heat of fusion, see Enthalpy of fusion

Heat of sublimation
   I$_2$-IV-VI$_3$ compounds, 256, 260
   IV-VI compounds, 338
   organic semiconductors, 394–395
   zinc sulfide crystals, 186
Heat of transition, 46
Heterojunction devices, 431, see also Photodiode
Hexagonal graphite, 43, see also Diamond
Hexagonal lattice, 57, 108, 123
Holes, see also Electrons
   bipolar junction transistors, 434–436
   boron crystals, 37
   distribution in p-n junction diode, 413–415
   germanium crystals, 69, 70, 76
   lifetime calculation, 31
   metal-insulator-semiconductor diodes, 444–445
   mobility
      aluminum antimonide crystals, 129
      aluminum arsenide crystals, 127
      aluminum nitride crystals, 123
      aluminum phosphide crystals, 125
      boron phosphide crystals, 119, 121
      cadmium selenide crystals, 202
      cadmium sulfide crystals, 199
      cadmium telluride crystals, 205–206
      elemental semiconducting alloys, 357
      gallium antimonide crystals, 150
      gallium arsenide crystals, 140–142, 144
      gallium nitride crystals, 132–133
      gallium phosphide crystals, 136
      germanium crystals, 71
      germanium-silicon alloys, 94
      gray tin crystals, 75, 80, 82
      IA-IB compounds, 296
      I$_2$-IV-VI$_3$ compounds, 264, 265
      II$_2$-IV compounds, 334
      II-IV-V$_2$ compounds, 239, 242
      III-V/III-V semiconductor alloys, 367
      IV-VI compounds, 323, 325, 327, 331
      V$_2$-VI$_3$ compounds, 339, 344, 345
      VIIIA-containing elements, 379–380
      indium antimonide crystals, 170
      indium arsenide crystals, 163
      indium nitride crystals, 154
      indium phosphide crystals, 160
      mercury telluride crystals, 212
      organic semiconductors, 395–396
      selenium crystals, 86
      silicon carbide crystals, 108, 109, 111
      silicon crystals, 60
      silver and copper chalcogenides, 324
      silver and copper oxides, 305–306
      tellurium crystals, 90
      zinc oxide crystals, 185
      zinc selenide crystals, 191
      zinc sulfide crystals, 187
      zinc telluride crystals, 196
   photodiodes, 427, 428, 430
   semiconductor crystals, 17, 19–20
      electrical conductivity, 23
      recombination, 25–26
   resistivity of diamond crystals, 44

Homojunction laser, 431, see also Photodiode
Horizontal melting zone, 213, see also Mercury telluride; Synthesis
Hot filament method, 54, see also Diamond; Synthesis
Hot-wall epitaxy, 206, see also Cadmium telluride; Synthesis
Hydrogen, atomic, 55, 56, see also Diamond; Synthesis

I

IC, see Integrated circuits
Icoshedra, face-sharing, 39, 40, see also Boron
Ideal crystal, 21–22, see also Electrical conductivity
Ideal solid insulators, 1
IGFET, see Insulated-gate field effect transistor
Impurities
  aluminum phosphide crystals, 125
  conductivity
    gallium antimonide crystals, 150, 153
    gallium arsenide crystals, 142, 146
  diamond crystals, 43, 44, 48, 50, 52
  distribution and calculation of resistance in semiconductor crystals, 29
  electrical conductivity, 18, 20–21
  electrical transport in silicon carbide crystals, 108
  germanium crystals, 72, 74
  III_2-VI_3 compounds, 279
  IV-VI compounds, 330
  indium antimonide crystals, 170
  p-n junction formation, 411, 412
    diode, 414
  Seebeck effect, 25
  silicon crystals, 60
  tunnel diodes, 424
Indium, 185, 412, see also Dopants/doping; p-n Junction; Zinc oxide
Indium antimonide, 165–171, see also Binary IV-IV and III-V semiconductors
Indium arsenide, 160–165, see also Binary IV-IV and III-V semiconductors
Indium nitride, 152–155, see also Binary IV-IV and III-V semiconductors
Indium phosphide, 155–160, see also Binary IV-IV and III-V semiconductors
Injection luminescence, 430, see also Photodiodes
Inorganic semiconductors
  binary, 7–11
  ternary, 12–13
Insulated-gate field effect transistors (IGFET), 443, see also Semiconductor devices
Insulators, 1, 16–17, 43
Integrated circuits (IC), 412, 442
Interatomic forces, 52–53, see also Diamond
Interface state, 28–29, 445
Intrinsic band gap, 41, see also Boron
Intrinsic carrier, 327, 347, see also Group IV-VI
Intrinsic conductivity, see also Extrinsic conductivity
  germanium-silicon alloys, 93
  gray tin crystals, 80
  I-VI compounds, 302
  indium antimonide crystals, 170
  semiconductor crystals, 18, 20
  silicon crystals, 59, 63
  tellurium crystals, 90
Intrinsic photoconductivity, 26
Ion implantation, 145, see also Gallium arsenide
Ion plantation method, 412, see also p-n Junction
Ionic bonding, 15, 16
Ionic conductivity
  copper bromide crystals, 216
  copper chloride crystals, 214–215
  copper iodide crystals, 218
  silver iodide crystals, 220
Ionic potential, 296, 298, see also Group IA-IB
Ionization energy
  diamond crystals, 44
  germanium, 66
  individual atoms in solid materials, 16
  phosphorus, 84
  selenium, 86
  silicon, 56
  tellurium, 89
Ionization potential, 37, see also Boron
Isotopes
  boron, 37
  germanium, 65
  phosphorus, 84
  selenium, 86
  silicon, 56
  tellurium, 88

J

JFET, see Junction field effect transistor
Junction field effect transistor (JFET), 439, 440

K

Kinetic energy, 16, 17
KKR method, see Kohn-Korringa-Rostoker method
Knoop microhardness, see also Hardness; Mechanical properties
  boron-carbon alloys, 91
  boron-silicon alloys, 92
  diamond crystals, 51
  gallium antimonide crystals, 148
  gallium arsenide crystals, 138
  gallium phosphide crystals, 134
  III-V/III-V semiconductor alloys, 365
  indium antimonide crystals, 168
  indium arsenide crystals, 161
  indium phosphide crystals, 158, 159
  silicon carbide crystals, 108
  zinc sulfide crystals, 186
Kohn-Korringa-Rostoker (KKR) method, 296, 298, see also Group IA-IB
k-Space, 23, see also Electrical conductivity
Kyropoulos method, 316, see also Group II_2-IV

L

Laser energy, 56, see also Diamond; Synthesis
Lattice energy, 267, see also Group I_3-VI_4 compounds
Lattice parameters

aluminum antimonide crystals, 128
aluminum arsenide crystals, 125–126
aluminum nitride crystals, 123
aluminum phosphide crystals, 125
boron crystals, 39
 -carbon alloys, 91
cadmium selenide crystals, 201–202
cadmium sulfide crystals, 198
cadmium telluride crystals, 203–204
copper bromide crystals, 216
copper chloride crystals, 214
copper fluoride crystals, 214
copper iodide crystals, 217
defect ternary compounds, 285–286
diamond crystals, 52–53
elemental semiconducting alloys, 360
gallium antimonide crystals, 145
gallium arsenide crystals, 138, 139
gallium nitride crystals, 131
gallium phosphide crystals, 134
germanium crystals, 66, 68
gray selenium crystals, 87
gray tin crystals, 79
IA-IB compounds, 295, 296, 300
$(IA)_3$-VB compounds, 299, 303
I-III-VI_2 compounds, 246–248
I-IV_2-V_3 compounds, 244, 247
I_2-IV-VI_3 compounds, 256, 259
I_3-VI_4 compounds, 266
II_2-IV compounds, 313, 330
II-IV-V_2 compounds, 230, 232–233
III_2-IV-VI compounds, 287
III-V/III-V semiconductor alloys, 362, 364, 383
III_2-VI_3 compounds, 276–277
IV_3-V_4 compounds, 274
IV-VI compounds, 318, 337
 temperature relation, 335, 354
V_2-VI_3 compounds, 339, 358
VIIIA-containing elements, 369–373
indium antimonide crystals, 165
indium arsenide crystals, 160–161
indium nitride crystals, 152
indium phosphide crystals, 155–156
mercury selenide crystals, 210
mercury sulfide crystals, 207–208
mercury telluride crystals, 211
organic semiconductors, 394–395
silicon crystals, 57
 carbide crystals, 106
silver and copper chalcogenides, 321
silver and copper oxides, 314
silver iodide crystals, 219
tellurium, 89
zinc oxide crystals, 183
zinc selenide crystals, 189
zinc sulfide crystals, 186
Lead, 35, see also Individual entries
Lead chalcogenides, 318, 320, 336, 338, see also Chalcogenides; Group IV-VI
Lead oxide, see also Group IV-VI
 electrical properties, 327, 346
 optical properties, 331, 333, 353
Lead selenide, 325, 330–331, 335, 336, 348, 356, see also Group IV-VI
Lead sulfide, see also Group IV-VI
 electrical conductivity, 327, 329, 330, 348
 intrinsic current, 327, 347
 magnetic properties, 331, 349
 optical properties, 333, 335, 336, 354, 356
Lead telluride, see also Group IV-VI
 applications, 338, 357
 electrical conductivity, 330–331, 336, 348
 optical properties, 333, 335, 336, 355, 357
 thermal conductivity, 331
Leakage current, 441, see also Field effect transistors
LED, see Light-emitting diodes
Lifetime, 31–32, 138, 430
Lifetime of excess minority charge carriers, 26
Light absorption, 85–86, see also Phosphorus
Light transmission, 42, see also Boron
Light-emitting diodes (LED), 431, see also Photodiode
Liquid nonmetals, 1
Liquid phase, 76, see also Gray tin
Liquid phase epitaxy (LPE), 412, see also p-n Junction
Liquid state
 aluminum antimonide, 131
 cadmium sulfide crystals, 198
 cesium auride, 296, see also IA-IB compounds
 gallium antimonide, 150
 gallium arsenide, 140–142, 145, 146
 germanium, 72
 I-III-VI_2 compounds, 250, 255
 II_2-IV compounds, 316, 335
 III_2-VI_3 compounds, 280–282
 IV-VI compounds, 331
 V_2-VI_3 compounds, 339, 344
 mobility of current carriers, 16
 selenium, 86, 87, 89
 silicon, 60
Litharge, 327, see also Lead oxide
Loellingite, 353, 376, see also Group VIIIA
LPE, see Liquid phase epitaxy
Luminescence spectra, 241, see also Group II-IV-V_2
Luminophors, 284, see also Defect ternary compounds

M

Magnesium, 313, see also Group II_2-IV compounds
Magnetic properties
 aluminum antimonide crystals, 131
 cadmium selenide crystals, 202
 cadmium sulfide crystals, 199
 cadmium telluride crystals, 206
 copper bromide crystals, 216
 copper chloride crystals, 215
 copper iodide crystals, 218
 gallium antimonide crystals, 150–151, 153
 gallium arsenide crystals, 142, 146
 gallium nitride crystals, 133
 gallium phosphide crystals, 136–137
 germanium crystals, 73
 gray tin crystals, 80

INDEX 463

I-III-VI$_2$ compounds, 251, 256
I$_2$-IV-VI$_3$ compounds, 265
I$_3$-VI$_4$ compounds, 268, 271
II$_2$-IV compounds, 316, 336
II-IV-V$_2$ compounds, 240, 244
III$_2$-VI$_3$ compounds, 280–282
IV-VI compounds, 331, 349
V$_2$-VI$_3$ compounds, 347, 365
VIIIA-containing elements, 355, 381–382
indium antimonide crystals, 170
indium arsenide crystals, 163, 165
indium nitride crystals, 154–155
indium phosphide crystals, 160
silicon crystals, 61
    carbide crystals, 108
silver and copper chalcogenides, 313
silver and copper oxides, 306
silver iodide crystals, 222
zinc oxide crystals, 185
zinc selenide crystals, 192
zinc sulfide crystals, 187
zinc telluride crystals, 197
Massicot, 327, see also Lead oxide
Mechanical properties
    aluminum antimonide crystals, 128
    aluminum arsenide crystals, 126
    aluminum nitride crystals, 123
    aluminum phosphide crystals, 125
    boron crystals, 42
    boron nitride crystals, 115, 116
    cadmium selenide crystals, 202
    cadmium sulfide crystals, 199
    cadmium telluride crystals, 204
    copper chloride crystals, 214
    diamond crystals, 51–52
    gallium antimonide crystals, 148
    gallium arsenide crystals, 138
    gallium phosphide, 134
    gray selenium crystals, 87
    gray tin crystals, 79
    I-III-VI$_2$ compounds, 247
    I$_2$-IV-VI$_3$ compounds, 256, 260
    I$_3$-VI$_4$ compounds, 266, 267
    II-IV-V$_2$ compounds, 235, 237
    IV-VI compounds, 338
    indium antimonide crystals, 168–169
    indium arsenide crystals, 161
    indium phosphide crystals, 158, 159
    mercury sulfide crystals, 208
    mercury telluride crystals, 211
    silicon carbide crystals, 108
    silver and copper chalcogenides, 307, 321
    silver and copper oxides, 315
    zinc oxide crystals, 183
    zinc selenide crystals, 189–190
    zinc sulfide crystals, 186
    zinc telluride crystals, 195
Melting point, see also Decomposition
    aluminum antimonide crystals, 128
    aluminum arsenide crystals, 126
    aluminum nitride crystals, 123
    boron crystals, 40

boron nitride crystals, 112
cadmium selenide crystals, 202
cadmium sulfide crystals, 198–199
cadmium telluride crystals, 204
copper bromide crystals, 216
copper chloride crystals, 214, 215
copper iodide crystals, 218
defect ternary compounds, 285–286
gallium antimonide crystals, 145
gallium arsenide crystals, 138
gallium nitride crystals, 131
gallium phosphide crystals, 134
gray selenium crystals, 87
IA-IB compounds, 296, 297, 300
(IA)$_3$-VB compounds, 299, 308
I-III-VI$_2$ compounds, 247, 249
I$_2$-IV-VI$_3$ compounds, 256, 260
I$_3$-VI$_4$ compounds, 266, 267
II$_2$-IV compounds, 315, 331
II-IV-V$_2$ compounds, 231
III$_2$-VI$_3$ compounds, 275, 279
IV-VI compounds, 338
IV$_3$-V$_4$ compounds, 274
V$_2$-VI$_3$ compounds, 339, 359
VIIIA-containing elements, 374–375
indium antimonide crystals, 166
indium arsenide crystals, 163
indium nitride crystals, 152
indium phosphide crystals, 156
mercury selenide crystals, 210
mercury sulfide crystals, 208, 209
mercury telluride crystals, 211
organic semiconductors, 394–395
selenium, 86
silicon, 60, 61
silver and copper chalcogenides, 321
silver and copper oxides, 315
silver iodide crystals, 220
zinc oxide crystals, 183
zinc selenide crystals, 189
zinc sulfide crystals, 186
zinc telluride crystals, 195
Melting process, 90, see also Tellurium
Mercury, 76, 87, see also Gray selenium; Gray tin
Mercury selenide, 210, see also Binary II-VI and I-VIII tetrahedral semiconductors
Mercury sulfide, 207–210, see also Binary II-VI and I-VIII tetrahedral semiconductors
Mercury telluride, 210–213, see also Binary II-VI and I-VIII tetrahedral semiconductors
MESFET, see Metal-semiconductor field effect transistor
Metal-insulator-semiconductor diode (MIS diode), 444–445, see also Multiterminal (multibarrier) devices
Metal-insulator-semiconductor field effect transistor (MISFET), 443, see also Field effect transistors; Multiterminal (multibarrier) devices
Metal-organic chemical vapor deposition (MOCVD), 206, see also Cadmium telluride; Chemical vapor deposition; Synthesis

Metal-oxide semiconductor field effect transistor
(MOSFET), 439, 442, 443, see also
Multiterminal (multibarrier) devices
Metal-semiconductor field effect transistor (MESFET),
439, 441, 442, see also Multiterminal
(multibarrier) devices
Metal-semiconductor rectifying contact, 415–416, see
also Semiconductor devices
Metallic bonding, 15
Metallization technique, 418, see also Ohmic contacts
Metallurgical methods, 43, see also Diamonds
Metals, 1, 15–17, 24, see also Individual entries
Methane, 55, 56, see also Diamond; Synthesis
Minority carriers, 20, 26
MIS diode, see Metal-insulator-semiconductor diode
Miscibility gap, 367, 376, see also Group II-VI/II-VI,
solid solutions
MISFET, see Metal-insulator-semiconductor field
effect transistor
Mobility, see also Electrons, mobility; Holes, mobility
  aluminum antimonide crystals, 129
  aluminum arsenide crystals, 127
  aluminum nitride crystals, 123
  aluminum phosphide crystals, 125
  boron crystals, 41
    -carbide alloys, 91
  boron nitride crystals, 116
  boron phosphide crystals, 119, 121
  cadmium selenide crystals, 202
  cadmium sulfide crystals, 199
  cadmium telluride crystals, 205–206
  current carriers in semiconductor crystals, 16
  defect ternary compounds, 288
  elemental semiconducting alloys, 357
  gallium antimonide crystals, 149–151
  gallium arsenide crystals, 140–142
  gallium nitride crystals, 132–133
  gallium phosphide crystals, 136
  germanium crystals, 70–71
    -silicon alloys, 94
  gray tin crystals, 75, 80, 82
  IA-IB compounds, 296
  I-III-VI$_2$ compounds, 247, 250
  I-IV$_2$-V$_3$ compounds, 244–245, 247
  I$_2$-IV-VI$_3$ compounds, 264, 265
  I-V binary systems, 297
  I$_3$-VI$_4$ compounds, 270
  II$_2$-IV compounds, 334
  II-IV-V$_2$ compounds, 239, 242
  II-VI/II-VI solid solutions, 378
  III-V/III-V semiconductor alloys, 367
  III$_2$-VI$_3$ compounds, 280
  IV-VI compounds, 323, 325, 327, 331
  V$_2$-VI$_3$ compounds, 339, 344
  VIIIA-containing elements, 379–380
  indium antimonide crystals, 169–170
  indium arsenide crystals, 163
  indium nitride crystals, 154–156
  indium phosphide crystals, 159–160
  mercury selenide crystals, 210
  mercury sulfide crystals, 208
  mercury telluride crystals, 212

  organic semiconductors, 395–396
  phosphorus crystals, 84, 85
  selenium crystals, 86
  silicon crystals, 59
    carbide crystals, 108, 111
  silver and copper chalcogenides, 324
  silver and copper oxides, 305–306, 318
  silver iodide crystals, 220
  tellurium crystals, 90
  zinc oxide crystals, 185
  zinc selenide crystals, 191
  zinc sulfide crystals, 187
  zinc telluride crystals, 196
MOCVD, see Metal-organic chemical vapor deposition
Mohs microhardness, see also Hardness; Mechanical
properties
  aluminum phosphide crystals, 125
  boron nitride crystals, 115, 116
  cadmium sulfide crystals, 199
  copper chloride crystals, 214
  copper iodide crystals, 218
  gallium antimonide crystals, 148
  gallium arsenide crystals, 138
  gallium phosphide crystals, 134
  I$_3$-VI$_4$ compounds, 267
  IV-VI compounds, 338
  mercury selenide crystals, 210
  mercury sulfide crystals, 208
  mercury telluride crystals, 211
  silicon carbide crystals, 108
  silver and copper chalcogenides, 321
  silver and copper oxides, 315
  silver iodide crystals, 220
  zinc oxide crystals, 183
  zinc sulfide crystals, 186
Monoclinic selenium, 86, 87, see also Selenium
MOSFET, see Metal-oxide semiconductor field effect
transistor
Multiphonon recombination, 31
Multiterminal (multibarrier) devices
  bipolar junction transistor, 434–436
  field-effect transistors, 439–443
  metal-insulator-semiconducting diode and charge-
coupled devices, 444–445
  thyristor, 436–438

N

Nernst effect, 30–31
Nernst vacuum calorimeter method, 33, see also
Specific heat
Neutral semiconductors, 25–26
Neutrality, electrical, 20
Neutron capture, 37, 42, see also Boron
Nitrides, 230, see also Group II-IV-V$_2$
Nitrogen, see also Diamond; Dopants/doping
  diamond impurity, 43, 44, 50, 52
  dopants and mobility in silicon carbide crystals, 108
Nonadamantine semiconductors
  binary compounds of group VIIIA, see Group VIIIA
  IA-IB compounds, 295–297
  I-V binary systems

INDEX

IA-IB compounds, 297, 298, 300, 314
$(IA)_3$-VB compounds, 297, 299, 301–302, 304, 308
I-VI compounds, see Group I-VI
II_4-IV compounds with antifluorite structure, see Group II_2-IV
IV-VI galenite-type compounds, see Group IV-VI
V_2-VI_3 compounds, see Group V_2-VI_3
solid solutions
   elemental alloys, 355–357, 360
   II-VI and III-V compounds, 380
   II-VI/II-VI, 367, 376, 378, 380
   III-V/III-V alloys, 362, 364–365, 367, 382, 383, 385
organic, 380, 384, 386–397
Nonmetal solids, 1
Nonstationary methods, 32–33, see also Thermal conductivity
n-Type conductivity, see also p-Type conductivity
   aluminum antimonide crystals, 129
   aluminum nitride crystals, 123
   boron crystals, 41
   boron nitride crystals, 116
   boron phosphide crystals, 121
   cadmium selenide crystals, 202
   cadmium telluride crystals, 205–206, 208
   gallium antimonide crystals, 150
   gallium arsenide crystals, 138, 140–141
   gallium nitride crystals, 132–133
   gray selenium crystals, 87
   I_2-IV-VI_3 compounds, 264
   I_3-VI_4 compounds, 270
   IA–$(IA)_3$-VB compounds, 301
   II_2-IV compounds, 316, 334
   III_2-VI_3 compounds, 280
   IV-VI compounds, 325–326, 330
   mercury telluride crystals, 212
   semiconductor crystals, 17, 20–21
N-type impurities, see Donors

O

Octahedral coordination, 2–3
Ohmic contact
   fabrication for semiconductors, 417–420
   gallium arsenide crystals, 143
   thyristors, 437
   unijunction transistors, 433
Ohmic mobility, 60, see also Silicon
Operation modes, 436, 437, see also Semiconductor devices
Optical absorption, see also Absorption coefficient
   diamond crystals, 50–52
   VIIIA-containing elements, 355
   tellurium crystals, 91
Optical devices, 43, see also Diamond
Optical interference dilatometers, 33, see also Coefficient of thermal expansion; Dilatometers
Optical properties
   aluminum antimonide crystals, 130–131
   boron arsenide crystals, 121

   boron nitride crystals, 115, 116
   cadmium selenide crystals, 202–203, 205
   cadmium sulfide crystals, 199
   cadmium telluride crystals, 206
   copper bromide crystals, 216–217
   copper chloride crystals, 215
   copper iodide crystals, 218
   defect ternary compounds, 284, 288
   diamond crystals, 52
   gallium antimonide crystals, 151, 154
   gallium arsenide crystals, 147
   gallium nitride crystals, 133
   gallium phosphide crystals, 137
   germanium crystals, 73–74, 78, 79
   gray selenium crystals, 87
   gray tin crystals, 80, 83
   IA-IB compounds, 296, 297, 299, 302
   $(IA)_3$-VB compounds, 301, 311, 312
   I-III-VI_2 compounds, 251, 254, 257, 258
   I-IV_2-V_3 compounds, 245
   II_2-IV compounds, 316, 336
   II-IV-V_2 compounds, 240–241, 245
   II-VI/II-VI solid solutions, 378
   III_2-VI_3 compounds, 281, 282
   III-V/III-V semiconductor alloys, 367
   IV-VI compounds, 331, 333, 335, 349–357
   V_2-VI_3 compounds, 348, 366–368
   VIIIA-containing elements, 355, 381–382
   indium antimonide crystals, 170
   indium arsenide crystals, 163, 165
   indium phosphide crystals, 160, 161
   mercury selenide crystals, 210
   mercury sulfide crystals, 209
   mercury telluride crystals, 213
   organic semiconductors, 386, 395–396
   phosphorus crystals, 84, 85
   silicon crystals, 61, 65–67
   silver and copper chalcogenides, 310–311, 323, 324, 326–329
   silver and copper oxides, 306–307, 320
   silver iodide crystals, 222
   tellurium crystals, 90–91
   zinc oxide crystals, 185–186
   zinc selenide crystals, 192
   zinc sulfide crystals, 187
   zinc telluride crystals, 197
Opticoupler, see Optisolator
Optisolator, 431, see also Photodiode
Orbit parameters, 16
Organic semiconductors, 380, 384, 386–397
Orthorhombic structure, 84, 128
Osmium, 354, 355, Group VIIIA
Overcooling, 86, see also Selenium
Oxidation, 105, 296
Oxides, see also Group I-VI
Oxygen, 133, 305, see also p-Type conductivity

P

Pauli's exclusion principle, 16
Peltier effect, 24
Periodic table

calculation of valency electrons, 2–3
element location and semiconductor properties, 5
spiral variant, 105, 106
Peritectic reactions, 231, 256, see also Group I_2-IV-VI_3; Group II-IV-V_2; Synthesis
Phase diagram
  copper chloride crystals, 214
  gallium arsenide crystals, 138, 140
  germanium-silicon alloys, 92
  V_2-VI_3 compounds, 339
  indium antimonide crystals, 166
  indium arsenide crystals, 165, 167
  indium phosphide crystals, 156, 157
  pseudobinary system, 284, 285
  silver iodide crystals, 222
Phase transition, 323, see also Group IV-VI
  temperatures, 107
Phonon dispersion, 41–42, 52–53, 72, see also Boron; Diamond; Germanium
Phonon impulse, 27, see also Photoconductivity
Phosphides, 230, see also Group II-IV-V_2
Phosphorus
  -boron chemical interactions, 92
  elemental semiconductor, 81–86
  p-n junction fabrication for semiconducting devices, 412
Photoconductive decay method, 31
Photoconductivity
  boron crystals, 41, 42
  gallium phosphide crystals, 136
  gray arsenic and antimony, 81–82
  gray tin crystals, 80
  I_2-IV-VI_3 compounds, 256, 261, 265
  I_3-VI_4 compounds, 268–269, 271, 272
  IA-IB compounds, 297, 301, 303
  $(IA)_3$-VB compounds, 301
  V_2-VI_3 compounds, 348
  selenium, 2
  semiconductor crystals, 26–28
  tellurium crystals, 91
  zinc selenide crystals, 191
Photocurrent gain, 430, see also Photodiodes
Photodiodes, 427–432, see also Two-electrode devices
Photoelectric properties
  I-V binary systems, 297
  $(IA)_3$-VB compounds, 301
  III_2-VI_3 compounds, 278
  II-IV-V_2 compounds, 241
  silver iodide crystals, 220
Photoemissive threshold, 301, 308, see also Group $(IA)_3$-VB
Photoluminescence, 191, 192, 380, see also Group II-VI and III-V alloys; Zinc selenide
Photon energy, 27, 61, see also Silicon
Photoresponse, 27
Photosensitivity, 284, see also Defect ternary compounds
Photovoltaic devices, 251, 254, see also Group I-III-VI_2
Photo-emf
  I_2-IV-VI_3 compounds, 256, 260–262, 265
  measurement in cadmium telluride crystals, 205
  semiconductor crystals, 26–28

Phthalocyanines, 386, see also Organic semiconductors
Physics of semiconductors
  chemical bonds, 15–18
  electrical conductivity, 18–24
  evaluation of basic properties, 29–35
  phenomena at interface of n- and p-type, 28–29
  photoconductivity and photo-emf, 26–28
  recombination of electrons and holes, 25–26
  thermoelectric phenomena, 24–25
Pnictogenides, 354, 381–382, see also Group VIIIA
p-n junction
  boron nitride crystals, 116
  gallium nitride crystals, 133
  semiconductor crystals, 28
  semiconductor devices, 411–412
    applications, 446
    bipolar junction transistors, 434
    equivalent circuit, 421
    field effect transistors, 439, 441
    photodiodes, 428
    thyristors, 437
p-n Junction diode, 413–415, see also Semiconductor devices
p-n transitions, 108, see also Silicon carbide
Poisson equation, 139, 148, 190, 414
Polymers, 386, 388–393, see also Organic semiconductors
Polytypes, see Allotropes
Potassium antimonide, 301, 304, see also Group $(IA)_3$-VB
Potentiometers, 43, see also Boron
Power conversion efficiency, 430, see also Photodiode
Power gain, 435, 436, see also Bipolar junction transistors
Pressure-temperature diagram, 43, 44, see also Carbon
Propane, 55, see also Diamond; Synthesis
Pseudobinary system, 284, see also Defect ternary compounds
Pseudodirect energy gap, 238–239, see also Group II-IV-V_2
Pseudopotential method
  boron arsenide crystals, 122
  cadmium sulfide crystals, 200
  cadmium telluride crystals, 205, 207
  gallium antimonide crystals, 150
  $(IA)_3$-VB compounds, 307, 308
  indium phosphide crystals, 159
  zinc selenide crystals, 190
p-Type conductivity, see also n-Type conductivity
  aluminum antimonide crystals, 129
  aluminum nitride crystals, 123
  boron crystals, 41
  boron nitride crystals, 116
  boron phosphide crystals, 121
  cadmium telluride crystals, 205–206, 209
  copper oxides, 305
  gallium arsenide crystals, 138
  gallium nitride crystals, 133
  gray selenium crystals, 87
  gray tin crystals, 80
  I_2-IV-VI_3 compounds, 262, 264
  I_3-VI_4 compounds, 270

(IA)$_3$-VB compounds, 301
II$_2$-IV compounds, 316, 334
III$_2$-VI$_3$ compounds, 280
IV-VI compounds, 323, 326, 327, 329
mercury telluride crystals, 212
semiconductor crystals, 17, 20–21
zinc telluride crystals, 196
P-type impurities, see Acceptors

Q

Quantum energy, 427, see also Photodiodes
Quantum theory, 16–18
Quartz dilatometers, 33–34, see also Coefficient of thermal expansion
Quasiparticles, 18

R

Radiative combination, 31
Raman spectroscopy, 53–56, see also Diamond
Recombination
   electrons/holes in semiconductor crystals, 25–27
   lifetime calculation, 31, 32
   rate in p-n junction diode, 415
Rectifiers, 2, 29
Rectifying effect, 302, see also Group I-VI compounds
Red phosphorus, 84, 85, see also Phosphorus
Reflectivity, see also Optical properties
   boron nitride crystals, 113, 116
   germanium crystals, 73, 74, 79
   I-IV$_2$-V$_3$ compounds, 245
   III-V/III-V semiconductor alloys, 367
   IV-VI compounds, 331, 350, 352
   V$_2$-VI$_3$ compounds, 348, 366
   VIIIA-containing elements, 355
   silver and copper chalcogenides, 313, 327, 328
   zinc oxide crystals, 186
Refraction index, see also Optical properties
   aluminum antimonide crystals, 130
   aluminum phosphide crystals, 125, 126
   boron crystals, 42
   boron nitride crystals, 115–116
   boron phosphide crystals, 120
   cadmium selenide crystals, 203, 205
   cadmium sulfide crystals, 199
   cadmium telluride crystals, 206
   copper bromide crystals, 216
   copper chloride crystals, 215
   copper iodide crystals, 218
   gallium antimonide crystals, 151, 154
   gallium arsenide crystals, 142–143, 147
   gallium nitride crystals, 133
   gallium phosphide crystals, 137
   germanium crystals, 73, 74
   gray tin crystals, 83
   I-III-VI$_2$ compounds, 251, 257
   (IA)$_3$-VB compounds, 301, 311, 312
   II$_2$-IV compounds, 316, 336
   II-IV-V$_2$ compounds, 240, 244
   III$_2$-VI$_3$ compounds, 281, 282
   III-V/III-V semiconductor alloys, 367
   IV-VI compounds, 331, 334–336, 349, 351, 353–355
   V$_2$-VI$_3$ compounds, 348, 366–368
   VIIIA-containing elements, 355, 381–382
   indium arsenide crystals, 163, 165
   indium nitride crystals, 155
   mercury selenide crystals, 210
   organic semiconductors, 395–396
   phosphorus crystals, 84
   silicon crystals, 61
      carbide crystals, 108, 111
   silver and copper chalcogenides, 313, 326, 328
   silver and copper oxides, 306
   silver iodide crystals, 222
   tellurium crystals, 91
   zinc oxide crystals, 185–186
   zinc selenide crystals, 192, 195
   zinc sulfide crystals, 187
   zinc telluride crystals, 197
Relative dilatometers, 33, see also Coefficient of thermal expansion; Dilatometers
Resistivity
   aluminum nitride crystals, 123
   boron crystals, 41
   boron nitride crystals, 113, 118
   calculation, 29
   copper chloride crystals, 214–215
   defect ternary compounds, 288
   depletion layer at p-n junction, 28
   diamond crystals, 44–46
   field effect transistors, 439
   gallium antimonide crystals, 150, 152
   gallium arsenide crystals, 140
   gallium nitride crystals, 133
   gallium phosphide crystals, 136
   gray tin crystals, 80, 83
   IA-IB compounds, 296
   (IA)$_3$-VB compounds, 299, 301, 309
   I-VI compounds, 302
   II$_2$-IV compounds, 316, 334
   III$_2$-VI$_3$ compounds, 280
   IV$_3$-V$_4$ compounds, 271
   IV-VI compounds, 323, 325, 344
   V$_2$-VI$_3$ compounds, 339, 344
   VIIIA-containing elements, 353
   mercury telluride crystals, 212
   phosphorus crystals, 85
   p-n junction diode, 414–415
   selenium, 86
   silicon carbide crystals, 108
   silver and copper oxides, 305
   silver and copper chalcogenides, 324
   solid electrolytes, 1
   thyristors, 437
   zinc oxide crystals, 185
   zinc selenide crystals, 191
Resistors, 43, see also Boron
Reverse bias, see also Bias
   bipolar junction transistors, 434
   photodiodes, 427
   p-n junction of semiconductor crystals, 28, 29
   thyristors, 437
RF plasma method, 54, 55, see also Diamond; Synthesis

rf-Magnetron sputtering, 155, see also Indium Nitride; Synthesis
Rheostats, 43, see also Boron
Rhombohedral structure
  boron crystals, 37, 39–42
  copper iodide crystals, 217
  phosphorus crystals, 84
  silicon carbide crystals, 105, 108
  zinc sulfide crystals, 186
Rock salt structure
  aluminum nitride crystals, 123
  cadmium selenide crystals, 202
  copper bromide crystals, 216
  copper chloride crystals, 214
  copper iodide crystals, 217
  gallium arsenide crystals, 138
  gallium phosphide crystals, 134
  indium arsenide crystals, 161
  mercury telluride crystals, 211
  zinc oxide crystals, 183
Rubidium antimonide, 301, see also Group $(IA)_3$-VB compounds
Rubidium auride, 296, 298, see also Group IA-IB compounds
Ruthenium, 354, 355, see also Group VIIIA

S

Saturation current, 415, 433
Scattering techniques, 23, 41–42, 150, 151
Schottky barrier, 441, 445, 446
  diode, 415–417
Schrödinger's equation, 21–22, see also Electrical conductivity
Seebeck coefficient
  boron phosphide crystals, 120
  boron-carbon alloys, 91
  elemental semiconducting alloys, 357
  I_2-IV-VI_3 compounds, 260, 262–265
  IA-IB compounds, 300
  II_2-IV compounds, 316, 334
  III_2-VI_3 compounds, 280–282
  IV-VI compounds, 323, 326, 329–330
  V_2-VI_3 compounds, 339, 342, 344, 345, 364
  VIIIA-containing elements, 379–380
  indium antimonide crystals, 169
  indium arsenide crystals, 162
  semiconductor crystals, 24, 25
  silver and copper chalcogenides, 309, 311, 312, 324, 326
  silver and copper oxides, 305
Selection rules, 237–239, see also Group II-IV-V_2
Selenium, 2, 86–88, 123
Self-diffusion coefficient, 74, see also Germanium
Semiconductor devices
  applications, 445–447
  manufacture from silicon crystals, 65
  multiterminal (multibarrier) devices
    bipolar devices, 434–438
    overview, 433–434
    unipolar devices, 438–445
  ohmic contact, 417–420
  p-n junction and schottky barrier, 411–417
  p-n junction and schottky-barrier two-electrode devices
    overview, 420–422
    photodiode, 427–432
    tunnel diodes, 424–427
    unijunction transistor, 433
    varistor and varactor, 422
    zener and avalanche diodes, 423–424
Semiconductor laser, 431, see also Photodiode
Semiconductors
  definition, history, systematization, 1–13
  free state of valence electrons, 16–17
Semimetals, 1
Senarmonite, 348, see also Group V_2-VI_3
Shear modulus
  gray tin crystals, 79
  gallium arsenide crystals, 139
  I_2-IV-VI_3 compounds, 256, 260
  I_3-VI_4 compounds, 267
  tellurium crystals, 89
Shockley analysis, 28, 440–441, see also Field effect transistors; p-n Junction
Silicon
  charge-coupled devices, 445, 446
  chemical interactions, 92–94
  criteria for ohmic contact and Schottky barriers, 417, 418
  dopants
    conductivity in boron nitride crystals, 116
    hardening in gallium arsenide crystals, 138–139
    synthesis of gallium arsenide crystals, 144, 145
  early use, 2
  elemental semiconductor, 56–65
    alloys, 355–357, 360
  field effect transistors, 441
  grown junction method in p-n junction fabrication, 412
  metal-insulator-semiconductor diodes, 444
  thermal expansion comparison with silicon carbide crystals, 106, 107
Silicon carbide, 105–110
Silver and copper chalcogenides, see Group I-VI
Silver and copper oxides, see Group I-VI
Silver iodide, 219–222, see also Binary II-VI and I-VIII tetrahedral semiconductors
Skutterudite structure, 353, 355, see also Group VIIIA
Sodium, 299, 301, 310, see also Group $(IA)_3$-VB
Solar cell, see Photodiode
Solid materials, 1, 16–17
Solid solutions
  elemental semiconducting alloys, 355–357, 360
  II-VI and III-V compounds semiconductor alloys, 380
  II-VI/II-VI semiconductor alloys, 367, 376, 378, 380
  III-V/III-V semiconductor alloys, 362, 364–365, 367, 382, 383, 385
  organic semiconductors, 380, 384, 386–397
Solid-state amorphization, 145, see also Gallium arsenide

INDEX 469

Solubility
  II-VI and III-V alloys, 387
  II-VI/II-VI solid solutions, 367, 376
  III-V/III-V semiconductor alloys, 365, 382, 384
Specific heat
  aluminum phosphide crystals, 125
  cadmium selenide crystals, 202
  cadmium sulfide crystals, 199
  cadmium telluride crystals, 204
  copper bromide crystals, 216
  copper chloride crystals, 214
  copper iodide crystals, 218
  elemental semiconducting alloys, 360
  gallium arsenide crystals, 138, 141
  I-III-VI_2 compounds, 249
  I_2-IV-VI_3 compounds, 256, 260
  I_3-VI_4 compounds, 266, 267
  II-IV-V_2 compounds, 235
  indium phosphide crystals, 156
  measurement in semiconductor crystals, 33
  mercury sulfide crystals, 208
  mercury telluride crystals, 211, 213
  silicon carbide crystals, 107
  silicon crystals, 58, 60
  silver iodide crystals, 220
  tellurium, 89
  zinc oxide crystals, 183
  zinc selenide crystals, 189
  zinc sulfide crystals, 186
  zinc telluride crystals, 195
Speed of ultrasound, 256, 260, 267, see also Group I_2-IV-VI_3; Group I_3-VI_4
Sphalerite structure
  aluminum antimonide crystals, 128
  aluminum arsenide, 125
  boron arsenide, 121
  boron nitride, 112, 113
  cadmium selenide crystals, 202
  cadmium sulfide crystals, 198
  cadmium telluride crystals, 203
  chalcopyrite structure comparison, 235–239
  copper bromide crystals, 216
  copper chloride crystals, 214
  copper fluoride crystals, 214
  copper iodide crystals, 217
  gallium antimonide crystals, 145
  gallium arsenide crystals, 138
  gallium phosphide crystals, 134
  I_2-IV-VI_3 compounds, 256
  II-IV-V_2 compounds, 230
  II-VI/II-VI solid solutions, 367, 376
  III_2-IV-VI compounds, 289
  III-V/III-V semiconductor alloys, 364, 385
  III_2-VI_3 compounds, 273
  indium antimonide crystals, 165, 169
  indium arsenide crystals, 160–161
  indium phosphide crystals, 155–156
  mercury selenide crystals, 210
  silver iodide crystals, 219, 222
  zinc selenide crystals, 189
  zinc sulfide crystals, 186
Spinel type structure, 278, see also Group III_2-VI_3

Stability
  aluminum phosphide crystals, 125
  boron arsenide crystals, 121
  boron nitride crystals, 112
  copper bromide crystals, 216
  germanium, 66
  gray tin, 76
  III_2-IV-VI compounds, 289
  mercury sulfide crystals, 208
  phosphorus crystals, 84
  silver and copper oxides, 303, 314
  silver iodide crystals, 219
  tellurium, 88
  zinc sulfide crystals, 186
Stationary methods, 32, see also Thermal conductivity
Strength, 15
Stress, 143, 275, see also Gallium arsenide; Group III_2-VI_3
  -strain relations, 138–139
Sulfur, 158, 159, see also Indium phosphide
Superconductivity, 65, 325, 327, 384, see also Group IV-VI; Organic semiconductors; Silicon
Superlattice, 105, see also Silicon carbide
Surface energy, 54, see also Diamond
Synthesis
  aluminum antimonide crystals, 131
  aluminum arsenide crystals, 127–128
  boron crystals, 41–43
  boron arsenide crystals, 121
  boron nitride crystals, 112–113
  boron phosphide crystals, 120
  cadmium selenide crystals, 203
  cadmium sulfide crystals, 198, 199
  cadmium telluride crystals, 206
  calculation of composition prediction, 2–3
  copper bromide crystals, 217
  copper chloride crystals, 215
  copper iodide crystals, 218
  defect ternary compounds, 284
  diamonds, 43, 54–56
  elemental semiconducting alloys, 360
  gallium arsenide crystals, 138, 143–145
  gallium nitride crystals, 134
  gray tin, 76
  IA-IB compounds, 295–296
  I-IV_2-V_3 compounds, 245
  I_2-IV-VI_3 compounds, 256
  II_2-IV compounds, 316
  III_2-IV-VI compounds, 284
  indium antimonide crystals, 170
  indium arsenide crystals, 165
  indium nitride crystals, 154
  indium phosphide crystals, 160
  mercury selenide crystals, 210
  mercury sulfide crystals, 209
  mercury telluride crystals, 213
  phosphorus crystals, 85
  silicon crystals, 61–65
    carbide crystals, 107
  silver iodide crystals, 222
  zinc sulfide crystals, 187
  zinc telluride crystals, 197–198

T

Tantalum, 87, see also Dopants/doping; Gray selenium
Tellurium
  dopants
    conductivity in gallium arsenide crystals, 138
    electrical transport in aluminum antimonide crystals, 129
    hardening in gallium arsenide crystals, 138–139
    elemental semiconductor, 88–91
Temperature, see also Individual entries
  conductivity
    indium phosphide crystals, 160, 161
    IV-VI compounds, 323, 344
  Debye characteristics in germanium crystals, 69, 71
  elastic properties of gallium antimonide crystals, 148, 149
  electrical conductivity
    aluminum nitride crystals, 123
    gray tin crystals, 80, 81
    I-III-VI$_2$ compounds, 250, 255
    I$_2$-IV-VI$_3$ compounds, 264, 265
    I$_3$-VI$_4$ compounds, 267–269
    II$_2$-IV compounds, 316, 335
    II-IV-V$_2$ compounds, 239, 243
    V$_2$-VI$_3$ compounds, 339, 344–345, 362, 363
    metals vs. nonmetals, 1
    semiconductor crystals, 17
    silver and copper chalcogenides, 312, 325
    semiconductor crystals, 15
  electrical conductivity/magnetic properties in III$_2$-VI$_3$ compounds, 280, 281
  glass transformation of selenium, 86
  II$_2$-IV compounds, 239, 241, 315
  IV-VI compounds
    intrinsic carrier, 327, 347
    mobility, 331
    refractive index, 335, 354
    resistivity, 323, 325, 344
  indium antimonide crystals, 170, 171
  lattice parameters
    cadmium sulfide crystals, 198
    III-V/III-V semiconductor alloys, 362, 364, 383
    indium antimonide crystals, 165, 167
    silicon carbide crystals, 106
  magnetic properties, 306, 313
  mobility, 16, 191, 193, 331
  optical properties
    IA-IB compounds, 296, 297, 299
    silver and copper chalcogenides, 313, 329
  resistivity
    diamond crystals, 44–46
    electrical conductivity of boron nitride crystals, 118
    (IA)$_3$-VB compounds, 301, 309, 310
    semiconductor crystals, 30
  Seebeck effect, 326, 344, 362, 364
  stability of boron crystals, 43
  thermal conductivity
    aluminum antimonide crystals, 130
    boron phosphide crystals, 119–120, 122
    cadmium selenide crystals, 202
    cadmium sulfide crystals, 199, 201
    cadmium telluride crystals, 206
    diamond crystals, 48, 49
    gallium antimonide crystals, 150, 153
    gallium arsenide crystals, 142, 146
    gallium nitride crystals, 133
    gallium phosphide crystals, 136, 137
    germanium crystals, 72, 77
    gray selenium crystals, 87, 89
    I-III-VI$_2$ compounds, 250–251
    I$_2$-IV-VI$_3$ compounds, 265
    I$_3$-VI$_4$ compounds, 268, 270
    II$_2$-IV compounds, 316, 336
    II-IV-V$_2$ compounds, 240, 243, 244
    V$_2$-VI$_3$ compounds, 347, 365
    indium arsenide crystals, 163, 164
    mercury selenide crystals, 210
    mercury telluride crystals, 213
    silicon crystals, 59
    silicon carbide crystals, 108, 112
    silver iodide crystals, 220, 222
    tellurium crystals, 90
    zinc oxide crystals, 185
    zinc selenide crystals, 191–192, 194
    zinc sulfide crystals, 187, 190
    zinc telluride crystals, 196–197
  thermal expansion
    aluminum antimonide crystals, 128
    cadmium telluride crystals, 204, 206
    gallium arsenide crystals, 138, 139
    gallium phosphide crystals, 134, 135
    gray tin crystals, 79, 80
    indium arsenide crystals, 161, 162
    indium phosphide crystals, 156, 157
    mercury telluride crystals, 211, 212
    silicon crystals, 57–59
    zinc selenide crystals, 189
    zinc telluride crystals, 193, 195
Temperature-composition
  copper chloride crystals, 214
  phase diagram
    gallium arsenide crystals, 138, 140
    indium arsenide crystals, 165
    indium phosphide crystals, 156, 157
Tensile strength, 42, see also Boron
Ternary adamantine semiconductors
  analogs of group II-VI
    I-III-VI$_2$ compounds, 246–257
    I$_2$-IV-VI$_3$ compounds, 256, 258, 260–265
    I$_3$-VI$_4$ compounds, 266–271
  analogs of group III-V
    II-IV-V$_2$ compounds, 230–243
    III-V, I-IV$_2$-V$_3$ compounds, 242, 244–246
  defect semiconductors
    III$_2$-VI$_3$ compounds, 273, 275–281
    IV$_3$-V$_4$ compounds, 271–274
    ternary compounds, 281–286, 288
    III$_2$-IV-VI compounds, 284, 287, 289
Ternary inorganic semiconductors, 12–13

Ternary valence compounds, 3–5
Tetradymite structure, 339, 351, 360, see also Group V_2-VI_3
Tetragonal structure
  boron crystals, 37, 39–41
  carbon, 43
  silicon crystals, 57
  copper chloride crystals, 214
Tetrahedral coordination, 2–3
Tetrahedral structure, 105, see also Silicon carbide
Tetrahedral semiconductors, see Binary II-VI and I-VII tetrahedral semiconductors
Tetrathiofulvalene, 397, see also Organic semiconductors
Thatogenides, 354, 381–382, see also Group II_2-IV; Group VIIIA
Thermal conductivity
  aluminum antimonide crystals, 130
  aluminum arsenide crystals, 127
  aluminum nitride crystals, 123
  aluminum phosphide crystals, 125
  boron crystals, 41–42
  boron nitride crystals, 112, 114, 116
  boron phosphide crystals, 119–120
  cadmium selenide crystals, 202
  cadmium sulfide crystals, 199, 201
  cadmium telluride crystals, 206
  calculation in semiconductor crystals, 32, 33
  copper bromide crystals, 216
  copper chloride crystals, 215
  copper iodide crystals, 218
  diamond crystals, 43, 46, 48–50
  elemental semiconducting alloys, 357, 360
  gallium antimonide crystals, 150, 153
  gallium arsenide crystals, 142, 146
  gallium nitride crystals, 133
  gallium phosphide crystals, 136, 137
  germanium crystals, 72, 77
  gray selenium crystals, 87, 89
  I-III-VI_2 compounds, 247, 250–251, 256
  I-IV_2-V_3 compounds, 245, 247
  I_2-IV-VI_3 compounds, 265
  I_3-VI_4 compounds, 268, 270
  II_2-IV compounds, 315–316, 334, 336
  II-IV-V_2 compounds, 240, 243, 244
  III_2-VI_3 compounds, 280
  III-V/III-V semiconductor alloys, 367
  IV-VI compounds, 323, 325, 331, 344
  V_2-VI_3 compounds, 339, 345, 347
  VIIIA-containing elements, 354, 355, 374–375
  indium antimonide crystals, 170, 171
  indium arsenide crystals, 163, 164
  indium phosphide crystals, 160, 161
  mercury selenide crystals, 210
  mercury telluride crystals, 213
  phosphorus crystals, 84
  silicon crystals, 59
    carbide crystals, 108, 112
  silver and copper chalcogenides, 324
  silver iodide crystals, 220, 222
  tellurium crystals, 89, 90

  zinc oxide crystals, 185
  zinc selenide crystals, 191–192, 194
  zinc sulfide crystals, 187, 190
  zinc telluride crystals, 195–197
Thermal diffusivity, 32, 114, 213, see also Boron nitride; Mercury telluride
Thermal emf test, 34–35, see also Coefficient of thermal expansion
Thermal oscillations, 17
Thermal properties, 44, see also Diamond
Thermodynamic equilibrium, 25
Thermodynamic properties
  cadmium sulfide crystals, 198
  diamond crystals, 45, 48, 50, 51
  germanium crystals, 69
  gray tin crystals, 80–81
  I-III-VI_2 compounds, 247, 249
  I_2-IV-VI_3 compounds, 256, 260
  I_3-VI_4 compounds, 266, 267
  silver and copper chalcogenides, 307, 321
  silver and copper oxides, 315
Thermoelectric power
  I_2-IV-VI_3 compounds, 262–264
  I-III-VI_2 compounds, 250, 255
Thermoelectric properties, 24–25, 41, see also Boron crystals
Thermoelements, 446, see also Semiconductor devices
Thin film deposition, 218, see also Copper iodide; Synthesis
Thiophene, 397
Thomson effect, 24
Three-dimensional defects, 145, see also Gallium arsenide
Threshold voltage, 443, see also Field effect transistors
Thyristors, 436–438, see also Multiterminal (multibarrier) devices
Tin, 35, 94
Tin chalcogenides, 331, 351, see also Chalcogenides; Group IV-VI
Tin selenide, 325–326, see also Group IV-VI
Tin sulfides, 319, 331, 351, see also Group IV-VI
Tin telluride, 327, 331, 352, see also Group IV-VI
Torsion modulus, 69, see also Germanium
Transition energy, 239, 241, see also Group II-IV-V_2
Transition temperature, 235, 236, see also Group II-IV-V_2
Transparency, 119, 124, see also Aluminum nitride; Boron phosphide
Traveling heater method, see also Synthesis
  copper chloride crystals, 214–216
  copper iodide crystals, 217, 218
Trichlorosilane, 62–63, see also Silicon; Synthesis
Tunnel diodes, 424–427
Tunnel diode/resistor amplifier, 425, 427
Two-dimensional defects, 144–145, see also Gallium arsenide
Two-electrode devices
  overview, 420–422
  photodiodes, 427–432
  tunnel diodes, 424–427
  unijunction transistors, 433

varistor and varactor, 422–423
zener and avalanche diodes, 423–424
Two-temperature method, 128, see also Aluminum arsenide; Synthesis
Type I diamonds, see Diamond
Type II diamonds, see Diamond

U

UHV, see Ultrahigh vacuum
UJT, see Unijunction transistors
Ultrahigh vacuum (UHV), 55, see also Diamond; Synthesis
Ultrasound speed, 235, 237, see also Group II-IV-V_2
Unijunction transistors (UJT), 433, see also Photodiode
Unipolar devices, 438–445, see also Multiterminal (multibarrier) devices
Unit cell
   diamond crystals, 52
   I-III-VI_2 compounds, 247, 249
   $(IA)_3$-VB compounds, 299
   II_2-IV compounds, 313, 330
   II-IV-V_2 compounds, 235, 237
   III_2-VI_3 compounds, 278, 279
   IIIB-VB compounds, 112, 113
   IV-VI compounds, 318
   selenium crystals, 87, 88
   silver and copper oxides, 303, 315
   tellurium, 88, 89

V

Vacancies, see also Holes
   boron crystals, 43
   copper oxides, 305
   defect ternary compounds, 281–282
   electron positions in semiconductor crystals, 17
   III_2-VI_3 compounds, 273, 278
   IV_3-V_4 compounds, 271
Valence band, 19, see also Electrical conductivity
Valence electrons, 2–3, see also Electrons
Valency, 2, 66
van der Pauw method, 30, see also Resistivity
van der Waals bonding, 15
Varactor, 422
Varistor, 422
Vegard's rule, 92, 355, 362, 367
Vickers microhardness, see also Hardness; Mechanical properties
   gallium antimonide crystals, 148
   gallium phosphide crystals, 134
   I-III-VI_2 compounds, 247
   I_3-VI_4 compounds, 267
   III_2-VI_3 compounds, 275, 279
   IV_3-V_4 compounds, 272–273, 275
   IV-VI compounds, 338
   indium antimonide crystals, 168
   indium arsenide crystals, 161
   indium phosphide crystals, 158
   selenium crystals, 86
   zinc selenide crystals, 189
Viscosity, 86, see also Selenium

Voltage, 1, 414–415, see also p-n Junction diode
Volume resistivity, 30
Voyager, 446–447

W

Welding torch, 54, see also Diamond; Synthesis
Wurtzite structure
   boron nitride, 112, 113
   cadmium sulfide crystals, 198
   copper bromide crystals, 216
   copper chloride crystals, 214, 215
   copper iodide crystals, 217
   gallium nitride crystals, 131
   I-III-VI_2 compounds, 247
   II-IV-V_2 compounds, 230
   II-VI/II-VI solid solutions, 367, 376
   III_2-IV-VI compounds, 287
   III_2-VI_3 compounds, 273
   indium nitride crystals, 152
   silver iodide crystals, 219, 221
   zinc oxide crystals, 183, 184
   zinc selenide crystals, 189
   zinc sulfide crystals, 186
   zinc telluride crystals, 193

Y

Yellow phosphorus, properties, 84, see also Phosphorus
Young's modulus
   diamond crystals, 50–51
   gallium antimonide crystals, 148
   gallium arsenide crystals, 139
   germanium crystals, 69
   gray selenium crystals, 87
   gray tin crystals, 79
   I-III-VI_2 compounds, 247
   I_2-IV-VI_3 compounds, 256, 260
   I_3-VI_4 compounds, 267
   II-IV-V_2 compounds, 235, 237
   IV_3-V_4 compounds, 273, 274
   indium arsenide crystals, 162
   indium phosphide crystals, 158
   silicon carbide crystals, 108
   silver and copper oxides, 315
   tellurium crystals, 89
   zinc selenide crystals, 190

Z

Zener diodes, 423–424
Zero method, 33, see also Thermal conductivity
Zinc, 87, 123
Zinc oxide, 183–186, see also Binary II-VI and I-VIII tetrahedral semiconductors
Zinc selenide, 187, 189–193, see also Binary II-VI and I-VIII tetrahedral semiconductors
Zinc sulfide, 186–187, see also Binary II-VI and I-VIII tetrahedral semiconductors
Zinc telluride, 193, 195–198, see also Binary II-VI and I-VIII tetrahedral semiconductors
Zone refining, 42, see also Boron